# On the Emergence Theme of Physics

# ON THE EMERGENCE THEME OF PHYSICS

**Robert Carroll**

University of Illinois at Urbana-Champaign, USA

World Scientific

NEW JERSEY • LONDON • SINGAPORE • BEIJING • SHANGHAI • HONG KONG • TAIPEI • CHENNAI

*Published by*

World Scientific Publishing Co. Pte. Ltd.
5 Toh Tuck Link, Singapore 596224
*USA office:* 27 Warren Street, Suite 401-402, Hackensack, NJ 07601
*UK office:* 57 Shelton Street, Covent Garden, London WC2H 9HE

**British Library Cataloguing-in-Publication Data**
A catalogue record for this book is available from the British Library.

**ON THE EMERGENCE THEME OF PHYSICS**



ISBN-13 978-981-4291-79-8
ISBN-10 981-4291-79-X
ISBN-13 978-981-4304-82-5 (pbk)
ISBN-10 978-4304-82-4 (pbk)

Printed in Singapore.

# Contents

# Preface

This book is designed as an investigation of certain aspects of quantum mechanics and general relativity. The approach is based largely on equations and mathematical structures from which one is able to extract some "meaning" in physics (whence the title). The mathematics can be regarded as a basically simple language (it can be self taught for example) but questions of physical meaning are often unclear, or inadequately developed in the mathematical context chosen. Partly for this reason mathematics is itself a developing language of course and it is often enlarged to handle emergent physical phenomena, as well as feeding on itself for development. In any event various features or aspects of physics are displayed, based on recent literature and the author's interpretation of these matters. Some original work is included, based on [**170, 171, 172, 173, 179, 185**], relating the Bohmian quantum potential to gravity and Ricci flow, but generally we will extract from known results and conjectures with explicit references to the sources (although we will not trace historical origins). Much of the material is taken from the electronic bulletin boards at arXiv.org and publication information can often be found there if it is omitted in the bibliography here; we hope to have portrayed a small amount of the range, versatility, and significance of net activity. The material is more or less up to date in a number of areas and hopefully will contribute to further discoveries. As this seems to be an age of emergence, this theme will be emphasized in various places, especially in the quantum-classical spirit, but also e.g. in the connections between gravity and thermodynamics. The book is written in the language of mathematics and the interaction with physics, along with the resulting emergent points of view, are best expressed via equations. Given that the Schrödinger equation can arise from e.g. subquantum oscillations, or diffusion, or hydrodynamics it still is not quantum mechanics since there remains a firm linking of quantum mechanics to eigenvalues and other algebraic notions (e.g. $C^*$ algebras and K-theory). There still seem to be many mysteries which we will not try to discuss involving the vacuum, the cosmological constant, dark energy and matter, etc. We will devote considerable attention to Bohmian mechanics and to semi-classical theory as well since these arenas interact with both classical and quantum regimes. In particular we will emphasize certain aspects of the quantum potential (QP) whose basic nature is best characterized via the quantum equivalence principle QEP) of Faraggi-Matone (cf. [**97, 170, 297, 589, 590**] and see also Section 1.3 and Chapter 8). We also discuss phase space techniques involving symplectic ideas which involve quantum mechanics via the metaplectic group for example. On the other hand the development of non-commutative geometry

(NCG), via $C^*$ algebras, K-theory, etc. involves a minimal coupling of quantum mechanics and gravity, and firmly establishes quantum mechanics with its algebraic structure as a seriously fundamental theory. We sketch this in Chapter 7 and indicate a little bit of the magnificent edifice involving the standard model and non-commutative geometry. Deep algebraic connections of quantum mechanics to number theory and algebraic geometry are known and flourishing and we will mention "arithmetic" at various times with the faintest suggestion that meaning (in a fundamental sense) really lies in number theory.

Thermodynamic ideas have assumed a definitive role in many areas of physics (including gravity) and suitable adjustments have to be made in partition functions for example when dealing with quantum particles (due to Bose-Einstein or Fermi-Dirac statistics, Pauli exclusion principles, etc.). Information theory can also be given a prominent role in many areas of physics and there is an emerging quantum information theory of possible use in computing as well (see e.g. [**146, 147, 197, 199, 547**]). In any event the SE applies to many situations where it can be modeled on momentum perturbations $\delta p = \nabla|\psi|/|\psi$ (where $\psi$ satisfies the SE) and corresponding position perturbations $\delta q$ will involve the Heisenberg inequality. This is lovely and can be obtained via Bohmian mechanics and there is no need to postulate any discrete spectrum of energies, or Hilbert space, or $C^*$ algebras, etc. So the Schrödinger and/or Heisenberg pictures have some meaning independent of the algebraic structure arising from atomic frequencies, etc. There is also the idea of quantum chaos which can be discussed in various manners and we consider semi-classical methods leading to trace formulas à la Gutzwiller (cf. [**396**]). This again suggests number theory, zeta functions, etc.

The work of Dürr, Goldstein, and Zanghi, in collaboration with e.g. Berndl, Dauman, Peruzzi, Struyve, and Tumulka, (see e.g. [**274, 364**]) has been very important in developing the deBroglie-Bohm theory, and we mention also fundamental contributions of Bacciagaluppi ([**56**]), Brown ([**132, 133**]), deGosson ([**369**]), Hiley ([**436**]), Holland ([**440**]), Valentini ([**848, 849, 850**]), etc. (cf. [**173**] for more references). Questions of ontology and epistimology will be largely ignored (except when citing results) but it is hoped that a philosophical flavor arises concerning "meaning" as arising from the mathematics of physics and perhaps the physics of mathematics. Generally it is clear that Bohmian mechanics is not the same as quantum mechanics and the quantum potential is best viewed via the QEP (see also comments below on trajectories) but the Bohmian approach has many extraordinary features applying to both the Schrödinger and Heisenberg pictures which lead to results of major interest (the Bell inequalities are a non-issue for us). We refer also to [**95, 137, 145, 187, 213, 406, 502, 597, 778, 903**] for information on time-energy uncertainty. In dealing with trajectories however there are striking differences between dBB trajectories and the classical ones and this is revealed beautifully in papers of Matzkin, Lombardi, and Nurok (cf. [**593, 594, 595**]). The dBB trajectories tend to follow the streamlines of the probability flow and are generically nonclassical but the Floydian type trajectories stemming from the QEP are generically quantum in nature (cf. [**177, 187, 297, 310, 311, 312, 590, 890**]).

Semiclassical systems are quantum systems that display the manifestation of classical trajectories; the wave function and the observable properties of such systems depend on the trajectories of the classical counterpart of the quantum system. Thus we will study semiclassical systems and phase space methods in some detail including the Gutzwiller trace formula (cf. also [**360, 361, 362**]). There is an interesting occurance of the Schwarzian derivative arising in [**595**] and the Schwarzian derivative is a fundamental object in the beautiful theory of Bertoldi, Farragi and Matone et al about the quantum equivalence principal (cf. [**97, 98, 177, 175, 176, 297, 589, 590**]). In this theory quantization (and eigenvalues) arise via the quantum stationary Hamilton Jacobi equation (QSHJE) which characterizes quantum solutions of the Schrödinger equation (SE). The Gutzwiller trace formula connects quantum energy eigenvalues with periodic orbits of the associated classical system and the techniques involve the metaplectic representation; however we do not try to cover the related arithmetic features of quantization which are entangled with the Riemann zeta function, the Selberg trace formula, etc. (see e.g. [**25, 128, 231, 233, 423, 505, 576, 749, 808, 809, 847**]) and recall the NCG approach leading also to zeta functions, etc. Given that Bohmian trajectories do not seem to correspond generally to classical ones while, via Gutzwiller, classical orbits are connected with the arithmetic of quantum mechanics (via eigenvalues), there may be no way to extract arithmetic information from the Bohmian theory (see [**36, 37, 360, 361, 362, 364, 557, 593, 594, 595, 597, 713, 781, 890, 903**] for more on trajectories). There is also a strong connection of the Bohmian quantum potential to Fisher information and thence to Riemannian geometry as developed in Chapter 2 and the question can arise of seeking suitable arithmetic in Riemannian geometry (cf. [**97, 98, 591, 592**]).

Chapter 2 contains some original work of the author on the quantum potential, Weyl geometry, and Ricci flow extracted from [**170, 172, 173, 185**] for example. There are also results in Chapter 1 concerning the quantum potential, Fisher information, and thermalization based on the author's work ([**173**]) as well as some comments on stability and trajectories. We remark that in the process of citing information from various sources we will often use the appelation "one" to refer to the author or authors of the source in question (instead of boringly repeating names or citations). The source is always clearly indicated and this practice involves no proprietary claim on my part to the information. Accurate rendering of mathematical or physical arguments can often involve repetition of words as well as equations. Chapter 3 (Sections 1-2) and the first two sections of Chapter 6 deal with some work of Elze, t'Hooft, Isidro, et al on emergence (cf. [**2, 290, 444, 445, 446, 456, 457, 458, 459, 461, 462**]). This becomes involved with Ricci flow and Perelman entropy (cf. [**170, 172, 506, 507, 621, 693, 836**]. Section 3 of Chapter 3 is deals with work of H. Yang et al involving gravity as a collective phenomenon emerging from gauge fields of electromagnetism living in a fuzzy space-time (cf. [**893, 894, 895, 896**]). Chapter 4 is about Kaluza-Klein and 5 dimensional physics and there is material on electrodynamics, induced matter, the zero point field, Klein-Gordon and Dirac equations, etc. following Mashoon, Ponce de Leon, Wesson, et al (cf. [**585, 586, 717, 718, 719, 868, 869, 870,**

**871, 872**] etc.). Chapter 5 (Sections 1-5) is about connections of thermodynamics and gravity and gives a derivation of the Einstein equations from entropy ideas following Padmanabhan et al; this is later expanded enormously in Sections 3-5 of Chapter 6 (see e.g. [**15, 16, 149, 178, 471, 474, 476, 523, 524, 525, 620, 659, 660, 661, 662, 663, 664, 665, 666, 667, 668, 669, 670, 671, 672, 673, 674, 675, 676, 677, 678, 776, 860, 861**]). The theory is developed beyond GR into Lanczos-Lovelock gravity for example and provides a gravitational connection to thermodynamical laws. In Sections 6-8 of Chapter 5 we sketch some work of L. Glinka et al on thermodynamics and quantum gravity (cf. [**353, 354, 355, 356, 357, 358**]). This involves bosonic strings and quantum field theory (QFT) and is speculative (but very interesting). Chapter 7 starts with some ideas about emergent time and the Connes-Rovelli approach (see [**234**]) via $C^*$ algebras compels us to pursue these themes via NCG to arrive at the Chamseddine-Connes-Kreimer-Marcolli universe where the standard model and gravity are united in NCG. Only a brief sketch of this is given for obvious reasons and we refer to [**205, 206, 207, 208, 230, 231, 232, 233, 236, 576**] for details. In Section 4 we sketch some work of Singh on noncommutative geometry and Klein-Gordon equations; this is followed by a sketch of recent results of Jejjala, Kavic, Minic, and Tze on quantum mechanics, gravity, time, and geometry. Chapter 8 is first a summary of material on gravity and the quantum potential (with some repetition from previous chapters). Then, following Matone [**590**] the QEP is shown to generate quantization via the QSHJE.

In this book I have tried to bring together many different "strings" of mathematics and physics involving many points of view and many authors (incidentally string theory and loop quantum gravity are not covered and it is assumed in a few places that very elementary string theory is known - see [**716**] for strings and [**509, 750, 831**] for quantum gravity). In particular, in various sections it has been indicated how many originally distinct concepts are related to one another at a fundamental level (often via mathematical identifications) or how a theory "emerges" or "arises" from another theory via some manipulations or some added factors (such as information loss). It is hoped that this exposure will stimulate new insights, combinations, and developments. The formulas and equations may seem daunting at times but the mathematical topics are easily available in books and university courses.

I would like to thank C. Castro, B. Roy Frieden, M. deGosson, J. Isidro, A. Kholodenko, M. Matone, and D. Schuch for helpful correspondence. I would also like to express my profound indebtedness to my mother Dorothy LeMonnier Carroll and my aunt Amelia LeMonnier, as well as to the extended family of Margaret and Morris Jacobs (living and dead).

The book is dedicated to my beloved wife Denise Rzewska Bredt Carroll.

CHAPTER 1

# SOME QUANTUM BACKGROUND

We assume basic quantum mechanics as in e.g. [**603**] (cf. also [**170, 177, 175**]).

## 1. DEBROGLIE-BOHM MECHANICS

This is covered e.g. in [**170, 440**] but we sketch matters here following [**170**]. There are many ways to derive or develop the dBB theory but for simplicity first look at a 1-D SE of the form (**1A**) $-(\hbar^2/2m)\partial_x^2\psi+V(x)\psi = i\hbar\partial_t\psi$. Then putting $\psi = Rexp(iS/\hbar)$ yields

$$\partial_t S + \frac{S_x^2}{2m} + V - \frac{\hbar^2 R_{xx}}{2mR} = 0;\ \partial_t(R^2) + \frac{1}{m}\partial_x(R^2 S_x) = 0 \tag{1.1}$$

which in e.g. three space dimensions can be written (with $P = R^2 = |\psi|^2$, and (**1B**) $Q = -(\hbar^2/2m)(\Delta R/R))$ is the quantum potential (QP)

$$\partial_t S + \frac{1}{2m}|\nabla S|^2 + V + Q = 0;\ \partial_t P + \frac{1}{m}\nabla\cdot(P\nabla S) = 0 \tag{1.2}$$

Writing $\rho = mP$ one has formulas

$$\frac{\Delta\sqrt{\rho}}{\sqrt{\rho}} = \frac{1}{4}\left[\frac{2\Delta\rho}{\rho} - \left(\frac{\nabla\rho}{\rho}\right)^2\right] \tag{1.3}$$

and for $\mathbf{v} = \dot{\mathbf{q}}$ with $p = mv = \partial_x S$ (cf. however p. 8.3 and [**177, 187, 310, 843**])

$$\partial_t\rho + \partial_x(\rho\dot{q}) = 0;\ \partial_t S + \frac{p^2}{2m} + V - \frac{\hbar^2}{3m}\frac{\partial_x^2\sqrt{\rho}}{\sqrt{\rho}} = 0 \tag{1.4}$$

Now we observe that ($P = 0$ for large $x$)

$$\int P\frac{\partial_x^2\sqrt{P}}{\sqrt{P}}dx = \frac{1}{4}\int\left(2P'' - \frac{(P')^2}{P}\right)dx = -\frac{1}{4}\int\frac{(P')^2}{P}dx = -\frac{1}{4}FI \tag{1.5}$$

FI is Fisher information or better Fisher channel capacity (cf. [**318, 321**]). Hence, writing Q via P,

$$\int PQdx = \frac{\hbar^2}{8m}\int\frac{(P')^2}{P}dx = \frac{\hbar^2}{8m}FI \tag{1.6}$$

For the differential entropy (**1C**) $\mathfrak{S} = -\int\rho log(\rho)dx$ one has

$$\partial_t\mathfrak{S} = -m\int[P_t log(P) + P_t]dx = -m\int P_t(log(P)+1)dx \tag{1.7}$$

From (1.1) we have $P_t = -(1/m)(PS')'$ ($P' \sim \partial_x P$ etc.) so

$$\partial_t \mathfrak{S} = \int (PS')'[log(P)+1]dx = -\int (PS')\frac{P'}{P}dx = -\int S'P'dx \tag{1.8}$$

At this point one often refers to diffusion or hydrodynamical processes and thinks of $S' = p = m\dot{q}$ (omitting boldface for vectors in 1-D) where $v = \dot{q}$ represents some collective velocity field related to a free for example with say $v = -(\hbar/2m)(\nabla P/P)$ involving an $v = -u$ (cf. [**170, 332, 333**]). Note that the typical Bohmian hypothesis $p \sim S'$ or $p \sim \nabla S$ is not valid for particle models and does not lead to correct quantum trajectories (cf. [**187, 297, 310, 311, 312, 589, 590**]) but we will continue to use this notation for hydrodynamical and diffusion processes essentially for historical purposes (cf. also [**843**] in this connection). Then (1.1) gives (**1D**) $P_t + (Pv)' = 0$ and it is interesting to note that this form of velocity involving $\nabla P/P$ or $\nabla\rho/\rho$ comes up in the form of momentum fluctuations or perturbations in work of Hall-Reginatto [**400, 401, 402, 736**] on the exact uncertainty principle and in Crowell [**246**] in a fluctuation context. Indeed in [**246**] one takes (**1E**) $\delta\rho = (\hbar/2i)(\nabla\rho/\rho)$ (think of $m = 1$ for convenience) with

$$< \delta\rho > = \int \rho\delta\rho dx = \frac{\hbar}{2i}\int \nabla\rho dx = 0;\ < (\delta\rho)^2 > = c\int \left[\frac{(\nabla\rho)^2}{\rho}\right] dx \sim FI \tag{1.9}$$

Thus from (1.3) and (**1B**) $Q$ can be written as

$$Q = -\frac{\hbar^2}{4m}\left[\frac{1}{2}(\delta p)^2 - \frac{\Delta\rho}{\rho}\right] = \frac{\hbar^2}{4m}\left[\nabla(\delta p) + \frac{1}{2}(\delta p)^2\right] \tag{1.10}$$

The exact uncertainty principle involves the characterization of quantum fluctuations as being generated by momentum fluctuations $\nabla P/P$ under certain quite general assumptions. This principle can be applied in many interesting circumstances (cf. [**170, 171, 400, 401, 402, 736**]).

We note also, going back to (1.8) with $P' \sim \nabla P$, when (as in the Brownian motion situation) $S' = -(\hbar/2m)(P'/P)$, then (**1F**) $\partial_t\mathfrak{S} = (\hbar/2m)\int[(P')^2/P]dx \sim \hat{c}FI$ showing that $\partial_t\mathfrak{S} \geq 0$ and exactly how it varies. This feature also arises in studying Ricci flow and information on this is provided in [**172, 506, 507, 621, 693, 836**]. Let $P(x)$ be a probability distribution on $\mathbf{R}$ or $\mathbf{R}^3$; we work in $\mathbf{R}$ for simplicity. Consider a differential entropy $\mathfrak{S} = -\int P\,log(P)dx$ and assume one wants to specify some value $\bar{A} = \int A(x)P(x)dx$. Then consider extremizing

$$\tilde{\mathfrak{S}} = -\int Plog(P)dx + \lambda(1 - \int Pdx) + \alpha(\bar{A} - \int APdx) \tag{1.11}$$

Recall $log(1+x) \sim x$ for $x$ small and write

$$\delta\tilde{\mathfrak{S}} = -\int [(P+\delta P)log(P+\delta P) - Plog(P)]\,dx - \lambda\int \delta Pdx - \alpha\int A\delta Pdx \tag{1.12}$$

$$= -\int \left\{(P+\delta P)log\left[P\left(1+\frac{\delta P}{P}\right)\right] - Plog(P)\right\} dx - \int (\lambda + \alpha A)\delta Pdx$$

$$\approx -\int \delta P(x)\,[log(P) + 1 + \lambda + \alpha A]\,dx$$

Since $\delta P$ is arbitrary one obtains (**1G**) $log(P)+1+\lambda+\alpha A=0$ or $P=exp[-1-\lambda-\alpha A]=(1/Z)exp(-\alpha A)$ where $Z=exp(1+\lambda)$ and $\int Pdx=1\Rightarrow Z=\int exp[-\alpha A(x)]dx$. Here a suitable candidate for A would be the total energy of a system and if S is determined by $P$ alone with $P\sim p=(1/m)S'$ as above then A would have to be the fluctuation energy determined by the quantum potential.

**1.1. SUBQUANTUM THERMODYNAMICS.** We go here to [**392**] for a derivation of the SE from vacuum fluctuations and diffusion waves in sub-quantum thermodynamics (cf. also [**93, 123, 329, 332, 333, 335, 334, 532, 878**] for background). Grössing's work in [**392**] specifies the energy in quantum mechanics arising from sub-quantum fluctuations via nonequilibrium thermodynamics. The ideas are motivated and discussed at great length in e.g. papers 1, 2, and 4 in [**392**] and we only summarize here following [**392**]-6. To each particle of nature is attributed an energy $E=\hbar\omega$ for some kind of angular frequency $\omega$ and one can generally assume that "particles" are actually dissipative systems maintained in a nonequilibrium steady-state by a permanent source of kinetic energy, or heat flow, which is not identical with the kinetic energy of the particle, but an additional contribution. Thus it is assumed here that (**1H**) $E_{tot}=\hbar\omega+[(\delta p)^2/2m]$ where $\delta p$ is the additional fluctuating momentum component of the particle of mass $m$. Similarly the particle's environment is considered to provide detection probability distributions which can be modeled by wave-like intensity distributions $I(x,t)=R^2(x,t)$ with $R$ being the wave's real valued amplitude; thus one assumes (**1I**) $P(x,t)=R^2(x,t)$ with $\int Pd^nx=1$ (note $x\sim\mathbf{x}$). In [**392**]-1 it was proposed to merge some results of nonequilibrium thermodynamics with classical wave mechanics in such a manner that the many microscopic degrees of freedom associated with the hypothesized sub-quantum medium can be recast into the more "macroscopic" properties that characterize the wave-like behavior on the quantum level. Thus one considers a particle as being surrounded by a large "heat bath" so that the momentum distribution in this region is given by the usual Maxwell-Boltzmann distribution. This corresponds to a "thermostatic" regulation" of the reservoir's temperature which is equivalent to saying that the energy lost to the thermostat can be regarded as heat. This leads to emergence at the equilibrium-type probability density ratio (**1J**) $[P(x,t)/P(x,0)]=exp[-(\Delta Q)/kT)]$ where T is the reservoir temperature and $\Delta Q$ the exchanged heat between the particle and its environment. The conditions (**1H**)-(**1J**) are sufficient to derive the SE. Thus first, via Boltzmann, the relation between heat and action is given via an action function $S=\int(E_{kin}-V)dt$ with $\delta S=\delta\int E_{kin}dt$ via

$$\Delta Q=2\omega\delta S=2\omega[\delta(S)(t)-\delta S(0)] \tag{1.13}$$

(cf. [**392**] for more details). Next the kinetic energy of the thermostat is $kT/2$ per degree of freedom and the average kinetic energy of an oscillator is $(1/2)\hbar\omega$ so equality of average kinetic energies demands (**1K**) $kT/2=\hbar\omega/2$ or $\hbar\omega=kT=1/\beta$. Combining (**1H**), (1.13), and (**1K**) yields then

$$P(x,t)=P(x,0)e^{-\frac{2}{\hbar}[\delta S(x,t)-\delta S(x,0)]} \tag{1.14}$$

leading to a momentum fluctuation

$$\delta p(x,t) = \nabla(\delta S(x,t)) = -\frac{\hbar}{2}\frac{\nabla P(x,t)}{P(x,t)} \tag{1.15}$$

and an additional kinetic energy term

$$\delta E_{kin} = \frac{1}{2m}\nabla(\delta S)\cdot\nabla(\delta S) = \frac{1}{2m}\left(\frac{\hbar}{2}\frac{\nabla P}{P}\right)^2 \tag{1.16}$$

The action integral then becomes

$$A = \int L d^n x dt = \int P(x,t)\left[\partial_t S + \frac{1}{2m}\nabla S\cdot\nabla S + \frac{1}{2m}\left(\frac{\hbar}{2}\frac{\nabla P}{P}\right)^2 + V\right] \tag{1.17}$$

We emphasize here that

$$(\bigstar\bigstar)\int P(\nabla S\cdot\delta p)d^n x = \int P(\nabla S\cdot\nabla(\delta S))d^n x = 0$$

(i.e. the fluctuations terms $\delta p$ are uncorrelated with the momentum $p \sim \nabla S$). Now one uses the Madelung form **(1L)** $\psi = Rexp[(i/\hbar)S]$ where $R = \sqrt{P}$ to obtain **(1M)** $[\nabla\psi/\psi]^2 = [\nabla P/2P]^2 + [\nabla S/\hbar]^2$ leading to (1.17) in the form

$$A = \int L dt = \int d^n x dt\left[|\psi|^2(\partial_t S + V) + \frac{\hbar^2}{2m}|\nabla\psi|^2\right] \tag{1.18}$$

(cf. [**848**] for $|\psi|^2 \sim P$). Then via $|\psi|^2\partial_t S = -(i\hbar/2)(\psi^*\dot{\psi} - \dot{\psi}^*\psi)$ one has **(1N)** $L = -(i\hbar/2)(\psi^*\dot{\psi} - \dot{\psi}^*\psi) + (\hbar^2/2m)\nabla\psi\cdot\nabla\psi^* + V\psi\psi^*$ leading to the SE **(1O)** $i\hbar\partial_t\psi = [-(\hbar^2/2m)\nabla^2 + V]\psi$ along with the "modified" Hamilton-Jacobi equation (cf. [**843**] for another version of the Madelung theory).

$$\partial_t S + \frac{1}{2m}(\nabla S)^2 + V + Q = 0;\ Q = -\frac{\hbar^2}{4m}\left[\frac{1}{2}\left(\frac{\nabla P}{P}\right)^2 - \frac{\Delta P}{P}\right] = -\frac{\hbar^2}{2m}\frac{\Delta R}{R} \tag{1.19}$$

Then define **(1P)** $\mathbf{u} = (\delta\mathbf{p}/m) = -(\hbar/2m)(\nabla P/P)$ and $k_{\mathbf{u}} = -(1/2)(\nabla P/P) = -(\nabla R/R)$ so that Q can be rewritten as

$$Q = \frac{m\mathbf{u}\cdot\mathbf{u}}{2} - \frac{\hbar}{2}(\nabla\cdot\mathbf{u}) = \frac{\hbar^2}{2m}(k_{\mathbf{u}}\cdot k_{\mathbf{u}} - \nabla\cdot k_{\mathbf{u}}) \tag{1.20}$$

Using (1.13) and (1.14) one can also write **(1Q)** $\mathbf{u} = (1/2\omega m)\nabla\mathcal{Q}$.

Generally a steady state oscillator in nonequilibrium thermodynamics corresponds to a kinetic energy at the sub-quantum level providing the necessary energy to maintain a constant oscillation frequency $\omega$ and some excess kinetic energy resulting in a fluctuating momentum contribution $\delta p$ to the momentum $p$ of the particle (note $p \sim \mathbf{p}$). Similarly a steady state resonator representing a "particle" in a thermodynamic environment will not only receive kinetic energy from it but in order to balance the stochastic influence of the buffeting momentum fluctuations it will also dissipate heat into the environment. There is a vacuum fluctuation theorem (VFT) from [**392**]-1 which proposes that larger energy fluctuations of the oscillating system correspond to higher probability of heat dissipated into the environment (rather that absorbed). The corresponding balancing velocity is called (after Einstein) the "osmotic" velocity. Thus recalling the stochastic "forward" movement $\mathbf{u} \sim (\delta\mathbf{p}/m)$, the current $\mathbf{J} = P\mathbf{u}$ has to be balanced by $-\mathbf{u}$, i.e. $\mathbf{J} = -P\mathbf{u}$. Putting (**1P**) into the definition of the "forward" diffusive current $\mathbf{J}$ and recalling the diffusivity $D = \hbar/2m$ one has (**1R**) $\mathbf{J} = P\mathbf{u} = -D\nabla P$ and when combined with the continuity equation $\dot{P} = -\nabla\cdot\mathbf{J}$ this gives (**1S**) $\partial_t P = D\nabla^2 P$. Here (**1R**) and (**1S**) are the first and second of the Fick laws of diffusion and $\mathbf{J}$ is called the diffusion current.

Returning now to (**1Q**) one defines $\Delta\mathcal{Q} = \mathcal{Q}(t) - \mathcal{Q}(0) < 0$ and maintaining heat flow as positive one writes $-\Delta\mathcal{Q}$ for heat dissipation and puts this in (**1Q**) to get the osmotic velocity (**1T**) $\bar{\mathbf{u}} = -\mathbf{u} = D(\nabla P/P) = -(1/2\omega m)\nabla\mathcal{Q}$ with osmotic current (**1U**) $\bar{\mathbf{J}} = P\bar{\mathbf{u}} = D\nabla P = -(P/2\omega m)\nabla\mathcal{Q}$. As a corollary to Fick's second law one has then

$$\partial_t P = -\nabla\cdot\bar{\mathbf{J}} = -D\nabla^2 P = \frac{1}{2\omega m}[\nabla P\cdot\nabla\mathcal{Q} + P\nabla^2\mathcal{Q}] \tag{1.21}$$

Next one looks for a thermodynamic meaning for the quantum potential Q. Take first $Q = 0$ and look at the osmotic velocity (**1T**) that represents the heat dissipation from the particle into its environment. From (1.20) one has (**1V**) $(\hbar/2)(\nabla\cdot\mathbf{u}) = (1/2)(m\mathbf{u}\cdot\mathbf{u})$. Inserting now (**1T**) instead of (**1Q**) yields the thermodynamic corollary of a vanishing QP as (**1W**) $\nabla^2\mathcal{Q} = (1/2\hbar\omega)(\nabla\mathcal{Q})^2$. Returning to (1.21) put first (**1K**) into (**1S**) to get (**1X**) $P = P_0 exp[-(\Delta\mathcal{Q}/\hbar\omega)]$ and one obtains then from (1.21)

$$\partial_t P = \frac{P}{2\omega m}\left[\nabla^2\mathcal{Q} - \frac{(\nabla\mathcal{Q})^2}{\hbar\omega}\right] \Rightarrow \partial_t P = -\frac{P}{2\omega m}\nabla^2\mathcal{Q} \tag{1.22}$$

(via (**1W**)). Now from (**1X**) one has also (**1Y**) $\partial_t P = -(P/\hbar\omega)\partial_t\mathcal{Q}$ so comparison of (1.22) and (**1X**) yields (**1Z**) $\nabla^2\mathcal{Q} - (1/D)\partial_t\mathcal{Q} = 0$ ($\tilde{\mathcal{Q}} = (1/\hbar\omega)\mathcal{Q}$ is inserted here in a second version of the last paper in [**392**]). This is nothing but a classical heat equation obtained by the requirement that the quantum potential $Q = 0$; it shows that even for free particles both in the quantum and classical case one can identify a heat dissipation process emanating from the particle. A non-vanishing quantum potential then is a means of describing the spatial and temporal dependencies of the corresponding thermal flow in the case that the particle is not free.

Various particular solutions to (**1Z**) are indicated for the case $Q = 0$ as well as when $Q \neq 0$. Examples are discussed with a view toward resolving a certain "particle in a box" problem of Einstein. In [**392**]-1 one concentrated on the momentum fluctuations $\delta p$ generated from the environment to the particle while in [**392**]-3 one develops the idea of excess energy developed as heat from the particle to its environment (which is described via the quantum potential). In fact one can rewrite (1.19) as

$$\partial_t S + \frac{1}{2m}(\nabla S)^2 + V + \frac{\hbar^2}{4m}\left[\nabla^2\tilde{\mathcal{Q}} - \frac{1}{D}\partial_t\tilde{\mathcal{Q}}\right] = 0 \tag{1.23}$$

where $\tilde{\mathcal{Q}} = \mathcal{Q}/\hbar\omega$. This provides considerable insight into the nature and role of the quantum potential.

Note that one has achieved a "thermalization" of the quantum potential (QP) in the form

$$Q = \frac{\hbar^2}{4m}\left[\nabla^2\tilde{\mathcal{Q}} - \frac{1}{D}\partial_t\tilde{\mathcal{Q}}\right] \tag{1.24}$$

where $\tilde{\mathcal{Q}} = \mathcal{Q}/\hbar\omega = \alpha\mathcal{Q}$ is an expression of heat and $D = \hbar/2m$ is a diffusion coefficient (note $\alpha = \beta$ as in (**1K**). In [**173**] we show that, as a corollary, one can produce a related thermalization of Fisher information (FI) which should have interesting consequences. Thus in (**1U**) one uses a formula

$$\frac{\nabla P}{P} = -\frac{1}{2\omega m D}\nabla\mathcal{Q} = -\frac{1}{\omega\hbar}\nabla\mathcal{Q} = -\beta\nabla\mathcal{Q} \tag{1.25}$$

and this leads one to think of $\nabla log(P) = -\alpha\nabla\mathcal{Q} = -\nabla(\alpha\mathcal{Q})$ with a possible solution

$$log(P) = -\alpha\mathcal{Q} + c(t) \Rightarrow P = exp[-\alpha\mathcal{Q} + c(t)] = \hat{c}(t)e^{-\alpha\mathcal{Q}} \tag{1.26}$$

(cf. also (1.13)). Now Fisher information (FI) is defined via ($dx \sim dx^3$ for example)

$$F = \int\frac{(\nabla P)^2}{P}dx = \int P\left(\frac{\nabla P}{P}\right)^2 dx \tag{1.27}$$

and one can write $Q$ as in (1.19). Consequently (since $\int\Delta P dx = 0$)

$$\int PQdx = -\frac{\hbar^2}{8m}\int\frac{(\nabla P)^2}{P}dx = -\frac{\hbar^2}{8m}F \tag{1.28}$$

Then as in [**173**] one can write formally, first using (1.24) and $\alpha = 1/\omega\hbar$

$$F = -\frac{8m}{\hbar^2}\int PQdx = -\frac{8m}{\hbar^2}\int P\frac{\hbar^2}{4m}\left[\nabla^2\tilde{\mathcal{Q}} - \frac{1}{D}\partial_t\tilde{\mathcal{Q}}\right]dx \tag{1.29}$$
$$= -2\alpha\int P\left[\nabla^2\mathcal{Q} - \frac{2m}{\hbar}\partial_t\mathcal{Q}\right]dx$$

and secondly, using (1.25) and (1.26) (recall also $\alpha = \beta$)

$$F = \int P\left(\frac{\nabla P}{P}\right)^2 dx = \beta^2\hat{c}(t)\int e^{-\beta\mathcal{Q}}(\nabla\mathcal{Q})^2 dx \tag{1.30}$$

In view of the thermal aspects of gravity theories now prevalent it may perhaps be suggested that connections of quantum mechanics to gravity may best be handled thermally. There may also be connections here to the emergent quantum mechanics of [**290, 444, 445, 446**].

Following Garbaczewski [**333**] one can look at a diffusion model with current velocity $\mathbf{v} = \mathbf{b} - \mathbf{u}$ where $u = D\nabla(log(\rho))$ is an osmotic velocity field and $D = \hbar/2m$. The continuity equation is a Fokker-Planck equation (♦) $\partial_t\rho = D\Delta\rho = \nabla\cdot(\mathbf{b}\rho)$ and assuming $\rho$, $\mathbf{v}\rho$, and $\mathbf{b}\rho$ vanish at spatial infinity (or at boundaries) leads to an entropy balance equation

$$\frac{d\mathfrak{S}}{dt} = \int\left[\rho(\nabla\cdot\mathbf{b}) + D\frac{(\nabla\rho)^2}{\rho}\right]d^3x \tag{1.31}$$

(cf. also [**170, 332, 334, 335**]). ■

**1.2. FLUCTUATIONS AND FISHER INFORMATION.** We go next to [**246**] which contains a rich lode of important material on quantum fluctuations and some penentrating insight into physics (but some proofreading seems indicated). We will try to rewrite some of this in a more complete manner. Thus consider a SE for $\psi = Rexp(iS/\hbar)$ with momentum operator $\hat{p}$ so that

$$\hat{p}\psi = p\psi = \left(\nabla S + \frac{\hbar}{i}\frac{\nabla R}{R}\right)\psi \Rightarrow p \sim <p> + \delta p \tag{1.32}$$

which identifies $(\hbar/i)(\nabla R/R)$ as a fluctuation $\delta p$. Recall now that the SE is (**2A**) $i\hbar\psi_t = -(\hbar^2/2m)\Delta\psi + V\psi$ and setting $\rho = R^2 = \psi^*\psi$ one has

$$\partial_t\rho = \frac{i\hbar}{2m}[\psi^*\Delta\psi - (\Delta\psi^*)\psi] = \frac{i\hbar}{2m}\nabla\cdot(\psi^*\nabla\psi - \nabla(\psi^*)\psi) \tag{1.33}$$

Now in polar form $\psi = Rexp(iS/\hbar)$, $\rho = R^2$ this becomes (calculating sometimes in 1-D for simplicity)

$$\nabla(Re^{iS/\hbar}) = \nabla Re^{iS/\hbar} + \frac{iR\nabla S}{\hbar}e^{iS/\hbar} \tag{1.34}$$

$$\psi^*\nabla\psi - (\nabla\psi^*)\psi = R\nabla R + \frac{iR^2\nabla S}{\hbar} - R\nabla R + \frac{iR^2\nabla S}{\hbar} = \frac{2i\rho\nabla S}{\hbar} \tag{1.35}$$

$$\Rightarrow \partial_t\rho = \frac{i\hbar}{2m}\nabla\left(\frac{2i\rho\nabla S}{\hbar}\right) = -\frac{1}{m}\nabla(\rho\nabla S)$$

Now the quantum potential is

$$Q = -\frac{\hbar^2}{2m}\frac{\Delta\rho^{1/2}}{\rho^{1/2}} = -\frac{\hbar^2}{4m}\left[\frac{1}{2}\left(\frac{\nabla\rho}{\rho}\right)^2 - \frac{\Delta\rho}{\rho}\right] = \frac{\hbar^2}{4m}\left[\frac{\Delta\rho}{\rho} - \frac{1}{2}\left(\frac{\nabla\rho}{\rho}\right)^2\right] \tag{1.36}$$

(cf. (1.1) and $\rho = R^2$ so $\rho' = 2RR'$ and $2(R'/R) = \rho'/\rho$; hence one has (**2B**) $\delta p \sim (\hbar/2i)(\nabla\rho/\rho)$. There is equivalent material about Fokker-Planck equations in [**170**] so we omit the discussion in [**246**]. One notes that (**2C**) $<\delta p> \sim \int \rho\delta p dx = (\hbar/2i)\int \nabla\rho dx = 0$ whereas $<(\delta p)^2> \sim c\int[(\nabla\rho)^2/\rho]dx \sim$ *Fisher information*. Here the quantum potential can also be written as

$$Q = -\frac{\hbar^2}{4m}\left[\frac{1}{2}(\delta p)^2 - \frac{\Delta\rho}{\rho}\right] = \frac{\hbar^2}{4m}\left[\nabla(\delta p) + \frac{1}{2}(\delta p)^2\right] \tag{1.37}$$

since $(\rho'/\rho)' = (\rho''/\rho) - [(\rho')^2/\rho^2] \sim \Delta\rho/\rho = \nabla(\delta p) + (\delta p)^2$. Note also that the exact uncertainty principle of Hall-Reginatto ([**400, 401**]), developed at length in [**170**], is based on momentum fluctuations $p = \nabla S + \delta p$ with $<\delta p> = 0$. If one writes $V(q) = V(<q>) + \nabla_q V(<q>)\delta q$ as a function of position and recalls that the quantum HJ equation has the form (**2D**) $S_t + (p^2/2m) + V + Q = 0$ (cf. [**170**]) then this can be rewritten in terms of fluctuations as

$$S_t + \frac{1}{2m} <p^2> + V(<q>) + \nabla_q V(<q>)\delta q + Q(p,\delta p); \tag{1.38}$$

$$Q = \frac{\hbar^2}{4m}\left(-i\hbar\nabla\cdot\delta p + \frac{1}{2}(\delta p)^2\right)$$

We go now to [**322**] with (**2E**) $\int dx\, p(x) = 1$ and (**2F**) $I[p] = \int dx F_I(p)$ where $F_I(p) = p(x)[(p')/p]^2$. Assume that there are known

$$< A_j >= \int dx A_j(x)p(x) \ \ (j = 1, \cdots, M) \tag{1.39}$$

and use the principle of extreme physical information (EPI) developed by B. Frieden et al (cf. [**318**]) to find the probability distribution $p = p_I$ extremizing $I[p]$ subject to prior conditions $< A_j >$. Jaynes used the Shannon functional $F = -plog(p)$ with (**2G**) $S[p] = -\int dxplog(p)$ but here one uses the Fisher extremization with

$$\delta_p\left[I[p] - \alpha < 1 > - \sum_1^M \lambda_i < A_i >\right] = 0 \tag{1.40}$$

$$\equiv \delta_p\left[\int dx\left(F_I[p] - \alpha p - \sum_i^M \lambda_i A_i p\right)\right] = 0$$

Variation leads to

$$\delta\int \frac{p'^2}{p^2}dx \approx \int\left[-\frac{p'^2}{p^2}\delta p + \frac{2p'}{p}\delta p'\right]dx \sim \int\left[-\frac{p'^2}{p^2} - \partial\left(\frac{2p'}{p}\right)\right]\delta p dx$$

$$\int dx\delta p\left[(p)^{-2}(p')^2 + \partial_x\left(\frac{2}{p}p'\right) + \alpha + \sum_1^M \lambda_i A_i\right] = 0 \tag{1.41}$$

which implies, via the arbitrary nature of $\delta p$

$$\left[(p)^{-2}(p')^2 + \partial_x\left(\frac{2}{p}p'\right) + \alpha + \sum_1^M \lambda_i A_i\right] = 0 \tag{1.42}$$

The normalization condition on $p$ makes $\alpha$ a function of the $\lambda_i$ and one lets $p_I(x, \lambda_i)$ be a solution of (1.42). Then the extreme Fisher information is (**2H**) $I = \int dx p_I^{-1}[p_I']^2$. Now one can simplify (1.42) via (**2I**) $G(x) = +\alpha + \sum_1^M \lambda_i A_i(x)$ and write

$$[\partial_x log(p_I)]^2 + 2\frac{\partial^2 log(p_I)}{\partial x^2} + G(x) = 0 \tag{1.43}$$

Then introduce $p_I = \psi^2$ and (**2J**) $v(x) = \partial_x log(\psi(x))$ so that (1.43) becomes (**2K**) $v'(x) = -[(1/4)G(x) + v^2(x)]$ which is a Riccati equation. Setting

$$u(x) = exp\left[\int^x dx\frac{dlog(\psi)}{dx}\right] = \psi \tag{1.44}$$

makes (1.43) into a Schrödinger-like equation

$$-\frac{1}{2}\psi''(x) - \frac{1}{8}\sum \lambda_i A_i(x)\psi(x) = \frac{\alpha}{8}\psi \tag{1.45}$$

where (**2L**) $U(x) = (1/8)\sum_1^M \lambda_i A_i(x)$ is an effective potential (cf. [**320, 322, 323, 641**]). Note that $\psi$ is defined here completely via $p_I$ so any quantum motion is automatically generated by fluctuation energy (i.e. $S \sim \delta S$).

Consider now a situation with one function $A_i$ **(2M)** $\bar{A} = \int pAdx$. Then from (1.40)-(1.43)

$$p^{-2}(p')^2 + \partial_x\left(\frac{2p'}{p}\right) + \alpha + \lambda A = 0 \tag{1.46}$$

Now, following [**322**], one translates the Legendre structure of thermodynamics into a Fisher context. Thus from (**2H**), integrating by parts implies

$$\frac{\partial I}{\partial\lambda} = \int dx\frac{\partial p_I}{\partial\lambda}\left[-p_I^{-2}\left(\frac{\partial p_I}{\partial x}\right)^2 - \frac{\partial}{\partial x}\left(\frac{2}{p_I}\frac{\partial p_I}{\partial x}\right)\right] \tag{1.47}$$

where $p_I$ is a solution of (1.46). Comparing (1.46) to (1.47) one has

$$\frac{\partial I}{\partial\lambda} = \int dx\frac{\partial p_I}{\partial\lambda}[\alpha + \lambda A] \tag{1.48}$$

Then on account of normalization ($\int p_I dx = 1$)

$$\frac{\partial I}{\partial\lambda} = \lambda\frac{\partial}{\partial\lambda}\int dx\, p_I A(x) \equiv \frac{\partial I}{\partial\lambda} = \lambda\frac{\partial}{\partial\lambda} < A > \tag{1.49}$$

which is a generalized Fisher-Euler theorem. The term $\int dx\alpha\partial_\lambda p_I \sim \alpha\partial_\lambda\int dx p_I = 0$ via $\int p_I dx = 1$. The thermodynamic counterpart of (1.49) is the derivative of the entropy with respect to mean values. Thus $I = I(\lambda)$, $p_I = p_I(\lambda)$, and via normalization $\alpha = \alpha(\lambda)$. Thus $\lambda$ and $< A >$ play reciprocal roles within thermodynamics and one introduces a generalized thermodynamic potential as a Legendre transform of $I$, namely

$$\Lambda = I(< A >) - \lambda < A > \tag{1.50}$$

Then, using (1.49)

$$\frac{\partial\Lambda}{\partial\lambda} = \frac{\partial I}{\partial < A >}\frac{\partial < A >}{\partial\lambda} - \lambda\frac{\partial < A >}{\partial\lambda} - < A > = - < A > \tag{1.51}$$

and one has a summary collection of formulas

$$\Lambda = I - \lambda < A >;\ \frac{\partial\Lambda}{\partial\lambda} = - < A >;\ \frac{\partial I}{\partial < A >} = \lambda; \tag{1.52}$$

$$\frac{\partial\lambda}{\partial < A >} = \frac{\partial^2 I}{\partial < A >^2};\ \frac{\partial < A >}{\partial\lambda} = -\frac{\partial^2\Lambda}{\partial\lambda^2}$$

and we recall (1.49) in the form (**2N**) $\frac{\partial I}{\partial\lambda} = \lambda\frac{\partial < A >}{\partial\lambda}$. Thus the Legendre transform structure of thermodynamics has been translated into the Fisher context (see here also Chapter 4 of [**318**] for more on this and for additional general information we cite e.g. [**308, 319, 580, 583, 584, 609, 687, 688, 689, 690, 691, 705, 706, 707, 708, 709, 722, 734**]).

On the other hand with $\bar{A}$ the sole constraint consider

$$\tilde{H} = FI + \alpha(\bar{A} - \int pAdx) \tag{1.53}$$

Then one finds directly

$$\delta\tilde{H} = \int dx\,\delta p\left[\frac{(p')^2}{p^2} + \partial_x\left(\frac{2p'}{p}\right) + \alpha A\right] \tag{1.54}$$

This means (as in (1.46))

$$\frac{(p')^2}{p^2} + \partial_x\left(\frac{2p'}{p}\right) + \alpha A = 0 \tag{1.55}$$

Now we can write $\partial_x(2p'/p) = (2p''/p) - [2(p')^2/p^2]$ which means, via (1.36), that

$$2\frac{\Delta p^{1/2}}{p^{1/2}} + \alpha A = 0 \tag{1.56}$$

and via (★) in Section 1.1 this means that the extreme probability $p_I$ directly determines a quantum potential $Q$ via

$$Q = -\frac{\hbar^2}{2m}\frac{\Delta p_I^{1/2}}{p_I^{1/2}} \Rightarrow Q = \frac{\hbar^2}{4m}[\alpha A] \tag{1.57}$$

However, although $A$ has not been specified we seem to have a result that a constraint $\bar{A}$ for which the Fisher thermodynamic procedure works with $I[p]$ as defined, requires $A$ to satisfy (2.26). This is in fact tautological since we are dealing with a situation where fluctuations based on $p_I$ are the only source of energy and one will have (**2O**) $\tilde{F} = \int PQdx = (\hbar^2/8m)\int[(\nabla p)^2/p]dx$ (cf. [**170**]) and $\tilde{F}$ corresponds to a fluctuation energy. In particular as indicated in [**170**]

$$\int p_I Q dx = \frac{\hbar^2}{4m}\int dx\, p[\alpha A] = \frac{\hbar^2}{4m}[\alpha\bar{A}] \tag{1.58}$$

Note that in general Fisher information $\tilde{F} = I[p]$ as in (**2G**) is an action term which can be added to a classical Hamiltonian in order to quantize it and thus $\tilde{F}$ is a "natural" constraint ingredient. Fixing $\bar{A}$ would mean fixing the contribution of the quantum potential (QP) Q or fixing the amount of quantization allowed. In some way this would also correspond to restraining the probability in order to achieve a fixed amount of quantization.

**REMARK 1.1.1.** Suppose we extremize $\tilde{\mathfrak{S}}$ as in (1.11) with $A \sim E =$ total energy to arrive at a probability (**2P**) $P = (1/Z)exp(-\gamma E)$ where $E \sim E(x,t)$, $Z = \int exp(-\gamma E)dx$. Then compute the Fisher information for this P which will be based upon (cf. (1.3))

$$P' = \frac{1}{Z}(-\gamma E')e^{-\gamma E};\ P'' = -\frac{\gamma}{Z}[E'' - \gamma(E')^2]e^{-\gamma E} \tag{1.59}$$

$$\frac{P'}{P} = -\gamma E';\ \frac{P''}{P} = -\gamma E'' + \gamma^2(E')^2 \tag{1.60}$$

$$Q = \frac{\gamma\hbar^2}{8m}[\gamma(E')^2 - 2E''] \tag{1.61}$$

$$FI = \frac{8m}{\hbar^2}\int PQdx = \frac{\gamma^2}{Z}\int(E')^2e^{-\gamma E}dx \tag{1.62}$$

On the other hand extremizing FI as in (1.53), with fluctuations as the only source of energy, leads to $P = p_I$ defined via A as in (1.56) which yields in paticular $Q = (\hbar^2/4m)[\alpha A]$ as in (2.26). Now recall for $P$ as in (**1J**)

(1) $\frac{P(x,t)}{P(x,0)} = e^{\frac{-\Delta\mathfrak{Q}}{kT}}$ from (**1J**)

(2) $\Delta\mathfrak{Q} = \mathfrak{Q}(t) - \mathfrak{Q}(0) = 2\omega[(\delta S)(t) - (\delta S)(0)]$ from (1.13) and $(2/\hbar)\delta S(t) \sim \beta\mathfrak{Q}(t)$ (cf. (•))

(3) $\delta p = \nabla(\delta S(x,t)) = -(\hbar/2)(\nabla P/P)$ as in (1.15) and from (1.25) $(\nabla P/P) = (1/\omega\hbar)\nabla\mathfrak{Q} = \beta\nabla\mathfrak{Q}$

(4) $\delta E_{kin} = (\hbar^2/8m)(\nabla P/P)^2 = (1/8m\omega^2)(\nabla\mathfrak{Q})^2$

(5) $P = \hat{c}(t)exp(-\alpha\mathfrak{Q})$ from (1.26) with $\alpha = 1/\omega\hbar \sim 1/kT \sim \beta$ - this can also be seen from (1.14) as $P \propto exp[-\beta\mathfrak{Q}(t)]$

For E = total energy as in (**2P**) it is clear that to apply this here we must think of $E = \hbar\omega + \delta E_{kin}$ and $\hbar\omega$ will only enter as a constant. The thermalization fluctuation energy for probabilities appears as $\hat{c}exp[-\delta S(x,t)] \sim \hat{c}exp[-\beta\mathfrak{Q}(t)]$ as indicated so $\delta E_{kin}$ is expressed in terms of $\mathfrak{Q}$. This means that (1.62) involving $(E')^2$ (or $(\nabla E)^2$) for FI corresponds to the $(\nabla\mathfrak{Q})^2$ formula of (1.30). In other words the fluctuation energy is equivalent to the thermal energy (modulo constants); see here also [**170, 369, 436**]. ■

We omit here mention of many related topics and work involving Fisher information, thermodynamics, and entropy based on results of Abe, Frieden, Garbaczewski, Gellman, Hall, Kaniadakis, Naudts, Pennini, A. Plastino, A.R. Plastino, Reginatto, Soffer, and Tsallis in particular; some references can be found in [**170, 278, 319, 318, 341, 647**]. For cogent discussions of many aspects of quantum mechanics see Nikolić [**635, 636, 637, 638, 639, 640**].

**REMARK 1.1.2.** We refer here to [**97, 170, 177, 175, 176, 297, 589, 851**], where the Schwarzian derivative arises in the quantum equivalence principle (QEP) of Faraggi-Matone, and to [**593, 594, 595**] where it comes up in semi-classical expansions (note also that there is a difference between semi-classical and quasi-classical - cf. [**606**]). Thus the stationary SE $(\hbar^2/2m)\psi'' - V\psi = E\psi$ can be written via $\psi = exp(iW/\hbar)$ with $W = s + (\hbar/i)log(A)$ as

$$(s')^2 - 2m(E - V) = \hbar^2\frac{A''}{A};\ 2A's' + As'' = 0 \tag{1.63}$$

Therefore (**2Q**) $A = c(s')^{-1/2}$ and

$$(s')^2 = 2m(E - V) + \frac{\hbar^2}{2}\left[\frac{3}{2}\left(\frac{s''}{s'}\right)^2 - \frac{s'''}{s'}\right] \tag{1.64}$$

which can be written as

$$\frac{1}{2m}(s')^2 + V + Q = E \tag{1.65}$$

Now from (1.1) we have $Q = -(\hbar^2/4m)(A''/A)$ (since $A = exp(log(A))$) and hence

$$Q = -\frac{\hbar^2}{2m}\left(\frac{[(s')^{-1/2}]''}{[s']^{-1/2}}\right) = \frac{\hbar^2}{4m}\left[\frac{s'''}{s'} - \frac{3}{2}\left(\frac{s''}{s'}\right)^2\right] = \frac{\hbar^2}{4m}\mathfrak{S}Z(s) \tag{1.66}$$

defines the quantum potential via a Schwarzian derivative $\mathfrak{S}Z$. The Schwarzian derivative (and Ermakov invariant) arise also in [**595**] in dealing with semiclassical theory for (**2R**) $\hbar^2\partial_x^2\psi(x) + p^2(x)\psi(x) = 0$ when writing the wave function as $\psi(x) = u(x)w(\xi(x))$ with the assumption (**2S**) $\hbar^2\partial_\xi^2 w(\xi) + R(\xi)w(\xi) = 0$. This leads to (**2T**) $\psi(x) = (\partial_x\xi)^{-1/2}w(\xi(x))$ and

$$R(\xi)(\partial_x\xi)^2 - p(x)^2 + \frac{\hbar^2}{2}\left[\frac{\xi'''}{\xi'} - \frac{3}{2}\left(\frac{\xi''}{\xi'}\right)^2\right] = 0 \tag{1.67}$$

where the last term can be written as (**2U**) $(\hbar^2/2) < \xi; x >= (\hbar^2/2)\mathfrak{S}Z(\xi)$ (cf. also [**785, 786**] where relations of Ricatti equations to the Wigner function and Ermakov invariants arise).

**1.3. THE QUANTUM EQUIVALENCE PRINCIPLE - QEP.** The postulate that physical states be equivalent under coordinate transformations leads to the quantum Hamilton-Jacobi equation (QSJE) which in turn gives the SE (cf. [**97, 297, 589**]). We have written about this before (cf. [**170, 177, 175, 176**]) but it is now imperative to put in more detail about the QEP if one hopes to understand anything about relations between classical and quantum mechanics. We will only deal with the non-relativistic situation here and refer to [**297**] (IJMPA - 2000) for a detailed discussion. Thus one looks for coordinate transformations (**TV**) $q \to q^v = v(q)$ with $S_0^v(q^v) = S_0((q^v))$ where $S_0$ corresponds to the reduced action from Hamilton's principal function $S = S_0 - Et$ in the stationary case. Then $p \to p_v = (\partial_q q^v)^{-1}p$ so $p$ transforms as $\partial_q$ and the formalism will be covariant. Now in classical mechanics there are problems with $p - q$ duality due to the manner in which time enters into the formalism and one looks for a formulation of dynamics with manifest $p - q$ duality via Legendre transforms

$$s = S_0(q) = pq - T_0(p);\ q = \frac{\partial T_0(p)}{\partial p} \tag{1.68}$$

There is now a canonical equation (cf. also Chapter 8)

$$\left(\frac{\partial^2}{\partial s^2} + \mathfrak{U}(s)\right) q\sqrt{p} = \left(\frac{\partial^2}{\partial s^2} + \mathfrak{U}(s)\right)\sqrt{p} = 0 \tag{1.69}$$

where $\mathfrak{U}(s) = \mathfrak{S}Z(s) = \{q; s\} = (q'''/q') - (3/2)(q''/q')^2$ ($' \sim d/ds$). This can be directly checked from (1.68) via e.g.

$$\partial_s\sqrt{p} = (1/2)p^{-1/2}p_s;\ \partial_s^2\sqrt{p} = -(1/4)p^{-3/2}p_s^2 + (1/2)p^{-1/2}p_{ss}; \tag{1.70}$$

$$p_s q_s + p q_{ss} = 0 \Rightarrow \frac{p_s}{p} = -\frac{q_{ss}}{q_s} \Rightarrow \frac{p_{ss}}{p} = -\frac{q_{sss}}{q_s} + 2\left(\frac{q_{ss}}{q_s}\right)^2$$

Hence e.g. $\partial_s^2\sqrt{p} = -\mathfrak{U}(s)\sqrt{p}$ and we note also that (**3A**) $(1/q\sqrt{p})(\partial^2(q\sqrt{p})/\partial s^2) = (1/\sqrt{p})(\partial^2\sqrt{p})/\partial s^2)$. Now $\mathfrak{U}(s)$ is considered as a sort of canonical potential (consistent with the identification (1.66)) and we note for a Möbius transformation ($GL(2, \mathbf{C})$ transformation)

$$q \to q^v = \frac{Aq + B}{Cq + D};\ p \to p_v = \rho^{-1}(Cq + D)^2 p \tag{1.71}$$

for $\rho = AD - BC$. For the Schwarzian derivative (in $x$) one has

(1.72) $$\{\gamma(h), x\} = \{h; x\};\ \gamma(h) = \frac{Ah + B}{Ch + D}$$

(cf. [**297**]). Then the $GL(2, \mathbf{C})$ invariance of the canonical equation (1.69) means that given $\mathfrak{U}(s)$ ($\sim Q$) with $y_1(s)$ and $y_2(s)$ linearly independent solutions of (1.69) it follows that

(1.73) $$q\sqrt{p} = Ay_1(s) + By_2(s);\ \sqrt{p} = Cy_1(s) + Dy_2(s)$$

This then leads to the dynamical coordinate $q$ via

(1.74) $$q = \frac{Ah(s) + B}{Ch(s) + D};\ h(s) = \frac{y_1(s)}{y_2(s)}$$

Inverting this one finds $S_0 = h^{-1} \circ \gamma^{-1}$ with $\gamma$ as in (1.72) so the solution of the dynamical problem is given via (**3B**) $S_0(q) = h^{-1}(\gamma^{-1}(q))$.

Now the formal $p - q$ duality of classical (Hamiltonian) mechanics is broken in the explicit solution of the equations of motion (due to the difference between the structure of the kinetic term and the potential $V(q)$ term). What is lacking is a formula in which $p$ and $q$ descriptions have the same structure. From a classical Hamiltonian $(1/2m)p^2 + V(q)$ and action $S = S_0^{cl} - Et$ for example one has classical equations (**3C**) $(1/2m)(\partial_q S_0^{cl})^2 + V - E = 0$ and the quantity (**3D**) $\mathfrak{W} = V - E$ has special importance in the theory (as indicated below). One is looking for a formulation of dynamics in which there always exists coordinate transformations connecting arbitrary systems. For this there should be a nonzero energy function for any structure of $V$, which would then avoid the degenerate situation where $S_0$ is a constant. The EP states in fact that

(1) For each pair $(\mathfrak{W}^a, \mathfrak{W}^b)$ there is a $v$-transformation $q^a \to q^b = v(q^a)$ such that $\mathfrak{W}^a(q^a) \to \mathfrak{W}^b(q^b)$.

However this stipulation cannot be consistently implemented in classical mechanics (CM). Indeed $\mathfrak{W}$ states transform as quadratic differentials under $v$-maps, i.e. (**3E**) $\mathfrak{W}^v(q^v)(dq^v)^2 = \mathfrak{W}(q)(dq)^2$ and a constant state $q_0$ corresponds to (**3F**) $\mathfrak{W}^0(q^0) \to (\partial_{q^v} q^0)^2 \mathfrak{W}^0(q^0) = 0$. This means that $\mathfrak{W}^0$ could not be connected to other states not equal to 0 and this rules out CM as an arena for an EP. The way out of such an impasse is then to look at quantum mechanics (QM) where $\mathfrak{W} \to \mathfrak{W} + Q(q)$ with $Q$ as in (1.66).

If $\mathfrak{Q}$ is the space of functions transforming as quadratic differentials then in fact $\mathfrak{W} \notin \mathfrak{Q}$ and $Q \notin \mathfrak{Q}$ but $\mathfrak{W} + Q \in \mathfrak{Q}$, i.e.

(1.75) $$\mathfrak{W}^v(q^v) + Q^v(q^v) = (\partial_{q^v} q)^2(\mathfrak{W}(q) + Q(q))$$

This is connected with an important cocycle condition which arises since $\mathfrak{W}^v(q^v)$ must have an inhomogeneous form

(1.76) $$\mathfrak{W}^v(q^v) = (\partial_{q^v} q)^2(\mathfrak{W}(q) + (q; q^v)$$

In particular via (1.75)

(1.77) $$(\partial_{q^v} q)^2 \mathfrak{W}(q) + (q; q^v) + Q^v(q^v) = (\partial_{q^v}(q)^2(\mathfrak{W}(q) + Q(q))$$

which means

$$Q^v(q^v) = (\partial_{q^v} q)^2 Q(q) - (q; q^v); \ \mathfrak{W}(q) = (q^0; q) \tag{1.78}$$

Thus all states originate from the map $q^0 \to q = v_0^{-1}(q^0)$ and a little calculation leads to the cocycle condition

$$(q^a; q^c) = (\partial_{q^c} q^b)^2 [(q^a; q^b) - (q^c; q^b)] \tag{1.79}$$

There is then further calculation in [**297**] (IJMPA-2000) involving an explicit calculation to show that (1.79) implies (★) $(q^a; q^b) = (\beta^2/4m)\{q^a, q^b\}$ (where of course $\beta \sim \hbar$ in quantum mechanics). The classical limit is associated as usual with $lim_{\beta\to 0} Q = 0$. For the SE one has then the HJ equation form (QSHJE)

$$\frac{1}{2m}\left(\frac{\partial S_0(q)}{\partial q}\right)^2 + V(q) - E + \frac{\beta^2}{4m}\{S_0, q\} = 0 \tag{1.80}$$

where (cf. (1.65))

$$\mathfrak{W}(q) = -\frac{\beta^2}{4m}\{e^{2iS_0/\beta}, q\}; \ Q(q) = \frac{\beta^2}{4m}\{S_0, q\} \tag{1.81}$$

One sees that (1.80) implies the SE and it also implies

$$e^{(2i/\beta)S_0} = \frac{\psi^D}{\psi} \tag{1.82}$$

where $\psi$ and $\psi^D$ are two linearly independent solutions of the stationary SE $[-(\beta^2/2m)\partial_q^2 + V(q)]\psi = E\psi$ with $\beta = \hbar$. In the language of [**297**] the SE is the equation which linearizes the QSHJE. If one writes (1.80) via $\psi = Rexp[(i/\hbar)\hat{S}_0]$ for real $R$ and $\hat{S}_0$ then

$$\frac{1}{2m}\left(\frac{\partial \hat{S}_0}{\partial q}\right)^2 + V - E - \frac{\hbar^2}{2m}\frac{\partial_q^2 R}{\partial q^2} = 0; \ \partial_q\left(R^2\frac{\partial \hat{S}_0}{\partial q}\right) = 0 \tag{1.83}$$

In general of course one should use the linear combination $R(Aexp[-(i/\hbar)\hat{S}_0] + Bexp[(i/\hbar)\hat{S}_0]$. In fact the QSHJE is more fundamental than the SE (cf. [**187, 297, 310, 589, 890**]). For now however consider (1.82) and then the derivation of connections between (1.83) and (1.80) implies a distinction between these cases. Thus let $\psi^D$ be a linearly independent from $\psi$ solution of the SE; then either $\bar{\psi} \propto \psi$ or not. If $\psi \not\propto \bar{\psi}$ then write (♦) $\psi^D = \bar{\psi}$ so by (1.82) (**3G**) $S_0 = \hat{S}_0 + \pi k\hbar$ for $k \in \mathbf{R}$. The continuity equation in (1.83) gives then $R \propto 1/\sqrt{S_0'}$ and hence by (1.82) and (♦) there results (**3H**) $Q = (\hbar^2/4m)\{S_0; q\} = -(\hbar^2/4mR)\partial_q^2 R$ and (1.83) corresponds to (1.80). The difference between (1.80) and (1.83) becomes relevant when $\bar{\psi} \propto \psi$. In particular (**3E**) implies that $\hat{S}_0$ must be a constant and since this can be absorbed by normalization one can choose $\hat{S}_0 = 0$ which implies (**3I**) $\psi = R$ and $\psi^D = R\int_{q_0}^q dx R^{-2}$ which via (1.82) yields

$$S_0 = \frac{\hbar}{2i}log\int_{q_0}^q dxR^{-2}; \ (\partial_q S_0)^2 + \frac{\hbar^2}{2}\{S_0; q\} = -\frac{\hbar^2}{R}\partial_q^2 R \tag{1.84}$$

It follows that (1.80) is then equivalent to (**3J**) $(\hbar^2/2mR)\partial_q^2 R = V - E$. On the other hand for $\hat{S}_0 = constant$ (1.83) degenerates giving rise to (**3J**) or (1.80) (see [**297**]-(IJMPA-2000)) for further discussion of the classical limit and related matters and [**310, 594, 890**] for more on trajectories)

## 2. ON THE GUTZWILLER TRACE FORMULA

We begin with some remarks about phase space techniques and we will only try to say a little bit in relation to Bohmian mechanics, semi-classical methods, and quantum mechanics. A preliminary list of references should include [**6, 36, 57, 63, 84, 90, 94, 100, 113, 116, 121, 122, 124, 132, 133, 141, 170, 173, 177, 176, 241, 244, 257, 261, 267, 273, 274, 294, 296, 297, 307, 310, 324, 337, 342, 360, 361, 362, 369, 370, 372, 373, 379, 381, 382, 390, 391, 394, 396, 399, 410, 411, 436, 440, 441, 442, 448, 457, 458, 460, 486, 487, 488, 520, 541, 542, 555, 556, 588, 593, 594, 595, 598, 618, 628, 630, 650, 654, 658, 686, 711, 728, 774, 785, 786, 789, 815, 823, 838, 890, 901**]. For connections of semi-classical (or WKB) methods to the SE we extract from [**515**] (cf. [**536, 537**] for fractal aspects). Consider a stationary SE (**1A**) $-(\hbar^2/2m)\Delta\psi = (E-V(x))\psi$ with $p^2 = \sqrt{2m(E-V(x))}$ and $\psi = exp(iS_0/\hbar)$ leading to a Riccati equation (**1B**) $-i\hbar\Delta S_0 + |\nabla S_0|^2 - p^2 = 0$. One assumes $p(x)$ does not change much over a deBroglie wavelength (**1C**) $\lambda = 2\pi\hbar/p(x)$, i.e. (**1D**) $\epsilon = (2\pi\hbar/p(x))|\nabla p/p| << 1$ (WKB condition). The limit $\hbar \to 0$ determines the lowest order approximation to the eikonal $S_0(x)$ via (**1E**) $|\nabla S_0|^2 - p^2 = 0$. Writing $S_0 = \phi_0 - i\hbar\phi_1 + (-i\hbar)^2\phi_2 + \cdots$ one finds the WKB equations

(2.1)

$$(\nabla S_0)^2 - p^2 = 0; \ (\nabla^2 S_0 + 2\nabla S_0 \cdot \nabla S_1 = 0, \cdots, \nabla^2 S_n + \sum_0^{n+1} \nabla S_m \cdot \nabla S_{n+1-m} = 0$$

where only the vectors $s_n = \nabla S_n$ are basic allowing for determination of the terms. In 1-D there results

$$s_0 = \pm p(x); \ s_1 = -\frac{p'}{2p}; \ s_2 = \mp\left[-\frac{p''}{4p^2} + \frac{3p''}{8p^3}\right] = \mp p(x)\epsilon(x); \ s_3 = \frac{1}{2}\epsilon'; \cdots \tag{2.2}$$

Thus

$$S_0(x) = \pm \int dx p(x)[1 + \hbar^2\epsilon(x)] + \frac{i\hbar}{2}[log(p(x)) + \hbar^2\epsilon(x)] \pm \cdots \tag{2.3}$$

Keeping only terms up to order $\hbar$ (which is reasonable if $\hbar^2|\epsilon(x)| << 1$) one writes for the WKB wave function

$$\psi_{WKB}(x) = \frac{1}{\sqrt{p}} e^{\pm(i/\hbar)\int^x p(\xi)d\xi} \tag{2.4}$$

In the classically accessible regime $V(x) \leq E$ this is an oscillating wave function but in the inaccessible regime $V(x) \geq E$ it decreases or increases exponentially. The transition between regimes is nontrivial since for $V(x) \simeq E$ the WKB approximation breaks down (however some connection rules exist - cf. [**515**]).

The most interesting approach however is via path integrals à la van Vleck which leads eventually to the Gutzwiller trace formula. Thus consider a path integral of the form (**1F**) $(x_b t_b|x_a t_a) = \int \mathcal{D}x \, exp(i\mathcal{A}[x]/\hbar)$. When $\hbar \to 0$ one has a sum of rapidly oscillating terms which will approximately cancel each other out. In this spirit the dominant contribution will come from the region where the oscillations are weakest, i.e. near the extremum of the action $\delta\mathcal{A}[x] = 0$. For a point particle with action $\mathcal{A}[x] = \int dt[(m\dot{x}^2/2) - V(x)]$ one would look at $m\ddot{x} = -V'(x)$ with $E = (1/2)m\dot{x}^2 + V(x) = constant$ and $p_{cl}(t) = m\dot{x}_{cl}(t)$. Then **(1G)** $\mathcal{A}[x_{cl}] = \int_{x_a}^{x_b} dxp(x) - (t_b - t_a)E$. Now for an integral **(1H)** with $a'(x_0) = 0$

$$\int \frac{dx}{\sqrt{2\pi i\hbar}} e^{ia(x)/\hbar} \to ce^{a(x_{cl})/\hbar} \tag{2.5}$$

as $\hbar \to 0$. This can be proved by expanding

$$a(x) = a)x_{cl}) + \frac{1}{2}a''(x_{cl})(\delta x)^2 + \cdots \tag{2.6}$$

($\delta x = x - x_{cl}$) and one can determine the constant in (2.5) via $c = 1/\sqrt{a''(x_{cl})}$ (saddle point approximation). The semiclassical aapproximation to the quantum path integral **(1F)** involves now

$$\mathcal{A}[x,\dot{x}] = \mathcal{A}[x_{cl}] + \int_{t_a}^{t_b} dt \frac{\delta\mathcal{A}}{\delta x(t)}\delta x(t) + \frac{1}{2}\int_{t_a}^{t_b} dtdt' \frac{\delta^2\mathcal{A}}{\delta x(t)\delta x(t')}\delta x(t)\delta x(t') + \cdots \tag{2.7}$$

For a point particle the quadratic term is

$$\frac{1}{2}\int_{t_a}^{t_b} dtdt' \frac{\delta^2\mathcal{A}}{\delta x(t)\delta x(t')}\delta x(t)\delta x(t') = \int_{t_a}^{t_b} dt\left[\frac{m(\delta\dot{x})^2}{2} + \frac{1}{2}V''(x_{cl}(t))(\delta x)^2\right] \tag{2.8}$$

Thus the fluctuations behave like those of a harmonic oscillator with a time dependent frequency **(1H)** $\Omega^2(t) = (1/m)V''(x_{cl}(t))$ with $\delta x = 0$ at the endpoints. Since $x(t)$ and $\delta x(t)$ differ only by $x_{cl}(t)$ one arrives at a semi-classical limit of amplitude **(1I)** $(x_b t_b|x_a t_a) = exp[iA(x_b, x_a; t_b - t_a)F_{sc}(x_b, x_a; t_b - t_a)$ with fluctuation factor

$$F_{sc} = \int \mathcal{D}\delta x(t) exp\left[\frac{i}{\hbar}\int_{t_a}^{t_b} dt\frac{m}{2}(\delta\dot{x}^2 - \Omega^2(t)\delta x^2)\right] \tag{2.9}$$

$$= \frac{1}{\sqrt{2\pi i\epsilon\hbar/m}} det(-\bar{\nabla}\nabla - \Omega^2(t))^{-1/2} = \frac{1}{[2\pi i\hbar(t_b - t_a)/m]^{1/2}}\sqrt{\frac{det(-\partial_t^2}{det[-\partial_t^2 - \Omega^2(t)]}}$$

Further calculation is abetted by results of Gelfand-Yaglom (cf. [**515**]) and one arrives at

(2.10)

$$(x_b t_b; x_a t_a) = (2\pi i\hbar)^{-D/2}[det_D(-\partial_b^{x^i}\partial_a^{x^j} A(x_b, x_a; t_b - t_a))]^{1/2} e^{iA(x_b, x_a, t_b - t_a)/\hbar}$$

where the $D \times D$ determinant is the van Vleck-Pauli-Morette determinant (cf. [**515**]) and (1.10) can be written as

$$(x_b t_b|x_a t_a) = (2\pi i\hbar)^{-D/2}\left[det_D\left(-\frac{\partial p_b}{\partial x_a}\right)\right]^{1/2} e^{iA(x_b, x_a; t_b - t_a)/\hbar} \tag{2.11}$$

(here one has **(1J)** $\partial_{x_b^i}\partial_{x_a^j} A(x_b, x_a; t_b - t_a) = \partial p_b^i/\partial x_a^j$). Note **(1K)** $A(x_b, x_a; t_b - t_a) \sim S(x_b, x_a; E) - (t_b - t_a)E$ where $\partial A/\partial x_{b,a} = \pm p(x_{b,a})$ with $\mathcal{A}(x_{cl}) \sim A$.

**2.1. PROPAGATORS, KERNELS, AND THE WIGNER FUNCTION.** Going to [**124**] consider a single particle Hamiltonian $\hat{H} \sim -(\hbar^2/2m)\nabla^2 + V(r)$ ($r \sim$ radius in say 3-D). There are bound states with $\hat{H}|n>= E_n|n>$ ($E_n > 0$) and wave functions $\psi_n(r)$ with (**1L**) $<n|m>= \delta_{mn}$ and $\sum \psi_n^*(r')\psi_n(r) = \delta(r'-r)$. The canonical single particle partition function is then (**1M**) $Z(\beta) = \int_0^\infty exp(-\beta E)g(E)dE = \sum_n exp(-\beta E_n)$ where $g(E) = \sum \delta(E-E_n)$ and the Bloch density is

$$C(r,r';\beta) = \sum \psi_n^*(r')\psi_n(r)e^{-\beta E_n} = <r|e^{-\beta\hat{H}}|r'>; \tag{2.12}$$

with (**1N**) $-\partial_\beta C(r,r',\beta) = \hat{H}_r C(r,r',\beta)$ ($\hat{H}_r$ acts on the variable $r$ with boundary conditions $C(r,r',\beta=0) = \delta(r-r')$). Due to the orthogonality of the states one can write

$$Z(\beta) = Tr(C) = \int C(r,r,\beta)d^3r \tag{2.13}$$

and there is a classical form of the local Bloch density in D dimensions leading to

$$C_{cl}(r,r',\beta) = \left(\frac{m}{2\pi\hbar^2\beta}\right)^{D/2} e^{-\beta V(r)} exp\left[-\frac{m}{2\hbar^2\beta}(r-r')^2\right] \tag{2.14}$$

If one replaces $\beta$ in (2.12) by an imaginary time interval $\beta \to i(t-t')/\hbar$ the Bloch density becomes the single particle propagator describing the propagation of the particle from $r'$ to $r$ in a time interval $t-t'>0$; thus

$$K(r,r';t-t') = \sum \psi_n^*(r')\psi_n(r)e^{-(i/\hbar)E_n(t-t')} = <r|e^{-(i/\hbar)\hat{H}(t-t')}|r'> \tag{2.15}$$

It follows that

$$\psi_n(r,t) = \hat{K}\psi_n(r',t') = \int d^3r' K(r,r',t-t')\psi_n(r',t') \tag{2.16}$$

where (**1O**) $(-i\hbar\partial_t + \hat{H}_r)K(r,r',t-t') = -i\hbar\delta(r-r')$. Via (**1L**) one then sees that

$$K(r,r';t-t') = \int K(r,r'',t-t'')K(r'',r',t''-t')d^3r'' \tag{2.17}$$

Taking the Fourier integral of $K$ one has then

$$-\frac{i}{\hbar}\int_0^\infty K(r,r';t)e^{(i/\hbar)Et}dt = -\frac{i}{\hbar}\sum \psi_n^*(r')\psi_n(r)\int_0^\infty e^{(i/\hbar)(E-E_n)t}dt \tag{2.18}$$

Consequently the Green's function in energy representation is
(2.19)

$$G(r,r';E) = -\frac{i}{\hbar}lim_{\epsilon\to 0}\int_0^\infty K(r,r't)e^{(i/\hbar)(E+i\epsilon)t}dt = \sum \psi_n^*(r')\psi_n(r)\frac{1}{E-E_n}$$

with (**1P**) $(E-\hat{H}_r)G(r,r';E) = \delta(r-r')$. In 3-D one obtains then for the free Green's function ($V=0$ and $\beta \to it/\hbar$)

$$G_0(r,r';E) = -\left(\frac{2m}{\hbar}\right)\frac{exp(ik|r-r'|)}{4\pi|r-r'|} \tag{2.20}$$

(in 2-D there is a Hankel function $H_0^+(k|r-r'|)$ and $k \sim \sqrt{2mE}/\hbar$ is the wave number).

The problem of finding a particle between $(x,p)$ and $(x+\delta x, p+\delta p)$ is classically

$$F_{cl}(x,p,\beta) = \frac{1}{\hbar}e^{-(\beta p^2/2m)}e^{-\beta V(x)} \tag{2.21}$$

The quantum probability can be a function of $x$ or $p$ alone with

$$\hat{\rho} = |\psi><\psi|;\ \rho(x,x') = <x|\hat{\rho}|x,> = \psi^*(x')\psi(x) \tag{2.22}$$

The Wigner transform of $\hat{\rho}$ is

$$\rho_W(x,p) = \frac{1}{2\pi\hbar}\int_{-\infty}^{\infty} dy <x-(y/2)|\hat{\rho}|x+(y/2)> e^{ipy/\hbar} \tag{2.23}$$
$$= \frac{1}{2\pi\hbar}\int_{-\infty}^{\infty} dy\psi^*(x+(y/2))\psi(x-(y/2))e^{(ipy/\hbar)}$$

and one defines the momentum space wave function via a formula (**1Q**) $\phi(p) = (1/\sqrt{2\pi\hbar})\int_{-\infty}^{\infty} dx\psi(x)exp(ipx/\hbar)$. The density matrix in the Gibbs ensemble is (**1R**) $\hat{C}_\beta = exp(-\beta\hat{H})$ and assuming a local potential $V(x)$ with $\hat{H} = -(\hbar^2/2m)\partial^2 + V$ one has

(2.24)
$$C(x,x';\beta) = <x|e^{-\beta\hat{H}}|x'> = e^{-\beta\hat{H}_x}\delta(x-x') = \frac{1}{2\pi\hbar}\int_{-\infty}^{\infty} dpe^{-ipx'/\hbar}e^{-\beta\hat{H}_x}e^{ipx/\hbar}$$

(where (**1S**) $\delta(x-x') = (1/2\pi\hbar)\int exp[ip(x-x')/\hbar]dp$. The ensemble average of an operator $\hat{Q}$ is $<\hat{Q}> = Tr(\hat{C}_\beta\hat{Q})/Tr(\hat{C}_\beta)$ and the Gibbs density matrix operator (**1R**) may then be written as (**1T**) $\hat{C}_\beta = \sum |n> exp[-\beta E_n] <n|$ leading to

$$C(x,x';\beta) = <x|\hat{C}_\beta|x'> = \sum \psi_n^*(x')e^{-\beta E_n}\psi_n(x) \tag{2.25}$$

which is called the Bloch density matrix. The Wigner transform $C_W(x,p;\beta)$ of the operator in (**1T**) is

$$C_W(x,p,\beta) = \frac{1}{2\pi\hbar}\int_{-\infty}^{\infty} dyC(x-(y/2), x+(y/2);\beta)e^{ipy'\hbar} \tag{2.26}$$
$$= \frac{1}{2\pi\hbar}\sum e^{-\beta E_n}\int_{-\infty}^{\infty} dy\psi_n^*(x+(y/2))\psi(x-(y/2))e^{ipy/\hbar}$$

**2.2. REMARKS ON THE TRACE FORMULA.** We refer here to [**65, 228, 248, 293, 340, 370, 375, 396, 555, 556, 588, 600, 623, 787, 811, 835**] and mention first [**787**] for additional information and details about the semi-classical path integral. Then we begin with [**623**] for the Gutzwiller trace formula, where it is shown that the energy spectrum of a generic (non-relativistic) quantum system can be expressed in terms of the invariant properties of the periodic orbits of the corresponding classical system via a series over all the periodic orbits of a corresponding classical system. This approach is not as illuminating mathematically as others (see e.g. [**811**]) but we prefer to keep matters as physical as possible here for various reasons. In any case we are not ultimately concerned with a rigorous proof but mainly want to understand how the trace formula links classical and quantum ideas. Thus begin with (**2A**) $i\hbar\partial_t\psi = \hat{H}\psi$ and write (**2B**)$H\psi = (1/2m)g^{\alpha\beta}(P_\alpha + A_\alpha)(P_\beta + A_\beta)\psi + U\psi$ where $A_\alpha$ and $U$ are functions of $q$ (and eventually $t$) and $g^{\alpha\beta}$ is the inverse of $g_{\alpha\beta}$ (the $P_\alpha$ are momentum operators given via $P_\alpha = -i\hbar\nabla - \alpha)$. Thus **(2C)** $P_\alpha A^\alpha\psi = (P_\alpha A^\alpha)\psi + A^\alpha P_\alpha\psi$ (covariant derivatives). The Hamiltonian acts on the space $L^2_M$ where the Schrödinger equation is self-adjoint with respect to the scalar product

$$(2.27)\qquad (\phi,\psi) = \int_M d^dq\sqrt{g(q)}(\phi^*\psi)(q,t);,\ g = det(g_{\alpha\beta})$$

If the potentials $A_\alpha$ and $U$ are time independent one can use the stationary SE **(2D)** $E\psi_E = H\psi_E$. Introduce now the forward time evolution operator or propagator with kernel $K$ via

$$(2.28)\qquad (i\hbar\partial_t - H)K(q,t|q',t') = -i\delta(t-t')\delta(q-q')\ (t\geq t');\ K = 0(t<t')$$

One stipulates then

$$(2.29)\qquad \psi(q,t) = \int_M d^dq\sqrt{g(q')}K(q,t|q't')\psi(q',t')$$

$$K(q,t|q',t') = \theta(t-t')\sum_n \psi_n(q)\psi_n^*(q')e^{-(i/\hbar)E_n(t-t')}$$

where $\theta(t-t') = 1\ t\geq t')$ and $\theta = 0\ (t<t')$. It follows that

$$(2.30)\qquad \int_0^\infty \frac{dt}{\hbar}e^{izt}K(q,t|q',0)\sum_n \frac{\psi_h(q)\psi_n^\dagger(q')}{z-E_n} = i(G(q|q',E)$$

defining the Green's function $G$ as the "resolvant" of the time independent problem (2.31)

$$(z-H)G(q|q',z) = \delta(q-q');\ G(q|q',z) = \int_0^\infty \frac{dt}{i\hbar}e^{izt/\hbar}K(q,t|q',t')\ \Im(z)>0$$

The poles of $G$ are real and coincide with the energy levels of the quantum system and using the Plemlj formula **(2E)** $lim_{\Im z\downarrow 0}[1/(z-E_n)] = PV[1/(E=E_n)] - i\pi\delta(E-E_n)$ one obtains a relation between energy density and the trace of $G$, namely **(2F)** $\rho(E) = -lim_{\Im z\downarrow 0}\Im(1/\pi)\int_M d^dq\sqrt{g(q)}G(q,q',z)|_{E=\Re z}$. Note that this means that the energy density is independent of the representation of the Hilbert space and this can be written in terms of the propagator via

$$(2.32)\qquad \rho(E) = lim_{\Im z\downarrow 0}\,\Im\int_0^\infty \frac{dt}{i\pi\hbar}e^{izt/\hbar}\int_M d^dq\sqrt{g(q)}K(q,t,q',0)\Big|_{E=\Re z}$$

A main result of semiclassical methods is the ability to express quantum observables in terms of classical objects. In classical mechanics a generic Hamiltonian exhibits a chaotic dynamics whereas this is not generally present in quantum behavior. The expectation here is that the quantum energy levels are associated with invariant sets of the classical dynamics. Recall that a classical Hamiltonian system is separable in $d$ dimensions if there are $d$ independent integrals of motion (including the Hamiltonian); in such a case one can produce action angle coordinates $(A_1,\cdots,A_d,\theta_1,\cdots,\theta_d)$ and quantization will involve **(2G)** $A_j = [n_j + (\mu/4)]$ for quantum numbers $(A_j,\mu_j)$. In any event for the case when $E$ is the only conserved quantity the Gutzwiller trace formula takes the form ($p.p.o \sim$ primative periodic orbits)

$$\rho(E) = \sum_n \delta(E - E_n) \simeq \int \frac{d^d q d^d p}{(2\pi\hbar)^d} \delta(E - H(p,q)) \tag{2.33}$$

$$+ \Im \left[ \sum_{o \in p.p.o} \frac{it_o}{\pi\hbar} \sum_1^\infty \frac{e^{i(r/\hbar)W_o(E) - (i\pi/2)\kappa_{o,r}}}{\sqrt{|det_\perp[I_{2d} - M_o^r]|}} \right]$$

The second term consists of a formal series ranging over all classical primitive periodic orbits of finite period $t_o$ and their representations $r$ ($\kappa_o^r$ is the Maslov index, $W_o$ is a reduced action, and $M_o$ denotes a monodromy matrix). The symbol $det_\perp$ refers to the eigendirections transversal to the orbit and in fact

$$\frac{1}{\sqrt{|det_\perp(I_{2d} - M(t))|}} \sim exp\left| -\frac{h_{KS}t}{2} \right| \tag{2.34}$$

where $h_{KS}$ is the Kolmogorov-Sinai entropy (sum of positive Lyapounov exponents of the orbit).

Thus periodic orbits are seen to affect the spectrum both individually and collectively and the collective contribution gives rise to major physical and mathematical difficulties (following [**623**]). For fixed values of the energy the number of periodic orbits is infinite and in a chaotic system the number of periodic orbits proliferates exponentially with the period $T$ and growth rate give by the topological entropy $h_T$ where (**2H**) $\#(periodic\ orbits) \sim exp[h_T T]$ ($T \uparrow \infty$). The topological entropy is the Rènyi entropy of order zero (cf. [**86**]) and if one assumes that on average the topological and KS entropies are equal then the diminishing amplitude of orbits of period $T$ is dominated by their proliferation. Thus the series consists effectively of terms with exponentially growing amplitudes and consequently a literal interpretation of the Gutzwiller trace formula is problematic. Nevertheless experimental and numerical evidence (cf. [**124**]) support the existence in some mathematical sense of a semiclassical approximation to the density of states related to the trace formula (cf. [**248, 396, 623**]).

**2.3. COHERENT STATES AND THE TRACE FORMULA.** It is clear that all this needs further clarification and explanation and we shift here to a somewhat "simpler" approach using coherent states (see [**170, 228, 370, 375, 600, 877**]). Since the point of view will be somewhat different there will be some repetition of ideas and we follow [**600**] for a framework with additional structure via [**370, 372, 369, 375**]. The Gutzwiller trace formula relates the density of states of a quantum system to periodic orbits of the corresponding classical Hamiltonian. It usually applies to systems with a discrete quantum spectrum and isolated, unstable, periodic orbits and the density of states may be expressed via

$$\rho(E) = \sum_n \delta(E - E_n) = \int_{-\infty}^{\infty} \frac{dt}{2\pi\hbar} e^{iEt/\hbar} Tr[\hat{U}]_t \tag{2.35}$$

The approach of [**600**] uses a coherent-state basis to evaluate the trace. The coherent states are labeled by a phase-space point $\alpha = (q,p)$ and may be thought of as wave packets positioned at $\alpha$. The trace of an operator can be expressed as an integral over coherent states, e.g. **(3A)** $Tr[\hat{U}_t] = \int [d\alpha/(2\pi\hbar)^d)] < \alpha|\hat{U}_t|\alpha >$. The contribution to the integral from a phase point $\alpha$ is clearly negligible unless $\alpha$ is very close to a periodic orbit of period close to $t$. The constructions of [**600**] build on semiclassical evolution of coherent states as described in [**555**] using Weyl-Heisenberg operators $\hat{Y}(\alpha)$. The deformation is described by a symplectic transformation based on metaplectic operators $\hat{R}(\tilde{S})$ (cf. also [**877**]) but we will extend this to the more elegant version by deGosson (cf. [**228, 369, 370, 372, 375**]). Generally the trace of a metaplectic operator is given via $Tr[\hat{R}(\tilde{S})] = exp(i\pi\nu/2)/\sqrt{|det(\tilde{S}-\tilde{I})|}$ where $\nu$ is an integer. Here the periodic orbit contributions to $Tr[\hat{U}_t]$ are proportional to $exp(iR_p/\hbar)Tr[\hat{R}(\tilde{M}_p)]$ where $R_p$ is the Hamiltonian action along the periodic orbit and $\tilde{M}_p$ is a corresponding stability matrix (see below and cf. [**96, 877**]).

Thus the metaplectic group $Mp(n)$ is a two fold covering of the symplectic group $Sp_n$ and there is a simple formula for the trace of a metaplectic operator of the form **(3B)** $Tr[\hat{R}(\tilde{S})] = exp(i\pi\nu/2)/\sqrt{|det(\tilde{S}-\tilde{I})|}$ (here $\tilde{S}$ is a symplectic matrix and we refer to [**369**] for general theory). One considers a quantum system with time-independent Hamiltonian $\hat{H}$ with a SE having energy levels $E_j$, eigenfunctions $\psi_j$ with $\hat{H}|\psi_j >= E_j|\psi_j >$ and **(3C)** $\rho(E) = \sum \delta(E - E_j)$. One considers densities

$$\rho_A(E) = \sum < \psi_j|\hat{A}|\psi_j > \delta(E - E_j); \tag{2.36}$$

$$S_A(E, \hbar\omega) = \sum | < \psi_j|\hat{A}|\psi_k > |^2 \delta(\hbar\omega - E_k + E_j)\delta(E - E_j)$$

where, or $\hat{U} = exp(-i\hat{H}t/\hbar)$

$$\rho(E) = \int_{-\infty}^{\infty} \frac{dt}{2\pi\hbar} e^{iEt/\hbar} Tr[\hat{U}_t]; \; \rho_A(E) = \int_{-\infty}^{\infty} \frac{dt}{2\pi\hbar} e^{iEt/\hbar} Tr[\hat{A}\hat{U}_t]; \tag{2.37}$$

$$S_A(E, \hbar\omega) = \int_{-\infty}^{\infty} \frac{dt}{2\pi\hbar} e^{iEt/\hbar} \int_{-\infty}^{\infty} \frac{ds}{2\pi\hbar} e^{i\omega s} Tr[\hat{A}_s \hat{A}\hat{U}_{t-s}]$$

In the explicitly time-dependent cases of a driving force with period $T$ one will consider the density of eigenphases $\theta_j$ of the Floquet operator $\hat{U}_T$ where **(3E)** $\rho(\theta) = \sum_{-\infty}^{\infty} \sum_{\ell=\infty}^{\infty} \sum_{j=1}^{N} \delta(\theta - \theta_j - 2\pi\ell) = (1/2\pi\hbar) \sum_{n=-\infty}^{\infty} \sum_{j=1}^{N} exp(in(\theta - \theta_j)$ (via the Poisson summation formula). Thus

$$\rho(\theta) = \frac{1}{2\pi\hbar} \sum_{-\infty}^{\infty} Tr[\hat{U}_T^n] e^{in\theta} = \frac{N}{2\pi\hbar} + \frac{1}{n\hbar} \Re \sum_{n>0} Tr[\hat{U}_T^n] e^{in\theta} \tag{2.38}$$

Given $\alpha = (q, p)$ in $d$-dimensions the Hamiltonian flow is given via

$$\dot{\alpha}_t = \tilde{J} \frac{\partial H}{\partial \alpha_t}; \; \tilde{J} = \begin{pmatrix} \tilde{0} & \tilde{I} \\ -\tilde{I} & \tilde{0} \end{pmatrix} \tag{2.39}$$

The separation $\delta\alpha_y$ of two nearby trajectories is given via the stability matrix $\tilde{M}_t$ where

$$(2.40)\qquad \delta\alpha_t = \tilde{M}_t\delta\alpha_0;\ \frac{d}{dt}\tilde{M}_t = \tilde{J}\tilde{K}_t\tilde{M}_t;\ [\tilde{K}_t]_{ij} = \left.\frac{\partial^2 H}{\partial\alpha_i\partial\alpha_j}\right|_{\alpha=g\alpha_y}$$

Following [**555**] one denotes coherent states via (**3F**) $|\alpha_0> = \hat{T}(\alpha_0)|0>$ where $|0>$ is the ground state of a harmonic oscillator and the Weyl-Heisenberg operator is given by (**3G**) $\hat{T}(\alpha_0) = exp[-i(\alpha_0 \wedge \hat{\alpha})/\hbar]$ where $\hat{\alpha} = (\hat{q}, \hat{p})$; here $\alpha_0 \wedge \alpha_1 = q_0 \cdot p_1 - q_1 \cdot p_0$. From the Baker-Cambell-Hausdorff formulas one has

(1) $\hat{T}^\dagger(\alpha_0)\hat{\alpha}\hat{T}(\alpha_0) = \hat{\alpha} + \alpha + 0$

(2) $\hat{T}(\alpha_0)\hat{T}(\alpha_1) = e^{-i[\alpha_0\wedge\alpha_1]/2\hbar}\hat{T}(\alpha_0 + \alpha_1)$ which implies that $\hat{T}^{-1}(\alpha_0) = \hat{T}(-\alpha_0) = \hat{T}^\dagger(\alpha_0)$.

(3) $i\hbar(d/dt)\hat{T}(\alpha_t) = \hat{T}(\alpha_t)[(1/2)[\dot{\alpha}_t \wedge \alpha_t] + [\dot{\alpha}_t \wedge \hat{\alpha}])$

By constructions, in order to represent deformations of wave packets one needs a representation of symplectic transformations $\tilde{S}$ on the Hilbert space of the system. For this one denotes representations of symplectic matrices $\tilde{S}$ by metaplectic operators in the form of unitary matrices (★) $\hat{R}(\tilde{S})\hat{T}(\alpha)\hat{R}^{-1}(\tilde{S}) = \hat{T}(\tilde{S}\alpha)$. Rather than basing this on configuration space representations one goes here to Weyl representations of the metaplectic operators $\hat{R}(\tilde{S})$ in the form

$$(2.41)\qquad \hat{R}(\tilde{S}) = \frac{exp(i\pi\nu/2}{\sqrt{|det(\tilde{S}-\tilde{I})|}}\int\frac{dy}{(2\pi\hbar)^d}exp\left[\frac{i}{2\hbar}y\cdot\tilde{A}y\right]\hat{T}(y)$$

where (**3H**) $\tilde{A} = (1/2)\tilde{J}(\tilde{S}+\tilde{I})(\tilde{S}-\tilde{I})^{-1}$. Actually following [**370**] the integer $\nu$ is compared to the Maslov index and we pick up the story now from [**370**] (math.SG 0411453).

Thus one looks at unitary operators $\hat{S} : L^2(\mathbf{R}^n) \to L^2(\mathbf{R}^n)$ which can be defined as follows. Let $S \in Sp(n)$ have no eigenvalue equal to 1 and associate to $S$ the Weyl operator (rewriting (1.41) in slightly different notation)

$$(2.42)\qquad \hat{R}(S) = \left(\frac{1}{2\pi}\right)^n\frac{i^\nu}{\sqrt{|det(S-I)|}}\int e^{(i/2)<M_S z_0, z_0>}\hat{T}(z_0)d^{2n}z_0$$

where $\hat{T}(z_0)$ is the Weyl-Heisenberg operator and (**3J**) $M_S = (1/2)J(S+I)(S-I)^{-1}$ ($I$ is the identity and $J$ the standard symplectic matrix). Then $\hat{R}(SS') = \pm\hat{R}(S)\hat{R}(S')$ (as above) where following [**370**] $\hat{R}(S)$ is a multiple by a scalar factor of modulus one of either of the two metaplectic operators $\pm\hat{S}$ associated to $Mp(n)$ (via the metaplectic covariance of the Weyl-Heisenberg operators - see below). The idea here is to make precise the work in [**600**] by comparing the integer $\nu$ to the Maslov index and in [**370**] one gives a semiclassical interpretation of $\hat{R}(S)$ in terms of the phase space wavefunctions. One denotes now by $\sigma$ the canonical symplectic form on $\mathbf{R}_z^{2n}$ via $\sigma(z,z') = <p,x'> - <p',x>$ if $z = (x,p)$ and $z' = (x',p')$; thus

$$(2.43)\qquad \sigma(z,z') = <Jz,z'>;\ J = \begin{pmatrix} 0 & I \\ -I & 0 \end{pmatrix}$$

Now via [**369**] every $S \in Mp(n)$ is the product of two Fourier transforms which are operators $S_{W,m}$ defined on $S(X)$ via

$$S_{W.m}f(x) = \left(\frac{1}{2\pi i}\right)^n i^m \sqrt{|det(L)|} \int e^{iW(x,x')} f(x') d^n x' \tag{2.44}$$

where $W$ is a quadratic form of the type

$$W(x,x') = \frac{1}{2} < Px, x' > - < Lx, x' > + \frac{1}{2} < Qx', x' > \tag{2.45}$$

with $P = P^T$, $Q = A^T$, and $det(L) \neq 0$. The integer in (2.44) corresponds to a choice of $arg(det(L))$, namely $m\pi \equiv arg(det(L))\ mod(2\pi)$ and hence to every $W$ there corresponds two different choices of $m$ modulo 4; if $m$ is one choice then $m+2$ is the other (reflecting the fact that $Mp(n)$ is a two fold covering of $Sp(n)$. The projection $\pi : Mp(n) \to Sp(n)$ is entirely specified by the datum of each $\pi(S_{W,m})$ and $\pi(S_{W,m}) = S_W$ where (**3J**) $(x,p) = S_W(x',p') \iff p = \partial_x W(x,x')$ and $p' = -\partial_{x'} W(x,x')$. In particular

$$(\bigstar\bigstar)\ S_W = \begin{pmatrix} L^{-1}Q & L^{-1} \\ PL^{-1}Q - L^T & PL^{-1} \end{pmatrix}$$

is the free symplectic automorphism generated by the quadratic form $W$ (note that $S_W(\ell_P \cup \ell_P) = 0$ for every $W$. The inverse $\hat{S}^{1}_{W,m} = \hat{S}^{*}_{W,m}$ is the operator $S_{W^*,m^*}$ where $W^*(x,x') = -W(x',x)$ and $m^* = n - m,\ mod(4)$. Note also that if $S$ is a free symplectic matrix

$$S_W = \begin{pmatrix} L^{-1}Q & L^{-1} \\ PL^{-1}Q - L^T & PL^{-1} \end{pmatrix} \tag{2.46}$$

then $S = S_W$ with $P = B^{-1}A$, $L = B^{-1}$, and $Q = DB^{-1}$.

For $z_0 = (x_0, p_0)$ one denotes by $T(z_0)$ the translation $z \to z + z_0$ acting on functions by push forward $T(z_0)f(z) = f(z - z_0)$. Let $\hat{T}(z_0)$ be the corresponding Weyl-Heisenberg operator so for $f \in \mathfrak{S}(\mathbf{R}^n)$ (Schwartz space) one has (**3K**) $\hat{T}(z_0) = exp[i < p_0, x > -(1/2) < p_0, x_0 >] f(x - x_0)$ and the operators $\hat{T}(z_0)$ then satisfy the metaplectic covariance formula (**3L**) $\hat{S}\hat{T}(z) = \hat{T}(Sz)\hat{S}$ $(S = \pi(\hat{S})$ for every $\hat{S} \in Mp(n)$ and $z$. In fact the metaplectic operators are the only unitary operators up to a factor in $S$ satisfying (**3K**) and one has

- For every $S \in Sp(n)$ there exists a unitary transformation $\hat{U}$ in $L^2(\mathbf{R}^n)$ satisfying (**3L**) and $\hat{U}$ is uniquely determined apart from a constant factor of modulus one.

The Weyl-Heisenberg operators satisfy in addition

$$\hat{T}(z_0)\hat{T}(z_1) = e^{-i\sigma(z_0,z_1)}\hat{T}(z_1)\hat{T}(z_0);\ \hat{T}(z + 0 + z_1) = e^{-(i/2)\sigma(z_0,z_1)}\hat{T}(z_0)\hat{T}(z_1) \tag{2.47}$$

Now let $a^w$ denote the Weyl operator with symbol $a$ so that

$$a^w f = \left(\frac{1}{2\pi}\right)^n \int e^{i(p,x-y)} a[(1/2)(x+y), p) f(y) d^n y d^n p \tag{2.48}$$

where $f \in \mathfrak{S}(\mathbf{R}^n)$; equivalently (**3M**) $a^w = \int a_{gs}(z_0)\hat{T}(z_0)d^n z_0$ where $a_\sigma$ is the symplectic Fourier transform $F_\sigma a$ defined via

$$(2.49) \qquad F_\sigma a(z) = \left(\frac{1}{2\pi}\right)^n \int e^{i\sigma(z,z')}a(z')d^{2n}z'$$

The kernel of $a^w$ is related to $a$ via

$$(2.50) \qquad a(x,p) = \int e^{-i<p,y>}K[x+(y/2),x-(y/2)]d^n y$$

and the Mehlig-Wilkinson (MW) operator (2.42) is the Weyl operator with twisted Weyl symbol

$$(2.51) \qquad a_\sigma(z) = \left(\frac{1}{2\pi}\right)^n \frac{i^\nu}{\sqrt{|det(S-I)|}} e^{(i/2)<M_S z_0,z_0>}$$

One recalls also a generalized Fresnel formula (for invertible M)

$$(2.52) \qquad \left(\frac{1}{2\pi}\right)^{n/2} \int e^{-i<p,x>}e^{(i/2)<Mx,x>}d^n x$$

$$= |det(M)|^{-1/2}e^{(i\pi/4)sgn(M)}e^{-(i/2)<M^{-1}x,x>}$$

The twisted Weyl symbol in the Mehlig-Wilkinson operators has the form (2.51) and one can provide two alternative formulations (cf. [**370**]). First one notes that $M_S = (1/2)J(S+I)(S-I)^{-1}$ is symmetric since (**3N**) $S \in Sp(n) \iff S^TJS = J \iff SJS^T = J$. Note that (**3I**) can be "solved" to get $S = (2M-J)^{-1}(2M+J)$ and one shows now in [**370**] that the operator

$$(2.53) \qquad \hat{R}(S) = \left(\frac{1}{2\pi}\right)^n \frac{i^\nu}{\sqrt{|det(S-I)|}} \int e^{(i/2)<M_S z_0,z_0>}\hat{T}(z_0)d^{2n}z_0$$

can be written in the following alternative two forms (for $det(S-I) \neq 0$)

$$(2.54) \qquad \hat{R}(S) = \left(\frac{1}{2\pi}\right)^n \frac{i^\nu}{\sqrt{|det(S-1)|}} \int e^{-(i/2)\sigma(Sz_0,z_0)}\hat{T}((S_I)z_0)d^{2n}z_0;$$

$$\hat{R}(S) = \left(\frac{1}{2\pi}\right)^n i^\nu\sqrt{|det(S-I)|}\int \hat{T}(Sz_0)\hat{T}(-z_0)d^{2n}z_0$$

To see this one notes that (**3O**) $(1/2)J(S+I)(S-I)^{-1} = *1/2)J + J(S-I)^{-1}$ and hence (**3P**) $< M_S z_0, z_0 > = < J(S-I)^{-1}z_0, z_0 > = \sigma((S-I)^{-1}z_0, z_0)$. Making a change of variables $z_0 \to (S-I)^{-1}z_0$ the right side of (1.53) becomes

$$(2.55) \qquad \int e^{(i/2)<M_S z_0,z_0>}\hat{T}(z)d^{2n}z_0 = \int e^{(1/2)\sigma(z_0,(S-I)z_0)}\hat{T}((S-I)z_0)d^{2n}z_0$$

$$= \int e^{-(1/2)\sigma(Sz_0,z_0)}\hat{T}((S-I)z_0)d^{2n}z_0$$

Hence (2.54)-1 holds and taking (1.47) into account one has (**3Q**) $\hat{T}((S-I)z_0) = exp[-(i/2)\sigma(Sz_0,z_0)]\hat{T}(Sz_0)\hat{T}(-z_0)$ leading to (1.54)-2. Consequently as a corollary there results $\hat{R}(S) = c_S\hat{S}_{W,m}$ where $|c| = 1$ (since (**3R**)$\hat{R}(S)\hat{T}(z_0) = \hat{T}(Sz_0)\hat{R}(S)$ (via (2.54)-2).

Next one shows that the Mehlig-Wilkinson operators coincide with the metaplectic operators $\hat{S}_{W,m}$ when $S = S_W$ and will determine the correct choice for $\nu$ (which is related to the usual Maslov index in [**376**]). One proves first that for a free symplectic matix as in (2.46)

$$(2.56)\qquad det(S_W - I) = det(B)det(B^{-1}A + DB^{-1} - B^{-1} - (B^T)^{-1}$$

Thus when $S$ is written as in (★★) then (**3S**) $det(S_W - I) = det(L^{-1})det(P + Q - L - L^T)$. To see this note that since $B$ is invertible $S - I$ can be written as

$$(2.57)\quad \begin{pmatrix} A-I & B \\ C & D-I \end{pmatrix} = \begin{pmatrix} 0 & B \\ I & D-I \end{pmatrix}\begin{pmatrix} C-(D-I)B^{-1}(A-I) & 0 \\ B^{-1}(A-I) & I \end{pmatrix}$$

hence (**3T**) $det(S_W - I) = det(B)det[C-(D-I)B^{-1}(A-I)]$. Since $S$ is symplectic one has $C - DB^{-1}A = -((B^T)^{-1}$ (using e.g. $S^TJS = SJS^T = J$) and hence

$$(2.58)\qquad C - (D-I)B^{-1}(A-I) = B^{-1}A + DB^{-1} = (B^T)^{-1}$$

Now let $S$ be a free symplectic matrix as in (2.46) and $\hat{R}(S)$ the corresponding MW operator. Then $\hat{R}(S) = \hat{S}_{W,m}$ provided that (**3U**) $\nu \equiv m - Inert(P + Q - L - L^T)\ mod(4)$ (here $Inert(M)$ is the number of eigenvalues $< 0$ of $M$). To see this recall that $\hat{R}(S) = c_S\hat{S}_{W,m}$ where $|c_S| = 1$. To determine $c_S$ let $\delta$ be the Dirac distribution centered at $x = 0$ and set

$$(2.59)\qquad C = \left(\frac{1}{2\pi}\right)^n \frac{i^\nu}{\sqrt{|det(S_W - I)|}}$$

Then by definition of $\hat{R}(S)$

$$(2.60)\qquad \hat{R}(S)\delta(x) = C\int e^{(i/2)<M_S z_0, z_0>} e^{i(,p_0,x>-(1/2(<p_0,x_0>)}\delta(x - x_0)d^{2n}z_0$$

$$= C\int e^{(i/2)<M_S(x,p_0),(x,p_0)>}e^{(i/2)<p,x>}\delta(x - x_0)d^{2n}z_0$$

Hence setting $x = 0$

$$(2.61)\qquad \hat{R}(S)\delta(0) = C\int e^{(i/2)<M_S(0,p_0),(0,p_0)>}\delta(-x_0)d^{2n}z_0$$

Since $\int\delta(-x_0)d^nx_0 = 1$ this yields

$$(2.62)\qquad \hat{R}(S)\delta(0) = \left(\frac{1}{2\pi}\right)^n \frac{i^\nu}{\sqrt{|det(S-I)|}}\int e^{(i/2)<M_S(0,p_0),(0,p_0)>}d^np_0$$

To calculate the scalar product $< M_S(0,p_0),(0,p_0) >= \sigma((S-I)^{-1}0,p_0),(0,p_0)$ note that $(x,p) = (S - I^{-1}(0,p_0))$ is equivalent to $S(x,p) = (x, p+p_0)$, i.e. to (**3V**) $p + p_0 = \partial_x W(x,x')$ and $p = -\partial_{x'}W(x,x')$. Using the explicit form (1.45) of $W$ with (1.56) implies then

$$(2.63)\qquad x = (P+Q-L-L^T)^{-1}p_0;\ p = (L-Q)(P+Q-L-L^T)^{-1}p_0$$

$$\Rightarrow < M_S(0,p_0),(0,p_0) >= - < (P+Q-L-L^T)^{-1}p_0, p_0 >$$

Applying then the Fresnel formula (1.52) gives (for $k_n = [1/2\pi]^n$)
(2.64)

$$k_n \int e^{(i/2)<M_S(0,p_0),(0,p_0)>} d^n p_0 = e^{-(i\pi/4)sgn(P+Q-L-L^T)}|det(P+Q-L-L^T)|^{1/2}$$

since (**3W**) $\sqrt{|det(S-I)|}^{-1/2} = |det(L)|^{1/2}|det(P+Q-L-L^T|^{-1/2}$ and in view of (**3S**) this leads to (**3X**) $\hat{R}(S)\delta(0) = k_n i^\nu exp[-(i\pi/4)sgn(P+Q-L-L^T)]|det(L)|^{1/2}$. Then by definition of $\hat{S}_{W,m}$ one has the formula (**3Y**) $\hat{S}_{W,m}\delta(0) = k_n i^{m-(n/2)}|det(L)|^{1/2}$ leading to (**3Z**) $i^\nu exp[-(i\pi/4)sgn(P+Q-L-L^T) = i^{m-(n/2)}$. Consequently

$$\nu - \frac{1}{2}sgn(P+Q-L-L^T) \equiv m - \frac{n}{2}\ mod(4) \tag{2.65}$$

For the general case recall from (**3S**) that (♦) $det(S_W - I) = det(L^{-1})det(P+Q-L-L^T)$ for all matrices $S_W \in Sp(n)$. Recall also that every $\hat{S} \in Mp(n)$ can be written in infinitely many ways as a product $\hat{S} = \hat{S}_{W,m}\hat{S}_{W',m'}$. One shows now that $\hat{S}_{W,m}$ and $\hat{S}_{W',m'}$ can be chosen such that $det(\hat{S}_{W,m} - I) \neq 0$ and $det(\hat{S}_{W',m'} - I) \neq 0$. For that purpose one recalls a factorization result from [**376**], namely for $W$ as in (1.45) one can write (•) $\hat{S}_{W,m}\hat{V}_P\hat{M}_{L,m}\hat{J}\hat{V}_Q$ where

$$\hat{V}_P f(x) = e^{(i/2)<Px,x>}f(x);\ \hat{M}_{L,m}f(x) = i^m\sqrt{|det(L)|}f(Lx); \tag{2.66}$$

$$\hat{J}f(x) = \left(\frac{1}{2\pi i}\right)^n \int e^{-i<x,x'>}f(x')d^n x'$$

Consequently one can state that every $\hat{S} \in Mp(n)$ is the product of two MW operators and these operators generate $Mp(n)$. To see this write $\hat{S} = \hat{S}_{W,m}\hat{S}_{W',m'}$ and apply (•) to each factor, leading to (••) $\hat{S} = \hat{V}_P\hat{M}_{L,m}\hat{J}\hat{V}_{-(P'+Q)}\hat{M}_{L',m'}\hat{J}\hat{V}_{Q'}$. One claims now that $\hat{S}_{W,m}$ and $\hat{S}_{W'm'}$ can be chosen so that $det(\hat{S}_{W,m} - I) \neq 0$ and $det(\hat{S}_{W',m'} - I) \neq 0$, i.e. (♦♦) $det(P+Q-L-L^T) \neq 0$ and $det(P'+Q'-L)-L'^T) \neq 0$. One refers here to (♦) and remarks first that the right side of (••) does not change if one replaces $P'$ by $P' + \lambda I$ and $Q$ by $Q - \lambda I$ for $\lambda \in \mathbf{R}$. Pick $\lambda$ not to be an eigenvalue of $P+Q-L-L^T$ and $-\lambda$ not an eigenvalue of $P'+Q'-L'-L'^T$; then (•••) $det(P+Q-\lambda I-L-L^T) \neq 0$ and $det(P'+\lambda I+Q'-L-L^T) \neq 0$.

**2.4. SOME MATHEMATICAL VARIATIONS.** The approach of [**228**] is somewhat more "mathematical" (i.e. complete and rigorous with theorems and proofs) and we sketch this as a prelude to a deeper study as in [**811, 847**] where the language of Fourier integral operators and microlocal analysis is used (cf. also [**229, 264, 313, 431, 432, 447**]). One considers a quantum system in $L^2(\mathbf{R}^n)$ with Hamiltonian $\hat{H} = -\hbar^2\Delta + V(x)$ with $V(x)$ real and $C^\infty$. The corresponding classical Hamiltonian is of course $H(q,p) = p^2 + V(q)$ (with mass suitable normalized via e.g. $2m = 1$) and for a given energy E on denotes by $\Sigma_E = \{(q,o) \in \mathbf{R}^{2n};\ H(q,p) = E\}$ (energy shell). More generally one considers Hamiltonians $\hat{H}$ obtained by the $\hbar$-Weyl quantization so that $\hat{H} = Op_\hbar^w(H)$ where

$$Op_\hbar^w(H)\psi(x) = (2\pi\hbar)^{-n}\int_{\mathbf{R}^{2n}} H\left(\frac{x+y}{2},\xi\right)\psi(y)e^{i(x-y)\cdot(\xi/\hbar)}dyd\xi \tag{2.67}$$

The Hamiltonian $H$ is assumed to be a smooth real-valued function of $z = (x, \xi) \in \mathbf{R}^{2n}$ which satisfies the global estimates ($< u > = (1 + |u|^2)^{1/2}$ for $u \in \mathbf{R}^m$)

(1) (**H.0**) There exist non-negative constants $C$, $m$ $C_\gamma$ such that ($i$) $|\partial_z^\gamma H(z)| \leq C_\gamma < H(z) > \ \forall z \in \mathbf{R}^{2n}$, $\forall \gamma \in \mathbf{N}^{2n}$
(2) (ii) $< H(z) > \leq C < H(z') >< z - z' >^m \ \forall z, z' \in \mathbf{R}^{2n}$

It follows that

(1) (iii) $H = p^2 + V(q)$ satisfies (**H.0**) if $V(q)$ is bounded below by some $a > 0$ and satisfies (**H.0**) in the variable $q$.
(2) (iv) The technical condition (**H.0**) implies in particular that $\hat{H}$ is essentially self adjoint on $L^2(\mathbf{R}^n)$ for $\hbar$ small enough and that $\chi(\hat{H})$ is an $\hbar$-pseudodifferential operator (PSDO) if $\chi \in C_0^\infty(\mathbf{R})$ (cf, [**431**]).

Now denote by $\phi_t$ the classical flow induced by Hamilton's equations with $H$ and by $S(q, p, t)$ the classical action along the trajectory starting at $(q, p)$ for $t = 0$ and evolving via (**4A**) $S(q, p, t) = \int_0^t (p_s \cdot \dot{Q}_s - H(q, p))ds$ where $(q_t, p_t) = \phi_t(q, p)$ (one writes $\alpha_t = \phi_t(\alpha)$ where $\alpha = (q, p)$ is a point in phase space). Let now (**4B**) $H''(\alpha_t) = (\partial^2 H/\partial\alpha^2)|_{\alpha=\alpha_y}$ be the Hessian of $H$ at $\alpha_t$ and $J$ be the standard symplectic matrix. Let $F(t)$ be the $2n \times 2n$ real symplectic matrix solution of the linear differential equation

$$\dot{F}(t) = JH''(\alpha_t)F(t);\ F(0) = \begin{pmatrix} I & 0 \\ 0 & I \end{pmatrix};\ J = \begin{pmatrix} 0 & I \\ -I & 0 \end{pmatrix} \tag{2.68}$$

Let now $\gamma$ be a closed orbit on $\Sigma_E$ with period $T_\gamma$ and denote by $F_\gamma$ the matrix $F_\gamma = F(T_\gamma)$ (monodromy matrix of $\gamma$). Evidently $F_\gamma$ depends on $\alpha$ but its eigenvalues do not since the monodromy matrix with a different initial point on $\gamma$ is conjugate to $F_\gamma$ and $F\gamma$ has 1 as an eigenvalue of algebraic multiplicity at least equal to 2. Then (**4C**) $\gamma$ is a nondegenerate orbit if the eigenvalue 1 of $F_\gamma$ has algebraic multiplicity 2. Then let $\sigma$ denote the standard symplectic form (**4D**) $\sigma(\alpha, \alpha') = p \cdot q' - p' \cdot q$; $\alpha + (q, p)$, $\alpha' = (q', p')$. Let $\{\alpha_1, \alpha_1'\}$ be the eigenspace of $F_\gamma$ belonging to the eigenvalue 1 and let $V$ be its orthogonal symplectic complement, i.e. (**4E**) $V = \{\alpha \in \mathbf{R}^{2n};\ \sigma(\alpha, \alpha_1) = \sigma(\alpha, \alpha_1') = 0\}$. In some cases the Hamiltonian flow will contain manifolds of periodic orbits with the same energy; this involves degenerate orbits but the techniques of [**228**] still apply. Now let $(\Gamma_E)_T$ be the set of all periodic orbits on $\Sigma_E$ with periods $T_\gamma$ with $0 < |T_\gamma| \leq T$ (including repetitions of primitive orbits and assigning negative periods to primitive orbits traced in the opposite direction). Then one requires:

(1) (**H.1**) There exists $\delta E > 0$ such that $H^{-1}([E - \delta E, E + \delta E])$ is a compact set of $\mathbf{R}^{2n}$ and $E$ is a noncritical value of $H$ (i.e. $H(z) = E \Rightarrow \nabla H(z) \neq 0$).
(2) (**H.2**) For any $T > 0$, $(\Gamma_E)_T$ is a discrete set with periods $-T \leq T_{\gamma_1} < \cdots < T_{\gamma_N} \leq T$. (**H.3**) All $\gamma$ in $(\Gamma_E)_T$ are nondegenerate, i.e. 1 is not an eigenvalue for the corresponding Poincaré map $P_\gamma$.

One recalls now the Gutzwiller trace formula as follows: Let $\hat{A} = Op_\hbar^w(A)$ be a quantum observable such that A satisfies:

(1) (**H.4**) There exists $\delta \in \mathbf{R}$ and $C_\gamma > 0$ ($\gamma \in \mathbf{N}^{2n}$) such that $|\partial_z^\gamma A(z)| \leq C_\gamma < H(z)^\delta$ ($\forall z \in \mathbf{R}^{2n}$).

(2) **(H.5)** $g \in C^\infty$ is a function whose Fourier transform $\hat{g}$ is of compact support with $Supp\, \hat{g} \subset [-T, T]$.
(3) For $\chi$ a smooth function with compact support contained in $]E-\delta E, E+\delta E[$, equal to 1 in a neighborhood of $E$, assume well defined the "regularized density of states" $\rho_A(E) = Tr[\chi(\hat{H})\hat{A}_\chi(\hat{H})g[(E-\hat{H})/\hbar]$. Note that **H.1** implies the spectrum of $\hat{H}$ is purely discrete in a neighborhood of E so that $\rho_A(E)$ is well defined.

**THEOREM 2.4.1.** Assume **(H.0)-(H.3)** for $H$, **(H.4)** for A, and **(H.5)** for $g$. Then the following asymptotic expansion holds modulo $O(\hbar^\infty)$

$$\rho_A(E) \equiv \pi^{-n/2}\hat{g}(0)\hbar^{-(n-1)} \int_{\Sigma_E} A(\alpha)d\sigma_E(\alpha) + \sum_{k\geq -n+2} c_k(\hat{g})\hbar^k \tag{2.69}$$

$$+ \sum_{\gamma\in(\Gamma_E)_T} (2\pi)^{(n/2)-1} \left\{ \hat{g}(T_\gamma)\frac{e^{i[(S_\gamma/\hbar)+\sigma_\gamma\pi/2)]}}{|det(I-P_\gamma|^{1/2}} \int_0^{T_\gamma^*} A(\alpha_s)ds + \sum_{j\geq 1} d_j^\gamma(\hat{g})\hbar^j \right\}$$

where $A(\alpha)$ is the classical Weyl symbol of $\hat{A}$, $T_\gamma^*$ is the primitive period of $\gamma$, $\sigma_\gamma$ is the Maslov index of $\gamma$ ($\sigma_\gamma \in \mathbf{Z}$), $S_\gamma = \oint_\gamma pdq$ is the classical action along $\gamma$, $c_k(\hat{g})$ are distributions in $\hat{g}$ with support in $\{0\}$, $d_j^\gamma$ are distributions in $\hat{g}$ with support $\{T_\gamma\}$ and $d\sigma_E$ is the Liouville measure on $\Sigma_E$, namely $d\sigma_E = d\Sigma_E/|\nabla H|$ (where $d\Sigma_E$ is Euclidean measure).

**REMARK 2.4.1.** One can include more general Hamiltonians depending explicitly on $\hbar$, namely $H = \sum_1^K \hbar^j H^{(j)}$ where $H^{(0)}$ satisfies **(H.0)** and for $j \geq 1$, $|\partial^\gamma H^{(j)}(z)| \leq C_{\gamma,j} < H^{(0)}(z) >$. This is useful since e.g. $H^{(0)} + \hbar H^{(1)}$ could involve a spin term. Then the formula in Theorem 1.4.1 is true with different coefficients. In particular the first term in the contribution of $T_\gamma$ is multiplied by $exp[-i\int_0^{T_\gamma^*} H^{(1)}(\alpha_s)ds]$.

**REMARK 2.4.2.** For Schrödinger operators one only needs smoothness of $V$. In this case the trace formula (2.69) is still valid without any assumptions at infinity for $V$ when one restricts the game to a compact energy surface, assuming $E < lim\,inf_{|x|\to\infty}V(x)$. Using exponential decrease of the eigenfunctions (cf. [**432**]) one can prove that, modulo an error term of order $\hbar^\infty$, the potential $V$ can be replaced by a potential $\tilde{V}$ satisfying items 3 and 4 after **(H.0)**.

To prove the theorem one makes use of "coherent states" which can be defined via **(4F)** $\psi_0(x) = (\hbar\pi)^{-n/4}exp[-(|x|^2/2\hbar)]$ as ground state with **(4G)** $T(\alpha) = exp[(i/\hbar)(p\cdot x - q\cdot \hbar D_x)]$ as the Weyl-Heisenberg operator of translation by $\alpha$ ($D_x = (1/i)\partial_x$). Then **(4H)** $\phi_\alpha = T(\alpha)\psi_0$ are the usual coherent states and it is known that any operator B with symbol decreasing sufficiently rapidly is in trace class (see [**313**]) and its trace is **(4I)** $Tr(B) = (2\pi\hbar)^{-n}\int < \phi_\alpha, B\phi_\alpha) > \alpha_0$. The regularized density of states $\rho_A(E)$ can then be rewritten as

$$\rho_A(E) = (2\pi)^{-n-1}\hbar^{-n}\int \hat{g}(t)e^{iEt/\hbar} < \phi_\alpha, \hat{A}_\chi U(t)\phi_\alpha > dtd\alpha \tag{2.70}$$

where $U(t)$ is the quantum unitary group **(4J)** $U(t) = exp[-it\hat{H}/\hbar]$ and $\hat{A}_\chi = \chi(\hat{H})\hat{A}_\chi(\hat{H})$ (sometimes the subscript $\chi$ is dropped in $A_\chi$). One can write **(4F)** as **(4L)** $\psi_0 = \Lambda_\hbar \tilde{\psi}_0$ where one assumes **(4M)** $(\Lambda_\hbar \psi)(x) = \hbar^{-n/4}\psi(x\hbar^{-1/2})$; $\tilde{\psi}_0(x) = \pi^{-n/4}exp(-|x|^2/2)$.

**LEMMA 2.4.1.** Assume that $A$ satisfies **(H.0)**; then

$$\hat{A}\phi_\alpha = \sum_\gamma \hbar^{|\gamma|/2}\frac{\partial^\gamma A(\alpha)}{\gamma!} + O(\hbar^\infty) \tag{2.71}$$

in $L^2(\mathbf{R}^n)$ where $\gamma \in \mathbf{N}^{2n}$, $|\gamma| = \sum_1^{2n} \gamma! = \prod_1^{2n} \gamma_j!$ and $\psi_{\gamma,\alpha} = T(\alpha)\Lambda_\hbar OP_1^w(z^\gamma)\tilde{\psi}_0$. Here $Op_1^w(z^\gamma)$ is the 1-Weyl quantization of the monomial $(x,\xi)^\gamma = x^{\gamma'}\xi^{\gamma''}$, $\gamma = (\gamma', \gamma'') \in \mathbf{N}^{2n}$.

The lemma is proved using a scaling argument and Taylor expansion for the symbol $A$ round the point $\alpha$. Thus $m(t,\alpha)$ is a linear combination of terms like **(4N)** $m_\gamma(\alpha,t) = <\psi_{\gamma,\alpha}, U(t)\phi_\alpha>$. Then one computes $U(t)\phi_\alpha$ using [**229**]. Recall that $F(t)$ is a time dependent symplectic matrix (Jacobi matrix) defined by a linear equation (2.68). Then, with $Met\,F$ denoting the metaplectic representation of the linearized flow F (cf. [**313**]) one defines the $\hbar$-dependent metaplectic flow via **(4O)** $Met_\hbar(F) = \Lambda_\hbar^{-1}Met(F)\Lambda_\hbar$. Also use the notation **(4P)** $\delta(\alpha,t) = \int_0^t p_s \cdot q_s - tH(\alpha) - \frac{1}{2}(p_t \cdot q_t - p \cdot q)$. From Theorem (3.5) of [**229**] and its proof one has the following estimation for the $L^2$ norm, namely for every $N \in \mathbf{N}$ and every $T > 0$ there exists $C_{N,T}$ such that

$$\|U(t)\phi_\alpha - e^{i\delta(\alpha,t)/\hbar}T(\alpha_t)Met_\hbar(F(t))\Lambda_\hbar P_N(x,D_x,t,\hbar)\tilde{\psi}_0\| \leq C_{N,T}\hbar^N \tag{2.72}$$

where $P_N(t,\hbar)$ is the differential operator defined via

$$P_N(x,D_x,t,\hbar) = I + \sum_{(k,j)\in I_N} \hbar^{(k/2)-j}p_{kj}^2(x,D,t) \tag{2.73}$$

where $I_N = \{(k,j) \in \mathbf{N}\times\mathbf{N},\ 1 \leq j \leq 2N-1,\ k \geq 3j,\ 1 \leq k-2j < 2N\}$. Here the differential operators $p_{kj}(x,D_x,t)$ are products of $j$ Weyl quantizations of homogeneous polynomials of degree $k_s$ with $\sum k_s = k$ $(1 \leq s \leq j)$. Consequently **(4Q)** $p_{kj}^2(x,D_x,t)\tilde{\psi}_0 = Q_{kj}(x)\tilde{\psi}_0(x)$ where $Q_{kj}(x)$ is a polynomial (with coefficients depending on $(\alpha,t)$) of degree $k$ having the same parity as $k$. Note that homogeneous polynomials have a definite parity and Weyl quatization behaves well with respect to symmetries - thus $Op^w(A)$ commutes with the parity operator $\Sigma f(x) = f(-x)$ if and only if $A$ is an even symbol and anticommutes for an odd symbol; further $\tilde{\psi}_0(x)$ is an even function. Consequently

$$m(\alpha,t) = \sum_{(j,k)\in I_N;\,|\gamma|\leq 2N} c_{k,j,\gamma}\hbar^{(1/2)(|\gamma|}e^{i\delta(\alpha,t)/\hbar} \tag{2.74}$$

$$\cdot <T(\alpha)\Lambda_\hbar Q_\gamma\tilde{\psi}_o, T(\alpha_t)\Lambda_\hbar Q_{k,j}Met(F(t)\tilde{\psi}_0> + O(\hbar^N)$$

where $Q_{k,j}$ (resp. $Q_\gamma$) are polynomials in $x$ with the same parity as k (resp. $|\gamma|$). This will be useful in proving that one has only even powers in $\hbar$ in (2.69) (although half integer powers appear naturally in the asymptotic propagation of coherent states).

We skip some sections now in [**228**] leading to the formulas

$$\rho_A(E) = \int dt \int_{\mathbf{R}^2n} \int_{\mathbf{R}^n} a(t,\alpha,y,\hbar) e^{(i/\hbar)\Phi_E(y,\alpha,t)} dy \tag{2.75}$$

(2.76)

$$\Phi_E(t,y,\alpha) = S(\alpha,t) + q\cdot p + (y-q_t)\cdot p_t + \frac{1}{2}(y-q_t)\cdot M(t)(y-q_t) + \frac{i}{2}|y-q|^2 - y\cdot + Et$$

where $\alpha = (q,p)$ and $\alpha_t = \phi_t(\alpha)$ as before and M arises via $F(t) = \begin{pmatrix} A & B \\ C & D \end{pmatrix}$ with $U = A + iB$, $V = C + iD$ and $M = VU^{-1}$. The procedure is now to prove Theorem 1.4.1 by expanding (2.75) by the method of stationary phase for which the background material can be found in [**447**]. The form needed here is contained in

**THEOREM 2.4.2.** Let $\mathcal{O} \subset \mathbf{R}^d$ be an open set and $a, f \in C^\infty(\mathcal{O})$ with $\Im(f) \geq 0$ in $\mathcal{O}$ and $supp(a) \subset \mathcal{O}$. Define $M = \{x = \in \mathcal{O},\ f'(x) = 0\}$ and assume $M$ is a smooth, compact, and connected submanifold of $\mathbf{R}^d$ of dimension $k$ such that for all $x \in M$ the Hessian $f''(x)$ is nondegenerate on the normal space $N_x$ to $M$ at $x$. Under these conditions the integral $J(\omega) = \int -\mathbf{R}^d exp[i\omega f(x)]a(x)dx$ has the following asymptotic expansion as $\omega \to \infty$

$$J(\omega) = \left(\frac{2\pi}{\omega}\right)^{(d-k)/2} \sum_{j\geq 0} c_j \omega^{-j}; \tag{2.77}$$

$$c_0 = e^{i\omega f(m_0)} \int_M \left[det\left(\frac{f''(m)|N_m}{i}\right)\right]_*^{-(1/2)} a(m) dV_M(m)$$

where $dV_M(m)$ is the canonical Euclidean volume in $M$, $m_0 \in M$ is arbitrary, and $[det(P)]_*^{-1/2}$ denotes the product of the reciprocals of square roots of the eigenvalues of $P$ chosen with positive real parts. Note that since $\Im(f) \geq 0$ the eigenvalues of $f''(m)|N_m/i$ lie in the closed right half plane.

A proof is sketched in [**228**]. Next one computes the stationary phase expansion of (2.75) with phase $\Phi_E$ given by (2.76). Note that $a(t,\alpha,y,\hbar)$ is actually, according to (2.74), a polynomial in $\hbar^{1/2}$ and $\hbar^{-1/2}$. Hence the stationary phase theorem (with $\hbar$ independent symbol $a$) applies to each coefficient of this polynomial. The first order derivatives of $\Phi_E(t,y,\alpha)$ (up to $O((y-q)^2, (\alpha-\alpha_t)^2)$) are given by

$$\partial_t \Phi_E = E - H(\alpha) + (y-q_t)\cdot \dot{p}_t - \dot{q}_t \cdot M(y-q_t); \tag{2.78}$$

$$\partial_y \Phi_E = p_t - p + i(y-q) + M(y-q)t); \ \partial_p \Phi_E = q - q_t + ({}^tD - {}^tBM - I)(y-q_t)$$

$$\partial_q \Phi_E = i(q-q_t) - P^tA(p-p_t)+)^tC - {}^tAM - iI)(y-q_t)$$

Moreover, since F is symplectic one has (**4R**) $2\Im(\Phi_E) = |y-q|^2 + |(A+iB)^{-1}(y-q_t)|^2$. This implies that $\Phi_E(y,\alpha,t)$ is critical on the set

$$C_E = \{(y,\alpha,t) \in \mathbf{R}^n_y \times \mathbf{R}^{2n}_\alpha \times \mathbf{R}_t :\ y = q_t;\ \alpha_t = \alpha; H(\alpha) = E\} \tag{2.79}$$

Thus each component $M_\gamma$ of $C_E$ has the form

$$M_\gamma = \{(y,\alpha,t) = (q,\alpha,T(\alpha)) :\ \alpha = (p,q) \in \gamma;\ \alpha_{T(\alpha)} = \alpha;\ H(\alpha) = E\} \tag{2.80}$$

One assumes now that each $\gamma$ is a smooth compact manifold and then the manifolds $\gamma$ are clearly unions of periodic classical trajectories of energy $E$. One assumes now a "clean intersection" hypothesis (see below). This will assure that **(4S)** $C_E = \{0\} \times \Sigma \cup \{\mathcal{M}_{\gamma_1}, \cdots, \mathcal{M}_{\gamma_N}\}$ where each $\mathcal{M}_{\gamma_k}$ has the form (2.80) wth $\gamma_k$ in the fixed point set of the mapping $\alpha \to \alpha_{T_k}$.

The first thing to check in order to apply the stationary phase theorem is that the support of $\alpha$ in (2.75) can be taken as compact, up to an error $O(\hbar^\infty)$. To see this one recalls some properties of $\hbar$-PSDO from [**264, 431**]. Thus the function $m(z) =< H(z) >$ is a weight function and in [**264**] it is proved that $\chi(\hat{H}) = \hat{H}_\chi$ where $H_\chi \in S(m^{-k})$ for every $k$. More precisely one has in the $\hbar$-asymptotic sense **(4T)** $H_\chi = \sum_{j\geq 0} H_{\chi j}\hbar^j$ and the support of $[H_{\chi,j}]$ is in a fixed compact set for every $j$ (cf. (bf H.5) and [**431**] for the computation of $H_{\chi,j}$). Recall also that the symbol space $S(m)$ is equipped with the family of semi-norms **(4U)** $sup_{z\in\mathbf{R}^{2n}} m^{-1}(z)|(\partial^\gamma/\partial z^\gamma u(z)|$. Then one can prove that there is a compact set K in $\mathbf{R}^{2n}$ such that for **(4V)** $m(\alpha,t) =< \hat{A}_\chi\phi_\alpha, U(t)\phi_\alpha >$ one has **(4W)** $\int_{\mathbf{R}^{2n}/K} |m(\alpha,t)|d\alpha = O(\hbar^\infty)$ uniformly in every bounded interval of $t$. We refer to [**228**] for proof.

Finally one computes the Hessian of $\Phi_E$ on a set $\mathcal{M}_{\gamma k}$. After some computation the Hessian $\Phi''_E$, with variables $(t, y, p, q)$ is the following $(1 + 3n) \times (1 + 3n)$ matrix

(2.81)
$$\begin{pmatrix} H_p \cdot (H_q + MH_p) & -H_q - H_p & -H_p(D - MB) & -H_p(C - MA) \\ -H_q - MH_p & M_iI & D - MB - I & C - MA - iI \\ -({}^tD - {}^tBM)H_p & {}^tD - {}^tBM - I & {}^tBMB - {}^tDB & {}^tBMA - {}^tBC \\ -({}^tC - {}^tAM)H_p & {}^tC - {}^tAM - iI & {}^tAMB - {}^tCB & {}^tAMA - {}^tCA + iI \end{pmatrix}$$

where $H_p$ (resp. $H_q$) denotes $\partial_p H|_{\alpha=\alpha_t}$ (resp. $\partial_q H|_{\alpha=\alpha_t}$ ${}^tA = A^T$ is the transpose of A, and $A$, $B$, $C$, $D$, $M$ are given in (2.76). One performs elementary row and column operations on (2.80) to compute the nullspace of $\Phi''_E$ and the determinant of $\Phi''_E$ restricted to the normal space to the critical manifold (cf. [**228**] for calculations) and the clean flow condition is stated as

**HYPOTHESIS C.** Assume that $D_E = \{(\alpha, t) \in \Sigma_E \times \mathbf{R}/\phi_t(\alpha) = \alpha\}$ is a submanifold of $\mathbf{R}^{1+2n}$. Then one says that $D_E$ satisfies the clean flow condition if for any $(\alpha, t) \in D_E$ the tangent space to $D_E$ is given by

$$(2.82)\quad T_{\alpha,t}D_E = \left\{(v, w, \tau) \in \mathbf{R}^{1+2n} : (F - I)\begin{pmatrix} v \\ w \end{pmatrix} + \tau\begin{pmatrix} H_p \\ -H_q \end{pmatrix} = 0;\right.$$
$$\left. H_q \cdot v + H_p \cdot w = 0\right\}$$

Since $C_E = \{(y, \alpha, t) : (\alpha, t) \in D_E \text{ and } y = q\}$ the tangent space $T_{y,\alpha,t}C_E$ is

$$(2.83)\quad \{(\tau, v, w, v) : (F - I)\begin{pmatrix} v \\ w \end{pmatrix} + \tau\begin{pmatrix} H_p \\ -H_q \end{pmatrix} = 0;\ H_q \cdot v + H_p \cdot w = 0\}$$

and in fact this equals the null space of $\Phi''_E$ (see [**228**] for details and further calculations checking determinants and Maslov indices). Thus the computations of [**228**] provide a proof for the existence of a Gutzwiller trace formula as in Theorem 1.4.1 (under Hypothesis C). However the calculations are only carried out for the case that $\gamma$ consists of a single trajectory and Hypothesis C reduces to the assumption (**H.3**) of isolated nondegenerate periodic orbits.

## 3. MORE ON METAPLECTIC TECHNIQUES

Having opened the door to symplectic and metaplectic ideas related to quantum mechanics in Section 1.2.3 we are "obliged" to develop this further, rather sooner than later. We will mainly draw upon formulations of deGosson (cf. [**369, 370, 372, 373, 374, 375, 376**] and mention also [**459**] by Isidro and deGosson where gerbes and gauge theory arise). Further important references are [**133, 307, 313, 394, 439, 588, 741, 763, 855, 885**] but we make no attempt for completeness here (with apologies for omissions). First we sketch from [**370**] (quant-ph 0808.2774) (cf. also [**377**]) where some general classical facts about quantum mechanics are reviewed. In particular deGosson indicates that:

(1) The SE can be autonomously be derived from the Hamilton equations of motion. Consequently the SE is equivalent to the Hamiltonian equations.
(2) The uncertainty principle of QM is already present formally in classical mechanics in the Hamiltonian formulation.

This sounds sacrilegious of course (but note there is an "arbitrary" parameter $\epsilon$ involved and no assurance that $\epsilon$ is related to the Planck constant $\beta$) and we will sketch some of the argument from [**370**] (cf. also [**377**]). First consider a system of $N$ particles in 3-D space with phase space evolution governed by the Hamiltonian equations

$$\dot{x}_j = \frac{\partial H}{\partial p_j};\ \dot{p}_j = -\frac{\partial H}{\partial q_j} \tag{3.1}$$

Setting $x = (x_1, \cdots, x_{3N})^T$ and $p = (p_1, \cdots, p_{3N})^T$ the solution at time $t$ is given via

$$\begin{pmatrix} x(t) \\ p(t) \end{pmatrix} = S_t \begin{pmatrix} x(0) \\ p(0) \end{pmatrix};\ S_t J S_y^T = S_t^T J S_t = J \tag{3.2}$$

where $S_t$ is a $6N \times 6N$ real matrix and $J$ is the standard symplectic matrix. In fact (**1A**) $H(x,p) = (1/2)Z^T M z$ for $z = (x\,p)^T$ with $S_t = exp(tJM)$. The set of matrices $S$ as in (**1A**) with $SJS^T = S^T JS = J$ is the standard symplectic group $Sp(6N)$ and $S_t$ will describe a curve $\Sigma$ in the symplectic group passing through the identity at time 0. The double cover of $Sp(6N)$ is the metaplectic group $Mp(6N)$ and int can be realized (in infinitely many ways) as a group of unitary operators acting on $L^2(\mathbf{R}^{3N})$. These groups are parametrized by a positive parameter and the choice $Mp^\epsilon(6N)$ will contain a Fourier like transform $\hat{F}^\epsilon$ defined via

$$\hat{F}^{gep}(\psi(p)) = \left(\frac{1}{2\pi i\epsilon}\right)^{3N/2} \int_{\mathbf{R}^{3N}} e^{(i/\epsilon)p\cdot x}\psi(x)dx \tag{3.3}$$

Then fix $\epsilon$ and via the "path lifting property" of covering groups one knows that the curve $\Sigma$ unambiguously induces a unique curve $\hat{\Sigma}$ in $Mp^\epsilon(6N)$ passing through the identity operator for $t = 0$. This curve is the unique curve having this property such that the projection of a point $\hat{S}_t$ of $\hat{\Sigma}$ down to $Sp(6N)$ is precisely $S_t$. Then letting $\hat{S}_t$ act on a smooth $L^2$ function $\psi_0$ defines a "wave function" **(1B)** $\psi(x,t) = \hat{S}_t\psi_0(x)$ satisfying the SE like equation $\imath\epsilon\partial_t\psi(x,t) = H(x, -i\epsilon\nabla_x)\psi(x,t)$ where $H$ is obtained from the Hamiltonian function via the symmetrized quantization rules $x_j \to \hat{x}_j$ and $p_j \to \hat{p}_j = -i\epsilon\partial_x$ and $x_jp_k \to (1/2)(\hat{x}_j\hat{p}_k + \hat{p}_k\hat{x})j)$. The choice $\epsilon = \hbar = h/2\pi$ then yields the SE **(1D)** $i\hbar\partial_t\psi(x,t) = H(x, -i\hbar\nabla_x)\psi(x,t)$ and hence "mathematically" one has an equivalence between the SE and the Hamilton equations. This is actually closely related to the fact that via Ehrenfest's equation (for a quadratic potential $V$)

$$m\frac{d^2 <x>}{dt^2} = -\left\langle\frac{\partial V}{\partial x}(x)\right\rangle \Rightarrow m\frac{d^2 <x>}{dt^2} = -\frac{\partial V}{\partial x}(<x>) \tag{3.4}$$

In 1-D assume now $H = p^2/2m$ so via general formulas for the metaplectic representation

$$S_t = \begin{pmatrix} 1 & t/m \\ 0 & 1) \end{pmatrix} \Rightarrow i\hbar\partial_t\psi(x,t) = -\frac{\hbar^2}{2m}\partial_x^2\psi(x,t) \tag{3.5}$$

with
(3.6)

$$\psi(x,t) = \int_{-\infty}^{\infty} K_t(x,y)\psi_0(y)dy;\ K = (e^{i\pi/4})^{sign(t)}\sqrt{\frac{m}{2\pi\hbar|t|}}exp\left[\frac{i}{\hbar}\frac{m(x-y)^2}{2t}\right]$$

Suppose next that $H$ is the harmonic oscillator Hamiltonian, for simplicity $m = \omega = 1$ so $H = (1/2)p^2 + x^2)$, in which case the solution of the SE

$$i\hbar\partial_t\psi(x,t) = \frac{1}{2}(-\hbar^2\partial_x^2 + x^2)\psi(x,t) \tag{3.7}$$

is given by (3.6) where now (for $t \neq n\pi$)

$$K_t(x,y) = i^{-[t/\pi]}\sqrt{\frac{1}{2\pi\hbar|Sin(t)|}}exp\left[\frac{i}{2\hbar}\frac{(x^2+y^2)Cos(t) - xy}{2Sin(t)}\right] \tag{3.8}$$

These Feynman integral type formulas are well known but they are restricted to quadratic Hamiltonians and only reconstruct the metaplectic representation in special cases. There is however a theorem due to Groenwald and van Hove that says one cannot use the metaplectic representation to construct solutions to the SE for general Hamiltonians (cf. [**369, 394**]). But this does not mean that there is no way to derive the SE from the Hamilton equations. The first step toward such a program is a rather straightforward extension of the quadratic case. Assume $H$ is a non-homogeneous polynomial of degree 2 in the position and momentum variables and following **(1A)** write **(1E)** $H(x,p) = (1/2)z^TMz + u^Tz$ for some vector $u$. The flow determined by the corresponding Hamilton equations involves now affine canonical transformations which again form a group, the inhomogeneous symplectic group $ISp(3N)$ (semi-direct product of the symplectic group and translations). One can repeat the previous arguments and show that for every $\epsilon > 0$ there is a 1-1 correspondence between continuous curves in ISP(3N) and curves in a group of unitary operators $IMp^\epsilon(3N)$ (inhomogeneous metaplectic group). Here $Mp^\epsilon(3N)$

consists of operators in $Mp^{\epsilon}(3N)$ composed on the left or right with Heisenberg operators

$$\hat{T}(x_0, p_0)\psi(x) = exp\left[\frac{i}{\epsilon}\left(p_0 \cdot x - \frac{1}{2}p_0 \cdot x_0\right)\right]\psi(x - x_0) \tag{3.9}$$

This is familiar from the Schrödinger representation of the Heisenberg group when $\epsilon = \hbar$ and Hamilton's equations are again mathematically equivalent to the SE associated with the non-homogeneous $H$. In general the following sketch seems to work (cf. [**555, 369, 370**]). One replaces $H$ by its Taylor series to second order around a point $z_t = f_t(z_0)$ where $z_0 = (x_0, p_0)$ is arbitrary. Thus write

$$H_{z_0}(z, t) = H(z_t) + \nabla_z H(z_t) \cdot (z - z_t) + \frac{1}{2}H''(z_t)(z - z_t) \cdot (z - z_t) \tag{3.10}$$

($H''$ is called the Hessian). The Hamilton equations for $H_{z_0}$ define a flow $f_{z_0,t}$ consisting of affine symplectic transformations (i.e. each $f_{z_0,t} \in ISp(3N)$) and when $t$ varies, $f_{z_0,t}$ is just $z_t$, the solution of the Hamilton equations with initial data at $t$. Thus every Hamiltonian trajectory comes from an affine flow, but this flow depends each time on the initial point. This is well known and has been used to construct short-time solutions for the SE with initial data a narrow wavepacket, by propagating the center of this wavepacket along the classical curve (cf. [**302, 386, 369, 370, 555**]); it suffices to lift as before the affine Hamiltonian flow to the inhomogeneous metaplectic group. Using the theory of Gabor frames from time-frequency analysis one can write down such short-time solutions for arbitrary wavepackets - valid up to some "Ehrenfest time". However asymptotic validity for short times is sufficient to construct exact solutions via a Lie-Trotter argument and one ends up with wavepackets obeying the SE.

Concerning the second point above suppose one has at time $t = 0$ a cloud of $N$ particles in phase space which could be assumed spherical and identified with a ball $B(r) : |x|^2 + |p|^2 \leq r^2$; the orthogonal projection of this ball on any plane of coordinates $(x_j, p_k)$ will be a circle of area $\pi r^2$. Given however a plane of conjugate coordinates $(x_j, p_j)$ however the phase cloud may distort and assume a vastly different shape but the projection on any such plane of conjugate coordinates will never decrease below $\pi r^2$ (note the total volume is constant via Liouville's theorem). However a plane of non-conjugate coordinates would be uncontrolled and the projection could become arbitrarily small. This was proved by Gromov (cf. [**387**]) and is reminiscent of the Heisenberg uncertainty principle. Indeed in [**692**] Penrose comes to the conclusion that phase space spreading with many degrees of freedom suggests that classical mechanics cannot be true of our world; however he adds that quantum effects can prevent this spreading. He adds that while phase space spreading a priori opens the door to classical chaos, quantum effects have a tendency to "tame" the behavior by blocking and excluding most of the classically allowed motions. The phenomena described above show that there is a similar taming in Hamiltonian mechanics itself preventing anarchy and chaotic spreading of the ball in phase space. One makes this more precise in [**370**] as follows (cf. also [**377**] where it is emphasized that one needs to justify the existence of a universal constant $\hbar$, valid for all physical systems, in order to claim a derivation of QM from CM. Consider an arbitrary region $\Omega \subset \mathbf{R}^{6N}$ and recall that the Gromov capacity of $\Omega$ is the (possibly infinite) number $c_{min}(\Omega)$ which is defined for given $B(r)$ as above and assume first that there is no canonical transformation sending that ball inside $\Omega$ (in which case $c_{min}(\Omega) = 0$). If there are canonical transformations sending $B(r)$ into $\Omega$ let $R$ (= symplectic radius) be the supremum of all radii for which this is possible and one defines the Gromov capacity of $\Omega$ by $c_{min} = \pi R^2$. Thus for $r < R$ one can find a canonical transformation sending $B(r)$ into $\Omega$ but no canonical transformation will send a ball with $r > R$ into $\Omega$. Thus (**1F**) $c_{min}(f(\Omega)) = c_{min}(\Omega)$ if $f$ is canonical and (**1G**) $c_{min} \leq c_{min}(\Omega')$ if $\Omega \subset \Omega'$. Further (**1H**) $c_{min}(\lambda\Omega) = gl^2 c_{in}(\Omega)$ for $\lambda$ constant. Most important here however is

$$c_{min}(B(R)) = \pi R^2 = c_{min}(Z_j(R)) \tag{3.11}$$

where $Z_j(R)$ is the phase space cylinder based on the plane of conjugate variables satisfying $x_j^1 + p_j^2 \leq R^2$ (see [**370, 377**]) for further discussion and proofs). Generally one calls symplectic capacity any function associating to subsets $\Omega$ of phase space a non-negative number $c(\omega)$ satisfying (**1F**), (**1G**), (**1H**), and (3.11) (cf. [**439**]). Then $c_{min}(\Omega) \leq c(\Omega)$ for all such $c$ and there is also a largest symplectic capacity (cf. [**370**]). Further all symplectic capacities agree on phase space ellipsoids and one looks now at the ellipsoid (**1I**) $(z - z_0)^T M(z - z_0) \leq 1$. The eigenvalues of $JM$ are the same as those of $M^{1/2}JM^{1/2}$ and hence purely imaginary (say $\pm i\lambda_j$ where $\lambda_j > 0$). Then (**1J**) $c(\Omega) = \pi/\lambda_{max}$. Actually a weaker form of symplectic capacity used in [**370**] is $c_{lin}$ defined as with $c_{min}$ by restricting the maps to be affine symplectic transformations. Thus $c_{lin} = \pi R^2$ is the supremum of numbers $\pi r^2$ such that the is an affine transformation in $ISp(3)$ sending the bal $B(r)$ into $\Omega$ and one replaces (**1F**) by $1K$ $c_{lin}(f(\Omega)) = c_{lin}(\Omega)$ for $f \in ISp(3N)$. If $\Omega$ is an ellipsoid then $c_{lin}(\Omega)$ is again given via (**1J**).

For the uncertainty principle we follow [**370**] but mention [**377**] for a more detailed treatment. To make things look quantum one writes $h = \pi\hbar$ and assumes $c_{min}(\Omega) \geq (1/2)h$. The convexity of $\Omega$ implies that there is a unique ellipsoid $\mathcal{J}_\Omega$ contained in $\Omega$ with maximal volume among other such ellipsoids (called the John ellipsoid after F. John - cf. [**66**]). Then one checks that $c_{lin}(\mathcal{J}_\Omega) \geq (h/2)$ and hence there is a positive definite $6N \times 6N$ matrix $\Sigma$ such that $\mathcal{J}_\Omega$ consists of all phase space points $z = (x,p)^T$ satisfying (**1L**) $(1/2)z^T\Sigma^{-1}z \leq 1$. The notation suggests that $\Sigma$ can be viewed as a statistical covariance matrix so one writes

$$\Sigma = \begin{pmatrix} \Sigma_{XX} & \Sigma_{XP} \\ \Sigma_{PX} & \Sigma_{PP} \end{pmatrix} \tag{3.12}$$

where the blocks $\Sigma_{XX}$, $\Sigma_{XP} = \Sigma_{PX}^T$ and $\Sigma_{PP}$ are $3N \times 3N$ matrices which can then be written as $\Sigma_{XX} = (Cov(X_j, X_k))_{j,k}$, $\Sigma_{XP} = (Cov(X_j, P_k))_{j,k}$, and $\Sigma_{PP} = (Cov(P_j, P_k))_{j,k}$. Then from [**369**] it follows that (**1M**) $(\Delta X_j)^2(\Delta P_j)^2 \geq (Cov(X_j, P_j)^2 + (1/4)\hbar^2)$ where $(\Delta x_j)^2 = Cov(X_j, X_j)$ etc. This is the strong form of the Heisenberg uncertainty principle due to Robertson and Schrödinger (see e.g. [**746**]) which implies the standard inequality upon neglecting covariances. Thus the inequalities (**1M**) are mathematically equivalent to the statement that $c(\mathcal{J}_\Omega) \geq (1/2)h$ for every symplectic capacity and this in turn is equivalent to the matrix condition (**1N**) $\Sigma + (i\hbar/2)J$ is positive semi-definite (well known from quantum optics (cf. [**369**]). The proof of the equivalence between (**1L**) and (**1M**) relies on elementary linear algebra, using (**1J**). Thus the inequalities (**1M**) are conserved in time under Hamiltonian evolution (cf. [**370**] for a sketch of the proof). The generalization to arbitrary Hamiltonian flows is somewhat harder and we refer to [**369, 369, 377, 574**].

CHAPTER 2

# SOME GEOMETRIC ASPECTS

## 1. INTRODUCTION

In [**170**] we surveyed many feature of the so called quantum potential (QP) (cf. also [**177, 178, 179, 171, 180, 181, 182, 183**]). Some matters were treated more thoroughly than others and we want to discuss here certain geometrical aspects in more detail, some connections to nonlinear Schrödinger type equations, and various phase space approaches. The latter two topics were not developed in [**170**] and we will try to make amends here. Some relations to electromagnetic (EM) theory will also be discussed. To set the stage we recall the Schrödinger equation (SE) in 1-D of the form (**1A**) $-(\hbar^2/2m)\psi'' + V\psi = i\hbar\psi_t$ so that for $\psi = Rexp(iS/\hbar)$ one has

$$(1.1) \qquad S_t + \frac{1}{2m}S_x^2 + V + Q = 0;\ Q = -\frac{\hbar^2 R''}{2mR};\ \partial_t(R^2) + \frac{1}{m}(R^2 S_x)_x = 0$$

Here $Q$ is the quantum potential (QP) and one can argue that Bohmian mechanics is simply classical symplectic mechanics using the Hamiltonian (**1B**) $H_q = H_c + Q = (1/2m)S_x^2 + V + Q$ from the Hamiltonian-Jacobi (HJ) equation (1.1) (cf. here [**116, 132, 369, 436**]). One can write $P = R^2 = |\psi|^2$ (a probability density) with $\rho = mP$ a mass density and obtain a hydrodynamical version of (1.1). Note in particular (**1C**) $Q = -(\hbar^2/2m)(\partial^2\sqrt{\rho}/\sqrt{\rho})$ and using $p = S_x = m\dot{q} = mv$ one obtains

$$(1.2) \qquad mv_t + mvv_x + \partial V + \partial Q = 0;\ \rho_t + (\rho\dot{q})_x = 0$$

leading to

$$(1.3) \qquad \partial_t(\rho v) + \partial(\rho v^2) + \frac{\rho}{m}\partial V + \frac{\rho}{m}\partial Q = 0$$

which has the flavor of an Euler equation (cf. [**170, 180, 256**]). There is however a missing pressure term from the hydrodynamical theory (cf. [**170**]) and looking at (1.2) one could imagine a pressure term supplied in the form (**1D**) $\partial Q = (1/R^2)\partial\mathfrak{P}$ (where $\mathfrak{P}$ denotes pressure). This suggests a hydrodynamical interpretation for Q, namely, going to 3-D for example, (**1E**) $\nabla\mathfrak{P} = R^2\nabla Q$ (cf. [**173**]). This will all be discussed in detail below and we make first a few background remarks about the QP.

**REMARK 1.1.1.** In [**177**] we considered given a function $Q \in L^\infty(\Omega)$ (for $\Omega$ a bounded domain) and looked for $R \in H_0^1(\Omega)$ satisfying $Q = -(\hbar^2/2m)(\Delta R/R) \equiv \Delta R + (2m/\hbar^2)QR = 0$. We showed that if $Q < 0$ ($\beta = (2m/\hbar^2)$) then there is a unique solution and if 0 is not in the countable spectrum of $\Delta R + \beta QR$ then

$\Delta R + \beta QR = 0$ has a unique solution for any $Q \in L^\infty$. The corresponding HJ equation **(1F)** $\partial_t S + (1/2m)(\nabla S)^2 + Q + V = 0$ and the continuity equation **(1G)** $\partial_t R^2 + (1/m)\nabla(R^2\nabla S) = 0$ must then be solved to obtain some sort of generalized quantum theory. Here a priori $V$ must be assumed unknown and there are then two equations for two unknowns $S$ and $V$, namely (in 1-D for simplicity)

$$S_t + \frac{1}{2m}S_x^2 + Q + V = 0;\ \partial_t R^2 + \frac{1}{m}(R^2 S_x)_x = 0 \tag{1.4}$$

the solution of which would yield a SE based on Q (see here Remarks 5.1.1 and 5.1.2 for more detail in this regard). ∎

**EXAMPLE** 1.1. Now $(1/2m)p^2 + V = E$ (classical Hamiltonian - recall $p \sim S_x$) so we could perhaps treat E as an unknown here and try to solve

$$S_t + Q + E = 0;\ \partial_t R^2 + \frac{1}{m}(R^2 S_x)_x = 0 \tag{1.5}$$

Consider first $R^2 S_x = -\int^x m\partial_t R^2 dx + f(t)$ from which

$$2RR_t S_x + R^2 S_{xt} = -\int^x m\partial_t^2 R^2 dx + f' \Rightarrow (Q_x + E_x)R^2 \tag{1.6}$$

$$= -R^2 S_{xt} = \frac{2R_t}{R}\left(-\int^x m\partial_t R^2 dx + f\right) + \int^x m\partial_t^2 R^2 dx - f'$$

Hence

$$R^2 S_x = -\int^x m\partial_t R^2 dx + f(t);\ R^2 E_x = -Q_x R^2 \tag{1.7}$$

$$+ \int^x m\partial_t^2 R^2 dx - f' + \frac{2R_t}{R}\left(-\int^x m\partial_t R^2 dx + f\right)$$

giving $S_x$ and $E_x$ modulo an arbitrary differentiable function $f(t)$. Note also

$$R^2 E_x = R^2 V_x + \frac{R^2}{2m}(S_x^2)_x = R^2 V_x + \frac{R^2}{m}S_x S_{xx} \tag{1.8}$$

and $S_{xx}$ can be determined via $(1/m)(R^2 S_x)_x = -\partial_t R^2$. Hence $R^2 V_x$ can be determined from $R^2 E_x$. Note here

$$\frac{R^2}{m}S_x S_{xx} = -\partial_t R^2 - \frac{2}{m}RR_x f + 2RR_x \int^x \partial_t R^2 dx \tag{1.9}$$

(see Remarks 5.1.1 and 5.1.2 for more details). ∎

We mention also two examples from [**116, 177**]

**EXAMPLE** 1.2. For a free particle in 1-D there are possibilities such as $\psi_1 = Aexp[i[px - (p^2t/2m))/\hbar]$ and $\psi_2 = Aexp[-i(px + (p^2t/2m))/\hbar]$ in which case $Q = 0$ for both functions but for $\psi = (1/\sqrt{2})(\psi_1 + \psi_2)$ there results $Q = p^2/2m$ ($p \sim \hbar k$ here). Hence $Q = 0$ depends on the wave function and cannot be said to represent a classical limit. Further we note that $S = \hbar kx - (\hbar^2k^2/2m)t$ in $\psi_1$ with $S_t = -\hbar^2k^2/2m \sim -E$, $S_x = \hbar k$, and $R = 1 \notin H_0^1$. For $\psi = (1/\sqrt{2})(\psi_1 + \psi_2)$ on the other hand

$$R = \sqrt{2}ACos(kx) \notin H_0^1;\ \frac{R''}{R} = -k^2;\ Q = \frac{k^2\hbar^2}{2m};\ S = -\frac{k^2\hbar^2 t}{2m}; \tag{1.10}$$

$$S_t = -\frac{k^2\hbar^2}{2m} \sim -E;\ S_x = 0$$

Thus the same SE can arise from different $Q$ (which is generally obvious of course) and $S$ varies with $Q$. ■

**EXAMPLE** 1.3. For $V = m\omega^2 x^2/2$ and a stationary SE one has solutions of the form $\psi_n(x) = c_n H_n(\xi x)exp(-\xi^2 x^2/2)$ where $\xi = (m\omega\hbar)^{1/2}$, $c_n = (\xi/\sqrt{\pi}2^n n!)$, and $H_n$ is a Hermite function. One computes that $Q = \hbar\omega[n+(1/2)]-(1/2)m\omega^2 x^2$ and hence $\hbar \to 0$ does not imply $Q \to 0$ and moreover $Q = 0$ corresponds to $x = \pm\sqrt{(2\hbar/m\omega)[n+(1/2)]}$ so not all systems in quantum mechanics have a classical limit. This example corresponds to $\Omega = \mathbf{R}$ and $\psi_n \in H_0^1$ is satisfied. ■

## 2. REMARKS ON WEYL GEOMETRY

Now we recall how in various situations the QP is proportional to a Weyl-Ricci curvature $R_w$ for example (cf. [**170, 179, 191, 770**]) and this can be interpreted in terms of a statistical geometry for example (cf. also [**53, 54**]). In general (see e.g. [**116, 179, 177**]) one knows that each wave function $\psi = Rexp(iS/\hbar)$ for a given SE produces a different QP as in (1.1) (which in higher dimensions has the form $-(\hbar^2/2m)(\Delta R/R)$). Thus for $Q \sim R_w$ to make sense we have to think of a given $R$ or $R^2 = P$ (or $\rho \sim mP$) as generating a (Weyl) geometry as in [**770**]. (cf. also [**170, 179, 191**].) This is in accord with having a Weyl vector (**2A**) $\phi_i \sim -\partial_i log(\hat{\rho})$ (where $\hat{\rho} = \rho/\sqrt{g}$ in [**770**] for a Riemannian metric $g$). Thus following [**770**] one assumes that the motion of the particle is given by some random process $q^i(t,\omega)$ in a manifold M ($\omega$ is the random process label) with a probability density $\rho(q,t)$ and satisfying a deterministic equation (**2B**) $\dot{q}^i(t,\omega) = (dq^i/dt)(t,\omega) = v^i(q(t,\omega),t)$ with random initial conditions $q^i(t_0,\omega) = q_0^i(\omega)$. The probability density will satisfy (**2C**) $\partial_t\rho + \partial_i(\rho v^i) = 0$ with initial data $\rho_0(q)$. Let $L(q,\dot{q},t)$ be some Lagrangian for the particle and define an equivalent Lagrangian via

$$L^*(q,\dot{q},t) = L(q,\dot{q},t) - \partial_t S + q^i\partial_i S \tag{2.1}$$

for some function S. The velocity field $v^i(q,t)$ yielding a classical motion with probability one can be found by minimizing the action functional

$$I(t_0,t_1) = E\left[\int_{t_0}^{t_1} L^*(q(t,\omega),\dot{q}(t,\omega),t)dt\right] \tag{2.2}$$

This leads to (**2D**) $\partial_t S + H(q,\nabla S,t) = 0$ and $p_i = (\partial L/\partial\dot{q}^i) = \partial_t S$ where $H \sim p_i\dot{q}^i - L$ with $v^i(q,t) = (\partial H/\partial p_i)(q,\nabla S(q,t),t)$. Now suppose that some geometric structure is given on M via $ds^2 = g_{ij}dq^i dq^j$ so that a scalar curvature $\mathcal{R}(q,t)$ is meaningful and write the acutal Lagrangian as (**2E**) $L = L_C + \gamma(\hbar^2/m)\mathcal{R}(q,t)$ where $\gamma$ will turn out to have the form $\gamma = (1/8)[(n-2)/(n-1)] = 1/16$ for $n = 3$. Assume that in a transplantation $q^i \to q^i + dq^i$ the length of a vector $\ell = (g_{ik}A^i A^k)^{1/2}$ varies according to the law (**2F**) $\delta\ell = \ell\phi_k dq^k$ where the $\phi_k$ are covariant components of an arbitrary vector of $M$ (this characterizes a Weyl geometry). One imagines that physics determines geometry so that the $\phi_k$ must be determined from some averaged least action principle yielding the motion of the particle; in particular the minimum now in (2.2) is to be evaluated with respect to the class of all Weyl geometries with fixed metric tensor. Since the only term containing the gauge vector $\vec{\phi} = (\phi_k)$ is the curvature term one requires $E[\mathcal{R}(q(t,\omega)t] = minimum$ ($\gamma > 0$ for $n \geq 3$). This minimization yields

$$\mathcal{R} = \dot{\mathcal{R}} + (n-1)\left[(n-2)\phi_i\phi^i - 2\left(\frac{1}{\sqrt{g}}\partial_i(\sqrt{g}\phi^i)\right)\right] \tag{2.3}$$

where $\phi^i = g^{ik}\phi_k$ and $\dot{\mathcal{R}}$ is the Riemannian curvature based on the metric. Note here that a Weyl geometry is assumed as the proper background for the motion. One shows that the quantity $\hat{\rho}(q,t) = \rho(q,t)/\sqrt{g}$ transforms as a scalar under coordinate changes and a covariant equation of the form **(2G)** $\partial_t\hat{\rho} + (1/\sqrt{g})\partial_i(\sqrt{g}v^i\hat{\rho}) = 0$ ensues ($g_{ik}$ is assumed time independent). Some calculation gives then a minimum when **(2H)** $\phi_i(q,t) = -[1/(n-2)]\partial_i log(\hat{\rho})]$. This shows that the transplantation properties of space are determined by the presence of matter and in turn this change in geometry acts on the particle via a "quantum" force $f_i = \gamma(\hbar^2/m)\partial_i\mathcal{R}$ depending on the gauge vector $\vec{\phi}$. Putting this $\vec{\phi}$ in (2.3) yields

$$R_w = \mathcal{R} = \dot{\mathcal{R}} + \frac{1}{2\gamma\sqrt{\hat{\rho}}}\left[\frac{1}{\sqrt{g}}\partial_i(\sqrt{g}g^{ik}\partial_k\sqrt{\hat{\rho}}\right] \tag{2.4}$$

along with a (HJ) equation

$$\partial_t S + H_C(q,\nabla S,t) - \gamma\left(\frac{\hbar^2}{m}\right)\mathcal{R} = 0 \tag{2.5}$$

and for certain Hamiltonians of the form **(2I)** $H_C = (1/2m)g^{ik}(p_i - A_i)(p_k - A_k) + V$ with arbitrary fields $A_k$ and V it is shown that the function $\psi = \sqrt{\hat{\rho}}exp[(i/\hbar)S(q,t)]$ satisfies a SE (omitting the $A_i$)

$$i\hbar\partial_t\psi = -\frac{\hbar^2}{2m}\frac{1}{\sqrt{g}}\left[\partial_i\left(\sqrt{g}g^{ik}\partial_k\right)\right]\psi + \left[V - \gamma\left(\frac{\hbar^2}{m}\right)\dot{\mathcal{R}}\right]\psi \tag{2.6}$$

This Hamiltonian is characteristic of a particle in an EM field and all Hamiltonians arising in nonrelativistic applications may be reduced to the above form with corresponding HJ equation

$$\partial_t S = \frac{1}{2m}g^{ik}\partial_i S\partial_k S + V - \gamma\frac{\hbar^2}{m}\mathcal{R} = 0 \tag{2.7}$$

**REMARK 2.1.1.** Note that indices are lowered or raised via use of $g_{ij}$ or its inverse $g^{ij}$. The most complete sources of notation for differential calculus on Riemannian manifolds seem to be [**28, 860**]. It is seen that $\hbar$ arises only via (2.5) and for $\dot{\mathcal{R}} = 0$ there is no $\hbar$ in the SE. If $\mathcal{R} = 0$ the quantum force is zero and "quantum mechanics" involves no $\hbar$; $\mathcal{R} = 0$ (with $\dot{\mathcal{R}} = 0$) involves (2.10) below giving $Q = 0$. ■

Now given (2.7), and comparing to (1.1) for example, we see that **(2J)** $Q \sim -\gamma(\hbar^2/m)\mathcal{R}$ with $\mathcal{R}$ given by (2.4) and $\gamma = 1/16$ for $n = 3$. Thus

$$Q \sim -\frac{\hbar^2}{16m}\left[\dot{\mathcal{R}} + \frac{8}{\sqrt{\hat{\rho}g}}\partial_i(\sqrt{g}g^{ik}\partial_k\sqrt{\hat{\rho}})\right] \tag{2.8}$$

and the SE (2.6) contains only $\dot{\mathcal{R}}$. Further from (**2H**) we have for the Weyl vector $\phi_i = -\partial_i log(\hat{\rho}) = -\partial_i\hat{\rho}/\hat{\rho}$ and there is an expression for $\mathcal{R}$ in the form (2.3) leading to

$$(2.9) \qquad Q \sim -\frac{\hbar^2}{16m}\left[\dot{\mathcal{R}} + 2\left\{\phi_i\phi^i - \frac{2}{\sqrt{g}}\partial_i(\sqrt{g}\phi^i)\right\}\right]$$

showing how $Q$ depends directly on the Weyl vector. When $\dot{\mathcal{R}} = 0$ (flat space) one sees that the SE is classical and **(2K)** $Q = -(\hbar^2/8m)[\phi_i\phi^i - (1/\sqrt{g})\partial_i(\sqrt{g}\phi^i)]$. Note that when $g = 1$ (so $\dot{\mathcal{R}} = 0$ automatically) and $\hat{\rho} = \rho$ we have then

$$(2.10) \qquad \phi_k\phi^k - 2\partial_k\phi^k \sim -\left(\frac{|\nabla\rho|^2}{\rho^2} - \frac{2\Delta\rho}{\rho}\right) = 4\frac{\Delta\sqrt{\rho}}{\sqrt{\rho}}$$

which means **(2L)** $Q = -(\hbar^2/2m)(\Delta\sqrt{\rho}/\sqrt{\rho})$ as in the desired **(1C)**.

**REMARK 2.1.2.** Thus starting with a manifold M with metric $g_{ij}$ and random initial conditions as indicated for a particle of mass $m$, the resulting classical statistical dynamics based on a probability distribution P with $\rho = mP$ can be properly phrased in a Weyl geometry in which the particle undergoes classical motion with probability one. The assumed Weyl geometry as well as the particle motion is determined via $\hat{\rho}(\rho, g)$ which says that given a different P there will be a different $\rho$ and $\hat{\rho}$ (since $g$ is fixed). Hence writing $\psi = \sqrt{\hat{\rho}}exp(iS/\hbar)$ one expects a different quantum potential and a different Weyl geometry. The SE will however remain unchanged and this may be a solution to the apparent problems illustrated in Section 1 about different quantum potentials being attached to the same SE. Another point of view could be that for $m$ fixed each $P$ (or equivalently $\rho$) determines a $P$-dependent motion via it Weyl geometry and each such motion can be described by a $P$-dependent wave function. The choice of $\hbar$ is arbitrary; here it arises via $\psi$ and any $\hbar$ will do. The identification with Planck's constant has to come from other considerations. ■

We go now to the second paper in [**770**] and sketch an interesting role of Weyl geometry in the Klein-Gordon (KG) equation (cf. also [**170, 183, 179, 191**] for discussion of this approach). The idea is to start from first principles, extended to gauge invariance relative to an arbitrary choice of spacetime calibration. Weyl geometry is not assumed but derived with the particle motion from a single average action principle. Thus assume a generic 4-D manifold with torsion free connection $\Gamma^\lambda_{\mu\nu} = \Gamma^\lambda_{\nu\mu}$ and a metric tensor $g$ with signature $(+,-,-,-)$; $\hbar = c = 1$ is taken for convenience (although this loses important information in the equations). The analysis will produce an integrable Weyl geometry with weights $w(g_{\mu\nu}) = 1$ and $w(\Gamma^\lambda_{\mu\nu}) = 0$ (cf. [**170, 465**] for Weyl geometry and Weyl-Dirac theory). One takes random initial conditions on a spacelike 3-D hypersurface and produces both particle motion and spacetime geometry via an average stationary action principle **(2M)** $\delta\left[E\int_{\tau_1}^{\tau_2} L(x(\tau), \dot{x}(\tau))d\tau\right] = 0$ where $\tau$ is an arbitrary parameter along the particle trajectory. Given $L$ positively homogeneous of first degree in $\dot{x}^\mu = dx^\mu/d\tau$ and transforming as a scalar of weight $w(L) = 0$ as well as a gauge invariant probability measure it follows that the action integral will be parameter invariant, coordinate invariant, and gauge invariant. A suitable Lagrangian is **(2N)** $L(x, dx) = (m^2 - (1/6)\mathcal{R})^{1/2}ds + A_\mu dx^\mu$

where $ds = (g_{\mu\nu}\dot{x}^\mu \dot{x}^\nu)^{1/2}d\tau$ and $w(m) = -1/2$ ($m$ = rest mass corresponds to a scalar Weyl field with no equation needed and the factor $(1/6)$ in $L$ is for convenience later). One writes **(2O)** $A_\mu = \bar{A}_\mu - \partial_\mu S$ where $\bar{A}_\mu \sim$ EM 4-potential in Lorentz gauge and $w(S) = w(\bar{A}_\mu) = 0$.

Omitting here the considerable details of calculation (which are given in [**770**] and sketched in [**170, 191**]) we work with a modified Lagrangian **(2P)** $\bar{L}(x, dx) = (m^2 - (1/6)\mathcal{R})^{1/2} + \bar{A}_\mu dx^\mu$. Variational methods lead to a 1-parameter family of hypersurfaces $S(x) = constant$ satisfying the HJ equation

$$g^{\mu\nu}(\partial_\mu S - \bar{A}_\mu)(\partial_\nu S - \bar{A}_\nu) = m^2 - (1/6)\mathcal{R} \tag{2.11}$$

and a congruence of curves intersecting this family given via

$$\frac{dx^\mu}{ds} = \frac{g^{\mu\nu}(\partial_\nu S - \bar{A}_\nu)}{[g^{\rho\sigma}(\partial_\rho - \bar{A}_\rho)(\partial_\sigma S - \bar{A}_\sigma)]^{1/2}} \tag{2.12}$$

The probability measure is determined by its probability current density $j^\mu$ where $\partial_\mu j^\mu = 0$ and **(2Q)** $j^\mu = \rho(\sqrt{-g}g^{\mu\nu}(\partial_\nu S - \bar{A}_\nu)$. Gauge invariance implies $w(j^\mu) = 0 = w(S)$ and $w(\rho) = -1$ so $\rho$ is the scalar probability density of the particle random motion. To find the connection the variational principle for **(2M)** is rephrased as

$$\delta\left[\int_\Omega d^4x[(m^2 - (1/6)\mathcal{R})(g_{\mu\nu}j^\mu j^\nu)]^{1/2} + A_\mu j^\mu\right] = 0 \tag{2.13}$$

Since the $\Gamma^\lambda_{\mu\nu}$ arise only in $\mathcal{R}$ this reduces to **(2R)** $\delta[\int_\Omega \rho\mathcal{R}\sqrt{-g}d^4x] = 0$ where (2.11) has been used. This leads to

$$\Gamma^\lambda_{\mu\nu} = \left\{\begin{array}{c}\lambda\\ \mu\nu\end{array}\right\} + \frac{1}{2}(\phi_\mu\delta^\lambda_\nu + \phi_\nu\delta^\lambda_\mu - g_{\mu\nu}g^{\lambda\rho}\phi_\rho);\ \phi_\mu = \partial_\mu log(\rho) \tag{2.14}$$

and shows that the connections are integrable Weyl connections with a gauge field $\phi_\mu$ ((**2A**) suggests here perhaps $\phi_i = -(1/2)\partial_i log(\rho)$). The HJ equation (2.11) and $\partial_\mu j^\mu = 0$ can be combined into a single equation for $S(x)$, namely

$$e^{iS}g^{\mu\nu}(iD_\mu - \bar{A}_\mu)(iD_\nu - \bar{A}_\nu)e^{-iS} - (m^2 - (1/6)\mathcal{R}) = 0 \tag{2.15}$$

with $D_\mu\rho = 0$ where (cf. [**53, 170**])

$$D_\mu T^\alpha_\beta = \partial_\mu T^\alpha_\beta + \Gamma^\alpha_{\mu\epsilon}T^\epsilon_\beta - \Gamma^\epsilon_{\mu\beta}T^\alpha_\epsilon + w(T)\phi_\mu T^\alpha_\beta \tag{2.16}$$

($D_\mu$ is called the double-covariant Weyl derivative and one notes that it is $\rho$ and not $m$, as in [**53**], which behaves as a constant under $D_\mu$). Then to any solution $(\rho, S)$ of these equations corresponds a particular random motion for the particle. One notes that (2.15)-(2.16) can be written in a familiar KG form

$$\left(\frac{i}{\sqrt{-g}}\partial_\mu\sqrt{-g} - \bar{A}_\mu\right)g^{\mu\nu}(i\partial_\nu - \bar{A}_\nu)\psi - (m^2 - (1/6)\dot{\mathcal{R}})\psi = 0 \tag{2.17}$$

where $\psi = \sqrt{\rho}exp(-iS)$ and $\dot{\mathcal{R}}$ is the Riemannian scalar curvature. We have also from [**770**]

$$\mathcal{R} = \dot{\mathcal{R}} - 3\left[\frac{1}{2}g^{\mu\nu}\phi_\mu\phi_\nu + \frac{1}{\sqrt{-g}}\partial_\mu\sqrt{-g}g^{\mu\nu}\phi_\nu\right] = \dot{\mathcal{R}} + \mathcal{R}_w \tag{2.18}$$

in keeping also with [**191**].

**REMARK 2.1.3.** Note here $g^{\mu\nu}\phi_\nu = \phi^\mu$ so (2.18) gives for the last term (**2S**) $\mathcal{R}_w = -3[(1/2)\phi_\mu\phi^\mu + (1/\sqrt{-g})\partial_\mu(\sqrt{-g}\phi^\mu)]$ whereas (2.3) suggests here (★) $-3[2\phi_\mu\phi^\mu - (2/\sqrt{-g})\partial_\mu(\sqrt{-g}\phi^\mu)]$ which is similar to paper 3 of [**770**] in having a minus sign in the middle; we remark that a change $\phi_\mu \to -2\phi_\mu$ would produce some agreement and will stay with (2.18) or equivalently (**2S**) due to calculations in Remark 2.1.5. ■

**REMARK 2.1.4.** We add here a few standard formulas involving derivatives; thus

$$\nabla_\mu\lambda^\nu = \partial_\mu\lambda^\nu + \Gamma^\mu_{\rho\nu}\lambda^\rho;\ \nabla_\mu\lambda^\mu = \partial_\mu\lambda^\mu + \Gamma^\mu_{\rho\mu}\lambda^\rho\ (divergence); \tag{2.19}$$

$$\nabla_\mu\lambda_\nu = \partial_\mu - \Gamma^\rho_{\nu\mu}\lambda_\rho;\ \Gamma^\mu_{\rho\mu} = \partial_\rho log(\sqrt{g});\ \nabla_m\lambda^m = \frac{1}{\sqrt{g}}\partial_m(\sqrt{g}\lambda^m)$$

Also from (2.17)

$$\Box \sim \frac{1}{\sqrt{-g}}\partial_\mu(\sqrt{-g}g^{\mu\nu}\partial_\nu) = \nabla_\mu g^{\mu\nu}\partial_\nu = \nabla_\mu\nabla^\mu \tag{2.20}$$

since $\nabla^\mu \sim \partial^\mu$ acting on functions (one could use $|g|$ instead of $\pm g$). ■

**REMARK 2.1.5.** We see that for $\bar{A}_\mu = 0$ the HJ equation (2.11) has the form (**2T**) $\partial_\mu S\partial^\mu S = m^2 - (1/6)\mathcal{R}$ and mention that it is shown in [**191**] that the 1/6 factor is essential if one wants a linear KG equation. We want now to identify $\mathcal{R}_w$ with a multiple of Q which should have a form like (**2U**) $Q \propto (1/\sqrt{\rho})\nabla^\mu\nabla_\mu(\sqrt{\rho})$. A crude calculation suggests

$$\partial_\mu\partial^\mu\sqrt{\rho} = \partial_\mu\left[\frac{1}{2}\rho^{-1/2}\partial^\mu\rho\right] = \frac{1}{2}\left[-\frac{1}{2}\rho^{-3/2}\partial_\mu\rho\partial^\mu\rho + \rho^{-1/2}\partial_\mu\partial^\mu\rho\right] \tag{2.21}$$

$$\Rightarrow \frac{\partial_\mu\partial^\mu\sqrt{\rho}}{\sqrt{\rho}} = \frac{1}{2}\left[-\frac{1}{2}\frac{\partial_\mu\rho\partial^\mu\rho}{\rho^2} + \frac{\partial_\mu\partial^\mu\rho}{\rho}\right]$$

and it is easy to check (cf. [**646**]) that $\nabla_m(fg^m) = (\nabla_m f)g^m + f(\nabla_m g^m)$. Hence $\nabla_m\nabla^m\sqrt{\rho}$ can be written out as in (2.21) to get

$$\frac{\Box(\sqrt{\rho})}{\sqrt{\rho}} = \frac{1}{2}\left[-\frac{1}{2}\frac{\nabla_\mu\rho\nabla^\mu\rho}{\rho^2} + \frac{\Box(\rho)}{\rho}\right] \tag{2.22}$$

and hence from (2.18)

$$\mathcal{R}_w = -3\left[\frac{1}{2}\frac{\nabla_\mu\rho\nabla^\mu\rho}{\rho^2} + \nabla_\mu\left(\frac{\nabla^\mu\rho}{\rho}\right)\right] = -3\left[\frac{1}{2}\frac{\nabla_\mu\rho\nabla^\mu\rho}{\rho^2}\right. \tag{2.23}$$

$$\left. + \frac{\nabla_\mu\nabla^\mu\rho}{\rho} - \frac{\nabla_\mu\rho\nabla^\mu\rho}{\rho^2}\right] = -6\frac{\Box(\sqrt{\rho})}{\sqrt{\rho}}$$

The formula for Q is then (**2V**) $Q = -[\Box(\sqrt{\rho})/\sqrt{\rho}] = (1/6)\mathcal{R}_w$. We remark that in various contexts formulas for Q arise here with multipliers $1/m^2$, $\hbar^2/2m$, etc. (cf. [**149**] and remarks below). ■

## 3. EMERGENCE OF $Q$ IN GEOMETRY

In [**170**] we have indicated a number of contexts where $Q$ arises in geometrical situations involving KG type equations and we review this here (cf. also [**183**]). We list a number of occasions (while omitting others).

(1) We omit here details for the Bertoldi-Faraggi-Matone (BFM) approach (see [**97, 170, 297**]) since it involves a whole philosophy (of considerable importance - see Section 1.3 and Chapter 8). Thus for $\eta^{\mu\nu} = diag(-1,1,1,1)$ and $q = (ct, q_1, q_2, q_3)$ one has

$$\frac{1}{2m}\eta^{\mu\nu}\partial_\mu S^{cl}\partial_\nu S^{cl} + \mathfrak{W}'_{rel} = 0; \tag{3.1}$$

with $\mathfrak{W}'_{rel} = \frac{1}{2mc^2}[m^2c^4 - V^2(q) - 2cV(q)\partial_0 S^{cl}]$ where V is some potential which we could take to be zero. The quantum version attaches $Q$ to (3.1) to get ($S^{cl} \to S$)

$$\frac{1}{2m}(\partial S)^2 + \mathfrak{W}_{rel} + Q = 0;\ \mathfrak{W}_{rel} = \frac{1}{2mc^2}[m^2c^4 - V^2 - 2cV\partial_0 S] \tag{3.2}$$

This involves then

$$\mathfrak{W}_{rel} = \left(\frac{\hbar^2}{2m}\right)\frac{\Box(Re^{iS/\hbar})}{Re^{iS/\hbar}};\ Q = -\frac{\hbar^2}{2m}\frac{\Box R}{R};\ \partial\cdot(R^2\partial S) = 0 \tag{3.3}$$

where one uses $\partial \sim \nabla$ when $g_{\mu\nu} = \eta_{\mu\nu}$.

(2) One can derive the SE, the KG equation, and the Dirac equation using methods of scale relativity (cf. [**170, 183, 201, 245, 643, 644, 671**]); here e.g. quantum paths are considered to be continuous nondifferentiable curves with left and right derivatives at any point. Using a "diffusion" coefficient $D = \hbar/2m$ as in the Nelson theory (cf. [**170, 183, 632**]) one defines "average" velocities $V = (1/2)[d_+x(t) + d_-x(t)]$ and $U = (1/2)[d_+x(t) - d_-x(t)]$. Then e.g. there is a SE $i\hbar\psi_t = -(\hbar^2/2m)\Delta\psi + \mathfrak{U}\psi$ with quantum potential $Q = -(m/2)U^2 - (\hbar/2)\partial U$ where $U = (\hbar/m)(\partial\sqrt{\rho}/\sqrt{\rho})$. The ideas should be extendible to a KG equation where $Q \sim (\hbar^2/m^2c^2)(\Box_g|\psi|/|\psi|)$ (see later sections).

(3) One can construct directly a KG theory following [**635**] in the form $(\partial_0^2 - \nabla^2 + m^2)\phi = 0$ where $\eta_{\mu\nu} = (1,-1,-1,-1)$. If $\psi = \phi^+$ with $\psi^* = \phi^-$ correspond to positive and negative frequency parts of $\phi = \phi^+ + \phi^-$ the particle current is $j_\mu = i\psi^*\overleftrightarrow{\partial}_\mu\psi$ and $N = \int d^3x j_0$ is the particle number. Trajectories have the form $d\mathbf{x}/dt = \mathbf{j}(t,\mathbf{x})/j_0(t,\mathbf{x})$ for $t = x_0$ and for $c = \hbar = 1$ one arrives at

$$\partial^\mu(R^2\partial_\mu S) = 0;\ \frac{(\partial^\mu S)(\partial_\mu S)}{2m} - \frac{m}{2} + Q = 0;\ Q = -\frac{1}{2m}\frac{\partial^\mu\partial_\mu R}{R} \tag{3.4}$$

(4) A covariant field theoretic version is also given in [**635**] using deDonder-Weyl theory (cf. also [**170, 183**]). One works with a real scalar field $\phi(x)$ and defines **(3A)** $\mathfrak{A} = \int d^4x\mathfrak{L};\ \mathfrak{L} = (1/2)(\partial^\mu\phi)(\partial_\mu\phi) - V(\phi)$ with **(3B)** $\pi^\mu = \partial\mathfrak{L}/\partial(\partial_\mu\phi) = \partial^\mu\phi,\ \partial_\mu\phi = \partial\mathfrak{H}/\partial\pi^\mu$, and $\partial_\mu\pi^\mu = -\partial\mathfrak{H}/\partial\phi$. One takes a preferred foliation of spacetime with $R^\mu$ normal to the leaf $\Sigma$ and writes $\mathfrak{R}([\phi],\Sigma) = \int_\Sigma d\Sigma_\mu R^\mu$ with $\mathfrak{S}([\phi],\Sigma) = \int_\Sigma d\Sigma_\mu S^\mu$ and

$\Psi = \mathfrak{R}exp(i\mathfrak{S}/\hbar)$. A covariant version of Bohmian mechanics ensues with

$$\frac{1}{2}\frac{dS_\mu}{d\phi}\frac{dS^\mu}{d\phi} + V + Q + \partial_\mu S^\mu = 0;\ \frac{dR^\mu}{d\phi}\frac{dS^\mu}{d\phi} + J + \partial_\mu R^\mu = 0 \tag{3.5}$$

$$Q = -\frac{\hbar^2}{2\mathfrak{R}}\frac{\delta^2\mathfrak{R}}{\delta_\Sigma \phi^2};\ J = \frac{\mathfrak{R}}{2}\frac{\delta^2\mathfrak{S}}{\delta_\Sigma \phi^2} \tag{3.6}$$

The nature of this approach as a covariant version of the Bohmian hidden variable theory is spelled out in the last paper of [**635**]. This is a significant extension of earlier classical field theoretic approaches and another lovely extension is described by Nikolic in [**636**] involving a covariant many fingered time Bohmian interpretation of quantum field theory (QFT).

We preface the next examples with a discussion of a formula $\mathfrak{M}^2 = m^2 exp(\mathfrak{Q}_{rel})$ used in [**797**] in an important manner and produced also in [**642**]. This formula differs from the result $\mathfrak{M} = mexp(\mathfrak{Q}_{rel})$ of [**800**] (which was abandoned in [**797**]) and in order to clarify this we write out in more detail the approach of [**642**]. Thus one is dealing with a Bohmian theory and for a Klein-Gordon (KG) equation a wave function $\psi = Rexp(iS/\hbar)$ this leads to (cf. also p. 8-18)

$$\partial_\mu(R^2\partial^\mu S) = 0;\ \partial_\mu S\partial^\mu S = \mathfrak{M}^2c^2\ (\sim m^2c^2(1+\mathfrak{Q}_{rel})) \tag{3.7}$$

where $\mathfrak{Q}_{rel} = (\hbar^2/m^2c^2)(\partial_\mu\partial^\mu R/R)$ and (temporarily now) $\partial_\mu\partial^\mu \sim \Box = (1/c^2)\partial_t^2 - \Delta$ where $\eta_{\alpha\beta} = diag(1,-1,-1,-1)$. Now $\mathfrak{M} = 1 + \mathfrak{Q}_{rel}$ is only an approximation (leading e.g. to tachyon problems) and a better formula for $\mathfrak{M}$ can be found as follows. Thus one knows (**3C**) $(dx^\mu(\tau)/d\tau) = (1/\mathfrak{M})\partial^\mu S$ and differentiating gives

$$\partial_\tau\partial^\mu S = \partial_\nu\partial^\mu S\frac{dx^\mu}{d\tau} = \partial_\nu\mathfrak{M}\frac{dx^\nu}{d\tau}\frac{dx^\mu}{d\tau} + \mathfrak{M}\frac{d^2x^\mu}{d\tau^2} \tag{3.8}$$

But via the formula (valid for $g_{ab} = \eta_{ab}$ constant)

$$\partial_b(\partial_a S\partial^a S) = (\partial_b\partial_a S)(\partial^a S) + (\partial_a S)(\partial_b\partial^a S); \tag{3.9}$$
$$\partial_a\partial_b\partial^a S = \partial_a S\eta^{ac}\partial_c\partial_b S = \partial^c S\partial_c\partial_b S$$

one has (**3D**) $\partial_\nu(\partial_\mu S\partial^\mu S) = 2(\partial^\mu S)(\partial_\mu\partial_\nu S)$ and therefore

$$\partial_\nu(\partial_\mu S\partial^\mu S) = \partial_\nu(\mathfrak{M}^2c^2) = 2\mathfrak{M}\partial_\nu\mathfrak{M}c^2 = 2(\partial^\nu S)(\partial_\mu\partial_\nu S) \tag{3.10}$$
$$= 2\mathfrak{M}\frac{dx^\mu}{d\tau}(\partial_\mu\partial_\nu S)$$

Hence (**3E**) $\partial_\nu\mathfrak{M}c^2 = (\partial_\mu\partial_\nu S)(dx^\mu/d\tau)$ which implies

$$\eta^{\alpha\nu}c^2\partial_\nu\mathfrak{M} = \eta^{\alpha\nu}\partial_\mu\partial_\nu S(dx^\mu/d\tau) = \partial_\mu\partial^\alpha S(dx^\mu/d\tau) \tag{3.11}$$

Consequently (3.8) becomes

$$\eta^{\alpha\nu}c^2\partial_\nu\mathfrak{M} = \partial_\nu\mathfrak{M}\frac{dx^\nu}{d\tau}\frac{dx^\alpha}{d\tau} + \mathfrak{M}\frac{d^2x^\alpha}{d\tau^2} \tag{3.12}$$
$$\equiv \mathfrak{M}\frac{d^2x^\alpha}{d\tau^2} = \left(c^2\eta^{\alpha\nu} - \frac{dx^\nu}{d\tau}\frac{dx^\alpha}{d\tau}\right)\partial_\nu\mathfrak{M}$$

and this is equation (9) of [**642**]. For $|\dot{x}^\alpha| << c$ one obtains then $\mathfrak{M}\ddot{x}^\alpha \sim c^2\partial^\alpha\mathfrak{M} \sim -c^2\partial_\alpha\mathfrak{M}$ and comparing with the nonrelativistic equation $m\ddot{x}^\alpha = -\partial_\alpha Q_{cl}$ implies $\mathfrak{M} \sim mexp(\mathfrak{Q}_{cl}/mc^2)$ and suggests that $\mathfrak{M} \sim mexp(\mathfrak{Q}_{rel}/2)$ (recall $\mathfrak{Q}_{cl} = -(\hbar^2/2m)(\nabla^2|\psi|/|\psi|)$).

Now one observes that the quantum effects will affect the geometry and in fact are equivalent to a change of spacetime metric

$$g_{\mu\nu} \to \tilde{g}_{\mu\nu} = (\mathfrak{M}^2/m^2)g_{\mu\nu} \tag{3.13}$$

(conformal transformation). The QHJE becomes $\tilde{g}^{\mu\nu}\tilde{\nabla}_\mu S\tilde{\nabla}_\nu S = m^2c^2$ where $\tilde{\nabla}_\mu$ represents covariant differentiation with respect to the metric $\tilde{g}_{\mu\nu}$ and the continuity equation is then $\tilde{g}_{\mu\nu}\tilde{\nabla}_\mu(\rho\tilde{\nabla}_\nu S) = 0$. The important conclusion here is that the presence of the quantum potential is equivalent to a curved spacetime with its metric given by (3.13). This is a geometrization of the quantum aspects of matter and it seems that there is a dual aspect to the role of geometry in physics. The spacetime geometry sometimes looks like "gravity" and sometimes reveals quantum behavior. The curvature due to the quantum potential may have a large influence on the classical contribution to the curvature of spacetime. The particle trajectory can now be derived from the guidance relation via differentiation as in (**3C**) again, leading to the Newton equations of motion

$$\mathfrak{M}\frac{d^2x^\mu}{d\tau^2} + \mathfrak{M}\Gamma^\mu_{\nu\kappa}u^\nu u^\kappa = (c^2g^{\mu\nu} - u^\mu u^\nu)\nabla_\nu\mathfrak{M} \tag{3.14}$$

Using the conformal transformation above (3.14) reduces to the standard geodesic equation.

We extract now from [**170, 797, 798, 799, 800, 801, 802, 803, 804, 805, 806**] with emphasis on the survey article [**797**]. This may seem overly repetitious but the material seems worthy of further emphasis. Thus a general "canonical" relativistic system consisting of gravity and classical matter (no quantum effects) is determined by the action

$$\mathcal{A} = \frac{1}{2\kappa}\int d^4x\sqrt{-g}\mathcal{R} + \int d^4x\sqrt{-g}\frac{\hbar^2}{2m}\left(\frac{\rho}{\hbar^2}\mathcal{D}_\mu S\mathcal{D}^\mu S - \frac{m^2}{\hbar^2}\rho\right) \tag{3.15}$$

where $\kappa = 8\pi G$ and $c = 1$ for convenience and $\mathcal{D}_\mu$ is the covariant derivative based on $g_{\mu\nu}$ ($\mathcal{D}_\mu \sim \nabla_\mu$). It was seen above that via deBroglie the introduction of a quantum potential is equivalent to introducing a conformal factor $\Omega^2 = \mathfrak{M}^2/m^2$ in the metric. Hence in order to introduce quantum effects of matter into the action (3.15) one uses this conformal transformation to get ($1 + Q \sim exp(Q)$ and $Q \sim (\hbar^2/c^2m^2)(\Box(\sqrt{\rho})/\sqrt{\rho})$ from [**797**] with $c = 1$ here).

$$\begin{aligned}\mathfrak{A} = \frac{1}{2\kappa}\int d^4x\sqrt{-\bar{g}}(\bar{\mathcal{R}}\Omega^2 - 6\bar{\nabla}_\mu\Omega\bar{\nabla}^\mu\Omega) \\ + \int d^4x\sqrt{-\bar{g}}\left(\frac{\rho}{m}\Omega^2\bar{\nabla}_\mu S\bar{\nabla}^\mu S - m\rho\Omega^4\right) \\ + \int d^4x\sqrt{-\bar{g}}\lambda\left[\Omega^2 - \left(1 + \frac{\hbar^2}{m^2}\frac{\bar{\Box}\sqrt{\rho}}{\sqrt{\rho}}\right)\right]\end{aligned} \tag{3.16}$$

where a bar over any quantity means that it corresponds to the nonquantum regime. Here only the first two terms of the expansion of $\mathfrak{M}^2 = m^2 exp(\mathfrak{Q})$ have been used, namely $\mathfrak{M}^2 \sim m^2(1+\mathfrak{Q})$. $\lambda$ is a Lagrange multiplier introduced to identify the conformal factor with its Bohmian value. We note that the definitive QP $\mathfrak{Q}$ is given in Remark 8.4.2 (via the QEP) as $\mathfrak{Q} = -(\hbar^2/2m)(\Box\sqrt{\rho}/\sqrt{\rho})$ and this gives at the same time a definitive form for $Q$ (see Remark 4.1.1 below). Since the form $Q$ is used in frequently in the literature we will refer to either $Q$ or $\mathfrak{Q}$ as a quantum potential. One uses here $\bar{g}_{\mu\nu}$ to raise or lower indices and to evaluate the covariant derivatives; the physical metric (containing the quantum effects of matter) is $g_{\mu\nu} = \Omega^2 \bar{g}_{\mu\nu}$. By variation of the action with respect to $\bar{g}_{\mu\nu}$, $\Omega$, $\rho$, $S$, and $\lambda$ one arrives at quantum equations of motion, including quantum Einstein equations (cf. [**170, 797**]). There is a generalized equivalence principle. The gravitational effects determine the causal structure of spacetime as long as quantum effects give its conformal structure. This does not mean that quantum effects have nothing to do with the causal structure; they can act on the causal structure through back reaction terms appearing in the metric field equations. The conformal factor of the metric is a function of the quantum potential and the mass of a relativistic particle is a field produced by quantum corrections to the classical mass. One has shown that the presence of the quantum potential is equivalent to a conformal mapping of the metric. Thus in different conformally related frames one "feels" different quantum masses and different curvatures. In particular there are two frames with one containing the quantum mass field and the classical metric while the other contains the classical mass and the quantum metric. In general frames both the spacetime metric and the mass field have quantum properties so one can state that different conformal frames are identical pictures of the gravitational and quantum phenomena. One "feels" different quantum forces in different conformal frames. The question then arises of whether the geometrization of quantum effects implies conformal invariance just as gravitational effects imply general coordinate invariance. One sees here that Weyl geometry provides additional degrees of freedom which can be identified with quantum effects and seems to create a unified geometric framework for understanding both gravitational and quantum forces. Some features here are: (i) Quantum effects appear independent of any preferred length scale. (ii) The quantum mass of a particle is a field. (iii) The gravitational constant is also a field depending on the matter distribution via the quantum potential. (iv) A local variation of matter field distribution changes the quantum potential acting on the geometry and alters it globally; the nonlocal character is forced by the quantum potential (cf. [**170, 797, 804**]).

Next (still following [**797**]) one goes to Weyl geometry based on the Weyl-Dirac action

$$\mathfrak{A} = \int d^4x\sqrt{-g}(F_{\mu\nu}F^{\mu\nu} - \beta^2\ {}^W\mathcal{R} + (\sigma+6)\beta_{;\mu}\beta^{;\mu} + \mathfrak{L}_{matter}) \tag{3.17}$$

Here $F_{\mu\nu}$ is the curl of the Weyl 4-vector $\phi_\mu$, $\sigma$ is an arbitrary constant and $\beta$ is a scalar field of weight $-1$. The symbol ";" represents a covariant derivative under general coordinate and conformal transformations (Weyl covariant derivative) defined as $X_{;\mu} = {}^W\nabla_\mu X - \mathcal{N}\phi_\mu X$ where $\mathcal{N}$ is the Weyl weight of $X$. The equations of motion are then given in [**170, 797**]. There is then agreement with the Bohmian theory provided one identifies

$$\beta \sim \mathfrak{M};\ \frac{8\pi\mathfrak{T}}{\mathcal{R}} \sim m^2;\ \frac{1}{\sigma\phi_\alpha\phi^\alpha - (\mathcal{R}/6)} \sim \alpha = \frac{\hbar^2}{m^2c^2} \tag{3.18}$$

Thus $\beta$ is the Bohmian quantum mass field and the coupling constant $\alpha$ (which depends on $\hbar$) is also a field, related to geometrical properties of spacetime. One notes that the quantum effects and the length scale of the spacetime are related. To see this suppose one is in a gauge in which the Dirac field is constant; apply a gauge transformation to change this to a general spacetime dependent function, i.e.

$$\beta = \beta_0 \to \beta(x) = \beta_0 exp(-\Xi(x));\ \phi_\mu \to \phi_\mu + \partial_\mu\Xi \tag{3.19}$$

Thus the gauge in which the quantum mass is constant (and the quantum force is zero) and the gauge in which the quantum mass is spacetime dependent are related to one another via a scale change. In particular $\phi_\mu$ in the two gauges differ by $-\nabla_\mu(\beta/\beta_0)$ and since $\phi_\mu$ is a part of Weyl geometry and the Dirac field represents the quantum mass one concludes that the quantum effects are geometrized which shows that $\phi_\mu$ is not independent of $\beta$ so the Weyl vector is determined by the quantum mass and thus the geometrical aspects of the manifold are related to quantum effects.)

**3.1. QUANTUM POTENTIAL AS A DYNAMICAL FIELD.** In [**797, 804**] (cf. also [**170**]) one can write down a scalar tensor theory where the conformal factor and the quantum potential are both dynamical fields but first we deal with (3.16). For the relativistic situation one will have e.g. $\mathfrak{Q} = (\hbar^2/m^2c^2)(\Box_g|\psi|/|\psi|)$ where $\Box_g|\psi| \sim \nabla_\alpha\nabla^\alpha|\psi| = g^{\alpha\beta}\nabla_\beta\nabla_\alpha|\psi|$ and the HJ equation is $\nabla_\mu S\nabla^\mu S = \mathfrak{M}^2c^2$ where $\mathfrak{M}^2 = m^2 exp(\mathfrak{Q})$. Equivalently $\tilde{g}^{\mu\nu}\tilde{\nabla}_\mu S\tilde{\nabla}_\nu S = m^2c^2$ where $g_{\mu\nu} = (\mathfrak{M}/m)^2\tilde{g}_{\mu\nu}$ and $\tilde{\nabla}_\mu$ is the covariant derivative with respect to $\tilde{g}_{\mu\nu}$. The corresponding geodesic equation is given via (3.14). We write $\Omega^2 = (\mathfrak{M}/m)^2$ and this leads to (3.16) based on the fundamental action (3.15). Recall here $exp(\mathfrak{Q}) \sim m^2(1+\mathfrak{Q})$ has been used for $\mathfrak{M}$ in the last term in (3.16). We recall also the fundamental equations determined by varying the action (3.16) with respect to $\bar{g}_{\mu\nu}$, $\Omega$, $\rho$, $S$, and $\lambda$ are (cf. [**170, 797**] and note the calculations do not involve Q explicitly)

(1) The equation of motion for $\Omega$

$$\bar{\mathcal{R}}\Omega + 6\,\bar{\Box}\,\Omega + \frac{2\kappa}{m}\rho\Omega(\bar{\nabla}_\mu S\bar{\nabla}^\mu S - 2m^2\Omega^2) + 2\kappa\lambda\Omega = 0 \tag{3.20}$$

(2) The continuity equation for particles $\bar{\nabla}_\mu(\rho\Omega^2\bar{\nabla}^\mu S) = 0$
(3) The equations of motion for particles

$$(\bar{\nabla}_\mu S\bar{\nabla}^\mu S - m^2\Omega^2)\Omega^2\sqrt{\rho} + \frac{\hbar^2}{2m}\left[\bar{\Box}\left(\frac{\lambda}{\sqrt{\rho}}\right) - \lambda\frac{\bar{\Box}\sqrt{\rho}}{\rho}\right] = 0 \tag{3.21}$$

(4) The modified Einstein equations for $\bar{g}_{\mu\nu}$

$$\Omega^2\left[\bar{\mathcal{R}}_{\mu\nu}-\frac{1}{2}\bar{g}_{\mu\nu}\bar{\mathcal{R}}\right]-[\bar{g}_{\mu\nu}\bar{\Box}-\bar{\nabla}_\mu\bar{\nabla}_\nu]\Omega^2-6\bar{\nabla}_\mu\Omega\bar{\nabla}_\nu\Omega+3\bar{g}_{\mu\nu}\bar{\nabla}_\alpha\Omega\bar{\nabla}^\alpha\Omega \tag{3.22}$$

$$+\frac{2\kappa}{m}\rho\Omega^2\bar{\nabla}_\mu S\bar{\nabla}_\nu S-\frac{\kappa}{m}\rho\Omega^2\bar{g}_{\mu\nu}\bar{\nabla}_\alpha S\bar{\nabla}^\alpha S+\kappa m\rho\Omega^4\bar{g}_{\mu\nu}$$

$$+\frac{\kappa\hbar^2}{m^2}\left[\bar{\nabla}_\mu\sqrt{\rho}\bar{\nabla}_\nu\left(\frac{\lambda}{\sqrt{\rho}}\right)+\bar{\nabla}_\nu\sqrt{\rho}\bar{\nabla}_\mu\left(\frac{\lambda}{\sqrt{\rho}}\right)\right]-\frac{\kappa\hbar^2}{m^2}\bar{g}_{\mu\nu}\bar{\nabla}_\alpha\left[\lambda\frac{\bar{\nabla}^\alpha\sqrt{\rho}}{\sqrt{\rho}}\right]=0$$

(5) The constraint equation $\Omega^2=1+(\hbar^2/m^2)[(\bar{\Box}\sqrt{\rho})/\sqrt{\rho}]$

Thus the back reaction effects of the quantum factor on the background metric are contained in these highly coupled equations. A simpler form of (3.20) can be obtained by taking the trace of (3.21) and using (3.20) which produces $\lambda=(\hbar^2/m^2)\bar{\nabla}_\mu[\lambda(\bar{\nabla}^\mu\sqrt{\rho})/\sqrt{\rho}]$. A solution of this via perturbation methods using the small parameter $\alpha=\hbar^2/m^2$ yields the trivial solution $\lambda=0$ so the above equations reduce to

$$\bar{\nabla}_\mu(\rho\Omega^2\bar{\nabla}^\mu S)=0;\ \bar{\nabla}_\mu S\bar{\nabla}^\mu S=m^2\Omega^2;\ \mathfrak{G}_{\mu\nu}=-\kappa\mathfrak{T}^{(m)}_{\mu\nu}-\kappa\mathfrak{T}^{(\Omega)}_{\mu\nu} \tag{3.23}$$

where $\mathfrak{T}^{(m)}_{\mu\nu}$ is the matter energy-momentum (EM) tensor and

$$\kappa\mathfrak{T}^{(\Omega)}_{\mu\nu}=\frac{[g_{\mu\nu}\Box-\nabla_\mu\nabla_\nu]\Omega^2}{\Omega^2}+6\frac{\nabla_\mu\Omega\nabla_\nu\Omega}{\omega^2}-2g_{\mu\nu}\frac{\nabla_\alpha\Omega\nabla^\alpha\Omega}{\Omega^2} \tag{3.24}$$

with $\Omega^2=1+\alpha(\bar{\Box}\sqrt{\rho}/\sqrt{\rho})$. Note that the second relation in (3.23) is the Bohmian equation of motion and written in terms of $g_{\mu\nu}$ it becomes $\nabla_\mu S\nabla^\mu S=m^2c^2$. Many examples with a lot of expansion is to be found in [**170, 797**] and references there.

## 4. OTHER GEOMETRIC ASPECTS

The quantum potential arises in many geometrical and cosmological situations and we mention a few of these here.

(1) We have written about the Wheeler-deWitt (WDW) equation and the QP in [**182**] at some length and in [**170**] have discussed the QP in related geometric situations following [**663, 664, 771, 797, 806**] in particular. For background information on WDW we refer to [**509**] for example and to [**353, 354, 355, 357**] for a newer version outlined in Chapter 5. One thinks of an ADM situation with (**4A**) $ds^2=-(N^2-h^{ij}H_iN_j)dt^2+2N_idx^idt+h_{ij}dx^idx^j$ and the deWitt metric (**4B**) $G_{ijk\ell}=(1/\sqrt{h})h_{ik}h_{j\ell}+h_{i\ell}h_{jk}-h_{ij}h_{k\ell})$. Given a wave function $\psi=\sqrt{P}exp(iS/\hbar)$ where $P$ corresponds to momentum fluctuations $(1/P)(\delta P/\delta h_{ij})$ one finds a quantum potential

$$Q=-\frac{\hbar^2}{2}P^{-1/2}\frac{\delta}{\delta h_{ij}}\left(G_{ijk\ell}\frac{\delta P^{1/2}}{\delta h_{k\ell}}\right) \tag{4.1}$$

This is related to an intimate connection between $Q$ and Fisher information based on techniques of Hall and Reginatto (cf. [**400, 401, 402,**

**403, 736**]. The WDW equation is

$$\left[-\frac{\hbar^2}{2}\frac{\delta}{\delta h_{ij}}G_{ijk\ell}\frac{\delta}{\delta h_{k\ell}}+V\right]\psi=0; \tag{4.2}$$

and there is a lovely relation

$$\int \mathfrak{D}hPQ=-\int \mathfrak{D}h\frac{\delta P^{1/2}}{\delta h_{ij}}G_{ijk\ell}\frac{\delta P^{1/2}}{\delta h_{k\ell}} \tag{4.3}$$

where the last term is Fisher information (cf. [**182, 319, 401, 402, 736**]).

(2) In [**806**] one uses again the attractive sandwich ordering of (4.2) (which is mandatory in (4.2)) and considers WDW in the form

$$\left[h^{-q}\frac{\delta}{\delta h_{ij}}h^q G_{ijk\ell}\frac{\delta}{\delta h_{k\ell}}+\sqrt{h}^{(3)}\mathcal{R}\right. \tag{4.4}$$
$$\left.+\frac{1}{2\sqrt{h}}\frac{\delta^2}{\delta\phi^2}-\frac{1}{2}\sqrt{h}h^{ij}\partial_i\phi\partial_j\phi-\frac{1}{2}\sqrt{h}V(\phi)\right]\psi=0$$

with momentum constraint (**4C**) $i[2\nabla_j(\delta/\delta h_{ij})-h^{ij}\partial_j\phi(\delta/\delta\phi)]\psi=0$ where $\phi$ is a matter field, $q$ is an ordering parameter, and $h=det(h_{ij})$. Putting this in "polar" form $\psi=\sqrt{\rho}exp(iS/\hbar)$ leads to

$$G_{ijk\ell}\frac{\delta S}{\delta h_{ij}}\frac{\delta S}{\delta h_{k\ell}}+\frac{1}{2\sqrt{h}}\left(\frac{\delta S}{\delta\phi}\right)^2-\sqrt{h}(^{(3)}\mathcal{R}-\mathcal{Q}_G) \tag{4.5}$$
$$+\frac{\sqrt{h}}{2}h^{ij}\partial_i\phi\partial_j\phi+\frac{\sqrt{h}}{2}(V(\phi)-\mathcal{Q}_M)=0$$

where the gravity and matter quantum potentials are given via

$$\mathcal{Q}_G=-\frac{1}{\sqrt{\rho h}}\left(G_{ijk\ell}\frac{\delta^2\sqrt{\rho}}{\delta h_{ij}\delta h_{k\ell}}+h^{-q}\frac{\delta h^q G_{ijk\ell}}{\delta h_{ij}}\frac{\delta\sqrt{\rho}}{\delta h_{k\ell}}\right);\ \mathcal{Q}_M=-\frac{1}{h\sqrt{\rho}}\frac{\delta^2\sqrt{\rho}}{\delta\phi^2} \tag{4.6}$$

There is a continuity equation

$$\frac{\delta}{\delta h_{ij}}\left[2h^q G_{ijk\ell}\frac{\delta S}{\delta h_{k\ell}}\rho\right]+\frac{\delta}{\delta\phi}\left[\frac{h^q}{\sqrt{h}}\frac{\delta S}{\delta\phi}\rho\right]=0 \tag{4.7}$$

and the momentum constraint leads to equations (**4D**) $2\nabla_j(\delta\sqrt{\rho}/\delta h_{ij})-h^{ij}\partial_j\phi(\delta\sqrt{\rho}/\delta\phi)=0$ and $2\nabla_j(\delta S/\delta h_{ij})-h^{ij}\partial_j\phi(\delta S/\delta\rho)=0$ while the Bohmian "guidance" equations are

$$\frac{\delta S}{\delta h_{ij}}=\pi^{k\ell}=\sqrt{h}(K^{k\ell}-h^{k\ell}K);\ \frac{\delta S}{\delta\phi}=\pi_\phi=\frac{\sqrt{h}}{N^\perp}\dot{\phi}-\sqrt{h}\frac{N^i}{N^\perp}\partial_i\phi \tag{4.8}$$

where $K^{ij}$ is the extrinsic curvature. Since in the WDW equation the wavefunction is in the ground state with zero energy the stability condition of the metric and matter field is (**4E**) $h^{ij}\partial_i\phi\partial_j\phi+V(\phi)-2^3\mathcal{R}+\mathcal{Q}_M+2\mathcal{Q}_G=0$ which is a pure quantum solution (this follows from (4.5) by setting all functional derivatives of $S$ to be zero). In [**806**] these equations are examined perturbatively and we refer to [**797, 803**] for discussion of the constraint algebra and related matters.

(3) In [**803**] one studies the constraint algebra and equations of motion based on a Lagrangian (**4F**) $\mathfrak{L} = \sqrt{-g}\mathcal{R} = \sqrt{h}N({}^{(3)}\mathcal{R} + Tr(K^2) - (TrK)^2)$ where ${}^{(3)}$is the 3-D Ricci scalar, $K_{ij}$ the extrinsic curvature, and $h$ the induced spatial metric. The canonical momentum of the 3-metric is given via (**4G**) $P^{IJ} = \partial\mathfrak{L}/\partial\dot{h}_{ij}) = \sqrt{h}(K^{ij} - h^{ij}TrK)$ and the classical Hamiltonian is (**4H**) $H = \int d^3x\mathfrak{H}$ with $mfH = \sqrt{h}(NC + N^iC_i)$. Here one has

$$C = -{}^{(3)}\mathcal{R} + \frac{1}{h}\left(Tr(p^2) - \frac{1}{2}(Trp)^2)\right) = -2G_{\mu\nu}n^{mu}n^{\nu}; \tag{4.9}$$

$$C_i = -2^{(3)}\nabla^j\left(\frac{p_{ij}}{\sqrt{h}}\right) = -2G_{\mu i}n^{\mu}$$

where $n^{\mu}$ is normal given via $n^{\mu} = (1/N, -\mathbf{N}/N)$. To get the quantum version one takes $H \to H + Q$ ($\mathfrak{H} \to \mathfrak{H} + \mathfrak{Q}$ (where $Q = \int d^3x\mathfrak{Q}$) and

$$\mathfrak{Q} = \hbar^2 NhG_{ijk\ell}\frac{1}{|\psi|}\frac{\delta^2|\psi|}{\delta h_{ij}\delta h_{k\ell}} \tag{4.10}$$

The classical constraints are then modified via $C \to C + (\mathfrak{Q}/\sqrt{h}N)$ and $C_i \to C_i$. We disregard the constraint algebra here and go some formulas for quantum Einstein equations. First there is an HJ equation

$$G_{ijk\ell}\frac{\delta S}{\delta h_{ij}}\frac{\delta S}{\delta h_{k\ell}} - \sqrt{h}({}^{(3)}\mathcal{R} - \mathfrak{Q}) = 0 \tag{4.11}$$

where $S$ is the phase of the wave function and this leads to Bohm-Einstein equations

$$\mathcal{G}^{ij} = -\kappa T^{ij} - \frac{1}{N}\frac{\delta(\mathfrak{Q}_G + \mathfrak{Q}_m}{\delta g_{ij}};\ \mathcal{G}^{0\mu} = -\kappa T^{0\mu} + \frac{\mathfrak{Q}_G + \mathfrak{Q}_m}{2\sqrt{-g}}g^{0\mu}; \tag{4.12}$$

$$\mathfrak{Q}_m = \hbar^2\frac{N\sqrt{H}}{2}\frac{\delta^2|psi|}{\delta\phi^2};\ \mathfrak{Q}_G = \hbar^2 NhG_{ijk\ell}\frac{1}{|\psi|}\frac{\delta^2|\psi|}{\delta h_{ij}\delta h_{k\ell}}$$

These are the quantum version of the Einstein equations and since regularization here only affects the quantum potential (cf. [**803**]) for any regularization the quantum Einstein equations are the same and one can write (**4I**) $\mathcal{G}^{\mu\nu} = -\kappa T^{\mu\nu} + \mathfrak{S}^{mu\nu}$ with

$$\mathfrak{S}^{0\mu} = -\frac{\mathfrak{Q}_G + \mathfrak{Q}_m}{2\sqrt{-g}}g^{0\mu} = \frac{\mathfrak{Q}}{2\sqrt{-g}}g^{0\mu};\ \mathfrak{S}^{ij} = -\frac{1}{N}\frac{\delta Q}{\delta g_{ij}} \tag{4.13}$$

(4) There are also developments of Bohmian theory and quantum geometrodynamics in [**73, 83, 108, 238, 526, 663, 664, 771, 807**] (cf. [**170**] for some survey and more references). In [**663**] for example one writes the WDW equation in the form

$$\left\{-\hbar^2\left[\kappa G_{ijk\ell}\frac{\delta}{\delta h_{ij}}\frac{\delta}{\delta h_{k\ell}} + \frac{1}{2}h^{-1/2}\frac{\delta^2}{\delta\phi^2}\right] + V\right\}\psi(h_{ij}, \phi) = 0; \tag{4.14}$$

$$V = h^{1/2}\left[-\kappa^{-1}(\mathcal{R}^{(3)} - 2\Lambda) + \frac{1}{2}h^{ij}\partial_i\phi\partial_j\phi + U(\phi)\right]$$

(questions of factor ordering and regularization are ignored here) with a constraint **(4J)** $-2h_{ij}\nabla_j(\delta\psi/\delta h_{ij}) + (\delta\psi)/\delta\phi)\partial_i\phi = 0$. Writing now $\psi = Rexp(iS/\hbar)$ **(4J)** leads to

(4.15)
$$-2h_{ij}\nabla_j(\delta S/\delta h_{ij}) + (\delta S/\delta\phi)\partial_i\phi = 0;\ -2h_{ij}\nabla_j(\delta R/\delta h_{ij}) + (\delta R/\delta\phi)\partial_i\phi = 0$$

and (4.14) yields

$$\kappa G_{ijk\ell}\frac{\delta S}{\delta h_{ij}}\frac{\delta S}{\delta h_{k\ell}} + \frac{1}{2}h^{-1/2}\left(\frac{\delta S}{\delta\phi}\right)^2 + V + Q = 0; \tag{4.16}$$

$$Q = -\frac{\hbar^2}{R}\left(\kappa G_{ijk\ell}\frac{\delta^2 R}{\delta h_{ij}h_{j\ell}} + \frac{h^{-1/2}}{2}\frac{\delta^2 R}{\delta\phi^2}\right);$$

$$\kappa G_{ijk\ell}\frac{\delta}{\delta h_{ij}}\left(R^2\frac{\delta S}{\delta h_{k\ell}}\right) + \frac{1}{2}\frac{\delta}{\delta\phi}\left(R^2\frac{\delta S}{\delta\phi}\right) = 0$$

(5) In [**799**] one picks up again the approach of (2) to find a pure quantum state leading to a static Einstein universe whose classical counterpart is flat spacetime. For WDW one uses a form of (4.4) (with $16\pi G = 1$ and $\mathcal{R}$ the 3-curvature scalar), namely

$$\hbar^2 h^{-q}\frac{\delta}{\delta h_{ij}}\left(h^q G_{ijk\ell}\frac{\delta\psi}{\delta h_{k\ell}}\right) + \sqrt{h}\mathcal{R}\psi + \frac{1}{\sqrt{h}}\mathcal{T}^{00}\left(\frac{-i\hbar\delta}{\delta\phi_a},\phi_a\right)\psi = 0; \tag{4.17}$$

with 3-diffeomorphism constraint **(4K)** $2\nabla_j(\delta/\delta h_{ij})\psi - \mathcal{T}^{i0}(\delta/\delta\phi_a,\phi_a)\pi = 0$. $\mathcal{T}^{\mu\nu}$ is the energy momentum tensor of matter fields $\phi_a$ in which the matter is quantized by replacing its conjugate momenta by $-i\hbar\delta/\delta\phi_a$. For the causal interpretation one sets again $\psi = Rexp(iS/\hbar)$ to obtain

$$G_{ijk\ell}\frac{\delta S}{\delta h_{ij}}\frac{\delta S}{\delta h_{k\ell}} - \sqrt{h}(\mathcal{R} - Q_G) + \frac{1}{\sqrt{h}}(\mathcal{T}^{00}(\delta S/\delta\phi_a,\phi_a) + Q_M) = 0; \tag{4.18}$$

$$\frac{\delta}{\delta h_{ij}}\left(2h^q G_{ijk\ell}\frac{\delta S}{\delta h_{k\ell}}R^2\right) + \sum\frac{\delta}{\delta\phi_a}\left(h^{q-(1/2)}\frac{\delta S}{\delta\phi_a}R^2\right) = 0 \tag{4.19}$$

$$Q_G = -\frac{\hbar^2}{\sqrt{h}R}\left(h^{-q}\frac{\delta}{\delta h_{ij}}h^q G_{ijk\ell}\frac{\delta R}{\delta h_{k\ell}}\right);\ Q_M - \frac{\hbar^2}{hR}\sum\frac{\delta^2 R}{\delta\phi_a^2} \tag{4.20}$$

$$2\nabla_j\frac{\delta R}{\delta h_{ij}} - \mathcal{T}^{i0}(\delta R/\delta\phi_a,\phi_a) = 0;\ 2\nabla_j\frac{\delta S}{\delta h_{ij}} - \mathcal{T}^{i0}(\delta S/\delta\phi_a,\phi_a) = 0 \tag{4.21}$$

One notes that all terms containing the second functional derivative are ill defined and can be regulated via $(\delta/\delta h_{ij}(x))\ (\delta/\delta h_{ij}(x) \to \int d^3x\sqrt{h}U(x-x')(\delta/\delta h_{ij}(x))(\delta/\delta h_{ij}(x'))$ where $U$ is the regulator. Finally the guidance equations are **(4L)** $\pi^{k\ell} = \sqrt{h}(K^{k\ell} - Kh^{k\ell}) = \delta S/\delta h_{k\ell}$ and $\pi_{\phi_a} = \delta S/\delta\phi_a$ where $K_{ij} = (1/2N)(\dot{h}_{ij} - \nabla_i N_j - \nabla_j N_i)$ is the extrinsic curvature. Using the quantum Hamilton-Jacobi-Einstein (HJE) equation one can define a limit, called the pure quantum limit, where the total quantum potential is of the same order as the total classical potential and they can cancel each other. In this case one has **(4M)** $\delta S/\delta h_{ij} = \delta S/\delta\phi_a = 0$ and the continuity is satisfied identically. The resulting trajectory is not similar to any classical solution and the quantum HJE equation for a pure quantum state is an equation for spatial dependence of the metric and matter fields in terms of the norm of the wave function. Explicit calculations are given for some special situations.

**REMARK 4.1.1.** Note that for $\eta_{ab} \sim (1,-1,-1,-1)$ and $\hbar = c = 1$ one has $\partial_0^2 - \nabla^2 \sim \Box$ and (**4N**) $(\nabla S)^2 = m^2[1 + (\Box R/m^2 R)]$ which therefore agrees with

$$(4.22) \qquad (\nabla S)^2 = \mathfrak{M}^2 c^2 (1+Q);\ Q = \frac{\hbar^2}{m^2c^2}\frac{\Box R}{R}$$

from Section 3.1 (F. and A. Shojai). For the BFM theory with $\eta_{ab} \sim (-1,1,1,1)$ one has (**4O**) $(1/2m)(\nabla S)^2 + (mc^2/2) - (\hbar^2/2m)(\Box R/R) = 0$ from (3.2) (cf. Remark 8.4.2 for clarification). The $Q$ of (4.22) differs from the definitive $\mathfrak{Q}$ of Remark 8.4.2 but is thereby defined and both are being referred to as quantum potentials. Now $\Box R \to -\Box R$ and $(\nabla S)^2 \sim \eta^{ab}\nabla_b S \nabla_a S \to -(\nabla S)^2$ for $\eta_{ab} \to -\eta_{ab}$. Hence in the $\eta_{ab} = (1,-1,-1,-1)$ notation one obtains $(\nabla S)^2 = m^2c^2[1 + (\hbar^2/m^2c^2)(\Box R/R)]$ as in (4.22). ∎

**REMARK 4.1.2.** There is a lot of motivation here for using the quantum potential as a generator of quantum gravity and also for considering the conformal factor $\mathfrak{M}^2/m^2$ as a generator of Ricci flow (cf. [**246, 378, 693**]). ∎

**REMARK 4.1.3.** One finds fascinating connections between Bohmian theory and phase space mechanics in [**132, 133, 369, 370, 373, 372, 628, 629, 630, 816**]. In [**132**] one argues that if the quantum potential (QP) reflects the quantum aspects of a sysem it should be possible to to identify such aspects within the QP and in particular one shows how the balance between localization and dispersion energies suggests a link between the QP and the Heisenberg uncertainty principle. Recall first that from the SE (**5A**) $i\hbar\partial_t\psi = [-(\hbar^2/2m)\nabla^2 + V]\psi$ there follows

$$(4.23) \qquad \partial_t S + \frac{(\nabla S)^2}{2m} - \frac{\hbar^2}{2m}\frac{\nabla^2 R}{r} + V = 0;\ \partial_t\rho + \nabla\cdot\left(\rho\frac{p}{m}\right) = 0$$

where $\psi = Rexp(iS/\hbar)$, $\rho = |\psi|^2$, and $p = \nabla S$. The QP is manifestly of the form $Q = -(\hbar^2/2m)(\nabla^2 R/R)$ and one writes $F = -\nabla(Q+V)$ and $v = j/\rho = p/m$. Now consider a more general derivation of the QP by writing the SE in the form (**5B**) $i\hbar\partial_t\psi = (T(\hat{p}) + V(\hat{x}))\psi$. Setting again $\psi = Rexp(iS/\hbar)$ one obtains

$$(4.24) \qquad \partial_t S + \Re\left(\frac{T\psi}{\psi}\right) + V(x) = 0;\ \partial_t\rho - \frac{2\rho}{\hbar}\Im\left(\frac{T\psi}{\psi}\right) = 0$$

Correspondingly in the momentum space with $\hat{x} = i\hbar\nabla_p$ and $\hat{p} = p$ the real and imaginary parts of the SE are

$$(4.25) \qquad \partial_t S + T(p) + \Re\left(\frac{V\psi}{\psi}\right) = 0;\ \partial_t\rho - \frac{2\rho}{\hbar}\Im\left(\frac{V\psi}{\psi}\right) = 0$$

Then expanding exponentials one writes

$$(4.26) \qquad \Re\left(\frac{\psi^* T\psi}{\rho}\right) = \Re\left(\frac{R[1 - (iS/\hbar) - \cdots)T(\hat{p})R(1 + (iS/\hbar) - \cdots]}{\rho}\right)$$

(note the formal equivalence $T\psi/\psi = \psi^* T\psi/\rho$). If now $T(\hat{p})$ is a general but analytic function of $\hat{p}$ one can expand in a power series in $\hat{p} = -i\hbar\nabla$ and the kinetic term may be separated into the sum of two parts

$$\Re\left(\frac{T\psi}{\psi}\right) = T_h(x) + T_0(x);\ T_0(x) = T(\nabla S) \tag{4.27}$$

where $T_h(x)$ is an expansion in even positive powers of $\hbar$ and $T_0(x)$ is independent of $\hbar$ and identifies $p = \nabla S$. The same line of argument allows the potential term of the HJ equation in (4.25) to be separated as

$$\Re\left(\frac{V\psi}{\psi}\right) = V_h(p) + V_0(p);\ V_0(p) = V(-\nabla_p S) \tag{4.28}$$

where $V_h$ is an expansion in even positive powers of $\hbar$ and $V_0(p)$ is independent of $\hbar$ and identifies $x = -\nabla_p S$. We pursue this further in succeeding chapters. ■

## 5. THE QUANTUM POTENTIAL AND GEOMETRY

We begin with [**797**] and recall some features of Weyl geometry (some of which are indicated already in previous sections). We remember first that vectors change in length and direction under translation via (**5A**) $\delta\ell = \phi_\mu \delta x^\mu \ell$ so $\ell = \ell_0 exp(\int \phi_\mu dx^\mu)$ where $\phi_\mu$ is the Weyl vector. Equivalently (**5B**) $g_{\mu\nu} \to exp(2\int \phi_\mu \delta x^\mu) g_{\mu\nu}$ which is a conformal transformation. Recall also that the metric is a Weyl covariant object of weight 2 and the Weyl connection is given via

$$\Gamma^\mu_{\nu\lambda} = \left\{ \begin{matrix} \mu \\ \nu\lambda \end{matrix} \right\} + g_{\nu\lambda}\phi^\mu - \delta^\mu_\nu \phi_\lambda - \delta^\mu_\lambda \phi_\nu \tag{5.1}$$

A gauge transformation (**5C**) $\phi_\mu \to \phi'_\mu = \phi_\mu + \partial_\mu \Lambda$ transforms $g_{\mu\nu} \to g'_{\mu\nu} = exp(2\Lambda) g_{\mu\nu}$ with $\delta\ell \to \delta\ell' = \delta\ell + (\partial_\mu \Lambda) dx^\mu \ell$. In remarks just before Section 3.1 we have seen how quantum effects are geometrized via the Dirac field $\beta$ and gauge transformations. Let us now make more explicit some direct relations between the quantum potential and geometric ideas via the Weyl vector. We recall from Section 3 (#2) that for the SE $Q = -(m/2)\mathbf{u}^2 - (\hbar/2)\partial\mathbf{u}$ where $\mathbf{u}$ is an osmotic velocity (see also [**170, 333**]). Similarly for the KG equation one has $Q = (\hbar^2/m^2c^2)(\Box_g|\psi|/|\psi|$ for example as in Remark 4.1 and we will also consider an appropriate osmotic velocity for this situation.

Consider now the situation $Q = 0$ for the SE which can be expressed in several forms.

(1) Defining the osmotic velocity as $\mathbf{u} = D\nabla log(\rho)$ with $D = \hbar/2m$ from [**170, 333**] (cf. also [**643, 644**]) one has then (**5A**) $(m/2)\mathbf{u}^2 + (\hbar/2)\nabla\mathbf{u} = 0$.
(2) Another form is directly (for $g = 1$) (**5B**) $\Delta\sqrt{\rho} = 0$.
(3) There is a general form for (**5C**) $\phi_i = -\partial_i log(\hat{\rho})$ with $\hat{\rho} = \rho/\sqrt{g}$, namely

$$\dot{\mathcal{R}} + 2\left[\phi_i\phi^i - \frac{2}{\sqrt{g}}\partial_i(\sqrt{g}\phi^i)\right] = 0 \tag{5.2}$$

When $\dot{\mathcal{R}} = 0$ with $\sqrt{g} = 1$ and $\hat{\rho} = \rho$ this becomes (**5D**) $\phi_i\phi^i - 2\partial_i\phi^i = 0$. Note $\phi^i \sim g^{ik}\phi_k = -g^{ik}\partial_k log(\hat{\rho}) = -\partial^k log(\hat{\rho})$.

(4) From (2.8) another form of (5.2) above is

$$\dot{\mathcal{R}} + \frac{8}{\sqrt{\rho}}\partial_i(\sqrt{g}g^{ik}\partial_k\sqrt{\hat{\rho}}) = 0 \tag{5.3}$$

(5) In view of [**170, 171, 246, 401, 402, 736**] one can say that the fundamental quantum fluctuation or perturbation in momentum has the form (**5E**) $\delta p \sim c(\nabla\rho/\rho)$ and this means (**5F**) $\delta p \sim \hat{c}\mathbf{u}$ or equivalently $\delta p \sim \tilde{c}\vec{\phi}$. We can assume that in a Weyl space situation an osmotic velocity $\mathbf{u} = Dlog(\hat{\rho})$ is meaningful. The "obligatory" nature of $\delta p \sim c(\nabla\rho/\rho)$ is made even more striking in the developments in [**409, 617**]. One shows there in particular that a classical momentum can be written as

$$\hat{p}_{cl} = \hat{p} + \left(\frac{i\hbar}{2}\right)\left(\frac{\nabla\rho}{\rho}\right) \Rightarrow \hat{p}_{cl} = -i\hbar\left(\nabla - \frac{1}{2}\frac{\nabla(\psi^*\psi)}{\psi^*\psi}\right) \tag{5.4}$$

It should now be possible to extract some analytic and geometric features of the situation $Q = 0$.

**EXAMPLE** 5.1. We think of $\psi = \sqrt{\rho}exp(iS/\hbar)$ with $\sqrt{\rho} = R$. Take #2 first and look for solutions of $\Delta R = 0$ in a finite region $\Omega$ with $R \in H_0^1(\Omega)$ (Sobolev space) for example (see [**184, 292**] for techniques and results in PDE). For a QM situation $R = 0$ on $\partial\Omega$ and $H_0^1(\Omega)$ is the natural setting with $R \in L^2(\Omega)$. However by Green's theorem $\int_\Omega R\Delta R dV = -\int_\Omega |\nabla R|^2 dS = 0$ which implies $\nabla R = R = 0$. this is consistent with the Example 1.2 where $\psi$ involves plane waves and $L^2$ solutions are meaningless. ∎

**EXAMPLE** 5.2. Consider next a situation $(m/2)|\mathbf{u}|^2 + (\hbar/2)\nabla\mathbf{u} = 0$ or in 1-D $\partial u + cu^2 = 0$ with $c > 0$. Then $u'/u^2 = -c \Rightarrow u = (\hat{c} + cx)^{-1}$ and setting $u = D\rho'/\rho$ yields $R^2 = \rho = k(\hat{c} + cx)^d$. This is not reasonable for $R = 0$ outside of a finite $\Omega$. ∎

**EXAMPLE** 5.3. Consider $\phi_i\phi^i - 2\partial\phi^i = 0$ or equivalently (in 1-D for convenience) $\phi'/\phi^2 = 1/2$ leading to $-(1/\phi) = (1/2)x + c$ and problems similar to those in Example 5.2. ∎

We can however think of $\rho$, $\vec{\phi}$, or $\mathbf{u}$ as functions of $Q$ so for each admisible $Q$ there will be in principle some well determined $R$, modulo spectral conditions as in Remark 1.1.

**REMARK 5.1.1.** In Remark 1.1 we saw that determining $R$ from $Q$ involved solving $\Delta R + \beta QR = 0$ $(\beta > 0)$ in say $H_0^1(\Omega)$. If $Q \leq 0$ this yields a unique solution while if 0 is not in the spectrum of $\Delta + \beta Q$ then (**5G**) $\Delta R + \beta QR = 0$ has a unique solution for say $Q \in L^\infty(\Omega)$. We also saw that modulo solvability of (1.4) one would obtain a "generalized" quantum theory based on $Q$. We can improve the statement of this in Remark 1.1 by saying that, given solutions V and $S$ of (1.4) (via a solution $R$ of (**5G**)), in converting this to a SE one eliminates $Q$ from the picture entirely. For $R$ unique $S(x,t)$ is determined up to a function $f(t)$ and a function $g(x)$ arising from

$$S = -\int^t (Q+V)dt - \frac{1}{2R^4}\left[f(t) - \int^x \partial_t R^2 dx\right]^2 + g(x) \tag{5.5}$$

However $V_x$ is known via (1.4) in terms of $S_{xt}$ which depends only on $Q$ (via $R$), $f$, and $f'$, hence only in terms of one function $f(t)$. If then $V = V(x)$ it may actually be almost determined and $Q$, instead of determining only one trajectory based on $R$ and $S$, actually could lead to the SE itself (modulo $f$) for $\psi = Rexp(iS/\hbar)$; if it were to be the case that $V = V(x)$ does not use $f(t)$ this means that $Q$ alone would determine a "generalized" quantum theory via the SE (but one would of course want $L^2$ solutions for QM). It would be worthwhile checking the equations to find such situations (see below). We note also that if $Q$ contains $t$ it is transmitted to $R$ as a parameter in solving the elliptic equation; if $Q$ is independent of $t$ then of course so is $R$ and this could conceivably simplify matters in determining $V = V(x)$. ■

We check this last idea in more detail now. Assume $Q$ is a function of $x$ alone, $Q = Q(x)$, and let it determine a unique $R \in H_0^1(\Omega)$ (normalized so that $\int_\Omega R^2 dx = 1$). Then look at (1.4), namely

$$S_t + \frac{1}{2m}S_x^2 + Q + V = 0;\ \partial_t R^2 + \frac{1}{m}(R^2 S_x)_x = 0 \tag{5.6}$$

The second equation becomes $(R^2 S_x)_x = 0$ which implies (**5H**) $R^2 S_x = f(t)$ for some "arbitrary" $f(t)$. Then (**5I**) $S_t + (1/2m)(f^2/R^4) + Q + V = 0$ and we can eliminate S from (**5H**) and (**5I**) via

$$R^2 S_{xt} = f_t;\ S_{xt} - \frac{2f^2 R_x}{mR^5} + Q_x + V_x = 0 \Rightarrow \frac{2f^2}{m}\frac{R_x}{R^3} - f_t = R^2(Q_x + V_x) \tag{5.7}$$

This determines $V_x$ in terms of $Q(x)$ and $f(t)$ so we ask whether $V = V(x)$ can occur (no $t$ dependence). In such a case the $t$ derivatives of the last term in (5.7) are zero yielding (**5J**) $f_{tt} = (2R_x/mR^3)\partial_t f^2$. This means

$$\frac{f_{tt}}{\partial_t f^2} = F(t) = \frac{2R_x}{mR} = \mathfrak{F}(x) \tag{5.8}$$

Consequently $F(t) = \mathfrak{F}(x) = c$ and (**5K**) $f_{tt} = c\partial_t f^2$ while $(R_x/R^3) = (cm/2)$ leading to

$$f_t = cf^2 + \hat{c};\ R^2 = \frac{1}{\tilde{c} - cmx} \tag{5.9}$$

Thus $x > (\tilde{c}/cm)$ but $R \notin H_0^1(\Omega)$ for any $\Omega$. This seem to preclude $V = V(x)$ (or perhaps $Q = Q(x)$). Hence from (5.6) one has at least (cf. also subsequent sections).

**PROPOSITION** 5.1. Given $Q = Q(x)$ determining a unique $R(x) \in H_0^1(\Omega)$ it follows that $V_x$ is determined up to an "arbitrary" function $f(t)$ via

$$V_x = \frac{1}{R^2}\left[\frac{2f^2 R_x}{mR^3} - f_t\right] - Q_x \tag{5.10}$$

This situation precludes $V$ being a function of $x$ alone. ■

**REMARK 5.1.2.** Note if $\int_\Omega R^2(x,t)dx = r^2(t)$ then to get a proper normalization one would take $\mathcal{R}(x,t) = (1/r(t))R(x,t)$ and note that Q computed on $\mathcal{R}$ is equal via (**5L**) $\mathcal{Q} = -(\hbar^2/2m)(\mathcal{R}_{xx}/\mathcal{R}) = -(\hbar^2/2m)(R_{xx}/R)$. Note also that $r$ is determined by $Q$ via $R$. We still think of $\psi \sim Rexp(iS/\hbar)$ so (5.5) applies and (**5H**) becomes (**5M**) $RS_x^2 = -m\int^x \partial_t R^2 dx + f(t) = A(f,Q)$ (since Q determines R). Then we are still essentially in the context of Remark 5.1 and (**5N**) $2RR_tS_x + R^2S_{xt} = \partial_t A$ while from (5.5) (**5O**) $S_{xt} + (1/m)S_xS_{xx} + Q_x + V_x = 0$. Now we eliminate $S_{xx}$ and $S_{xt}$ to get $V_x$ in terms of $Q$ and $f$. First from (**5N**) one has

$$S_{xt} = \frac{1}{R^2}\left\{\partial_t A - 2\frac{R_t}{R}A\right\} \tag{5.11}$$

while (**5P**) $R^2S_{xx} + 2RR_x = \partial_x A \Rightarrow S_{xx} = (1/R^2)(\partial_x A - 2RR_x)$. Hence one arrives at

$$\frac{1}{R^2}\left[\partial_t A - \frac{2AR_t}{R}\right] = -Q_x - V_x - \frac{1}{m}\frac{A}{R^4}(\partial_x A - 2RR_x) \tag{5.12}$$

and we can state

**PROPOSITION** 5.2. Defining $A(f,Q) = f(t) = m\int^x \partial_t R^2 dx$ with $f$ "arbitrary" one can determine $V_x$ via (5.12) as $V_x(Q,f)$. Hence $V = \int^x V_x dx + h(t)$ for $h$ "arbitrary" provides a potential $V(Q,f,h)$ and the associated SE is determined completely by $V$. If choices $f = h = 0$ are "natural" one can say that $Q$ determines a natural SE and a corresponding "generalized" quantum theory. ∎

**REMARK 5.1.3.** Consider the stationary case (cf. [**170**]) (**5Q**) $(1/2m)S_x^2 + Q + V - E = 0$ with $(R^2S_x)_x = 0$ where $R = R(x)$ is say uniquely determined via $Q = Q(x)$ (note however that both $R$ and $Q$ must contain $E$ as a parameter). Then

$$S_x = \frac{c}{R^2};\ \frac{1}{2m}\left(\frac{c^2}{R^4}\right) + Q + V - E = 0 \tag{5.13}$$

This means (**5R**) $1 = \partial_E Q - (c^2/mR^5)\partial_E R$ (since $V$ does not depend on $E$) and hence

$$\frac{2c^2R_E}{mR^5} + 1 = \frac{\hbar^2 R'' R_E}{2mR^2} - \frac{\hbar^2 R''_E}{2mR} \tag{5.14}$$

Viewed in terms of $\rho = R^2$ this mean that $\rho = \rho(E,x)$ and the corresponding Weyl geometry based on $\vec{\phi} = -\nabla log(\rho)$ will depend on $E$ (as will $Q$ of course). We refer here also to the quantum mass idea of Floyd, namely $m_Q = m(1 - \partial_E Q)$ for stationary situations (this is sketched in [**177**] for example and we refer to [**170**] for more details). Some further ideas about this are sketched later. In particular one knows that $Q \sim -(\hbar^2/2m)\mathcal{R}$ (from Section 2.2) where $\mathcal{R}$ is the Ricci-Weyl curvature with $\hbar$ essentially put in by hand to conform to the wave function idea and operator QM. We see that the geometry of the space in which a trajectory transpires is thereby determined by $E$ (not surprisingly) which seems to say that the probability distribution $\rho$ is the basic unknown here (and in the time dependent situation). Once one has a probability distribution one can posit a wave function and insert $\hbar$. In fact (given V) the two equations $(1/2m)(c^2/R^2) + Q + V - E = 0$ and $Q = -(\hbar^2/2m)(R''/R)$ determine $R = R(E,x)$ directly ($\hbar$ being gratuitously inserted). ∎

## 6. CONFORMAL AND QUANTUM MASS

We collect here some results of the author from [**172**]. The first item clarifies the relation between conformal mass and quantum mass. We show that conformal general relativity (GR) is an integrable Weyl geometry and then construct an explicit identification of this with a Bohmian Weyl-Dirac theory with Dirac field $\beta \sim \mathcal{M}$. This confirms (3.19) in Chapter 2 and also enhances the result $\mathcal{M} = \hat{m}$ of [**117**] relating the conformal mass $\hat{m}$ to the Bohmian mass $\mathcal{M}$. We show that the conformal mapping $\Omega^2$ can be written as **(6A)** $\Omega^2 = \mathcal{M}^2/m^2$ where $\mathcal{M} \sim exp(Q)$ with Q being the quantum potential (cf. [**117, 172, 642, 797**] and note that the first paper of [**797**] appears in the book [**527**], pp. 59-98). For the details we extract from [**172**] and refer to Section 2.2 for remarks on Weyl geometry.

One has an Einstein form for GR of the form

$$S_{GR} = \int d^4x\sqrt{-g}(R - \alpha|\nabla\psi|^2 + 16\pi L_M) \tag{6.1}$$

(cf. [**117, 729**]) whose conformal form (conformal GR) is an integrable Weyl geometry based on

$$\hat{S}_{GR} = \int d^4x\sqrt{-\hat{g}}e^{-\psi}\left[\hat{R} - \left(\alpha - \frac{3}{2}\right)|\hat{\nabla}\psi|^2 + 16\pi e^{-\psi}L_M\right] \tag{6.2}$$
$$= \int d^4x\sqrt{-\hat{g}}\left[\hat{\phi}\hat{R} - \left(\alpha - \frac{3}{2}\right)\frac{|\hat{\nabla}\hat{\phi}|^2}{\hat{\phi}} + 16\pi\hat{\phi}^2 L_M\right]$$

where $\Omega^2 = exp(-\psi) = \phi$ with $\hat{g}_{ab} = \Omega^2 g_{ab}$ and $\hat{\phi} = exp(\psi) = \phi^{-1}$ (note $(\hat{\nabla}\psi)^2 = (\hat{\nabla}\hat{\phi})^2/(\hat{\phi})^2$). One sees also that (6.2) is the same as the Brans-Dicke (BD) action when $L_M = 0$, namely (using $\hat{g}$ as the basic metric)

$$S_{BD} = \int d^4x\sqrt{-\hat{g}}\left[\hat{\phi}\hat{R} - \frac{\omega}{\hat{\phi}}|\hat{\nabla}\hat{\phi}|^2 + 16\pi L_M\right]; \tag{6.3}$$

which corresponds to (6.2) provided $\omega = \alpha - (3/2)$ and $L_M = 0$. For (6.2) we have a Weyl gauge vector $w_a \sim \partial_a\psi = \partial_a\hat{\phi}/\hat{\phi}$ and a conformal mass $\hat{m} = \hat{\phi}^{-1/2}m$ with $\Omega^2 = \hat{\phi}^{-1}$ as the conformal factor above. Now in (6.2) we identify $\hat{m}$ with the quantum mass $\mathfrak{M}$ of [**797**] where for certain model situations $\mathfrak{M} \sim \beta$ is a Dirac field in a Bohmian-Dirac-Weyl theory as in (6.8) below with quantum potential $Q$ determined via $\mathfrak{M}^2 = m^2 exp(Q)$ (cf. [**170, 642, 797**] and note that $m^2 \propto T$ where $8\pi T^{ab} = (1/\sqrt{-g})(\delta\sqrt{-g}\mathfrak{L}_M/\delta g_{ab})$). Then $\hat{\phi}^{-1} = \hat{m}^2/m^2 = \mathfrak{M}^2/m^2 \sim \Omega^2$ for $\Omega^2$ the standard conformal factor of [**797**]. Further one can write **(6B)** $\sqrt{-\hat{g}}\hat{\phi}\hat{R} = \hat{\phi}^{-1}\sqrt{-\hat{g}}\hat{\phi}^2\hat{R} = \hat{\phi}^{-1}\sqrt{-g}\hat{R} = (\beta^2/m^2)\sqrt{-g}\hat{R}$. Recall here from [**170**] that for $g_{ab} = \hat{\phi}\hat{g}_{ab}$ one has $\sqrt{-g} = \hat{\phi}^2\sqrt{-\hat{g}}$ and for the Weyl-Dirac geometry we give a brief survey following [**170, 465**].

(1) Weyl gauge transformations: $g_{ab} \to \tilde{g}_{ab} = e^{2\lambda}g_{ab}$; $g^{ab} \to \tilde{g}^{ab} = e^{-2\lambda}g^{ab}$ - weight e.g. $\Pi(g^{ab}) = -2$. $\beta$ is a Dirac field of weight $-1$. Note $\Pi(\sqrt{-g}) = 4$.
(2) $\Gamma^c_{ab}$ is Riemannian connection; Weyl connection is $\hat{\Gamma}^c_{ab}$ and $\hat{\Gamma}^c_{ab} = \Gamma^c_{ab} = g_{ab}w^c - \delta^c_b w_a - \delta^c_a w_b$.
(3) $\nabla_a B_b = \partial_a B_b - B_c\Gamma^c_{ab}$; $\nabla_a B^b = \partial_a B^b + B^c\Gamma^b_{ca}$

(4) $\hat{\nabla}_a B_b = \partial_a B_b - B_c \hat{\Gamma}^c_{ab}$; $\hat{\nabla}_a B^b = \partial_a B^b + B^c \hat{\Gamma}^b_{ca}$
(5) $\hat{\nabla}_\lambda g^{ab} = -2g^{ab} w_\lambda$; $\hat{\nabla}_\lambda g_{ab} = 2g_{ab} w_\lambda$ and for $\Omega^2 = exp(-\psi)$ the requirement $\nabla_c g_{ab} = 0$ is transformed into $\hat{\nabla}_c \hat{g}_{ab} = \partial_c \psi \hat{g}_{ab}$ showing that $w_c = -\partial_c \psi$ (cf. [**117**]) leading to $w_\mu = \hat{\phi}_\mu / \hat{\phi}$ and hence via $\beta = m\hat{\phi}^{-1/2}$ one has $w_c = 2\beta_c/\beta$ with $\hat{\phi}_c/\hat{\phi} = -2\beta_c/\beta$ and $w^a = -2\beta^a/\beta$.

Consequently, via $\beta^2 \hat{R} = \beta^2 R - 6\beta^2 \nabla_\lambda w^\lambda + 6\beta^2 w^\lambda w_\lambda$ (cf. [**170, 185, 265, 465**]), one observes that $-\beta^2 \nabla_\lambda w^\lambda = -\nabla_\lambda(\beta^2 w^\lambda) + 2\beta \partial_\lambda \beta w^\lambda$, and the divergence term will vanish upon integration, so the first integral in (6.2) becomes

$$I_1 = \int d^4x \sqrt{-g} \left[ \frac{\beta^2}{m^2} R + 12\beta \partial_\lambda \beta w^\lambda + 6\beta^2 w^\lambda w_\lambda \right] \tag{6.4}$$

Setting now $\alpha - (3/2) = \gamma$ the second integral in (6.2) is

$$I_2 = -\gamma \int d^4x \sqrt{-\hat{g}} \hat{\phi} \frac{|\hat{\nabla}\hat{\phi}|^2}{|\hat{\phi}|^2} \tag{6.5}$$

$$= -4\gamma \int d^4x \sqrt{-\hat{g}} \hat{\phi}^{-1} \hat{\phi}^2 \frac{|\hat{\nabla}\beta|^2}{\beta^2} = -\frac{4\gamma}{m^2} \int d^4x \sqrt{-g} |\hat{\nabla}\beta|^2$$

while the third integral in (6.2) becomes (**6C**) $16\pi \int \sqrt{-g} d^4x L_M$. Combining now (1.4), (1.5), and (**6C**) gives then
(6.6)

$$\hat{S}_{GR} = \frac{1}{m^2} \int d^4x \sqrt{-g} \left[ \beta^2 R + 6\beta^2 w^\alpha w_\alpha + 12\beta \partial_\alpha \beta w^\alpha - 4\gamma |\hat{\nabla}\beta|^2 + 16\pi m^2 L_M \right]$$

We will think of $\hat{\nabla}\beta$ in the form (**6D**) $\hat{\nabla}_\mu \beta = \partial_\mu \beta - w_\mu \beta = -\partial_\mu \beta$. Putting then $|\hat{\nabla}\beta|^2 = |\partial\beta|^2$ (6.6) becomes (recall $\gamma = \alpha - (3/2)$)

$$\hat{S}_{GR} = \frac{1}{m^2} \int d^4x \sqrt{-g} \left[ \beta^2 R + (3 - 4\alpha)|\partial\beta|^2 + 16\pi m^2 L_M \right] \tag{6.7}$$

One then checks this against some Weyl-Dirac actions. Thus, neglecting terms $W^{ab}W_{ab}$ we find integrands involving $dx^4\sqrt{-g}$ times

$$-\beta^2 R + 3(3\sigma + 2)|\partial\beta|^2 + 2\Lambda\beta^4 + \mathfrak{L}_M \tag{6.8}$$

(see e.g. [**170, 185, 465, 797**]); the term $2\Lambda\beta^4$ of weight $-4$ is added gratuitously (recall $\Pi(\sqrt{-g}) = 4$). Consequently, omitting the $\Lambda$ term, (6.8) corresponds to (6.7) times $m^2$ for $\mathfrak{L}_M \sim 16\pi L_M$ and (**6E**) $9\sigma + 4\alpha + 3 = 0$. Hence one can identify conformal GR (without $\Lambda$) with a Bohmian-Weyl-Dirac theory where conformal mass $\hat{m}$ corresponds to quantum mass $\mathfrak{M}$.

## 7. RICCI FLOW AND THE QUANTUM POTENTIAL

Certain aspects of Perelman's work on the Poincaré conjecture have applications in physics and we want to suggest a few formulas in this direction (cf. [**170, 172**]). We go first to [**217, 491, 506, 507, 508, 621, 693, 836**] and simply write down a few formulas from [**621, 836**] here with minimal explanation. Thus one has Perelman's functional ($\dot{\mathcal{R}}$ is the Riemannian Ricci curvature)

$$\mathfrak{F} = \int_M (\dot{\mathcal{R}} + |\nabla f|^2) exp(-f) dV \tag{7.1}$$

and a so-called Nash entropy (**1A**) $N(u) = \int_M ulog(u)dV$ where $u = exp(-f)$. One considers Ricci flows with $\delta g \sim \partial_t g = h$ and for (**1B**) $\Box^* u = -\partial_t u - \Delta u + \dot{\mathcal{R}}u = 0$ (or equivalently $\partial_t f + \Delta f - |\nabla f|^2 + \dot{\mathcal{R}} = 0$) it follows that $\int_M exp(-f)dV = 1$ is preserved and $\partial_t N = \mathfrak{F}$. Note the Ricci flow equation is $\partial_t g = -2Ric$. Extremizing $\mathfrak{F}$ via $\delta\mathfrak{F} \sim \partial_t \mathfrak{F} = 0$ involves $Ric + Hess(f) = 0$ or $R_{ij} + \nabla_i \nabla_j f = 0$ and one knows also that

$$\partial_t N = \int_M (|\nabla f|^2 + \dot{\mathcal{R}})exp(-f)dV = \mathfrak{F}; \tag{7.2}$$

$$\partial_t \mathfrak{F} = 2\int_M |Ric + Hess(f)|^2 exp(-f)dV$$

Now referring to [**97, 170, 178, 186, 171, 179, 191, 193, 297, 319, 333, 332, 401, 402, 637, 638, 681, 770, 797, 884**] for details we note first the important observation in [**836**] that $\mathfrak{F}$ is in fact a Fisher information functional. Fisher information has come up repeatedly in studies of the Schrödinger equation (SE) and the Wheeler-deWitt equation (WDW) and is connected to a differential entropy correspondingto the Nash entropy above (cf. [**178, 186, 333, 332**]). The basic ideas involve (using 1-D for simplicity) a quantum potential Q such that $\int_M PQdx \sim \mathfrak{F}$ arising from a wave function $\psi = Rexp(iS/\hbar)$ where $Q = -(\hbar^2/2m)(\Delta R/R)$ and $P \sim |\psi|^2$ is a probability density. In a WDW context for example one can develop a framework

$$Q = cP^{-1/2}\partial(GP^{1/2});\ \int QP = c\int P^{1/2}\partial(GP^{1/2})\mathfrak{D}hdx \tag{7.3}$$

$$\to -c\int \partial P^{1/2} G\partial P^{1/2}\mathfrak{D}hdx$$

where $G$ is an expression involving the deWitt metric $G_{ijk\ell}(h)$. In a more simple minded context consider a SE in 1-D $i\hbar\partial_t\psi = -(\hbar^2/2m)\partial_x^2\psi + V\psi$ where $\psi = Rexp(iS/\hbar)$ leads to the equations

$$S_t + \frac{1}{2m}S_x^2 + Q + V = 0;\ \partial_t R^2 + \frac{1}{m}(R^2 S_x)_x = 0:\ Q = -\frac{\hbar^2}{2m}\frac{R_{xx}}{R} \tag{7.4}$$

In terms of the exact uncertainty principle of Hall and Reginatto (see [**400, 402, 401, 736**] and cf. also [**170, 182, 186, 681**]) the quantum Hamiltonian has a Fisher information term $c\int dx(\nabla P\cdot\nabla P/2mP)$ added to the classical Hamiltonian (where $P = R^2 \sim |\psi|^2$) and a simple calculation gives

$$\int PQd^3x \sim -\frac{\hbar^2}{8m}\int\left[2\Delta P - \frac{1}{P}|\nabla P|^2\right]d^3x = \frac{\hbar^2}{8m}\int\frac{1}{P}|\nabla P|^2 d^3x \tag{7.5}$$

In the situation of (7.3) the analogues to Section 1 involve ($\partial \sim \partial_x$)

(7.6)

$$P \sim e^{-f};\ P' \sim P_x \sim -f'e^{-f};\ Q \sim e^{f/2}\partial(G\partial e^{-f/2});\ PQ \sim e^{-f/2}\partial(G\partial e^{-f/2});$$

$$\int PQ \to -\int \partial e^{-f/2} G\partial e^{-f/2} \sim -\int \partial P^{1/2} G \partial P^{1/2}$$

In the context of the SE in Weyl space developed in [**53, 54, 170, 171, 186, 191, 193, 770, 874**] one has a situation $|\psi|^2 \sim R^2 \sim P \sim \hat{\rho} = \rho/\sqrt{g}$ with a Weyl vector $\vec{\phi} = -\nabla log(\hat{\rho})$ and a quantum potential

$$Q \sim -\frac{\hbar^2}{16m}\left[\dot{\mathcal{R}} + \frac{8}{\sqrt{\hat{\rho}}}\frac{1}{\sqrt{g}}\partial_i\left(\sqrt{g}g^{ik}\partial_k\sqrt{\hat{\rho}}\right)\right] = -\frac{\hbar^2}{16m}\left[\dot{\mathcal{R}} + \frac{8}{\sqrt{\hat{\rho}}}\Delta\sqrt{\hat{\rho}}\right] \quad (7.7)$$

(recall $divgrad(U) = \Delta U = (1/\sqrt{g})\partial_m(\sqrt{g}g^{mn}\partial_n U)$. Here the Weyl-Ricci curvature is **(1C)** $\mathcal{R} = \dot{\mathcal{R}} + \mathcal{R}_w$ where

$$\mathcal{R}_w = 2|\vec{\phi}|^2 - 4\nabla\cdot\vec{\phi} = 8\frac{\Delta\sqrt{\hat{\rho}}}{\sqrt{\hat{\rho}}} \quad (7.8)$$

and $Q = -(\hbar^2/16m)\mathcal{R}$. Note that

$$-\nabla\cdot\vec{\phi} \sim -\Delta log(\hat{\rho}) \sim -\frac{\Delta\hat{\rho}}{\hat{\rho}} + \frac{|\nabla\hat{\rho}|^2}{\hat{\rho}^2} \quad (7.9)$$

and for $exp(-f) = \hat{\rho} = u$

$$\int \hat{\rho}\nabla\cdot\vec{\phi}dV = \int\left[-\Delta\hat{\rho} + \frac{|\nabla\hat{\rho}|^2}{\hat{\rho}}\right]dV \quad (7.10)$$

with the first term in the last integral vanishing and the second providing Fisher information again. Comparing with the beginning of this section we have analogues **(1D)** $G \sim (R + |\vec{\phi}|^2)$ with $\vec{\phi} = -\nabla log(\hat{\rho}) \sim \nabla f$ to go with (7.6). Clearly $\hat{\rho}$ is basically a probability concept with $\int \hat{\rho}dV = 1$ and quantum mechanics (QM) (or rather perhaps Bohmian mechanics) seems to enter the picture through the second equation in (2.2), namely **(1E)** $\partial_t\hat{\rho} + (1/m)div(\hat{\rho}\nabla S) = 0$ with $p = mv = \nabla S$, which must be reconciled with **(1B)** (i.e. $(1/m)div(u\nabla S) = \Delta u - \dot{\mathcal{R}}u$). In any event the term $G = \dot{\mathcal{R}} + |\vec{\phi}|^2$ can be written as **(1F)** $\dot{\mathcal{R}} + \mathcal{R}_w + (|\vec{\phi}|^2 - \mathcal{R}_w) = \alpha Q + (4\nabla\cdot\vec{\phi} - |\vec{\phi}|^2)$ which leads to **(1G)** $\mathfrak{F} \sim \alpha\int_M QPdV + \beta\int|\vec{\phi}|^2PdV$ putting Q directly into the picture and suggesting some sort of quantum mechanical connection.

**REMARK 7.1.1.** We mention also that Q appears in a fascinating geometrical role in the relativistic Bohmian format following [**97, 297, 770, 797**] (cf. also [**170, 186**] for survey material). Thus e.g. one can define a quantum mass field via

$$\mathfrak{M}^2 = m^2 exp(Q) \sim m^2(1+Q);\ Q \sim \frac{-\hbar^2}{c^2m^2}\frac{\Box(\sqrt{\rho})}{\sqrt{\rho}} \sim \frac{\alpha}{6}\mathcal{R}_w \quad (7.11)$$

where $\rho$ refers to an appropriate mass density and $\mathfrak{M}$ is in fact the Dirac field $\beta$ in a Weyl-Dirac formulation of Bohmian quantum gravity. Further one can change the 4-D Lorentzian metric via a conformal factor $\Omega^2 = \mathfrak{M}^2/m^2$ in the form $\tilde{g}_{\mu\nu} = \Omega^2 g_{\mu\nu}$ and this suggests possible interest in Ricci flows etc. in conformal Lorentzian spaces (cf. here also [**246**]). We refer to [**97, 170, 177, 297**] for another fascinating form of the quantum potential as a mass generating term and intrinsic self energy. ∎

**7.1. EXPANSION.** In [**186, 170**] we indicated some relations between Weyl geometry and the quantum potential, between conformal general relativity (GR) and Dirac-Weyl theory, and between Ricci flow and the quantum potential. We would now like to develop this a little further. First we consider simple Ricci flow as in [**621, 836**]. Thus from [**621**] we take the Perelman entropy functional as (**1H**) $\mathfrak{F}(g,f) = \int_M (|\nabla f|^2 + R)exp(-f)dV$ (restricted to $f$ such that $\int_M exp(-f)dV = 1$) and a Nash (or differential) entropy (**1I**) $N(u) = \int_M ulog(u)dV$ where $u = exp(-f)$ (M is a compact Riemannian manifold without boundary). One writes $dV = \sqrt{det(g)}\prod dx^i$ and shows that if $g \to g + sh$ $(g, h \in \mathcal{M} = Riem(M))$ then (**1J**) $\partial_s det(g)|_{s=0} = g^{ij}h_{ij}det(g) = (Tr_g h)det(g)$. This comes from a matrix formula of the form (**1K**) $\partial_s det(A+B)|_{s=0} = (A^{-1} : B)det(A)$ where $A^{-1} : B = a^{ij}b_{ji} = a^{ij}b_{ij}$ for symmetric B ($a^{ij}$ comes from $A^{-1}$). If one has Ricci flow (**1L**) $\partial_s g = -2Ric$ (i.e. $\partial_s g_{ij} = -2R_{ij}$) then, considering $h \sim -2Ric$, one arrives at (**1M**) $\partial_s dV = -RdV$ where $R = g^{ij}R_{ij}$ (more general Ricci flow involves (**1N**) $\partial_t g_{ik} = -2(R_{ik} + \nabla_i\nabla_k\phi)$). We use now $t$ and $s$ interchangeably and suppose $\partial_t g = -2Ric$ with $u = exp(-f)$ satisfying $\Box^* u = 0$ where $\Box^* = -\partial_t - \Delta + R$. Then $\int_M exp(-f)dV = 1$ is preserved since (**1O**) $\partial_t \int_M udV = \int_M (\partial_s u - Ru)dV = -\int_M \Delta u dV = 0$ and, after some integration by parts,

$$(7.12)\quad \partial_t N = \int_M [\partial_t u(log(u)+1)dV + ulog(u)\partial_t dV = \int_M (|\nabla f|^2 + R)e^{-f}dV = \mathfrak{F}$$

In particular for $R \geq 0$ N is monotone as befits an entropy. We note also that $\Box^* u = 0$ is equivalent to (**1P**) $\partial_t f = -\Delta f + |\nabla f|^2 - R$.

It was also noted in [**836**] that $\mathfrak{F}$ is a Fisher information functional (cf. [**170, 318, 319**]) and we showed in [**172**] that for a given 3-D manifold M and a Weyl-Schrödinger picture of quantum evolution based on [**770**] (cf. also [**53, 54, 55, 170, 186, 171, 179, 191, 193, 875**]) one can express $\mathfrak{F}$ in terms of a quantum potential $Q$ in the form (**1Q**) $\mathfrak{F} \sim \alpha \int_M QPdV + \beta \int_M |\vec{\phi}|^2 PdV$ where $\vec{\phi}$ is a Weyl vector and $P$ is a probability distribution associated with a quantum mass density $\hat{\rho} \sim |\psi|^2$. There will be a corresponding Schrödinger equation (SE) in a Weyl space as in [**170, 172**] provided there is a phase $S$ (for $\psi = |\psi|exp(iS/\hbar)$) satisfying (**1$\mathfrak{R}$**) $(1/m)div(P\nabla S) = \Delta P - RP$ (arising from $\partial_t\hat{\rho} - \Delta\hat{\phi} = -(1/m)div(\hat{\rho}\nabla S)$ and $\partial_t\hat{\rho} + \Delta\hat{\rho} - R\hat{\rho} = 0$ with $\hat{\rho} \sim P \sim u \sim |\psi|^2$ - we denote this by Gothic $\mathfrak{R}$ since it will be referred to again in other sections). In the present work we show that there can exist solutions $S$ of (**1$\mathfrak{R}$**) and this establishes a connection between Ricci flow and quantum theory (via Fisher information and the quantum potential). Another aspect is to look at a relativistic situation with conformal perturbations of a 4-D semi-Riemannian metric $g$ based on a quantum potential (defined via a quantum mass). Indeed in a simple minded way we could perhaps think of a conformal transformation $\hat{g}_{ab} = \Omega^2 g_{ab}$ (in 4-D) where following [**170, 172**] we can imagine ourselves immersed in conformal general relativity (GR) with metric $\hat{g}$ and (**1S**) $exp(Q) \sim \mathfrak{M}^2/m^2 = \Omega^2 = \hat{\phi}^{-1}$ with $\beta \sim \mathfrak{M}$ where $\beta$ is a Dirac field and $Q$ a quantum potential $Q \sim (\hbar^2/m^2c^2)(\Box_g\sqrt{\rho})/\sqrt{\rho})$ with $\rho \sim |\psi^2|$ referring to a quantum matter density. The theme here (as developed in [**170, 172**]) is that Weyl-Dirac action with Dirac field $\beta$ leads to $\beta \sim \mathfrak{M}$ and is equivalent to conformal GR (cf. also [**179, 642, 797, 798, 804**] and see [**378**] for ideas on Ricci flow gravity).

**REMARK 7.1.2.** For completeness we recall (cf. [**170, 860**]) for $\mathfrak{L}_G = (1/2\chi)\sqrt{-g}R$

$$\delta\mathfrak{L} = \frac{1}{2\chi}\left[R_{ab} - \frac{1}{2}g_{ab}R\right]\sqrt{-g}\delta g^{ab} + \frac{1}{2\chi}g^{ab}\sqrt{-g}\delta R_{ab} \tag{7.13}$$

The last term can be converted to a boundary integral if certain derivatives of $g_{ab}$ are fixed there. Next following [**117, 186, 172, 368, 729, 730, 731**] the Einstein frame GR action has the form

$$S_{GR} = \int d^4x\sqrt{-g}(R - \alpha(\nabla\psi)^2 + 16\pi L_M) \tag{7.14}$$

(cf. [**117**]) whose conformal form (conformal GR) is

$$\hat{S}_{GR} = \int d^4x\sqrt{-\hat{g}}e^{-\psi}\left[\hat{R} - \left(\alpha - \frac{3}{2}\right)(\hat{\nabla}\psi)^2 + 16\pi e^{-\psi}L_M\right] \tag{7.15}$$

$$= \int d^4x\sqrt{-g}\left[\hat{\phi}\hat{R} - \left(\alpha - \frac{3}{2}\right)\frac{(\hat{\nabla}\hat{\phi})^2}{\hat{\phi}} + 16\pi\hat{\phi}^2 L_M\right]$$

where $\hat{g}_{ab} = \Omega^2 g_{ab}$, $\Omega^2 = exp(\psi) = \phi$, and $\hat{\phi} = exp(-\psi) = \phi^{-1}$. If we omit the matter Lagrangians, and set $\lambda = (3/2) - \alpha$, (1.4) becomes for $\hat{g}_{ab} \to g_{ab}$

$$\tilde{S} = \int d^4x\sqrt{-g}e^{-\psi}[R + \lambda(\nabla\psi)^2] \tag{7.16}$$

In this form on a 3-D manifold $M$ we have *exactly* the situation treated in [**170, 172**] with an associated SE in Weyl space based on (**7**$\mathfrak{R}$). ■

Consider now (**1**$\mathfrak{R}$) : $(1/m)div(P\nabla S) = \Delta P - RP$ for $P \sim \hat{\rho} \sim |\psi|^2$ and $\int P\sqrt{|g|}d^3x = 1$ (in 3-D we will use here $\sqrt{|g|}$ for $\sqrt{-g}$). One knows that $div(P\nabla S) = P\Delta S + \nabla P \cdot \nabla S$ and

$$\Delta\psi = \frac{1}{\sqrt{|g|}}\partial_m(\sqrt{|g|}\nabla\psi);\ \nabla\psi = g^{mn}\partial_n\psi;\ \int_M div\mathbf{V}\sqrt{|g|}d^3x = \int_{\partial M}\mathbf{V}\cdot\mathbf{ds} \tag{7.17}$$

(cf. [**171**]). Recall also $\int P\sqrt{|g|}d^3x = 1$ and

$$Q \sim -\frac{\hbar^2}{8m}\left[\left(\frac{\nabla P}{P}\right)^2 - 2\left(\frac{\Delta P}{P}\right)\right];\ <Q>_\psi = \int PQd^3x \tag{7.18}$$

Now in 1-D an analogous equation to (**1**$\mathfrak{R}$) would be (**1T**) $(PS')' = P' - RP = F$ with solution determined via

$$PS' = P' - \int RP + c \Rightarrow S' = \partial_x log(P) - \frac{1}{P}\int RP + cP^{-1} \tag{7.19}$$

$$\Rightarrow S = log(P) - \int\frac{1}{P}\int RP + c\int P^{-1} + k$$

which suggests that solutions of (**1**$\mathfrak{R}$) do in fact exist in general. We approach the general case in Sobolev spaces à la [**51, 52, 184, 292**]. The volume element is defined via $\eta = \sqrt{|g|}dx^1 \wedge \cdots \wedge dx^n$ (where $n = 3$ for our purposes) and $* : \wedge^p M \to \wedge^{n-p}M$ is defined via

$$(*\alpha)_{\lambda_{p+1}\cdots\lambda_n} = \frac{1}{p!}\eta_{\lambda_1\cdots\lambda_n}\alpha^{\lambda_1\cdots\lambda_p};\ (\alpha,\beta) = \frac{1}{p!}\alpha_{\lambda_1\cdots\lambda_p}\beta^{\lambda_1\cdots\lambda_p}; \tag{7.20}$$

$$*1 = \eta;\ **\alpha = (-1)^{p(n-p)}\alpha;\ *\eta = 1;\ \alpha \wedge (*\beta) = (\alpha,\beta)\eta$$

One writes now $< \alpha, \beta >= \int_M (\alpha,\beta)\eta$ and, for $(\Omega, \phi)$ a local chart we have **(1U)** $\int_M f dV = \int_{\phi(\Omega)} (\sqrt{|g|}f) \circ \phi^{-1} \prod dx^i$ ($\sim \int_M f\sqrt{|g|}\prod dx^i$). Then one has **(1V)** $< d\alpha, \gamma >=< \alpha, \delta\gamma >$ for $\alpha \in \wedge^p M$ and $\gamma \in \wedge^{p+1}M$ where the codifferential $\delta$ on $p$-forms is defined via **(1W)** $\delta = (-1)^p *^{-1} d*$. Then $\delta^2 = d^2 = 0$ and $\Delta = d\delta + \delta d$ so that $\Delta f = \delta df = -\nabla^\nu\nabla_\nu f$. Indeed for $\alpha \in \wedge^p M$

$$(\delta\alpha)_{\lambda_1,\cdots,\lambda_{p-1}} = -\nabla^\gamma \alpha_{\gamma,\lambda_1,\cdots,\lambda_{p-1}} \tag{7.21}$$

with $\delta f = 0$ ($\delta : \wedge^p M \to \wedge^{p-1}M$). Then in particular **(1X)** $< \Delta\phi, \phi >=< \delta d\phi, \phi >=< d\phi, d\phi >= \int_M \nabla^\nu\phi\nabla_\nu\phi\eta$.

Now to deal with weak solutions of an equation in divergence form look at an operator **(1Y)** $Au = -\nabla(a\nabla u) \sim (-1/\sqrt{|g|})\partial_m(\sqrt{|g|}ag^{mn}\nabla_n u) = -\nabla_m(a\nabla^m u)$ so that for $\phi \in \mathcal{D}(M)$

$$\int_M Au\phi dV = -\int [\nabla_m(ag^{mn}\nabla_n u)]\phi dV \tag{7.22}$$

$$= \int ag^{mn}\nabla_n u\nabla_m\phi dV = \int a\nabla^m u\nabla_m\phi dV$$

Here one imagines $M$ to be a complete Riemannian manifold with Soblev spaces $H_0^1(M) \sim H^1(M)$ (see [**51, 52, 184, 352, 424, 833**]). The notation in [**51**] is different and we think of $H^1(M)$ as the space of $L^2$ functions $u$ on $M$ with $\nabla u \in L^2$ and $H_0^1$ means the completion of $\mathcal{D}(M)$ in the $H^1$ norm $\|u\|^2 = \int_M [|u|^2 + |\nabla u|^2]dV$. Following [**424**] we can also assume $\partial M = \emptyset$ with $M$ connected for all $M$ under consideration. Then let $H = H^1(M)$ be our Hilbert space and consider the operator $A(S) = -(1/m)\nabla(P\nabla S)$ with

$$B(S,\psi) = \frac{1}{m}\int P\nabla^m S\nabla_m\psi dV \tag{7.23}$$

for $S, \psi \in H_0^1 = H^1$. Then $A(S) = RP - \Delta P = F$ becomes **(1Z)** $B(S,\psi) = < F, \psi >= \int F\psi dV$ and one has (•) $|B(S,\psi)| \le c\|S\|_H\|\psi\|_H$ and $|B(S,S)| = \int P(\nabla S)^2 dV$. Now $P \ge 0$ with $\int PdV = 1$ but to use the Lax-Milgram theory we need here $|B(S,S)| \ge \beta\|S\|_H^2$ ($H = H^1$). In this direction one recalls that in Euclidean space for $\psi \in H_0^1(\mathbf{R}^3)$ there follows (♦) $\|\psi\|_{L^2}^2 \le c\|\nabla\psi\|_{L^2}^2$ (Friedrich's inequality - cf. [**833**]) which would imply $\|\psi\|_H^2 \le (c+1)\|\nabla\psi\|_{L^2}^2$. However such Sobolev and Poincaré-Sobolev inequalities become more complicated on manifolds and (♦) is in no way automatic (cf. [**51, 424, 833**]). However we have some recourse here to the definition of $P$, namely $P = exp(-f)$, which basically is a conformal factor and $P > 0$ unless $f \to \infty$. One heuristic situation would then be to assume (★) $0 < \epsilon \le P(x)$ on $M$ (and since $\int exp(-f)dV = 1$ with $dV = \sqrt{|g|}\prod_1^3 dx^i$ we must then have $\epsilon\int dV \le 1$ or $vol(M) = \int_M dV \le (1/\epsilon)$). Then from (•) we have (••) $|B(S,S)| \ge \epsilon\|(\nabla S)^2\|$ and for any $\kappa > 0$ it follows that ($\bullet$) $|B(S,S)| + \kappa\|S\|^2_{L^2} \geq min(\epsilon,\kappa)\|S\|^2_{H^1}$. This means via Lax-Milgram that the equation

$$A(S) + \kappa S = -\frac{1}{m}\nabla(P\nabla S) + \kappa S = F = RP - \Delta P \tag{7.24}$$

has a unique weak solution $S \in H^1(M)$ for any $\kappa > 0$ (assuming $F \in L^2(M)$). Equivalently ($\blacklozenge\blacklozenge$) $-\frac{1}{m}[P\Delta S + (\nabla P)(\nabla S)] + \kappa S = F$ has a unique weak solution $S \in H^1(M)$. This is close but we cannot put $\kappa = 0$. A different approach following from remarks in [**424**], pp. 56-57, leads to a genuine weak solution of ($\mathfrak{R}$). Thus from a result of Yau [**897**] if $M$ is a complete simply connected 3-D differential manifold with sectional curvature $K < 0$ one has (cf. [**424**], p. 56)

$$\int_M |\psi| dV \leq 2\sqrt{-K}\int_M |\nabla\psi| dV \Rightarrow \int_M |\psi|^2 dV \leq c\int_M |\nabla\psi|^2 dV \tag{7.25}$$

Hence ($\blacklozenge$) holds and one has $\|\psi\|^2_{H^1} \leq (1+c)\|\nabla\psi\|^2$ so that ($\bigstar\bigstar$) $|B(S,S)| \geq \epsilon\|(\nabla S)^2\| \geq (c+1)^{-1}\epsilon\|S\|^2_{H^1}$ and this leads via Lax-Milgram again to

**THEOREM 7.1.1.** Let M be a complete simply connected 3-D differential manifold with sectional curvature $K < 0$. Then there exists a unique weak solution of (**1**$\mathfrak{R}$) in $H^1(M)$.

**REMARK 7.1.3.** One must keep in mind here that the metric is changing under the Ricci flow and assume that estimates involving e.g. $K$ are considered over some time interval. ■

**REMARK 7.1.4.** Some local results for (**1**$\mathfrak{R}$) could be generated by lemma 4.9, p. 65 in [**424**] for $M$ a complete 3-D differential manifolds with $Ric \geq kg$ for $k \in \mathbf{R}$ whose injectivity radii satisfy $inj \geq i$ for $i > 0$ (see [**218**] for a clear definition of cut locus and injectivity radius). ■

## 8. REMARKS ON THE FRIEDMAN EQUATIONS

In [**170, 172**] we indicated some connections between geometry, quantum mechanics, Ricci flow, and gravity. We began with the Perelman entropy functional (**1A**) $\mathfrak{F} = \int_M (R + |\nabla f|^2) exp(-f) dV$ and recalled that $\mathfrak{F}$ is in fact a Fisher information (following [**836**]). One shows in [**621, 836**] that the $L^2$ gradient flow of $\mathfrak{F}$ is determined by evolution equations

$$\partial_t g_{ij} = -2(R_{ij} + \nabla_i\nabla_j f);\ \partial_t f = -\Delta f - R \tag{8.1}$$

and this is equivalent to the decoupled family

$$\partial_t g_{ij} = -2R_{ij};\ \partial_t f = -\Delta f + |\nabla f|^2 - R \tag{8.2}$$

via a time dependent diffeomorphism (cf. also [**506, 507, 693**]). We note from [**836**] that $\mathfrak{F}$ is invariant under such diffeomorphisms and hence we could deal with (1.1) as a generalized Ricci flow if desired. Note the equations involving $\psi$ are independent of choosing (1.1) or (1.2) since $\mathfrak{F}$ is invariant and this accounts for using a Ricci flow (and (**1C**) below) to obtain $P_t = -\Delta P - RP$ leading to ($\bigstar$) $(1/m) div(P\nabla S) = \Delta P - RP$. By developments in [**170, 179**] going back to Santamato's discussion of the Schrödinger equation (SE) in a Weyl space (cf. [**170, 179, 191, 193, 770**]) and using the exact uncertainty principle of Hall-Reginatto (cf. [**170, 179, 171, 400**]) we could envision a Schrödinger wave function $\psi = |\psi|exp(iS/\hbar)$ with $P = |\psi|^2 \sim exp(-f)$ a probability density and (**1B**) $Q = -(\hbar^2/16m)[R + (8/\sqrt{P})\Delta\sqrt{P}]$ a quantum potential. Here $R \sim \dot{\mathcal{R}}$ is the Ricci curvature associated with the space metric $g_{ij}$ and $\vec{\phi} = -\nabla log(P)$ is the Weyl vector where $P \sim \hat{\rho} = \rho/\sqrt{|g|}$ for a matter density $\rho$ related to a mass $m$ (note $R$ will then go into the SE in a standard manner - cf. [**170**]). Then in [**172**] we recalled the flow equation (**1C**) $\partial_t f + \Delta f - |\nabla f|^2 + R = 0$ under Ricci flow with $\int_M exp(-f)dV = 1$. Independently we showed that $\int_M PQdV \sim \mathfrak{F}$ (where $Q = -(\hbar^2/2m)(\Delta|\psi|/|\psi|))$ with $S$ entering the picture via (**1D**) $\partial_t P + (1/m)div(P\nabla S) = 0$ leading to (**1E**) $\mathfrak{F} = \int_M (R + |\nabla f|^2)exp(-f)dV$ (studied in [**172**]). Thence via (**1C**) and $u = exp(-f) = P$ in the form (**1F**) $P_t = -\Delta P + RP$ (Ricci flow situation) we were led to (**1$\mathfrak{R}$**) of [**172**], namely $(1/m)div(P\nabla S) = \Delta P - RP$. The constructions were largely heuristic and preliminary in nature but they suggested a number of possibly interesting connections to eigenvalue problems and lower bounds on the first eigenvalue of the Laplacian for a Riemannian manifold $M$ (of three dimensions for convenience here). One could also consider generalized Ricci flow and $\partial_t f = -R - \Delta f$ but this seems awkward. There are many references for this latter topic and we mention in particular [**51, 52, 210, 211, 424, 425, 546, 551, 767, 768, 833, 897**] without specific attribution. In any event inequalities of the form (**1G**) $\int_M |\nabla\phi|^2 dV \geq \lambda_D \int |\phi|^2 dV$ hold for Dirichlet eigenfunctions $\phi$ of the Laplacian. More generally for functions in $H_0^1(M)$ with $M$ bounded and $\partial M$ reasonable (e.g. weakly convex) one expects (**1H**) $\int_M |\phi|^2 dV \leq c\int_M |\nabla\phi|^2 dV$ (cf. [**767**]) leading to $\|\phi\|^2_{H_0^1} \leq (1+c)\|\nabla\phi\|_{L^2}$ as in [**172**]. This in turn leads to solutions of (**1$\mathfrak{R}$**) derived under the assumption (**1H**) with $M$ bounded and $0 < \epsilon \leq P(x)$ on $M$. Thus $P \sim |\psi|^2 \geq \epsilon$ on $M$ and $S = 0$ on $\partial M$ which seems to involve a permissible quantum wave function $\psi = \sqrt{P}exp(iS/\hbar)$. One is also assuming some finite time interval (with time treated as a parameter in the $S$ equation).

Another feature of the Ricci flow is developed in [**378**] in the form of Ricci flow gravity. In [**378**] one develops the theory from a geometric Lagrangian (**1I**) $\mathfrak{L} = \Omega(R + \lambda(\nabla\phi)^2)$ in a "volumetric" manifold $M_4$ where $\Omega = exp(-\phi)\omega$ with $\omega \sim \sqrt{|g|}\prod_1^4 dx^i$. One notes here that this is precisely the Lagrangian for conformal general relativity (GR), with no ordinary mass term (cf. (**2D**)), and this suggests looking at $M_4 \sim M \times T$ with a line element $ds^2 = d\sigma^2 - dt^2$. We refer to [**170, 172**] for a discussion of Brans-Dicke (BD) theory and conformal GR. Here $\phi$ in (**1I**) plays the role of $f$ in (**1G**) so we can identify the Perelman functional $\mathfrak{F}$ with some kind of constrained conformal GR action. Further we showed in [**185**] that (in 4-D) this involves an integrable Weyl geometry with conformal mass $\hat{m} = exp(f/2)m$, which can be identified with a quantum mass $\mathfrak{M}$ in a corresponding Dirac-Weyl theory, as well as a corresponding quantum potential $\tilde{Q} = (\hbar^2/m^2)(\Box\sqrt{\tilde{\rho}}/\sqrt{\tilde{\rho}}$ $(c = 1)$ where $|\psi| \sim \tilde{\rho}$ (cf. [**170, 185, 797, 804**]). Here $\mathfrak{M} \sim \beta$ where $\beta$ is the Dirac field and $\mathfrak{M}^2/m^2 = exp(\tilde{Q})$ (cf. [**185, 642**]).

# 9. ENHANCEMENT

The theme of Ricci flow gravity in [**378**] has a number of attractive features but we would like to approach it eventually via techniques of Bohmian "quantum gravity" following [**170, 179, 185, 616, 765, 797, 798, 804**]. Thus a correct relativistic quantum equation of motion for a spin zero particle should have the form (**2A**) $\nabla^\mu S\nabla_\mu S = \mathfrak{M}^2c^2$ where $M^2 = m^2exp(\tilde{Q})$ ($\tilde{Q} = (\hbar^2/m^2c^2)(\Box|\psi|/|\psi|)$ with (**2B**) $\nabla_\mu(\rho\nabla^{mu}S) = 0$ where $\rho \sim |\psi|^2$ (see e.g. [**179, 170, 797**]). Writing (**2C**) $g_{\mu\nu} \to \tilde{g}_{\mu\nu} = (\mathfrak{M}^2/m^2)g_{\mu\nu}$ (conformal transformation) one arrives at

$$\tilde{g}^{\mu\nu}\tilde{\nabla}_\mu S\tilde{\nabla}_\nu S = m^2c^2;\ \tilde{g}^{\mu\nu}\tilde{\nabla}_\mu(\rho\tilde{\nabla}_\nu S) = 0 \tag{9.1}$$

This indicates that the presence of a quantum potential is equivalent to having a curved spacetime with (**2C**) and can be considered as a geometrization of the quantum effects of matter. In order to compare with [**378**] we go to [**797, 801, 804**] and look at a scalar-tensor theory with action

$$\mathfrak{A} = \int \sqrt{-g}d^4x\left[\phi R - \frac{w}{\phi}\nabla^\mu\phi\nabla_\mu\phi + 2\Lambda\phi\right] \tag{9.2}$$

with no ordinary matter Lagrangian. This is a Brans-Dicke form without $\mathfrak{L}_M$ and is equivalent to conformal GR in the form

$$\hat{\mathfrak{A}} = \int d^4x\sqrt{-\hat{g}}e^{-\psi}\left[\hat{R} - \left(\alpha - \frac{3}{2}\right)|\hat{\nabla}\psi|^2\right] \tag{9.3}$$

(cf. [**185**] - we omit $\Lambda$ for simplicity and think of $\hat{g}_{ab} = \Omega^2 g_{ab}$ for $g_{ab} \sim GR$ with $\Omega^2 = exp(\psi) = \phi = \hat{\phi}^{-1}$ and $[\alpha - (3/2)] \sim w$). Then $\hat{g}_{ab} \sim (\mathfrak{M}^2/m^2)g_{ab}$ is involved as in (**2C**) and (**2D**) $\mathfrak{L} = exp(-\psi)\sqrt{-\hat{g}}\prod_1^4 dx^i(\hat{R} - w|\nabla\psi|^2)$ is exactly the $\mathfrak{L}$ of Graf in (**1I**) with $\lambda \sim -w$, $\psi \sim \phi$, and $\hat{g} \sim g$.

**REMARK 9.1.1** The field equations arising from (9.2) (with $\Lambda = 0$) in [**797, 798**] are

$$R + \frac{2w}{\phi}\Box\phi + \frac{w}{\phi^2}\nabla^\mu\phi\nabla_\mu\phi = 0;\ G^{\mu\nu} \tag{9.4}$$

$$= -\frac{1}{\phi^2}T^{\mu\nu} - \frac{1}{\phi}\left[\nabla^\mu\nabla^\nu - g^{\mu\nu}\Box\right]\phi + \frac{w}{\phi^2}\nabla^\mu\phi\nabla^\nu\phi - \frac{w}{2\phi^2}g^{\mu\nu}\nabla^\alpha\phi\nabla_\alpha\phi$$

where $G_{\mu\nu} = R_{\mu\nu} - (1/2)Rg_{\mu\nu}$ and $T^{\mu\nu} = 0$ in the absence of ordinary matter. However if we want a quantum contribution in this situation the matter must be introduced and in [**797, 804**] this is done by assuming a matter Lagrangian (**2E**) $L_M = (\rho/m)\nabla_\mu S\nabla^\mu S - \rho m$ with interaction between the scalar field $\phi$ and the matter field expressed via (**2F**) $L'_M = (\rho/m)\phi^\alpha\nabla^\mu S\nabla_\mu S - m\rho\phi^\beta - \Lambda(1+Q)^\gamma$ (with suitable choice of $\alpha$, $\beta$, $\gamma$ and $1 + Q \sim exp(Q)$); this leads to a theory incorporating both quantum and gravitational effects of matter. However one can apparently examine quantum effects directly via conformal transformation with $\Omega^2 = \mathfrak{M}^2/m^2 = exp(\tilde{Q})$ and the use of Weyl-Dirac theory with the natural Dirac field $\beta \sim \mathfrak{M}$ (cf. [**170, 179, 797**]). ■

Thus we recall that conformal GR (without $L_M$) is basically equivalent to Weyl-Dirac theory as shown in [**172**] with conformal mass $\hat{m}$ corresponding to

$\beta = \mathfrak{M}$ where $\mathfrak{M}$ is in turn generated by a "classical" mass $m$ with $\mathfrak{M}^2 = m^2 exp(\tilde{Q})$ and $\tilde{Q} = (\hbar^2/m^2c^2)(\Box\sqrt{\rho}/\sqrt{\rho})$. In fact this could serve as a definition of $\rho$ via $(\sqrt{\rho} = \chi)$

$$\text{(9.5)} \qquad \frac{\mathfrak{M}^2}{m^2} = exp(\tilde{Q}) \Rightarrow \tilde{Q} = 2log(\mathfrak{M}/m)$$

$$= \frac{\hbar^2}{m^2c^2}\frac{\Box\chi}{\chi} \Rightarrow 2\chi log(\beta/m) = \frac{\hbar^2}{m^2c^2}\Box(\chi)$$

and one could speculate that this approach allow one to geometrize mass in treating it only via its quantum effect. However it seems more natural to include a matter Lagrangian $L_M$ as well. Recall also the Weyl vector $w_\mu = 2\partial_\mu log(\beta)$ and in fact (cf. [**172**])

$$\text{(9.6)} \qquad \hat{\mathfrak{A}} = \int d^4x\sqrt{-\hat{g}}e^{-\tilde{Q}}\left[\hat{R} - \left(\alpha - \frac{3}{2}\right)(\hat{\nabla}\tilde{Q})^2 + 16\pi e^{-\tilde{Q}}L_M\right]$$

This couples the ordinary and quantum effects of matter directly and gives $\tilde{Q}$ a role as a scalar field (cf. [**797**] where $Q$ is made dynamical in a different manner).

Now in [**378**] one computes from (**1I**) the terms (**2G**) $(\delta\mathfrak{L}/\delta g_{ik}) \sim P_{ik}$ with $(\delta\mathfrak{L}/\delta\phi) \sim \mathcal{Q}$ and writes a Nöther identity

$$\text{(9.7)} \qquad div(\tilde{P}^i_k\xi^k) = P^{ik}L_\xi g_{ik} + \mathcal{Q}L_\xi\phi$$

followed by a discussion of various situations. Such an approach is developed systematically for variational frameworks in e.g. [**243, 298, 345, 346, 519, 775, 780**] and some concrete examples are given in [**159, 616**] for FRW spaces. The FRW framework is very helpful in general and we mention a few details following [**16, 157, 295, 388**]. Thus in general one has a metric (**2H**) $ds^2 = -dt^2 + a^2(t)\gamma_{ij}dx^i dx^j$ with $\gamma_{ij} = [dr^2/(1-kr^2)] + r^2 d\Omega_2^2$ in 4-D, where $d\Omega_2^2$ is the metric for a 2-D sphere with unit radius. If $R_{ij}$ is the Ricci tensor for $\gamma_{ij}$ then (**2I**) $R_{ij} = 2k\gamma_{ij}$ and $k = 1, 0, -1$ corresponding to positive, zero, or negative curvature for the 3-D hypersurface (note $d\Omega_2^2 \sim d\theta^2 + Sin^2(\theta)d\phi^2$).

The Jordan-Brans-Dicke action (with $\Lambda = 0$) is then given by (9.2) with $L_M = \int d^4x\sqrt{-g}L_M$ added and the equations of motion can be developed as follows (cf. [**157, 295**]). First $\phi$ is to depend only on $t$ so $\nabla^c\phi\nabla_c\phi = -(\dot{\phi})^2$ and

$$\text{(9.8)} \qquad \Box\phi = -(\ddot{\phi} + 3H\dot{\phi}) = -\frac{1}{a^3}\frac{d}{dt}(a^3\dot{\phi})$$

where $H = (\dot{a}/a)$ is the Hubble parameter. Further

$$\text{(9.9)} \qquad T_{ab} = -\frac{2}{\sqrt{-g}}\frac{\delta}{\delta g^{ab}}(\sqrt{-g}L_M) = (P+\rho)u_a u_b + Pg_{ab};\ T = 3P - \rho$$

where $\rho =$ energy density and $P$ is the pressure of a "cosmic fluid" moving with velocity $u^\mu$. The field equations are

$$\text{(9.10)} \qquad G_{ab} = \frac{8\pi}{\phi}T_{ab} + \frac{w}{\phi^2}\left(\nabla_a\phi\nabla_b\phi - \frac{1}{2}g_{ab}\nabla^c\phi\nabla_c\phi\right)$$

$$+\frac{1}{\phi}(\nabla_a\phi\nabla_b\phi - g_{ab}\Box\phi) - \frac{V}{2\phi}g_{ab}$$

and

$$\frac{2w}{\phi}\Box\phi + R - \frac{w}{\phi^2}\nabla^c\phi\nabla_c\phi - \frac{dV}{d\phi} = 0 \tag{9.11}$$

One obtains then from (9.10)

$$R = -\frac{8\pi T}{\phi} + \frac{w}{\phi^2}\nabla^c\phi\nabla_c\phi + \frac{3\Box\phi}{\phi} + \frac{2V}{\phi} \tag{9.12}$$

which implies (using $R = 6[\dot{H} + 2H^2 + (k/a^2)]$)

$$\dot{H} + 2H^2 + \frac{k}{a^2} = -\frac{4\pi T}{3\phi} - \frac{w}{6}\left(\frac{\dot{\phi}}{\phi}\right)^2 + \frac{1}{2}\frac{\Box\phi}{\phi} + \frac{V}{3\phi} \tag{9.13}$$

Further from (9.10) one has

$$H^2 = \frac{8\pi}{3\phi}\rho + \frac{w}{6}\left(\frac{\dot{\phi}}{\phi}\right)^2 - H\frac{\dot{\phi}}{\phi} - \frac{k}{a^2} + \frac{V}{6\phi} \tag{9.14}$$

which provides a first integral. Further from (9.11) and (9.12) follows also

$$\Box\phi = \frac{1}{2w+3}\left[8\pi T + \phi\frac{dV}{d\phi} - 2V\right] \tag{9.15}$$

and putting together (9.14) and (9.15) with $T = 3P - \rho$ yields then

$$\dot{H} = \frac{-8\pi}{(2w+3)\phi}[(w+2)\rho + wP] - \frac{w}{2}\left(\frac{\dot{\phi}}{\phi}\right)^2 \tag{9.16}$$

$$+2H\frac{\dot{\phi}}{\phi} + \frac{k}{a^2} + \frac{1}{2(2+3)\phi}\left(\phi\frac{dV}{d\phi} - 2V\right)$$

Moreover from (9.15)

$$\ddot{\phi} + 3H\dot{\phi} = \frac{1}{2w+3}\left[8\pi(\rho - 3P) - \phi\frac{dV}{d\phi} + 2V\right] \tag{9.17}$$

A remark might be appropriate here.

**REMARK 9.1.2.** In [**157**] one uses units with $8\pi G = 1 = c = \hbar$ and instead of a standard $L_M$ an inflaton field Lagrangian $L(\psi) = (1/2)\partial_\mu\psi\partial^\mu\psi - U(\psi)$ is used (also $k = 1$ is assumed). This leads to a conservation equation (**2J**) $\dot{\rho}+3H(\rho+P) = 0 \equiv \ddot{\psi} + 3H\dot{\psi} = -\partial_\psi U$ with (**2K**) $\rho = (1/2)\dot{\psi}^2 + U(\psi)$ and $P = (1/2)\dot{\psi}^2 - U(\psi)$.∎

What we want now is the Ricci curvature determined by the space metric $\gamma_{ij}$ as in (**2H**) and we refer to [**188, 353, 745**] for input. In [**185**] for example one uses a line element

$$ds^2 = -dt^2 + a^2(t)\left[\frac{dr^2}{1-\kappa r^2} + r^2 d\Omega^2\right] \tag{9.18}$$

Here $\kappa$ can take any value but it is related to $(+,0,-)$ curvatures according to sign. One sets $\dot{a} = da/dt$ and computes Christoffel symbols to be

$$\Gamma^0_{11} = \frac{a\dot{a}}{1-\kappa r^2};\ \Gamma^1_{11} = \frac{\kappa r}{1-\kappa r^2};\ \Gamma^0_{22} = a\dot{a}r^2;\ \Gamma^0_{33} = a\dot{a}r^2 Sin^2(\theta); \tag{9.19}$$

$$\Gamma^1_{01} = \Gamma^2_{02};\ \Gamma^3_{03} = \frac{\dot{a}}{a};\ \Gamma^1_{22} = -r(1-\kappa r^2);\ \Gamma^1_{33} = -r(1-\kappa r^2)Sin^2(\theta);$$

$$\Gamma^2_{12} = \Gamma^3_{13} = \frac{1}{r};\ \Gamma^2_{33} = -Sin(\theta)Cos(\theta);\ \Gamma^3_{23} = Ctn(\theta)$$

leading to

$$R_{00} = -3\frac{\ddot{a}}{a};\ R_{11} = \frac{a\ddot{a} + 2\dot{a}^2 + 2\kappa}{1-\kappa r^2}; \tag{9.20}$$

$$R_{22} = r^2[a\ddot{a} + 2\dot{a}^2 + 2\kappa];\ R_{33} = r^2[a\ddot{a} + 2\dot{a}^2 + 2\kappa]Sin^2(\theta)$$

with Ricci scalar

$$^4R = 6\left[\frac{\ddot{a}}{a} + \left(\frac{\dot{a}}{a}\right)^2 + \frac{\kappa}{a^2}\right] = {}^3R + 3\frac{\ddot{a}}{a} \tag{9.21}$$

Thus if we write now $g_{00} = -1$ with $\tilde{g}_{ii}$ given via (9.19) we consider Ricci flow for the space metric $\tilde{g}_{ij} = a^2(t)\gamma_{ii}$ in the form $\partial_t\tilde{g}_{ii} = -2\,{}^3R_{ii}$ $(i = 1,2,3)$ to arrive at **(2L)** $2a\dot{a}\gamma_{ii} = -2^3R_{ii}$. Since $\gamma^{ii}\gamma_{ii} = 3$ we obtain **(2M)** $6a\dot{a} = -2\,{}^3R$ which can be written as

$$6a\dot{a} = -2\left[3\frac{\ddot{a}}{a} + 6\left(\frac{\dot{a}}{a}\right)^2 + \frac{6\kappa}{a^2}\right] \tag{9.22}$$

**PROPOSITION** 9.1. Ricci flow in the form $\partial_t\tilde{g}_{ii} = -2\,{}^3R_{ii}$ with $\tilde{g}_{ij} = a^2(t)\gamma_{ij}$ as in (9.18) leads to (9.22) as a stipulation about $a(t)$.

Now to involve the Einstein equations $R_{\mu\nu} = 8\pi G(T_{\mu\nu} - (1/2)g_{\mu\nu}T)$ the $\mu\nu = 00$ equation gives **(2N)** $-3(\ddot{a}/a) = 4\pi G(\rho + 3P)$ and the $\mu\nu$ equations give

$$\frac{\ddot{a}}{a} + 2\left(\frac{\dot{a}}{a}\right)^2 + 2\frac{\kappa}{a^2} = 4\pi G(\rho - P) \tag{9.23}$$

Due to isotropy there is only one distinct equation from $\mu\nu = ij$ and one obtains

$$\left(\frac{\dot{a}}{a}\right)^2 = \frac{8\pi G}{3}\rho - \frac{\kappa}{a^2};\ \frac{\ddot{a}}{a} = -\frac{4\pi G}{3}(\rho + 3P) \tag{9.24}$$

which are known as the Friedman equations (note $P < -(1/3)\rho$ implies repulsive gravitation). Recalling the Hubble parameter $H = \dot{a}/a$ these can be written as

$$H^2 = \frac{8\pi G}{3}\rho - \frac{\kappa}{a^2};\ \dot{H} = -4\pi G(\rho + P) + \frac{\kappa}{a^2} \tag{9.25}$$

We will discuss (9.22) and (9.24) in Section 3.

**REMARK 9.1.3.** From **[388]** we note that for a universe in a state of accelerated expansion ($\ddot{a} > 0$) the second Friedman equation in (9.24) implies $\rho + 3P < 0$ and for a vacuum $P = -\rho$ so $\rho + 3P = -2\rho < 0$ (repulsive gravity).

The Friedman equations with a cosmological constant for a homogeneous isotropic universe models with pressure free matter are

$$\frac{\ddot{a}}{a} = \frac{\Lambda}{3} - \frac{4\pi G}{3}\rho;\ H^2 = \left(\frac{\dot{a}}{a}\right)^2 = \frac{\Lambda}{3} - \frac{k}{a^2} + \frac{8\pi G}{3}\rho \tag{9.26}$$

(see Section 3 for more on this). ∎

**REMARK 9.1.4.** In [**509**], p. 210, one has $ds^2 = -N^2(t)dt^2 + a^2(t)d\Omega_3^2$ with $d\Omega_3^2 = d\chi^2 + Sin^2(\chi)(d\theta^2 + Sin^2(\theta)d\phi^2)$, second fundamental form (**2O**) $K_{ab} = -(1/2N)(\partial h_{ab}/\partial t) = -(\dot{a}/aN)h_{ab}$, and $K = K_{ab}h^{ab} = -3\dot{a}/aN$ (cf. also [**353**]). Given the EH action

$$S_{EH} = \frac{c^4}{16\pi G}\int_M d^4x\sqrt{-g}(R - 2\Lambda) \pm \frac{c^4}{8\pi G}\int_{\partial M} d^3x\sqrt{h}K \tag{9.27}$$

we obtain for the surface term (**2P**) $2\int d^3xK\sqrt{h} = -6\int d^3x(\dot{a}/Na)\sqrt{h}$ with $\sqrt{h} = a^3Sin^2(\chi)Sin(\theta)$. This surface term is cancelled after partial integration and one obtains (setting $2G/3\pi = 1$)

$$S_g = \frac{1}{2}\int dtN\left(-\frac{a\dot{a}^2}{N^2} + a - \frac{\Lambda a^3}{3}\right) \tag{9.28}$$

The matter action is given by (after a rescaling $\phi \to \phi/\sqrt{2}\pi$) a formula (**2Q**) $S_m = (1/2)\int dtNa^3[(\dot{\phi}^2/N^2) - m^2\phi]$ leading to a minisuperspace action ($q^1 \sim a$ and $q^2 \sim \phi$)

$$S = S_q + S_m = \int dtN\left[\frac{1}{2}G_{AB}\frac{\dot{q}^A\dot{q}^B}{N^2} - V(q)\right];\ G_{AB} = \begin{pmatrix} -a & 0 \\ 0 & q^3 \end{pmatrix} \tag{9.29}$$

with $\sqrt{-G} = a^2$. Following general techniques (cf. [**509**]) this leads to a Wheeler-deWitt (WDW) equation

$$\frac{1}{2}\left(\frac{\hbar^2}{a^2}\partial_a(a\partial_a) - \frac{\hbar^2}{a^3}\frac{\partial^2}{\partial\phi^2} - a + \frac{\Lambda a^3}{3} + m^2a^3\phi^2\right)\psi(a,\phi) = 0 \tag{9.30}$$

and this has a simpler form upon writing $\alpha = log(a)$, namely

$$\frac{e^{-3\alpha}}{2}\left(\hbar^2\frac{\partial^2}{\partial\alpha^2} - \hbar^2\frac{\partial^2}{\partial\phi^2} - e^{4\alpha} + e^{6\alpha}\left[m^2\phi^2 + \frac{\Lambda}{3}\right]\right)\psi = 0 \tag{9.31}$$

This has the form of a Klein-Gordon equation with potential $V(\alpha,\phi) = -exp(4\alpha) + exp(6\alpha)[m^2\phi^2 + (\Lambda/3)]$. ∎

## 10. SOME HEURISTIC RESULTS

For general background we refer to [**188, 225, 388, 418, 743, 860, 864**] and for Bohmian aspects see [**170, 179, 172, 185, 616, 642, 765, 797, 798, 801, 804**]. Some of this is mentioned in Section 2 and we add here first a few comments.

The two most popular examples of cosmological fluids are known as matter and radiation. Matter involves collisionless, nonrelativistic particles having essentially zero pressure $P_M = 0$ (e.g. stars and galaxies or dust); the energy density in matter falls off as (**3A**) $\rho_M \propto a^{-3}$. For electromagnetic energy one has (**3B**) $T^{\mu\nu} = F^{\mu\lambda}F^{\nu}_{\lambda} - (1/4)g^{\mu\nu}F^{\lambda\sigma}F_{\lambda\sigma}$ with trace 0; since this must equal $T = -\rho + 3P$ one has $P_R = (1/3)\rho_R$. A universe dominated by radiation is then postulated to have (**3C**) $\rho_R \propto a^{-4}$; for a perfect fluid with $p_\Lambda = -\rho_\Lambda$ one assumes $\rho_\Lambda \propto a^0$ (vacuum dominated).

There is now a density parameter (**3D**) $\Omega = (8\pi G/3H^2)\rho = \rho/\rho_{crit}$ and from (2.24) one has (**3E**) $\Omega - 1 = \kappa/H^2a^2$. Thus the sign of $\kappa$ is determined via

(1) $\rho < \rho_{crit} \sim \Omega < 1 \sim \kappa < 0$ (*open*)
(2) $\rho = \rho_{crit} \sim \Omega = 1 \sim \kappa = 0$ (*flat*)
(3) $\rho > \rho_{crit} \sim \Omega > 1 \sim \kappa > 0$ (*closed*)

We refer to [**185**] for more discussion.

**REMARK 10.1.1.** We gather here a few observations from [**388**]. Note first from the Friedman equations that one can write (**3F**) $\dot{\rho} + 3(\dot{a}/a)(\rho + P) = 0 \equiv (d/dt)(\rho a^3) + P(d/dt)a^3 = 0$. Let then $V \sim a^3$ correspond to the volume of an expanding region together with the "cosmic fluid" (a comoving volume). The energy is then $U = \rho a^2$ so that (**3F**) becomes $dU + PdV = 0$. The first law of thermodynamics states that (**3G**) $TdS = dU + PdV$ for a fluid in equilibrium and here $dS = 0$ corresponds to an adiabatic process. If in addition $P = w\rho$ one obtains from (**3F**) the relation (**3H**) $(d/dt)(\rho a^3) + w\rho(d/dt)a^3 = 0 \Rightarrow \rho a^{3(w+1)} = \rho_0$. Lorentz invariant vacuum energy (LIVE) involves $w = -1$ whereas for dust $w = 0$ and radiation involves $w = 1/3$. For $P = w\rho$ the first Friedman equation becomes

$$\left(\frac{\dot{a}}{a}\right)^2 = \frac{8\pi G}{3}\frac{\rho_0}{a^{3(1+w)}} - \frac{\kappa}{a^2} \tag{10.1}$$

and we refer to [**388**] for a discussion of phenomena depending on $w > -(1/3)$, $w < -(1/3)$, or $w = -1/3$ (as well as many other matters). ■

**REMARK 10.1.2.** From [**388**] again the Einstein equations with a cosmological constant are

$$R_{\mu\nu} - \frac{1}{2}g_{\mu\nu}R + \Lambda g_{\mu\nu} = 8\pi G T_{\mu\nu} \tag{10.2}$$

and for a vacuum ($T_{\mu\nu} = 0$) a homogeneous isotropic model with $\Lambda > 0$ involves equations

$$3\frac{\dot{a}^2 + k}{a^2} - \Lambda = 0; \; -2\frac{\ddot{a}}{a} - \frac{\dot{a}^2 + k}{a^2} + \Lambda = 0 \tag{10.3}$$

(this is just (2.26) rewritten for $\rho = 0$). The first equation is $\dot{a}^2 - \omega^2 a^2 = -k$ with solutions

$$a(t) = \begin{cases} \frac{\sqrt{k}}{\omega}Cosh(\omega t) & (k > 0) \\ exp(\omega t) & (k = 0) \\ \frac{\sqrt{|k|}}{\omega}Sinh(\omega t) & (k < 0) \end{cases} \tag{10.4}$$

This gives a deSitter model and it is instructive to draw graphs for (3.20). For $k = 0$ one has $H = (\dot{a}/a) = \omega$ and $a^2 = exp(2Ht)$ with $ds^2 = -dt^2 + exp(2Ht)(dx^2 + dy^2 + dz^2)$. We refer to [**388**] for further discussion about event and particle horizons (cf. also [**225, 418, 743**] for information about such horizons). ■

Consider now Ricci flow via (2.22) in conjunction with the Friedman equations (2.24). One obtains first

$$3\frac{\ddot{a}}{a}+6\left(\frac{\dot{a}}{a}\right)^2=-4\pi G(\rho+3P)+6\left[\frac{8\pi G}{3}\rho-\frac{\kappa}{a^2}\right]=12\pi G(\rho-P)-\frac{6\kappa}{a^2} \tag{10.5}$$

which means that

**PROPOSITION** 10.1. Ricci flow plus the Friedman equations (2.24) implies

$$6a\dot{a}=-2[12\pi G(\rho-P)]\Rightarrow\frac{d}{dt}a^2=8\pi G(P-\rho) \tag{10.6}$$

(independently of $\kappa$).

**REMARK 10.1.3.** We assume here that the Ricci flow persists in time, ignoring much of the Hamilton-Perelman analysis (cf. [**218, 506, 507, 621, 693, 836**]). Then (3.6) says that $a^2$ is increasing (resp. decreasing) for $P>\rho$ (resp. $P<\rho$). This gives a somewhat different perspective in comparison with that arising from the Friedman equations (2.24). We note that (2.24) (or (2.26)) do not specify a sign for $\dot{a}$. In particular for $P=0$ or $P=-\rho$ (3.6) implies that $\dot{a}<0$ for $\rho>0$. ■

**REMARK 10.1.4.** From (2.25) (resp. (2.26)) in order to have $H^2\geq 0$ one requires

$$\frac{8\pi G}{3}\rho\geq\frac{\kappa}{a^2}\ \mathit{or}\ \frac{8\pi G}{3}+\frac{\Lambda}{3}\geq\frac{\kappa}{a^2} \tag{10.7}$$

We have then from (2.26) and (3.7)

$$2\frac{\ddot{a}}{a}=\frac{2\Lambda}{3}-\frac{8\pi G}{3}\rho\geq\frac{2\Lambda}{3}-\left(\frac{\kappa}{a^2}-\frac{\Lambda}{3}\right)\geq\Lambda-\frac{\kappa}{a^2} \tag{10.8}$$

which seems to predict repulsive gravity for $\Lambda-(\kappa/a^2)>0$. ■

One can of course ask for an existence (uniqueness) theorem for the ordinary differential equation (ODE) (2.22) on some time interval. This can be rewritten in terms of H via $\dot{H}=(\ddot{a}/a)-H^2$ to obtain

$$\dot{H}+3H^2+\frac{2\kappa}{a^2}-a^2H=0 \tag{10.9}$$

Set then $\chi=a^3$ and $\tau=3t$ so that $H=\chi_\tau/\chi$ (note $\chi_\tau=(1/3)\dot{\chi}=(1/3)3a\dot{a}$ and $\chi_\tau/\tau=(a^2\dot{a}/a^3)=H$). Then (3.9) becomes a Ricatti type equation (note $\dot{H}=3H_\tau$)

$$H_\tau+H^2+\frac{2\kappa}{3a^2}-\frac{a^2}{3}H=0\equiv\chi_{\tau\tau}-\frac{a^2}{3}\chi_\tau+\frac{2\kappa}{3a^2}\chi=0 \tag{10.10}$$

Consequently

$$\left[\chi_\tau e^{-\int_0^\tau(a^2/3)ds}\right]_\tau+\frac{2\kappa}{3a^2}\chi=0\Rightarrow\chi_\tau=e^{-\int_0^\tau ds}\left[c+\frac{2\kappa}{3}\int_0^\tau\frac{\chi}{a^2}ds\right] \tag{10.11}$$

$(c \sim \chi_\tau(0))$ leading to

$$2a^2 a_\tau = e^{-\int_0^\tau (a^2/3)ds} \left[ c + \frac{2\kappa}{3} \int_0^\tau ads \right] \tag{10.12}$$

and consequently

$$a^3 = a_0^3 + \int_0^T e^{-\int_0^\alpha (a^2/3)ds} \left[ c + \frac{2\kappa}{3} \int_0^\alpha ads \right] d\alpha \tag{10.13}$$

**PROPOSITION** 10.2. Equation (3.12) is an ODE of the form **(3I)** $a_\tau = J(a, \tau)$ with $J$ Lipschitz in any region $0 < \epsilon \leq a \leq A$ (and say $0 \leq \tau \leq T$). Hence via standard theorems (see e.g. [**184**]) there is a unique local solution.

There is then a question of the behavior of $a$ as a solution of the integrodifferential equation (3.13) or the ODE (3.12) and this will depend in particular on the signs of $c$ and $\kappa$.

CHAPTER 3

# ASPECTS OF EMERGENCE

## 1. INTRODUCTION

There have been many papers and books written on the idea of emergence related to classical and quantum mechanics, going in both directions, and this is discussed already at some length in Chapters 1 and 2. We have seen how a quantum potential $Q$ arises in the Bohmian context which can be modeled on momentum perturbations $\delta p = \nabla P/P$ where $P$ is a probability density ($P = R^2$ for a wave function $\psi = Rexp(iS/\hbar)$. In the end however classical quantum mechanics (QM) appears as a magic machine based on Hilbert space and operator algebras where the significant results arise almost trivially from the mathematics. This led e.g. to non-commutative geometry and the magnificent edifice connecting the standard model in physics to mathematics developed by A. Connes and collaborators (cf. [**175, 231, 233, 505, 569**]). This cannot be the whole story however since e.g. we have also looked at statistics and thermodynamics and their connections to QM and we would like to understand better the general relations among various types of perturbations. We do not enter here into questions of the emergence of practically everything from something else but restrict attention to topics in mathematical physics and in this direction mention e.g. [**2, 101, 105, 109, 116, 121, 129, 161, 166, 167, 170, 171, 172, 173, 222, 246, 290, 307, 318, 319, 332, 333, 342, 360, 369, 371, 372, 373, 381, 392, 400, 401, 402, 435, 436, 440, 444, 445, 446, 456, 457, 458, 459, 460, 517, 535, 547, 577, 628, 630, 632, 643, 650, 654, 736, 756, 757, 758, 759, 795, 815, 838, 843, 855, 873, 885, 901, 911**]. We mainly consider a theory as emergent if it contains or reduces to another theory in a significant manner or if its laws are tied to those of another theory via mathematical connections. As an example one shows in [**757, 758**] (cf. also [**173, 215, 560**]) that a "stable" Hamiltonian system generates a quantum potential like object $Q$ involving $\psi = Aexp(ikS)$ with

$$Q = -\frac{1}{2mk^2}\frac{\Delta A}{A};\ \frac{i}{k}\partial_t\psi = -\frac{1}{2k^2}\sum \partial_i(g_{ij}\partial_j\psi) + U\psi \tag{1.1}$$

where $\partial_i \sim \partial/\partial q_i$. Thus starting from a stable Hamiltonian system

$$\dot{q}_i = \frac{\partial H}{\partial p_i};\ \dot{p}_i = -\frac{\partial H}{\partial q_i} \tag{1.2}$$

(not a priori related to a SE) one arrives at a SE (for $k = 1/\hbar$)

$$i\hbar\partial_t\psi = -\left(\frac{\hbar^2}{2m}\right)\Delta\psi + U\psi \tag{1.3}$$

and a quantum mechanical context. The machinery here uses Poincaré-Chetaev equations, Lyapunov stability ideas, and a somewhat heuristic "least action principle" attributed to Chetaev (which is not clearly stated but looks hopeful).

We also want to mention several other points of view involving classical and quantum systems:

(1) The approach via Lax-Phillips theory (see e.g. [**72, 85, 285, 309, 449, 450, 451, 540, 830**]).
(2) Ideas based on Hooft's approach (see e.g. [**105, 106, 444**] and remarks below).
(3) Further aspects of Bohmian approaches (see e.g. [**132, 133, 170, 171, 172, 173, 222, 274, 342, 403, 436, 440, 635, 654, 815, 838**]).
(4) Phase space methods (see e.g. [**273, 901**]).

Perhaps the most "intuitive" picture comes from [**444, 445, 446**] (cf. also [**105, 107, 106, 290**]) where, with apologies to 't Hooft for brevity, in [**444**] for example it is suggested that the ontological quantum states arise from equivalence classes of deterministic paths with energy eigenstates corresponding to limit cycles of the deterministic model; periodicity of deterministic subsystems is connected to information loss and the ensuing theory ensures that the Hamiltonian operator $\hat{H}$ be bounded below. We note how this hypothesis fits in with the Gutzwiller trace formula and we will return to this in Chapter 6. We should add that almost everything in physics seems to be emergent from or related to information theory in some manner; we omit any sort of complete referencing but mention e.g. [**109, 129, 146, 147, 170, 172, 173, 194, 195, 197, 318, 319, 332, 333, 400, 401, 402, 403, 547, 911**].

There is still another "emergence" with origins in string theory and noncommutative (NC) geometry which leads from quantum fields in a NC space-time to gravity (see e.g. [**69, 270, 493, 494, 495, 554, 566, 622, 744, 788, 821, 824, 893**]). This material seems to involve a virtual paradigm shift and we will sketch some details in Section 3 following [**893, 894**]. Here NC field theory is the formulation of QFT on NC space-time where (**1A**) $[y^a, y^b]_* = i\theta^{ab}$ which arises from introducing a symplectic structure (**1B**) $B = (1/2)B_{ab}dy^a \wedge dy^b$ and then quantizing with Poisson structure $\theta^{ab} = (B^{-1})^{ab}$ to obtain a quantum phase space. Translation in NC directions involves inner automorphisms of an NC $C^*$-algebra $\mathfrak{A}_\theta$, i.e. (**1C**) $exp(ik\cdot y) * \hat{f}(y) * exp(-ik\cdot y) = \hat{f}(y+\theta\cdot k)$ for $\hat{f}(y) \in \mathfrak{A}_\theta$ or equivalently $[y^a, \hat{f}(y)]_* = i\theta^{ab}\partial_b\hat{f}(y)$. Then, from string theory, when a D-brane world volume $M$ supports a symplectic structure $B$ the Darboux theorem provides local symplectic coordinates and the electromagnetic (EM) forces can be locally "eliminated". This provides a form of equivalence principle for geometrization of EM and is the crux of an emergent gravity (and "Einstein" equations) arising from the NC gauge fields (see Section 4 for more detail). Further background information also appears later.

## 2. SOME CLASSICAL QUANTUM RELATIONS

We indicate here some relations between quantum and classical mechanics and connections to Bohmian mechanics (cf. [**116, 132, 133, 170, 403, 635**] and see also [**318, 319, 332, 333, 342, 392, 400, 401, 402, 403, 440, 628, 630, 635, 650, 654, 699, 736, 815, 838, 843, 890**]). We begin with [**635**] where it is shown that one can consider classical mechanics (CM) as probabilistic and/or quantum mechanics (QM) as deterministic. This can be reconciled e.g. by considering both from a Bohmian point of view. To see this just look at the standard Schrödinger equation (SE)

$$\left(\frac{\hat{\mathbf{p}}^2}{2m}+V(\mathbf{x},t)\right)\psi(\mathbf{x},t)=i\hbar\partial_t\psi(\mathbf{x},t);\ \hat{\mathbf{p}}=-i\hbar\nabla;\ \psi=Re^{iS/\hbar} \tag{2.1}$$

leading to (cf. [**170, 436, 440, 635**])

$$\frac{|\nabla S|^2}{2m}+V+Q=-\partial_t S;\ \partial_t\rho+\nabla\left(\rho\frac{\nabla S}{m}\right)=0; \tag{2.2}$$

$$\mathbf{v}=\frac{d\mathbf{x}}{dt}=\frac{1}{m}\nabla S;\ \rho=R^2;\ Q=-\frac{\hbar^2}{2m}\frac{\nabla^2|\psi|}{|\psi|}$$

Thus all QM uncertainties arise from lack of knowledge of $\mathbf{x}(t_0)$. It is then immediate that for a statistical ensemble $\rho=R^2$ with $\psi=Rexp(iS/\hbar)$ the nonlinear SE

$$\left(\frac{\hat{\mathbf{p}}^2}{2m}+V-Q\right)\psi=i\hbar\partial_t\psi \tag{2.3}$$

corresponds to the standard conservation equation

$$\partial_t\rho+\nabla\left(\rho\frac{\nabla S}{m}\right)=0\equiv\left[\partial_t+\left(\frac{\nabla S}{m}\right)\nabla+\left(\frac{\nabla^2 S}{2m}\right)\right]R=0 \tag{2.4}$$

along with

$$\frac{1}{2m}|\nabla S|^2+V=-\partial_t S \tag{2.5}$$

where the insertion of the quantum object $\hbar$ (via $Q$) corresponds to Bohr quantization (**2A**) $mvr=n\hbar$ when $\psi$ is required to be single valued. In a situation where $S=s-Et$ for example one has $\nabla S=\nabla s$ and $(1/2m)|\nabla s|^2+V=E$ but this does not lead immediately to a single nonlinear equation for $R$. Problems of measurement for nonlinear QM based on (2.3) are discussed in [**635**] along with the emergence of classical physics (statistics and trajectories). One argues in paper 1 of [**635**] that all measurable properties of CM can be predicted by the QM based on the nonlinear SE (2.3). In paper 2 the arguments are expanded and various conceptual choices are indicated, one of which is to accept the Bohmian deterministic interpretation of QM (cf. also [**682**]). The discussions in [**635**] are quite profound and worth reading from many points of view.

**REMARK 2.1.1.** Going to the situation of [**757, 758**], studied also in [**173**], given a Hamiltonian system

$$\frac{dq_j}{dt} = \frac{\partial H}{\partial p_j};\ \frac{dp_j}{dt} = -\frac{\partial H}{\partial q_j} \tag{2.6}$$

we note that there is no a priori probability distribution present. Hence there is no direct argument based on [**635**] to establish a quantum framework. However once a quantum potential $Q = -(\hbar^2/2m)(\Delta A/A)$ has been produced as in [**757, 758**] based on a stability analysis for the perturbed system one arrives at a Bohmian formulation for the system. In other words the stability analysis of a stable classical Hamilonian system (with potential $V(x)$) leads to a Bohmian Hamilton-Jacobi equation for a wave function phase function based on the perturbation of the space potential $V(x)$ by a related quantum potential $Q(x)$. ■

We go next to a treatment of interacting classical and quantum systems in [**400, 403**]. Thus in [**168, 764**] one has specified two minimal requirements for a consistent classical-quantum formulation of mechanics, namely

(1) A Lie bracket may be defined on the set of observables.
(2) The Lie bracket is equivalent to the classical Poisson bracket for any two classical observables and to $(i\hbar)^{-1}$ times the quantum commutator for any two quantum observables.

A third requirement is also developed in [**403**], namely

(1) The classical configuration is invariant under any canonical transformation applied to the quantum component and vice-versa.

It is also shown that the configuration ensemble approach of [**403**] is consistent with thermodynamics and implies that classical statistical mechanics may be given a Hamilton-Jacobi formulation. Thus given a probability density $P$, whose dynamics satisfies an action principle, there is a canonical conjugate $S$ and an ensemble Hamiltonian $\tilde{H}[P,S]$ such that

$$\partial_t P = \frac{\delta \tilde{H}}{\delta S};\ \partial_t S = -\frac{\delta \tilde{H}}{\delta P} \tag{2.7}$$

For a continuous configuration space with position coordinate $\xi$ and a functional $L[f] = \int d\xi F(f, \nabla f, \xi)$ one has (**2B**) $\delta L/\delta f = \partial_f F - \nabla \cdot [\partial F/\partial(\nabla f)]$ and as an example the ensemble Hamiltonian

$$\tilde{H}_Q = \int dq P\left[\frac{|\nabla S|^2}{2m} + \frac{\hbar^2}{8m}\frac{|\nabla P|^2}{P^2} + V(q)\right] \tag{2.8}$$

describes a quantum spin-zero particle of mass $m$ moving under a potential $V$. The equations of motion (2.7) reduce then to the real and imaginary parts of the SE (**2C**) $i\hbar\partial_t\psi = [-(\hbar^2/2m)\nabla^2 + V]\psi$ where $\psi = P^{1/2}exp(iS/\hbar)$. In the limit $\hbar \to 0$ these equations reduce to the Hamilton-Jacobi (HJ) and continuity equations for an ensemble of classical particles. One notes also that $\tilde{H}_Q[P,S] = \int dq\psi^*[-(\hbar^2/2m)\nabla^2 + V]\psi$ (cf. [**170**]). Observe that the conjugate pair $(P,S)$ allows a Poisson bracket to be defined for functionals $A[P,S]$ and $B[P,S]$, namely

$$\{A,B\} = \int d\xi\left(\frac{\delta A}{\delta P}\frac{\delta B}{\delta S} - \frac{\delta B}{\delta P}\frac{\delta A}{\delta S}\right) \tag{2.9}$$

Thus the equations of motion are $\partial_t P = \{P, \tilde{H}\}$ and $\partial_t S = \{S, \tilde{H}\}$ and in general **(2D)** $dA/dt = \{A, \tilde{H}\} + \partial_t A$ for $A = A[P, S, t]$. Note however that in general an arbitrary functional $A[P, S]$ is not allowable as an observable and a fundamental condition here is (cf. [**400**]) **(2E)** $A[P, S+c] = A[P, S]$ and $\delta A/\delta S = 0$ if $P(\xi) = 0$. Thus the observables should be chosen as some set of functionals satisfying the normalisation and positivity constraints **(2E)** that is closed under (2.9).

Now consider a mixed quantum-classical ensemble indexed via $\xi = (q, x)$ ($q$ could be the position of a quantum system or some complete set of kets $\{|q>\}$). The mixed ensemble will be described by two conjugate quantities $P(q, x, t)$ and $S(q, x, t)$. For any real classical phase space function $f(x, k)$ define the corresponding classical observable via

$$C_f = \int dqdxPf(x, \nabla_x S);\;\; C_f = < f(x, k) > \tag{2.10}$$

Here $\nabla_x S$ plays the role of a momentum associated with the configuration $(q, x)$ and

$$\{C_f, C_g\} = \int dqdx[-f\nabla_x \cdot (P\nabla_k g) + g\nabla_x \cdot (P\nabla_k f)] \tag{2.11}$$

$$= \int dqdxP(\nabla_x f \cdot \nabla_k g - \nabla_x g \cdot \nabla_k f) = C_{\{f,g\}}$$

where all quantities are evaluated at $k = \nabla_x S$ and $\{f, g\}$ is the usual Poisson bracket for phase space functions. Hence the Lie bracket for classical observables is equivalent to the usual phase space Poisson bracket. Next one notes that for any Hermitian operator M acting on the Hilbert space spanned by the kets $\{|q>\}$ one can define the corresponding quantum observable via
(2.12)

$$Q_M = \int dqdx\psi^*(q, x)M\psi(q, x) = \int dqdq'dx(PP')^{1/2}e^{i(S-S')/\hbar} < q'|M|q >$$

where $\psi = P(q, x)^{1/2}exp[iS(q, x)/\hbar]$ etc. In fact then $Q_M \equiv < M >$ and $Q_M$ satisfies the normalisation and positivity conditions **(2F)**. To evaluate the Poisson brackets of two quantum observables $Q_M$ and $Q_N$ one notes that for any real functional $A[P, S]$

$$\frac{\delta A}{\delta P} = \frac{\partial\psi}{\partial P}\frac{\delta A}{\delta\psi} + \frac{\partial\psi^*}{\partial P}\frac{\delta A}{\delta\psi^*} = \frac{1}{\psi\psi^*}\Re\left[\psi\frac{\delta A}{\delta\psi}\right] \tag{2.13}$$

$$\frac{\delta A}{\delta S} = \frac{\partial\psi}{\partial S}\frac{\delta A}{\delta\psi} + \frac{\partial\psi^*}{\partial S}\frac{\delta A}{\delta\psi^*} = -\frac{2}{\hbar}\Im\left[\psi\frac{\delta A}{\delta\psi}\right]$$

Hence, since $-ad + bc = \Im[(a + ib)(c - id)]$ one has

$$\{A, B\} = \frac{2}{\hbar}\Im\left[\int dqdx\frac{\delta A}{\delta\psi}\frac{\delta B}{\delta\psi^*}\right] \tag{2.14}$$

Since $M$ and $N$ are Hermitian one can replace $\psi^* M\psi$ by $(M\psi)^*\psi$ in (2.12) leading to

$$\{Q_M, Q_N\} = \frac{2}{\hbar}\Im\left[\int dqdx(M\psi)^* N\psi\right] = Q_{[M,N]/i\hbar} \tag{2.15}$$

where $[M,N] = MN - NM$. Thus (2.9), (2.11), and (2.15) are the main results so far.

Some further results are also discussed in [**403**]. Thus consider a mixed quantum-classical ensemble corresponding to a quantum particle of mass $m$ interacting with a classical particle of mass $M$ via a potential $V(q,x)$ (take 1D for simplicity). There is an ensemble Hamiltonian

$$(2.16)\qquad \tilde{H}_{QC}[P,S] = \int dqdxP\left[\frac{\partial_q S)^2}{2m} + \frac{\hbar^2}{8m}\frac{(\partial_q P)^2}{P^2} + \frac{(\partial_x S)^2}{2M} + V(q,x)\right]$$

From (2.10) one has

$$(2.17)\qquad <x>= C_x = \int dqdxPx;\ <k>= C_k = \int dqdxP\partial_x S;$$

$$<q>= Q_q = \int dqdxPq;\ <p>= Q_p = \int dqdxP\partial_q S$$

Following (**1D**) and (2.16) one obtains a generalization of the Ehrehfest relations

$$(2.18)\qquad \frac{d}{dt}<x>= M^{-1}<k>;\ \frac{d}{dt}<k>= -<\partial_x V>;$$

$$\frac{d}{dt}<q>= m^{-1}<p>;\ \frac{d}{dt}<p>= -<\partial_q V>$$

which implies that the centroid of a narrow initial probability density $P(q,x)$ will evolve classically for short timescales at least.

Thus properties (1) and (2) above hold and one shows that property also is valid in the form that a measurement of the classical configuration cannot detect whether or not a transformation has been applied to the quantum system and vice versa when the components are noninteracting. The latter is shown by a demonstration that

$$(2.19)\qquad \{C_{g(x)}, Q_M\} = 0 = \{Q_{G(q)}, C_f\}$$

It is not true that $\{C_f, Q_M\} = 0$ for arbitrary classical and quantum observables but it does hold when $P(q,x) = P_Q(q)P_C(x)$ and $S(q,x) = S_Q(q) + S_C(x)$. It is also shown that all classical phase space ensembles have a counterpart in the configuration ensemble approach and this approach can deal with thermal ensembles in a manner compatible with both classical and quantum thermodynamics. We refer to [**400, 401, 402, 403**] for more on all this.

It turns out that there is a great deal of information available about chaos (instability), entropy, information, quantum mechanics, and Bohmian trajectories and we sketch here a brief survey.

(1) The paper [**789**] defines Lyapounov exponents and Kolmogorov-Sinai (KS) entropy in quantum dynamics. For classical dynamics the largest Lyapounov number $\Lambda$ describes the asymptotic rate of exponential separation $d(t)$ between two initially close trajectories at a distance $d(0)$ via (**2G**) $\Lambda = lim(1/t)log[d(t)/d(0)]$ as $d(0) \to 0$ and $t \to \infty$. The KS entropy $h$ is defined for regions of connected stochasticity $M$ via

(**2H**) $h = \int_M (\sum_{\Lambda_+>0} \Lambda_+(x))d\mu$ where $d\mu$ is an element of the invariant volume (measure) in M and $x$ stands for canonical coordinates and momenta. In a single region of connected stochasticity the $\Lambda^{'s}$ are independent of $x$ and the integral over $d\mu$ gives simply (**2I**) $h = \sum_{\Lambda_+} \Lambda_+$. The KS entropy $h$ provides a measure of the rate of loss of information in predicting the future course of the trajectory; a dynamical system is said to be chaotic if $h$ is positive definite. Then in order to quantify quantum phase space dynamics one can use the Bohmian form of QM to define Lyapounov characteristic numbers and the associated KS entropy. Thus e.g. for a particle of mass $m$ in an external potential (**2J**) $i\hbar\psi_t = \hat{H}\psi$ with $\hat{H} \sim -(\hbar^2/2m)\Delta + V$ and one obtains as usual for $\psi = Rexp(iS/\hbar)$ and $P = R^2$

$$\partial_t S + \frac{(\nabla S)^2}{2m} + V + Q = 0;\ \partial_t P + \nabla \cdot \left(\frac{P\nabla S}{m}\right) = 0 \tag{2.20}$$

where $Q = -(\hbar^2 \Delta R/2mR)$. The dynamics is then governed by the trajectory $\mathbf{r}(t)$ where (**2K**) $\dot{\mathbf{r}} = (1/m)\nabla S(\mathbf{r}, t)$ or equivalently by (**2L**) $\ddot{\mathbf{r}}(t) = -(1/m)\nabla[V+Q]$. One notes that a fundamental property of the quantum phase space is the non-negative distribution function (**2M**) $f(\mathbf{r}, \mathbf{p}, t) = P(\mathbf{r}, t)\delta(\mathbf{p} - \nabla S(\mathbf{r}, t))$. From the solution of (**2K**) or (**2L**) the average rate of divergence of two neighboring trajectories in phase space can be determined and hence the KS entropy via the Lyapounov numbers. This is illustrated via a quantum delta-kicked rotor with numbers and graphs (see also [**22, 277, 278, 280, 324, 341, 538, 539, 561, 679, 680, 699, 777, 866, 879, 880, 881, 882**] for more in this direction) and cf. [**470**] for a fascinating study of entropic geometry for crowd dynamics.

(2) We refer also to [**253, 772, 773, 774**] for trajectory based discussions of interference, vorticity, decoherence, fractals, and related phenomena.

(3) In the papers [**169, 452, 486, 487, 529**] one looks at quantum chaos, renormalization, and quantum action. Here the idea is that for a given classical action (**2N**) $S[x] = \int (1/2)m\dot{x}^2 dt - V(x)$ there is a quantum action (**2O**) $\tilde{S} = \int (1/2)\tilde{m}\dot{x}^2 - \tilde{V}(x)$ which takes into account the quantum effects by a renormalized $\tilde{m}$ and $\tilde{V}$. The quantum action can be computed nonperturbatively and methods of classical chaos theory (Lyapounov exponents, Poincaré sections, etc.) can be applied using the quantum action.

(4) We mention without details that a number of papers deal with complex action $\hat{S}$ where $\psi \sim exp(i\hat{S}/\hbar)$ with

$$\partial_t \hat{S} + \left[\frac{1}{2m}|\nabla|\nabla\hat{S}|^2 + V\right] = \frac{i\hbar}{2m}\Delta\hat{S} \tag{2.21}$$

There are various techniques and results along with comparisons to complex WKB methods and we cite [**91, 360, 361, 488, 541, 568, 713, 774, 782, 890**] (cf. also [**116, 132, 133, 170, 177, 267, 297, 310, 440**] for general results on trajectories and the quantum potential). The complex action is very useful in treating problems where there are nodes or when bound states are considered.

## 3. NC FIELDS AND GRAVITY

A survey of Yang's approach to this subject appears in [**893, 894, 896**] and in connection with this point of view we mention also [**12, 67, 68, 69, 512, 622, 566, 625, 821, 895**]. Related ideas can be found in e.g. [**40, 41, 42, 204, 240, 270, 455, 493, 494, 495, 744, 788, 791, 792, 824, 869**] and see Section 2 for related work of deGosson and Isidro using gerbes. The long survey articles in [**894**] are sometimes confusing so we will extract liberally from the last preprint (hep-th 0902.0035, v2) and refer also to the short articles in IJMPA and MPA referenced in [**896**]. The main idea is that gravity is a collective phenomenon emerging from gauge fields of electromagnetic (EM) type, living in a fuzzy space-time. For motivation here one looks at the Einstein equations (★) $R_{ab}-(1/2)g_{ab}R = 8\pi GT_{ab}$ for a flat space where the left side vanishes identically. Hence $T_{ab}$ should also vanish so a flat spacetime would seem to have no "price". But an empty space-time is full of quantum fluctuations and the vacuum would be heavy of roughly Planck mass $\rho_{vac} \sim M_P^4$ and this apparent conflict between quantum field theory and relativity is related to the cosmological constant (CC) problem where a matter Lagrangian is shifted by a factor $2\Lambda$ with $T_{ab} \to T_{ab} - \Lambda g_{ab}$. After some convincing argument it is then inferred that a flat spacetime may not be free of "cost" but a result of Planck energy condensation in a vacuum. One arrives at the problem of formulating a physically viable theory (i.e. a background independent theory) to correctly explain the dynamical origin of flat spacetime. For this Yang et al develop a theory of emergent gravity from non-commutative (NC) geometry. Note that time will actually be considered emergent in a somewhat different manner (see Remark 1.1.1 below) so that one should consider the manifolds $M$ as Riemannian (not Lorentzian) and remarks about space-time should be taken heuristically.

Thus consider electromagnetism (EM) on a D-brane whose data are given by $(M, g, B)$ where $M$ is a smooth manifold with a metric $g$ and a symplectic structure $B$ (for strings see e.g. [**87, 384, 511, 716, 912**] and for branes see also [**489, 575, 825**]). The dynamics of $U(1)$ gauge fields on the D-brane is described by open string field theory whose low energy effective action on the D-brane is given via Dirac-Born-Infeld (DBI) action. In particular the DBI action predicts two important results, namely: (**3A**) The triple $(M, g, B)$ comes only in the combination $(M, g+\kappa B)$ which embodies a generalized geometry continuously interpolating between symplectic geometry ($|\kappa Bg^{-1}| >> 1$) and Riemannian geometry ($|\kappa Bg^{-1}| << 1$ - see here [**200, 393, 438**] for generalized complex geometries). (**3B**) The EM force $F = dA$ appears only as the deformation of symplectic structure $\Omega(x) = (B+F)(x)$. Then including the $U(1)$ gauge fields the data of "D-manifold" are given via $(M, g, \Omega) = (M, g+\kappa\Omega)$.

Consider now another D-brane whose D-manifold is described by different data $(N, G, B) = (N, G+\kappa B)$ and ask whether there is a diffeomorphism $\phi : N \to M$

such that (**3C**) $\phi^*(g+\kappa\Omega) = G+\kappa B$. Note here that a D-brane whose world volume $M$ supports a symplectic structure $B$ also respects (in addition to $Diff(M)$) the so-called $\Lambda$ symmetry (**3D**) $(B, A) \to (B - d\Lambda, A + \Lambda)$ where $\Lambda$ is a 1-form on $M$. Thus consider the symmetry transformation which is a combination of the $\Lambda$-transformation (**3D**) followed by a diffeomorphism $\phi:\ N \to M$. The action transforms the "DBI metric" $g+\kappa B$ on $M$ according to (**3E**) $g+\kappa B \to \phi^*(g+\kappa\Omega)$ and the main point here is that there always exists a diffeomorphism $\phi:\ N \to M$ such that $\phi * (\Omega) = B$ (via the Darboux theorem providing local coordinates e.g. $dq^j \wedge dp_j$). Therefore two different DBI metrics $g + \kappa\Omega$ and $G + \kappa B$ are diffeomorphic to each other, e.g. (**3F**) $\phi^*(g + \kappa\Omega) = G + \kappa B$ where $G = \phi^*(g)$. This property holds for any pair $(g, B)$ of Riemannian metric metric and symplectic structure $B$.

Since $B$ is nondegenerate at any point $y \in M$ one can invert this map to obtain the map (**3G**) $\theta \equiv B^{-1}:\ T^*M \to TM$ and the cosymplectic structure (**3H**) $\theta \in \wedge^2 TM$ is called the Poisson structure of $M$ (defining a Poisson bracket $\{\cdot,\cdot\}_\theta$). A NC spacetime then arises from quantizing the symplectic manifold $(M, B)$ with its Poisson structure, treating it as a quantum phase space, i.e.

$$< B_{ab} >_{vac} = (\theta^{-1})_{ab} \iff [y^a, y^b]_* = i\theta^{ab} \tag{3.1}$$

Thus there always is a coordinate transformation to locally eliminate the EM force $F = dA$ as long as $M$ supports a symplectic structure $B$, i.e. $M$ is an NC space. This is a novel form of the equivalence principle, corresponding to the Darboux transformation, for the geometrization of the electromagnetism. Since it is always possible to find a coordinate transformation $\phi \in Diff(M)$ such that $\phi^*(B+F) = B$ the relation (**3I**) $\phi^*(g_\kappa(B+F) = G+\kappa B$ shows that the EM fields in the DBI metric $g + \kappa(B + F)$ now appear as a new metric (**3J**) $G = \phi^*(g)$. One can also consider the inverse relation $\phi_*(G+\kappa B) = g+\kappa(B+F)$ showing that a nontrivial metric $G$ in a vacuum (3.1) can be interpreted as an inhomogeneous condensation of gauge fields on a D-manifold with metric $g$. Note that the relationship in the case (**3K**) $\kappa = 2\pi\alpha'$ (where $\alpha'$ is the $T$-duality number $R \to \alpha'/R$ of string theory) is the familiar diffeomorphism in Riemannian geometry and says nothing spectacular.

Concerning the Darboux theorem as providing a diffeomorphism between two different DBI metrics in terms of local coordinates $\phi:\ y \to x = x(y)$ one has

$$(g + \kappa\Omega)_{\alpha\beta}(x) = \frac{\partial y^a}{\partial x^\alpha}(G_{ab}(y) + \kappa B_{ab}(y))\frac{\partial y^b}{\partial x^\beta};\ G_{ab} = \frac{\partial x^\alpha}{\partial y^a}\frac{\partial x^\beta}{\partial y^b} g_{\alpha\beta}(x) \tag{3.2}$$

Equation (3.2) shows how NC gauge fields manifest themselves as a space-time geometry. To expose the intrinsic connection between NC gauge fields and space-time geometry one can write the coordinate transformation (1.2) as

$$x^a(y) = y^a + \theta^{ab}\hat{A}_b(y) \tag{3.3}$$

Note that $F(x) = 0$ or equivalently $\hat{A}_a(y) = 0$ corresponds to $G_{ab} = g_{ab}$ as it should. Clearly the second equation in (3.2) shows how the metric on M is deformed by the presence of NC gauge fields. Next one shows how the Darboux theorem in symplectic geometry materializes as a novel form of the equivalence principle such that EM in NC space-time can be regarded as a theory of gravity. Note first that for a given Poisson algebra $(C^\infty(M), \{\cdot,\cdot\}_\theta)$ there is a natural map $C^\infty(M) \to TM : f \to X_f$ between smooth functions in $C^\infty(M)$ and vector fields in TM such that (**3L**) $X_f(g) = \{g, f\}_\theta$ for any $g \in C^\infty(M)$. Also the 1-1 correspondence between $C^\infty(M)$ and $\Gamma(TM)$ is a Lie algebra homomorphism in the sense that

$$X_{\{f,g\}_\theta} = -[X_f, X_g] \tag{3.4}$$

Now let vector fields be denoted by $V_a \in \Gamma(TM)$ and their dual 1-forms as $D^a \in \Gamma(T^*M)$. Then $V_a$ and $D^a$ are related to the orthogonal frames $E_a \in \Gamma(TM)$ and $E^a \in \Gamma(T^*M)$ by (**3M**) $V_a = \lambda E_a$ and $E^a = \lambda D^a$ respectively where $\lambda^2 = det(V_a^b)$. This exhibits how the Darboux theorem implements a deep principle to realize a Riemannian manifold as an emergent geometry from NC gauge fields (3.3) through the correspondence (**3L**) whose metric is given by

$$ds^2 = g_{ab}E^a \otimes E^b = \lambda^2 g_{ab} D^a_\mu D^b_\nu dy^\mu \otimes dy^\nu \tag{3.5}$$

One remarks that the emergent gravity can be generalized to full NC fields in two different manners (which are dual). The first point is that the correspondence (**3L**) between a Poisson algebra $(C^\infty(M), \{\cdot,\cdot\}_\theta)$ and vector fields in $\Gamma(TM)$ can be generalized to a NC $C^*$-algebra $(A_\theta(M), \{\cdot,\cdot\}_\theta)$ by considering an adjoint action of NC gauge fields $\hat{D}_a(y) \in A_\theta$ as

$$ad_{\hat{D}_a}[\hat{f}](y) = -i[\hat{D}_a(y), \hat{f}(y)]_* = V_a^\mu \frac{\partial f(y)}{\partial y^\mu} + O(\theta^3) \tag{3.6}$$

where the leading term recovers the vector fields in (**3L**). Secondly every NC space can be represented as a theory of operators in a Hilbert space $H$, which consists of a NC $C^*$-algebra $A_\theta$ and any operator in $A_\theta$ or any NC field can be represented as a matrix whose size is determined by the dimension of $H$. For the Moyal NC space as an example of (3.1) one gets $N \times N$ matrices in the $N \to \infty$ limit. One considers now this Moyal NC space in (3.1) with $M = \mathbf{R}^{2n}$ and $B$ constant for simplicity. Note here

**REMARK 3.1.1** For a general space-time the emergent gravity based on NC geometry endows a natural concept of "emerged time" since a symplectic manifold $(M, B)$ always admits a Hamiltonian dynamical system (and hence a time) on M defined by a Hamiltonian vector field $X_H$, i.e. $i_{X_H}B = dH$ described by $(df/dt) = X_H(f) = \{f, H\}_\theta$ for any $f \in C^\infty(M)$. In a general case where the symplectic structure is changing along the dynamical flow as in (3.2) the dynamical evolution must be locally defined on every local Darboux chart. We refer also to Chapter 7 for another idea of emergent time. ∎

Now for the Moyal NC space the background NC space is described by $x^a(y) = y^a$ in (3.3) and the corresponding vector fields are given by $V_a = \partial/\partial y^a$ according to (**3L**) of (3.6) and hence for the background (3.1) is flat, i.e. $g_{ab} = \eta_{ab}$. Treating emergent time as in Remark 3.1.1 one has an emergent flat space-time, but this is not coming from an empty space but from a uniform condensation of gauge fields in a vacuum (3.1). Since gravity emerges from NC gauge fields the parameters $g_{YM}^2$ and $|\theta|$ should be related to the Newton constant $G$ in emergent gravity and some analysis shows that $G\hbar^2/c^2 \sim g_{YM}^2|\theta|$ (arising from $\rho_{vac} \sim |B_{ab}|^2 \sim M_P^4$ where $M_P = (8\pi G)^{1/2} \sim 10^{18}GeV$. This indicates that a flat space does not come free of cost but is based on the huge Planck mass $M_P$. Further estimations based on this picture lead to an estimate of vacuum fluctuations $\delta\rho \sim 1/L_P^2 L_H^2$ where $L_H$ refers to IR fluctuations of a typical scale; this estimate seems possibly related to dark energy.

**3.1. REVIEW OF SOME IDEAS.** We go here to various papers for a review of some of the material arising in Section 1 above (cf. [**40, 41, 42, 175, 493, 494, 495, 554, 566, 788, 792, 821, 824, 869**]. Going to [**566**] for example consider a NC algebra $\mathcal{A}$ and a NC space with coordinates $\hat{x}^i$ and (canonical) relations (**3M**) $\mathcal{R}: [\hat{x}^i, \hat{x}^j] = i\theta^{ij}$ with $\mathcal{A}_x = \mathbf{C}[\hat{x}^1, \cdots, \hat{x}^N]/\mathcal{R}$ and fields $\psi(\hat{x}) \in \mathcal{A}_x$ with covariant transformation law (•) $\delta\psi(\hat{x}) = i\alpha(\hat{x})\psi(\hat{x})$. One looks at infinitesimal gauge transformations such that $\delta\hat{x}^i = 0$ and introduces covariant coordinates $\hat{X}^i = \hat{x}^i + A^i(\hat{x})$ $(A_i(\hat{x}) \in \mathcal{A}_x)$ with (**3N**) $\delta(\hat{X}^i\psi) = i\alpha\hat{X}^i\psi$. Here $\alpha(\hat{x}) \in \mathcal{A}_x$ for Abelian gauge transformations and (**3O**) $\delta A^k = i[\alpha, A^k] - i[\hat{x}^k, \alpha]$. A tensor (**3P**) $T^{kj} = [\hat{X}^k, \hat{X}^j] - i\theta^{kj}$ can also be defined and one verifies that these are covariant tensors via

$$T^{ij} = [A^i, \hat{x}^j] + [\hat{x}^i, A^j] + [A^i, A^j]; \tag{3.7}$$

$$\delta T^{ij} = [\delta A^i, \hat{x}^j] + [\hat{x}^i, \delta A^j] + [\delta A^i, A^j] + [A^i, \delta A^j]$$

Inserting $\delta A^i$ from (**3O**) and using the Jacobi identity there results (**3Q**) $\delta T^{kj} = i[\alpha, T^{kj}]$.

Using the Fourier transform one defines then a Weyl transform via

$$\tilde{f}(k) = \frac{1}{(2\pi)^{n/2}} \int d^n x e^{-ik_j x^j} f(x); \; W(f) = \frac{1}{(2\pi)^{n/2}} \int d^n k e^{ik_j \hat{x}^j} \tilde{f}(k) \tag{3.8}$$

From this one can a find a function $f \diamond g$ with (**3R**) $W(f)W(g) = W(f \diamond g)$ which can be written out as

$$W(f)W(g) = \frac{1}{(2m)^n} \int d^n k d^n p e^{ik_m \hat{x}^m} e^{ip_j \hat{x}^j} \tilde{f}(k)\tilde{g}(p) \tag{3.9}$$

If the product of the exponentials can be calculated by the BCH formula to give an exponential of a linear combination of the $\hat{x}^i$ this will determine $f \diamond g$. This can be done for our cananonical structure (**3M**) via (**3S**) $exp(ik_m\hat{x}^m)exp(ip_j\hat{x}^j) = exp[i(k_j + p_j)\hat{x}^j - (i/2)k_m p_j \theta^{mj}]$ leading to the Weyl-Moyal star product (cf. [**177, 175**])

$$f \diamond g = f \bigstar g = exp\left[\frac{i}{2}\frac{\partial}{\partial x^i}\theta^{ij}\frac{\partial}{\partial y^j}\right] f(x)g(y)|_{y \to x} \tag{3.10}$$

The Weyl quantization allows the representation of an element of $\mathcal{A}_x$ by a classical function of $x$. Thus first $c \sim c$ and $\hat{x}^i \sim x^i$ and e.g.

$$\hat{x}^i\hat{x}^j = W(x^i)W(x^j) = W(x^i \diamond x^j) \Rightarrow \hat{x}^i\hat{x}^j \sim x^i \diamond x^j \tag{3.11}$$

In particular (**3T**) $W(x^ix^j) = (1/2)(\hat{x}^i\hat{x}^j + \hat{x}^j\hat{x}^i)$. For a classical function $\psi(x)$ the gauge transformation (•) yields (**3U**) $\delta_\alpha\psi(x) = i\alpha(x)\diamond\psi(x)$ and this implies

$$(3.12)\qquad (\delta_\alpha\delta_\beta - \delta_\beta\delta_\alpha)\psi(x) = i(\beta\diamond\alpha - \alpha\diamond\beta)\diamond\psi$$

The transformation law of $A^i(x)$ is (**3V**) $\delta A^j = i[\alpha \overset{\diamond}{,} A^j] - i[x^j \overset{\diamond}{,} \alpha]$ and for tensors $\delta T^{mj} = i[\alpha \overset{\diamond}{,} T^{mj}]$ where $T^{ij}$ is defined as in (**3P**); note also (**3W**) $T^{ij} = [A^i \overset{\diamond}{,} x^j] + [x^i \overset{\diamond}{,} A^j] + [A^i \overset{\diamond}{,} A^j]$.

Now for NC gauge theories and relations to NC Yang-Mills we write (**3X**) $\delta A^j = -i[\hat{x}^j, \alpha] + i[\alpha, A^j]$ and due to (**3M**) the commutator $[\hat{x}^j, \cdot]$ acts as a derivation on elements $f \in \mathcal{A}_x$ via (**3Y**) $[\hat{x}^j, f] = i\theta^{jk}\partial_k f$ (note $\partial_j(fg) = (\partial_j f)g + f(\partial_j g)$ and $\partial_j\hat{x}^j = \delta^i_j$ so the right side of (**3Y**) is a derivation for $\theta$ constant). One can write now (for $\theta$ nondegenerate)

$$(3.13)\qquad \delta A^j = \theta^{jk}\partial_k\alpha + i[\alpha, A^j];\ A^j = \theta^{jk}\hat{A}_k;\ \delta\hat{A}_j = \partial_j\alpha + i[\alpha, \hat{A}_j]$$

and an explicit expression for the tensor T of (3.7) is (**3Z**) $T^{jk} = i\theta^{jm}\partial_m A^k - i\theta^{kn}\partial_n A^j + [A^j, A^k]$. Up to a factor of $i$ the relation to the NC Yang-Mills field $\hat{F}$ is then (**3AA**) $T^{jk} = i\theta^{jm}\theta^{kn}\hat{F}_{mn}$. Assuming again nondegenerate $\theta$ one has also (**3AA**) $\hat{F}_{k\ell} = \partial_k\hat{A}_\ell - \partial_\ell\hat{A}_k - i[\hat{A}_k, \hat{A}_\ell]$. This can be considered as the field strength of an Abelian gauge potential in a NC geometry and except for the definition of the bracket it has the same form as a NC gauge field strength in a commutative geometry. Since $\theta^{ij} \in \mathbf{C}$, $\hat{F}$ is a tensor and (**3AB**) $\delta\hat{F}_{k\ell} = i[\alpha, \hat{F}_{k\ell}]$. The relation to NC Yang-Mills is clearer if one represents the elements of $\mathcal{A}_x$ by functions of $x^i$ and uses the Moyal-Weyl star product (3.10) so that (**3Y**) becomes (**3AC**) $x^j\bigstar f - f\bigstar x^j = i\theta^{jk}\partial_k f$ where $f = f(x)$ and $\partial_k f = \partial f/\partial x^k$. This follows from (3.10) and the identifications (3.13) have the same form as before. The relevant equations in terms of star products become

$$(3.14)\qquad \delta A^k = \theta^{kj}\partial_j\alpha + i\alpha\bigstar A^k - iA^k\bigstar\alpha;\ \delta T^{mj} = i\alpha\bigstar T^{mj} - iT^{mj}\bigstar\alpha;$$
$$T^{mj} = i\theta^{mk}\partial_k A^j - i\theta^{j\ell}\partial_\ell A^m + A^m\bigstar A^j - A^j\bigstar A^m;$$
$$\delta\hat{A}_j = \partial_j\alpha + i\alpha\bigstar\hat{A}_j - i\hat{A}_j\bigstar\alpha;\ \delta\hat{F}_{k\ell} = i\alpha\bigstar\hat{F}_{k\ell} - i\hat{F}_{k\ell}\bigstar\alpha;$$
$$\hat{F}_{k\ell} = \partial_k\hat{A}_\ell - \partial_\ell\hat{A}_k - i\hat{A}_k\bigstar\hat{A}_\ell + i\hat{A}_\ell\bigstar\hat{A}_k;\ \delta_\alpha\delta_\beta - \delta_\beta\delta_\alpha = \delta_{(\beta\bigstar\alpha - \alpha\bigstar\beta)}$$

This all generalizes to $A^i$, $\alpha$, $\hat{A}_j$, $\hat{F}_{k\ell}$ that are Hermitian $n\times n$ matrices; note also that (**3AD**) $\hat{X}^i = \hat{x}^i + \theta^{ij}\hat{A}_j$.

A brief sketch of the Seiberg-Witten (SW) map is now given via an infinitesimal gauge parameter $\epsilon$ and ordinary gauge potentials $a_i$ (cf. [**177, 175, 622, 792**]). Thus one has (**3AE**) $\delta_\epsilon a_j = \partial_i\epsilon + i[\epsilon, a_j]$ which is to be compared with the gauge transformation (**3V**) $\delta A^j = i[\alpha \overset{\diamond}{,} A^j] - i[x^j \overset{\diamond}{,} \alpha]$. One can write then (cf. [**175, 522**]) (**3AF**) $f\diamond g = fg + \sum_{n\geq 1} h^n B_n(f,g)$ where the $B_n$ are bilinear differential operators and $h$ is an expansion parameter. In our canonical case

$$(3.15)\qquad f\bigstar g = fg + \sum_{n\geq 1}\frac{1}{n!}\left(\frac{i}{2}\right)^n \theta^{j_1k_1}\cdots\theta^{j_nk_n}(\partial_{j_1}\cdots\partial_{j_n}f)(\partial_{k_1}\cdots\partial_{k_n}g)$$

Working up to second order in $h$ one has (**3AG**) $[f \overset{\bigstar}{,} \alpha] = i\theta^{jk}(x)\partial_j f\partial_k g + O(\theta^3)$ (there are no second order terms). Hence in (**3V**) one obtains (**3AH**) $[x^j \overset{\bigstar}{,} \alpha] =$

$i\theta^{jk}\partial_k\alpha$ (for all orders in $\theta$) and combining this with (**3AG**) one has (**3AI**) $\delta A^j = \theta^{jk}\partial_k\alpha - \theta^{jk}\partial_j f\partial_k g + O(\theta^3)$. Following [**792**] one can construct local expressions for $A$ and $\alpha$ in terms of $a$, $\epsilon$, $\theta$ via the Ansatz

$$A^k = \theta^{kj}a_j + G^k(\theta, a, \partial a, \cdots) + O(\theta^3); \tag{3.16}$$

$$\alpha = \epsilon + \gamma(\theta, \epsilon, \partial\epsilon, \cdots, a, \partial a, \cdots) + O(\theta^3)$$

One demands that $\delta A$ in (3.16) with the infinitesimal parameter $\alpha$ be obtained from the variation (**3AE**) of $a$. This is true to first order in $\theta$ due to (3.16). In second order one gets an equation for $G^k$ and $\gamma$, namely

$$\delta_\epsilon G^k = \theta^{kj}\partial_j\gamma - \frac{1}{2}\theta^{m\ell}[\partial_m\epsilon\partial_\ell(\theta^{kj}a_j] + \partial_m(\theta^{kj}a_j)\partial_\ell\epsilon) + i[\epsilon, G^k] + i[\gamma, \theta^{kj}a_j] \tag{3.17}$$

This equation has the following solution

$$G^k = -\frac{1}{4}\theta^{m\ell}\{a_m, [\partial_\ell(\theta^{kj}a_j) + \theta^{mj}F_{\ell j}]\};\ \gamma = \frac{1}{4}\theta^{\ell m}\{\partial_\ell\alpha, a_m\} \tag{3.18}$$

where $F_{ij}$ is the classical field strength $F_{mj} = \partial_m a_j - \partial_j a_m + i[a_m, a_j]$. The proof involves the Jacobi identity for $\theta^{m\ell}(x)$ and in the "canonical" case $\theta^{m\ell}$ constant it is the same result as in [**792**] if one uses (3.13).

## 4. GERBES AND STRUCTURE

The idea of combining a Riemannian and a symplectic structure to discuss QM has also been treated extensively in various ways in work of Isidro, deGosson, et al (cf. [**2, 456, 457, 458, 459, 460, 464, 461, 462, 567**]); for example it is shown in [**372, 458**] that symplectic covariance of the SE on phase space is equivalent to gauge invariance under a $U(1)$ gerbe on phase space (cf. also work of Klauder et al ([**513, 514**]). We sketch here the gerbe context following [**458, 459**] (cf. also [**45, 46, 126, 141, 316, 369, 372, 373, 374, 605, 624**]). The preceding section has clearly indicated that the string theory language and constructions should be explored more fully. In particular we will need the language of gerbes and corresponding connections to Hamiltonian mechanics. One can also pursue the theory of NC spaces and deformation quantization in a geometric context. We will begin with a discussion of work by M. de Gosson and J. Isidro et al (cf. [**369, 372, 373, 374, 457, 458, 459, 461**] and for perspective let us mention here related geometrical approaches in [**45, 46, 126, 141, 316, 605, 624, 779, 910**]. Starting out with [**458, 459, 461**] one notes first that for a differential manifold M with metric given via (**4A**) $ds^2 = g_{\mu\nu}dx^\mu dx^\nu$ it is more or less mandatory today that there will be an associated NC space with coordinates $\hat{x}^\mu$ and (**4B**) $[\hat{x}^\mu, \hat{x}^\nu] = ia\theta^{\mu\nu}$ with $\theta^{\mu\nu}$ constant (to begin with). Also as $a \to 0$ one expects the bracket in (**4B**) to vanish with (•) $a \sim L_P^2$ for Planck length $L_P$. Note that in [**659**] it has been argued that the existence of a fundamental length scale $L_P$ implies modifying the spacetime metric via (♦) $ds^2 \to ds^2 + L_P^2$ so that $L_P$ becomes the shortest possible distance. It is also indicated that modifying the spacetime interval in this manner is equivalent to requiring invariance of a field theory under the exchange of short and long distances. On the other hand in [**461**] one shows that the existence of a minimal length scale $L_P$ is equivalent to the exchange (★) $S/\hbar \to \hbar/S$ in Feynman's exponential of the action integral $exp(iS/\hbar)$ leading to $exp(i\hbar/S)$. One refers to this exchange (★) as semiclassical versus strong-quantum duality and it is shown in [**458**] (paper 1) that (**4B**) and (★) are equivalent (cf. also [**461**]). This is done in the context of gerbes and the B-field arises in [**461**] (for comparison with Section 1).

For gerbes we refer generally to [**141, 369, 458, 459, 461**]. Thus a unitary line bundle on a base manifold $M$ is a 1-cocycle (**4C**) $\lambda \in H^1(M, C^\infty(U(1))$ (first Čech cohomology group of $M$ with coefficients in the sheaf of germs of smooth $U(1)$ valued functions. Let $U_\alpha$ be a "good" cover of $M$ by open sets $U_\alpha$ so that the bundle is determined by a collection of $U(1)$ valued transition functions defined on 2-fold overlaps (**4D**) $\lambda_{\alpha_1\alpha_2} : U_{\alpha_1} \cap U_{\alpha_2} \to U(1)$ with (cf. [**458**])

$$\lambda_{\alpha_2\alpha_1} = \lambda^{-1}_{\alpha_1\alpha_2};\ \lambda_{\alpha_1\alpha_2}\lambda_{\alpha_2\alpha_3}\lambda_{\alpha_3\alpha_1} = 1\ (on\ U_{\alpha_1} \cap U_{\alpha_2} \cap U_{\alpha_3}) \tag{4.1}$$

A gerbe is a 2-cocycle $g \in H^2(M, C^\infty(U(1))$ which means there is a collection $\{g_{\alpha_1\alpha_2\alpha_3}\}$ of maps

$$g_{\alpha_1\alpha_2\alpha_3} : U_{\alpha_1} \cap U_{\alpha_2} \cap U_{\alpha_3} \to U(1); \tag{4.2}$$

$$g_{\alpha_1\alpha_2\alpha_3} = g^{-1}_{\alpha_2\alpha_1\alpha_3} = g^{-1}_{\alpha_1\alpha_3\alpha_2} = g^{-1}_{\alpha_3\alpha_2\alpha_1}$$

as well as the 2-cocycle condition (**4E**) $g_{\alpha_2\alpha_3\alpha_4}g^{-1}_{\alpha_1\alpha_3\alpha_4}g_{\alpha_1\alpha_2\alpha_4}g^{-1}_{\alpha_1\alpha_2\alpha_3} = 1$ on $U_{\alpha_1} \cap U_{\alpha_2} \cap U_{\alpha_3} \cap U_{\alpha_4}$. Now $g$ is a 2-coboundary in Čech cohomology whenever (••) $g_{\alpha_1\alpha_2\alpha_3} = \tau_{\alpha_1\alpha_2}\tau_{\alpha_2\alpha_3}\tau_{\alpha_3\alpha_1}$ where the trivializations satisfy $\tau_{\alpha_2\alpha_1} = \tau^{-1}_{\alpha_1\alpha_2}$. Note that any two trivializations differ by a unitary line bundle since $\tau'_{\alpha_1\alpha_2}/\tau_{\alpha_1\alpha_2}$ satisfies the 1-cocycle condition in (4.1) but this is not a transition function so a gerbe is not a manifold. Moreover on a gerbe specified by a 2-cocycle $g_{\alpha_1\alpha_2\alpha_3}$ a connection is specified by forms $A$, $B$, $H$ satisfing

$$H_{U_\alpha} = dB_\alpha;\ B_{\alpha_2} - B_{\alpha_1} = dA_{\alpha_1\alpha_2}; \tag{4.3}$$

$$A_{\alpha_1\alpha_2} + A_{\alpha_2\alpha_3} + A_{\alpha_3\alpha_1} = g^{-1}_{\alpha_1\alpha_2\alpha_3}dg_{\alpha_1\alpha_2\alpha_3}$$

The 3-form $H$ is the curvature of the gerbe connection and if $H = 0$ it is called flat. The connections to the $B$ field of Section 1 should be apparent.

Now in [**458**] (paper 1) or [**461**] (for phase space) one explicitly constructs a gerbe based on path integrals. Thus on a spacetime ((**4F**) $M \sim \mathbf{R} \times \mathbf{F}$ with $\mathbf{F}$ a configuration space take coordinates $x^\mu_\alpha \sim (t_\alpha, q^j_\alpha)$ with $j = 1, \cdots, n-1$. Take any two points $q_{\alpha_i}$, $q_{\alpha_2}$ on $\mathbf{F}$ with local charts $U_{\alpha_1}$, $U_{\alpha_2}$ around them and $L_{\alpha_1,\alpha_2}$ an oriented path connecting the two points. Define for all such trajectories (**4G**) $\tilde{a}_{\alpha_1\alpha_2} \sim \int DL_{\alpha_1\alpha_2} exp\left(\frac{i}{\hbar}S(L_{\alpha_1\alpha_2}\right)$ ($\sim$ means $\propto$ here). Any normalization factors will cancel out later since one is only interested in quotients as in (**4H**), (4.5), (4.7), etc.). The argument of the exponential involves the action $S$ evaluated along the path $L_{\alpha_1\alpha_2}$ so $\tilde{a}_{\alpha_1\alpha_2}$ is proportional to the probability amplitude for the particle to start at $q_{\alpha_1}$ and finish at $q_{\alpha_2}$. Consider then (**4H**) $a_{\alpha_1\alpha_2} = \tilde{a}_{\alpha_1\alpha_2}/|\tilde{a}_{\alpha_1\alpha_2}|$ (i.e. the $U(1)$ valued phase of the path integral (**4G**)). Assume (**4I**) $U_{\alpha_1\alpha_2} = U_{\alpha_1} \cap U_{\alpha_2} \neq \phi$ and define for $q_{\alpha_{12}} \in U_{\alpha_1\alpha_2}$

$$\tau_{\alpha_1\alpha_2} : U_{\alpha_1\alpha_2} \to U(1) : \tau_{\alpha_1\alpha_2}(q_{\alpha_{12}}) = a_{\alpha_1\alpha_{12}}(q_{\alpha_{12}})a_{\alpha_{12}\alpha_2}(q_{\alpha_{12}}) \tag{4.4}$$

Thus $\tau_{\alpha_1\alpha_2}$ is the $U(1)$ valued phase of the probability amplitude for the transition $q_{\alpha_1} \to q_{\alpha_{12}} \to q_{\alpha_2}$ and this qualifies as a gerbe trivialisation on **F** which can be expressed as

$$\tau_{\alpha_1\alpha_2}(q_{\alpha_{12}}) = \frac{\tilde{\tau}_{\alpha_1\alpha_2}(q_{\alpha_{12}})}{|\tilde{\tau}_{\alpha_1\alpha_2}(q_{\alpha_{12}})|} \tag{4.5}$$

$$\tilde{\tau}_{\alpha_1\alpha_2}(q_{\alpha_{12}}) \sim \int DL_{\alpha_1\alpha_2}(\alpha_{12})exp\left[\frac{i}{\hbar}S(L_{\alpha_1\alpha_2}(\alpha_{12}))\right]$$

where the integral extends over all paths meeting the requirements of (4.4).

Now consider three points with their charts (**4J**) $q_{\alpha_1} \in U_{\alpha_1}$, $q_{\alpha_2} \in U_{\alpha_2}$, and $q_{\alpha_3} \in U_{\alpha_3}$ with (**2K**) $U_{\alpha_1} \cap U_{\alpha_2} \cap U_{\alpha_3} \neq \phi$ and Once the trivialization (4.4) is known the 2-cocycle $g_{\alpha_1\alpha_2\alpha_3}$ defining a gerbe on **F** is given via

$$g_{\alpha_1\alpha_2\alpha_3} : U_{\alpha_1\alpha_2\alpha_3} \to U(1); \tag{4.6}$$

$$g_{\alpha_1\alpha_2\alpha_3}(q_{\alpha_{123}}) = \tau_{\alpha_1\alpha_2}(q_{\alpha_{123}})\tau_{\alpha_2\alpha_3}(q_{\alpha_{123}})\tau_{\alpha_3\alpha_1}(q_{\alpha_{123}})$$

Here one is talking about the $U(1)$-phase with a path $q_{\alpha_1} \to q_{\alpha_{123}} \to q_{\alpha_2} \to q_{\alpha_{123}} \to q_{\alpha_3} \to q_{\alpha_{123}} \to q_{\alpha_1}$ with (**4L**) $q_{\alpha_{123}} \in U_{\alpha_1\alpha_2\alpha_3}$ and (**4M**) $L_{\alpha_1\alpha_2\alpha_3}(\alpha 123) = L_{|ga_1\alpha_2}(\alpha 123 + L_{\alpha_2\alpha_3}(ga123 + L_{\alpha_3\alpha_1}(\alpha 123)$. The 2-cocycle can be expressed via

$$g_{\alpha_1\alpha_2\alpha_3}(q_{\alpha_{123}}) = \frac{\tilde{g}_{\alpha_1\alpha_2\alpha_3}(q_{\alpha_{123}})}{|\tilde{g}_{\alpha_1\alpha_2\alpha_3}(q_{\alpha_{123}})|} \tag{4.7}$$

$$\tilde{g}_{\alpha_1\alpha_2\alpha_3}(q_{\alpha_{123}}) \sim \int DL_{\alpha_1\alpha_2\alpha_3}(\alpha_{123})exp\left[\frac{i}{\hbar}S(L_{\alpha_1\alpha_2\alpha_3}(\alpha_{123}))\right]$$

The integral extends over all paths $L_{\alpha_1\alpha_2\alpha_3}(\alpha_{123})$ meeting the requirements indicated as in (**4K**) and (**4L**).

Given a closed loop $L$ set $\mathbf{S} \subset \mathbf{F}$ be a two-dimensional surface with boundary $\partial\mathbf{S} = L$ so that by Stokes' theorem (**4N**) $S(L) = \int_L \mathfrak{L}dt = \int_{\partial\mathbf{S}} \mathfrak{L}dt = \int_{\mathbf{S}} d\mathfrak{L} \wedge dt$. Any surface **S** such that $\partial\mathbf{S} = L$ will satisfy (**4N**) since $d\mathfrak{L} \wedge dt$ is closed; hence choose **S** to bound a closed loop $L_{\alpha_1\alpha_2\alpha_3}(\alpha_{123})$ as in (4.7) - with (**2M**). Consider the first half leg $L_{\alpha_1\alpha_2}(\alpha_{123})$, denoted by $(1/2)L_{\alpha_1\alpha_2}(\alpha_{123})$, running from $\alpha_1 \to \alpha_{123}$ with second half $(1/2')L_{\alpha_3\alpha_1}(\alpha_{123})$, running from $\alpha_{123} \to \alpha_1$. The sum of these two half legs (**4O**) $(1/2)L)_{\alpha_1\alpha_2}(\alpha_{123}) + (1/2')L_{\alpha_3\alpha_1}(\alpha_{123})$ will as a rule enclose an area $\mathbf{S}_{\alpha_1}(\alpha_{123})$ so one can write (**4P**) $\partial\mathbf{S}_{\alpha_1}(\alpha_{123}) = (1/2)L)\alpha_1\alpha_2(\alpha_{123}) + (1/2')L_{\alpha_3\alpha_1}(\alpha_{123})$. Similar considerations hold for the other half legs leading to

$$\partial\mathbf{S}_{\alpha_2}(\alpha_{123}) = (1/2)L_{\alpha_2\alpha_3}(\alpha_{123}) + (1/2')L_{\alpha_1\alpha_2}(\alpha_{123}); \tag{4.8}$$

$$\partial\mathbf{S}_{\alpha_3})\alpha_{123}) = (1/2)L_{\alpha_3\alpha_1})\alpha_{123}) + (1/2')L_{\alpha_2\alpha_3}(\alpha_{123})$$

The boundaries of the three surfaces all pass through the variable midpooint $\alpha_{123}$ and one defines their connected sum (**4Q**) $\mathbf{S}_{\alpha_1\alpha_2\alpha_3} = \mathbf{S}_{\alpha_1} + \mathbf{S}_{\alpha_2} + \mathbf{S}_{\alpha_3}$ so that

$$L_{\alpha_1\alpha_2\alpha_3} = \partial\mathbf{S}_{\alpha_1\alpha_2\alpha_4} = \partial\mathbf{S}_{\alpha_1} + \partial\mathbf{S}_{\alpha_2} + \partial\mathbf{S}_{\alpha_3} \tag{4.9}$$

For convenience one assumes here that at least one of the three surfaces on the right in (**4Q**) does not degenerate into a curve.

In general one cannot compute the integral (4.7) exactly but insight is gained via a steepest descent approximation (cf. [**461**], paper 2) where one obtains

$$g^0_{\alpha_1\alpha_2\alpha_3}(q_{\alpha_{123}}) = exp\left[\frac{i}{\hbar}S(L^0_{\alpha_1\alpha_2\alpha_3}(\alpha_{123}))\right] \tag{4.10}$$

where the superindex "0" means evaluation at the extremal (which is the path minimizing the action S under the requirements after (**4L**)). Then summarizing, via (4.7), (**4N**), (**4Q**), (4.9), and (4.10) one can write the steepest descent approximation to the 2-cocycle as

$$g^0_{\alpha_1\alpha_2\alpha_3} = exp\left[\frac{i}{\hbar}\int_{S^0_{\alpha_1\alpha_2\alpha_3}} d\mathfrak{L}\wedge dt\right] \tag{4.11}$$

where $S^0_{\alpha_1\alpha_2\alpha_3}$ is a minimal surface for the integrand $d\mathfrak{L}\wedge dt$. One can use (4.10) and (4.11) to compute the connection (to the same accuracy as the 2-cocycle) via

$$A_{\alpha_1\alpha_2} = \frac{i}{\hbar}(\mathfrak{L}dt)_{\alpha_1\alpha_2};\ H|_{U_\alpha} = dB_\alpha; \tag{4.12}$$

$$B_{\alpha_2} - B_{\alpha_1} = dA_{\alpha_1\alpha_2} = \frac{i}{\hbar}(d\mathfrak{L}\wedge dt)_{\alpha_1\alpha_2}$$

Note here that the potential $A$ is supposed to be a 1-form on $\mathbf{F}$ where the gerbe is defined. Here in (2.12) it is a 1-form on $\mathbf{F}\times\mathbf{R}$ so let $i:\mathbf{F}\to\mathbf{F}\times\mathbf{R}$ be the natural inclusion and the 1-form $A$ in (4.12) is understood to be the pullback $i^*(\mathfrak{L}dt)$ onto $\mathbf{F}$ (written as $\mathfrak{L}dt$). Thus the machinery of quantum mechanics encoded in the Feynman exponential $exp(iS/\hbar)$ has been packaged in the geometric language of gerbes on $\mathbf{F}$ which has produced a Neveu-Schwartz $B$ field. Recall also that a non-vanishing $B_{\mu\nu}$ field across the $q^\mu$, $q^\nu$ plane induces the space noncommutativity (**4B**) with $\theta^{\mu\nu} = aB^{-1}_{\mu\nu}$.

**REMARK 4.1.1.** The presentation of gerbes above seems (perhaps necessarily) complicated and one can find a simpler approach in [**605**] for example (cf. also [**459, 461**] where some of the details of the quantum connections are repeated in a phase space context and the important idea of quantum mechanics as a spontaneously broken gauge theory on a $U(1)$ gerbe is developed). ∎

We will rephrase much of the above now in a phase space context following [**459**] which will bring it closer to Section 1 (cf. also [**369**]). Thus consider a phase space $P$ as a finite dimensional symplectic manifold with symplectic form $\omega$ and Darboux coordinates $(q,p)$. The coordinate free reexpression of Heisenberg's principle (in WKB formulation) is simply

$$\frac{1}{2\pi\hbar}\int S\omega \in \mathbf{Z};\ \partial S = 0 \tag{4.13}$$

(i.e. symplectic area is quantized in units of $\hbar$) and one constructs a $U(1)$ gauge theory of quantum mechanics on phase space. For gerbes one has (4.2) as before with (**4E**) and (••) along with $H$, $B$, $A$ as in (4.3). Now a phase space $P$ may be the cotangent bundle to a configuration manifold M on which is given a mechanical action

$$S = \int_I dt\mathfrak{L} \tag{4.14}$$

for a Lagrangian $\mathfrak{L}$ over a time interval I. On an open set $U_\alpha \subset P$ one picks Darboux coordinates $q_\alpha^j, p_j^\alpha$ such that

$$\omega|_{U_\alpha} = \sum_1^d dq_\alpha^j \wedge dp_j^\alpha \equiv \omega = \sum_{j=1}^d dq^j \wedge dp_j \tag{4.15}$$

The canonical 1-form $\theta$ on $P$ is defined as

$$\theta = -\sum_1^d p_j dq^j;\ d\theta = \omega \tag{4.16}$$

and one also uses here the integral invariant of Poincaré-Cartan denoted by $\lambda$ in the form

$$\lambda = \theta + Hdt \tag{4.17}$$

where H is the Hamiltonian. This means the action (4.14) can be written in terms of the line integral of $\lambda$, namely as (**4R**) $S = -\int_I \lambda$, and on constant energy submanifolds of $P$ (or for fixed time), one has (**4S**) $d\lambda = \omega$ (H=constant). The index $j$ is dropped now while maintaining the index $\alpha$ of Čech cohomology and given 3 points $(q_{\alpha_i}, p_{\alpha_i})$ on $P$ covered by charts $U_{\alpha_i}$ with (**4T**) $\cap U_{\alpha_i} \neq \phi$ let (**4U**) $(q_{\alpha_{123}}, p_{\alpha_{123}})$ be a variable point in the triple overlap. Also as before let $L_{\alpha_1\alpha_2\alpha_3}(\alpha_{123})$ be a closed loop

$$L_{\alpha_1\alpha_2\alpha_3}(\alpha_{123}) = L_{\alpha_1\alpha_2}(\alpha_{123}) + L_{\alpha_2\alpha_3}(\alpha_{123}) + L_{\alpha_3\alpha_1}(\alpha_{123}) \tag{4.18}$$

Dropping the point $\alpha_{123}$ for convenience one has a 2-cocycle defining a $U(1)$ gerbe on $P$ given via (cf. (4.7))

$$g_{\alpha_1\alpha_2\alpha_3} = \frac{\tilde{g}_{\alpha_1\alpha_2\alpha_3}}{|\tilde{g}_{\alpha_1\alpha_2\alpha_3}|}; \tag{4.19}$$

$$\tilde{g}_{\alpha_1\alpha_2\alpha_3} \sim \int DL_{\alpha_1\alpha_2 g\alpha_3} exp\left[-\frac{i}{\hbar}\int_{L_{\alpha_1\alpha_2\alpha_3}} \lambda\right]$$

The integral (4.19) extends over all the closed trajectories of the type specified in (4.18). Consider now the sum of surfaces (**4V**) $S_{\alpha_1\alpha_2\alpha_3} = S_{\alpha_1\alpha_2} + S_{\alpha_1\alpha_3} + S_{\alpha_3\alpha_1}$ (dropping again the expression $\alpha_{123}$) and the closed trajectory (4.18) bounds the surface (**2V**), i.e. $L_{\alpha_1\alpha_2\alpha_3} = \partial S_{\alpha_1\alpha_2\alpha_3}$. Picking $S_{\alpha_1\alpha_2\alpha_3}$ to be a constant energy surface in $P$ or else using fixed values for the time one has by (**2S**) and Stokes' theorem

$$\tilde{g}_{\alpha_1\alpha_2\alpha_3} \sim \int DS_{\alpha_1\alpha_2\alpha_3} exp\left[-\frac{i}{\hbar}\int_{S_{\alpha_1\alpha_2\alpha_3}} \omega\right] \tag{4.20}$$

with integral extending over all surfaces (**4V**). One computes (4.19) in the stationary-phase approximation ($\hbar \to 0$) as in [**907**]. Then the stationary phase 2-cocycle

$g^0_{\alpha_1\alpha_2\alpha_3}$ defining a $U(1)$-gerbe on $P$ is (cf. [**461**])

$$g^0_{\alpha_1\alpha_2\alpha_3} = exp\left[-\frac{i}{\hbar}\int_{L^0_{\alpha_1\alpha_2\alpha_3}} \omega\right] \tag{4.21}$$

Alternatively one can express $g^0_{\alpha_1\alpha_2\alpha_3}$ in terms of an integral over an extremal surface as in (4.20) in the form

$$g^0_{\alpha_1\alpha_2\alpha_3} = exp\left[-\frac{i}{\hbar}\int_{S^0_{\alpha_1\alpha_2\alpha_3}} \omega\right] \tag{4.22}$$

Thus (4.21)-(4.22) give the stationary-phase approximation $g^0$ to the 2-cocycle $g_{\alpha_1\alpha_2\alpha_3}$ which is a function of the variable midpoint (**4U**) and one drops now the superindex 0 on everything with the understanding to work in the stationary-phase approximation (which is equivalent to the quantum mechanical WKB approximation). Using (4.3) and (4.18) one finds now for the 1-form A (**4W**) $A = -(i/\hbar)\lambda$ and on constant energy submanifolds of $P$ (**4X**) $B_{\alpha_2} - B_{\alpha_1} = -(i/\hbar)\omega_{\alpha_1\alpha_2}$ (where $\omega_{\alpha_1\alpha_2}$ is the restriction to a non-empty $U_{\alpha_1} \cap U_{\alpha_2}$); finally (**4Y**) $H = dB$. In the WKB approximation one knows that the symplectic area of any open surface $S_{\alpha_1\alpha_2\alpha_3}$ is quantized via (cf. [**907**])

$$\frac{1}{\hbar}\int_{S_{\alpha_1\alpha_2\alpha_3}} \omega = 2\pi\left(n_{\alpha_1\alpha_2\alpha_3} + \frac{1}{2}\right);\ n_{\alpha_1\alpha_2\alpha_3} \in \mathbf{Z} \tag{4.23}$$

Consider now two open, constant energy, symplectically minimal surfaces $S^i \subset P$ such that $\partial S^1 = -\partial S^2$; join them along their common boundary to form the closed surface $S = S^1 - S^2$ which bounds a 3-D volume V. Then following [**461**] (4.23) can be recast as a quantization condition (**4Z**) $(1/2\pi i)\int_V H \in \mathbf{Z}$ $(\partial V = S)$ under certain conditions. (**4Z**) is an equivalent coordinate-free rendering of the Heisenberg uncertainty principle making no use of Darboux coordinates on $P$; but it is not an equation in deRham cohomology because the volumes $V$ have a boundary. Since gerbes are characterized by integral deRham classes on compact manifolds there are problems in that $P$ is not compact and hence the gerbe constructed above cannot be identified with $[H]/2\pi i$.

**REMARK 4.1.2.** In (**4R**) one can write (♦♦) $\lambda \to \lambda + df$ $(f \in C^\infty(P))$ and since the action is given via (**4R**) this transformation amounts to shifting $S$ by a constant $C$, i.e. $S \to S + C$, $C = -\int df$. In quantum mechanics the WKB approximation involves $\psi_{WKB} = Rexp(iS/hbar)$ and (♦♦) multiplies any wave function by a constant phase factor $exp(iC/\hbar)$, i.e. $\psi \to exp(iC/\hbar)\psi$. Hence (★★) $\psi =\!\to \psi_f = exp(-if/\hbar)\psi$ with $f$ an arbitrary phase space function with dimensions of action. ∎

**REMARK 4.1.3.** One can refer to objects such as $\psi_f(q,p)$ as probability distributions and from [**369, 901**] one knows that the usual SE on $M$ (•••) $H(q, -i\hbar\partial_q)\psi(q) = E\psi(q)$ implies the Schrödinger like equation

$$H\left(\frac{q}{2} + i\hbar\partial_p, \frac{p}{2} - i\hbar\partial_q\right)\psi(q,p) = E\psi(q,p) \tag{4.24}$$

Moreover the quantum operators $Q_{A'_0} = (q/2) + i\hbar\partial_p$ and $P_{A'_0} = (p/2) - i\hbar\partial_q$ satisfy the usual Heisenberg equation $[Q_{A'_0}P_{A'_0}] = i\hbar$ so (2.24) can be written as

$$H(Q_{A'_0}, P_{A'_0})\psi(q,p) = E\psi(q,p) \tag{4.25}$$

and a computation shows that $\psi(q,p)$ in (4.24) and in ($\bullet\bullet\bullet$) are related as in ($\bigstar\bigstar$) where the argument $f(q,p)$ has the form $f_{A'_0}(q,p) = (1/2)pq = (1/2)p_j q^j$. Thus the SE (4.24) follows from ($\bullet\bullet\bullet$) if and only if

$$\psi(q,p) = exp\left(-\frac{i}{2\hbar}pq\right)\psi(q) \tag{4.26}$$

A straightforward computation shows that (4.26) corresponds to the choice $\phi = (2\pi\hbar)^{d/2}\delta(q)$ in [**373**], paper 1 where it is also shown that $|\psi|^2$ is a probability distribution for $\hbar \to 0$; indeed if $\Psi = U_\phi\psi$ for $\phi(q) = [1/(\pi\hbar)^{d/4}]exp[-(1/2\hbar)|q|^2]$ then

$$lim_{\hbar\to 0}\int|\Psi(q,o)|^2 dp = |\psi(q)|^2;\ lim_{\hbar\to 0}\int|\Psi(q,p)|^2 dq = |\hat{\psi}(p)|^2 \tag{4.27}$$

(where $\hat{\psi}$ denotes the Fourier transform). The reason for the subindex $A'_0$ above involves the symplectic exterior derivative on $P$, $d' = -dq\partial_q + dp\partial_p$. Thus consider the connection $A'_0$ on $P$ given via $A'_0 = -(i/\hbar)df_{A'_0} = (1/2i\hbar)(pdq + qdp)$. Then covariantize $d'$ as $d' \to D'_{A'_0} = d' + A'_0$ from which one sees that $Q$, $P$ based on $A'_0$ are the result of gauging the symplectic derivative $d'$ by the connection $A'_0$, i.e.

$$i\hbar D'_{A'_0} = dq\left(\frac{p}{2} - i\hbar\partial_q\right) + dp\left(\frac{q}{2} + i\hbar\partial_p\right) \tag{4.28}$$

Such covariantizing is equivalent to the symplectic transformation considered in [**373**] for example which renders the quantum theory manifestly symmetric under the symplectic exchange of $q$ and $p$. We refer to [**459**] for more in this spirit. ∎

CHAPTER 4

# KALUZA-KLEIN AND COSMOLOGY

## 1. INTRODUCTION

Referring to [**473**] one recalls the Maxwell equations

$$\nabla\cdot\mathbf{D}=\rho;\ \nabla\times\mathbf{H}=\mathbf{J}+\partial_t\mathbf{D};\ \nabla\cdot\mathbf{B}=0;\ \nabla\times\mathbf{E}+\partial_t\mathbf{B}=0 \tag{1.1}$$

These are traditionally accompanied by a scalar potential $\phi$ and a vector potential $\mathbf{A}$ with

$$\mathbf{B}=\nabla\times\mathbf{A};\ \nabla\times(\mathbf{E}+\partial_t\mathbf{A})=0 \tag{1.2}$$

leading to (**1A**) $\mathbf{E}+\partial_t\mathbf{A}=-\nabla\phi$ and consequently

$$\nabla^2\phi+\partial_t(\nabla\cdot\mathbf{A})=-\frac{\rho}{\epsilon_0};\ \nabla^2\mathbf{A}-\frac{1}{c^2}\partial_t^2\mathbf{A}-\nabla\left(\nabla\cdot\mathbf{A}+\frac{1}{c^2}\partial_t\phi\right)=-\mu_0\mathbf{J} \tag{1.3}$$

($\mathbf{D}=\epsilon_0\mathbf{E}$ and $\mathbf{B}=\mu_0\mathbf{H}$). There are gauge transformations

$$\mathbf{A}\to\mathbf{A}'=\mathbf{A}+\nabla\Lambda;\ \phi\to\phi'=\phi-\partial_t\Lambda \tag{1.4}$$

and one can satisfy the Lorentz condition (**1B**) $\nabla\cdot\mathbf{A}'+(1/c^2)\partial_t\phi'=0$ which uncouples (1.3) to yield

$$\nabla^2\phi-\frac{1}{c^2}\partial_t^2\phi=-\frac{1}{c^2}\partial_t\phi';\ \nabla^2\mathbf{A}-\frac{1}{c^2}\partial_t^2\mathbf{A}=-\mu_0\mathbf{J} \tag{1.5}$$

Thus given a gauge function $\Lambda$ such that

$$\nabla^2\Lambda-\frac{1}{c^2}\partial_t^2\Lambda=-\left(\nabla\cdot\mathbf{A}+\frac{1}{c^2}\partial_t\phi\right) \tag{1.6}$$

the new potentials $\mathbf{A}'$, $\phi'$ will satisfy the Lorentz condition and give rise to (1.5).

**REMARK 1.1.1** Mathematically one often writes the Maxwell equations as $dF=0$ and $d*F=0$ (cf. [**543**]) with Einstein-Maxwell equations in the form $dF=0$, $d*F=0$, and $[R_{jk}+F\circ F]_0=0$ where $R_{jk}$ is the Ricci tensor of $g$ and $(F\circ F)_{jk}=F_j^\ell F_{\ell k}$. These latter equations are understood as the Euler-Lagrange equations for the functional $\int_M(R+|F|^2)d\mu_g$ where $F$ varies over all closed 2-forms in a given deRham class and $g$ varies over Riemannian metrics of some fixed volume $V$. In dimension 4 (and only there) the Einstein-Maxwell equations imply that the scalar curvature $R$ is constant (see [**543**] for this and many other fascinating results. ■

We go now to [**389**] for some basic axiomatics (see also [**256, 426, 427, 428, 429, 430, 469, 533, 695, 720**] for more detailed expositions).

(1) Axiom 1 involves a charge density $\rho$ (3-form) with (**1C**) $Q = \int_V \rho dV$ and via the Poincaré lemma for suitably simple spaces (**1D**) $div(\mathfrak{D}) = \rho$. This is artificial of course since $\mathfrak{D}$ is not uniquely determined. One defines also a connective derivative $D/Dt$ such that $DQ/dt = 0$ and writes

$$\frac{DQ}{Dt} = \frac{D}{Dt}\int_{V(t)} \rho dV = \int_{V(t)} \partial_t \rho dV + \oint_{\partial V(t)} \rho u^i da_i \tag{1.7}$$
$$= \int_{V(t)} (\partial_t \rho + \partial_i(\rho u^i))dV$$

Then via (**1D**) one has (**1E**) $DQ/Dt = 0 \equiv \partial_t + \partial_i J^i = 0 \Rightarrow \partial_i(\partial_t D^i + J^i) = 0$ and consequently in suitably simple spaces

$$\partial_t D^i + J^i = \epsilon^{ijk}\partial_j H_k \equiv \mathfrak{J} = curl(\mathbf{H}) - \dot{\mathfrak{D}} \tag{1.8}$$

(2) Axiom 2 postulates a Lorentz force (**1F**) $F_q(E_i + \epsilon_{ijk}u^j B^k)$ (for electrical field strength $E_i$ and magnetic field strength $B^i$) acting on a small charge $q$ moving with velocity $u^j$. Some argument yields $F_i = 0$ so (**1G**) $E_i = -\epsilon_{ijk}u^j B^k$.

(3) For Axiom 3 one digresses first to hydrodynamics and relates vortex lines to magnetic flux lines. Some argument then leads to (**1H**) $div(\mathbf{B}) = 0$ (assuming no magnetic charge density). Further (**1I**) $\partial_t B^i + \epsilon^{ijk}\partial_j E_k = 0 \equiv \dot{\mathbf{B}} + curl(\mathbf{E}) = 0$ (Faraday induction law) reflects magnetic flux conservation).

(4) Axiom 4 relates field strengths and excitations via

$$D^i = \epsilon_0\sqrt{g}g^{ij}E_j;\ H_i = (\mu_0\sqrt{g})^{-1}g_{ij}B^j \tag{1.9}$$

and this connects the electromagnetic field to the metric structure of space-time (via $g_{ij}$). In fact, given no birefringence, one can derive the metric from propagation properties of the electromagnetic field (cf. [**426, 533**]). We omit here any discussion of skewons or axions which arise in connection with constitutive equations in matter genealizing (1.9).

There are 2 gauge potentials $\phi$ and $\mathbf{A}$ where

$$E_i = -\partial_i\phi - \partial_t A_i;\ B^i = \epsilon^{ijk}\partial_j A_k \tag{1.10}$$

and these define covariant derivatives

$$D_t^\phi = \partial_t + \frac{q}{\hbar}\phi;\ D_i^A = \partial_i - \frac{q}{\hbar}A_i \tag{1.11}$$

The gauge approach does not however reflect properties of space-time. Generally $(E_i, B^i)$ (field strengths) and $(D^i, H_i)$ (excitations) are grouped together.

**REMARK 1.1.2.** The axiomatics for electromagnetism seem to be in good shape but there is considerable doubt expressed in [**819**] for example concerning axiomatic formulations for general relativity (GR). We quote from [**819**] in their concluding remarks: "In conclusion, even though significant progress has been made on the front of gravitation theories, one cannot help but notice that it is still unclear how to relate principles and experiments in order to form simple theoretical viability criteria expressed in a mathematical way. Our inability to enunciate these criteria, as well as several of our very basic definitions, in a representation-invariant way seem to have played a crucial role in this lack of progress. However, this seems to be a critical obstacle to overcome, if we want to go beyond a trial-and-error approach when it comes to gravitational theories." ■

## 2. KALUZA-KLEIN THEORY

We begin with material from [**656, 868**] to which we refer for background, philosophy, motivation, explanations, and details. Consider a perfect isotropic fluid with density $\rho$ and pressure p (i.e. no viscosity and the pressure is equal in the 3 spatial directions). Then the energy-momentum tensor is (**2A**) $T_{ab} = (p+\rho c^2)u_a u_b - p g_{ab}$ where $u_a$ are the 4-velocities. This is constructed to have zero divergence and the equation of continuity and the equations of motion for the 3 spatial directions are derived from the 4 components of $\nabla_a T^{ab} = 0$. The full field equations can be written as

$$R_{ab} - \frac{1}{2}R g_{ab} + \Lambda g_{ab} = \frac{8\pi G}{c^4}[(p+\rho c^2)u_a u_b - p g_{ab}] \tag{2.1}$$

It is often convenient to move $\Lambda$ to the other side and incorporate it into $T_{ab}$ via (**2B**) $\rho_v = \Lambda c^2/8\pi G$ and $p_v = -\Lambda c^4/8\pi G$. The roles of pressure and density can be appreciated via the Raychaudhuri equations

$$\ddot{a}\left(\frac{3}{a}\right) = 2(\omega^2 - \sigma^2) - \frac{4\pi G}{c^2}(3p + \rho c^2);\ \dot{\rho}c^2 = -(p+\rho c^2)\frac{3\dot{a}}{a} \tag{2.2}$$

Here $a$ is the scale factor of a region of fluid with vorticity $\omega$, shear $\sigma$, and uniform pressure and density (a dot denotes the time derivative). Note that for mass to be attractive one needs (**2C**) $(3p + \rho c^2) > 0$ (gravitational energy density) and for stability (**2D**) $(p + \rho c^2) > 0$ is required (inertial energy density). For a homogeneous and isotropic fluid without viscosity or shear Einstein's equations reduce to the Friedman equations

$$8\pi G\rho = \frac{3}{a^2}(kc^2 + \dot{a}^2) - \Lambda c^2;\ \frac{8\pi G p}{c^2} = -\frac{1}{a^2}(kc^2 + \dot{a}^2 + 2a\ddot{a}) + \Lambda c^2 \tag{2.3}$$

where $k = \pm 1$ or 0 is a curvature constant describing the departure of the 3-D part of the spacetime from Minkowski space ($\eta_{ab} = diag(1,-1,-1,-1)$). The 3-D isotropic and non-isotropic forms of the metric are

$$ds^2 = c^2dt^2 - \frac{a^2(t)}{[1+(kr^2/4)]^2}[dr^2 + r^2 d\Omega^2]; \tag{2.4}$$

$$ds^2 = c^2dt^2 - a^2(t)\left[\frac{dr^2}{(1-kr^2)} + r^2 d\Omega^2\right]$$

where $d\Omega^2 = d\theta^2 + Sin^2(\theta)d\phi^2$. A photon moving radially is defined by $ds = 0$ with $d\theta = d\phi = 0$ leading to

$$\frac{dr}{ds} = \pm\frac{c(1-kr^2)^{1/2}}{a(t)} \tag{2.5}$$

and of course the speed is not $c$. FRW models have event or particle horizons (or both) and the distance to the particle horizon defines the size of that part of the universe which is in causal communication with us. In terms of the Hubble parameter $H_0 = \dot{a}_0/a_0$ and the deceleration parameter $q_0 = -a_0\ddot{a}_0/\dot{a}_0^2$ the distances are

$$d_{k=1} = \frac{c}{H_0(2q_0-1)^{1/2}} Cos^{-1}\left(\frac{1}{q_0}-1\right),\ q_0 > \frac{1}{2}; \tag{2.6}$$

$$d_{k=0} = \frac{2c}{H_0} = 3ct_0,\ q_0 = \frac{1}{2};\ d_{k=-1} = \frac{c}{H_0(1-2q_0)^{1/2}} Cosh^{-1}\left(\frac{1}{q_0}-1\right)$$

(the latter for $q_0 < \frac{1}{2}$). Even for the middle case (the Einstein-deSitter model with flat 3-space sections) the distance to the horizon is not $ct_0$ showing that in relativity the purpose of $c$ is merely to transpose time to a length. In particle physics with a finite rest mass the motion is along paths with $s$ a minimum and assuming $m$ constant with $ds^2 = g_{ab}dx^a dx^b$ the minimum principle leads to **(2E)** $(du^c/ds) + \Gamma^c_{ab}u^a u^b = 0$ (geodesic equation).

With some repetition now classical electromagnetism is described via a 4-potential $A_\alpha$ and a 4-current $J^\alpha$ with Maxwell equations contained in the tensor relations

$$\frac{\partial F^{\alpha\beta}}{\partial x^\alpha} = \frac{4\pi}{c}J^\beta;\ F_{\alpha\beta} = \frac{\partial A^\beta}{\partial x^\alpha} - \frac{\partial A^\alpha}{\partial x^\beta};\ \frac{\partial F_{\alpha\beta}}{\partial x^\gamma} + \frac{\partial F_{\beta\gamma}}{\partial x^\alpha} + \frac{\partial F_{\gamma\alpha}}{\partial x^\beta} = 0 \tag{2.7}$$

These can be obtained by using the Lagrangian

$$L = -\frac{1}{16\pi}F^{\alpha\beta}F_{\alpha\beta} - \frac{1}{c}J^\alpha A_\alpha \tag{2.8}$$

for the Euler-Lagrange equations. One recalls that for quantum mechanics (QM) the rule $p \to (\hbar/i)\nabla$ and $E \to (i\hbar)\partial/\partial t$ applied to the non-relativistic energy equation $(p^2/2m)+V = E$ yields the Schrödinger equation (SE) **(2F)** $-(\hbar^2/2m)\nabla^2\psi + V\psi = i\hbar\partial_t\psi$. For a particle with charge $q$ moving with a 3-velocity $(dx/dt) << c$ in an EM field the Lagrangian is $L = (m/2)(\dot{x}^2) - (q/c)A_\alpha\dot{x}^\alpha$ with path action $S = \int_{t_0}^{t_1} Ldt$. Alternatively one could redefine the Lagrangian as **(2G)** $L = (m/2\hbar)\dot{x}^2 = (q/c\hbar)A_\alpha\dot{x}^\alpha$ and one sees the importance of $q^2/\hbar$ and $m/\hbar$ (cf. Wesson [**868**] for details). The relativistic energy equation $E^2 - p^2c^2 = m^2c^4$ leads (for a freely moving particle with $V = 0$) **(2H)** $-(1/c^2)\partial_t^2\phi + \nabla^2\phi = (mc/\hbar)^2\phi$ with Lagrangian **(2J)** $L = (1/2)[(\phi_t/c)^2 - (\nabla\phi)^2] - (1/2)(mc/\hbar)^2\phi^2$ (flat spacetime). For spin 1/2 one has an energy equation $p_\alpha P^\alpha = m^2c^2$ which can be factorized via $\gamma$-matrices $\gamma^a\gamma^b + \gamma^b\gamma^a = 2\eta^{ab}$ to (•) $i\hbar\gamma^\alpha\partial_\alpha\psi - mc\psi = 0$ for a bi-spinor field $\psi$; the Lagrangian is **(2K)** $L = i\hbar c\bar{\psi}\partial_\alpha\psi - mc^2\bar{\psi}\psi$ (where $\bar{\psi} = \psi^T\gamma^0$). In order to achieve gauge invariance under $\psi \to exp(i\theta(x))\psi$ one replaces this by **(2L)** $L = i\hbar c\bar{\psi}\gamma^\alpha\partial_\alpha\psi - mc^2\bar{\psi}\psi - q\bar{\psi}\gamma^\alpha\psi A_\alpha$ (where $A_\alpha \to A_\alpha + \partial_\alpha\lambda$ under gauge transformations - $\lambda(x^\alpha)$ is a scalar function). This should in fact be further extended by including a "free" term for the gauge field to

$$L = i\hbar c\bar{\psi}\partial_\alpha\psi - mc^2\bar{\psi}\psi - \frac{1}{16\pi}F^{\alpha\beta}F_{\alpha\beta} - q\bar{\psi}\gamma^\alpha\psi A_\alpha \tag{2.9}$$

This is OK for massless photons but for spin 1 particles such as massive photons one would use the Proca equation **(2M)** $\partial_\alpha F^{\alpha\beta} + (mc/\hbar)^2A^\beta = 0$ with **(2N)** $L = -(1/16\pi)F^{\alpha\beta}F_{\alpha\beta} + (1/8\pi)(mc/\hbar)^2A^\alpha A_\alpha$.

We omit Yang-Mills and quantum chromodynamics here (cf. [**868**]) and go next to Kaluza-Klein. Thus for 5-D there are 15 dimensionless potential $g_{AB}$ (where $A, B = 0, 1, 2, 3, 4$). For simplicity take first $g_{AB} = g_{AB}(x^\alpha)$ with $g_{44} = -\Phi^2(x^\alpha)$ so that

$$g_{AB} = \begin{bmatrix} (g - \alpha\beta - \kappa^2\Phi^2 A_\alpha A_\beta) & -\kappa\Phi^2 A_\alpha \\ -\kappa\Phi^2 A_\beta & -\Phi^2 \end{bmatrix} \tag{2.10}$$

The field equations are then

$$G_{\alpha\beta} = \frac{\kappa^2\Phi^2}{2}T_{\alpha\beta} - \frac{1}{\Phi^2}(\nabla_\alpha\nabla_\beta\Phi - g_{\alpha\beta}\Box\Phi); \tag{2.11}$$

$$\nabla^\alpha F_{\alpha\beta} = -3\frac{\nabla^\alpha\Phi}{\Phi}F_{\alpha\beta};\ \Box\Phi = -\frac{\kappa^2\Phi^3}{4}F_{\alpha\beta}F^{\alpha\beta}$$

with $T_{\alpha\beta} = [(1/4)g_{\alpha\beta}F_{\gamma\delta}F^{\gamma\delta} - F^\gamma_\alpha F_{\beta\gamma}]/2$ and $\Box = g^{\alpha\beta}\nabla_\alpha\nabla_\beta$. Kaluza's case $g_{44} = -\Phi^2 = -1$ with $\kappa = (16\pi G/c^4)^{1/2}$ yields for (2.11)

$$G_{\alpha\beta} = \frac{8\pi G}{c^4}T_{\alpha\beta};\ \nabla^\alpha F_{\alpha\beta} = 0 \tag{2.12}$$

which are the Einstein and Maxwell equations in 4-D but derived from the vacuum in 5-D. Note they involve from (2.11) the choice of EM gauge $F_{\alpha\beta}F^{\alpha\beta} = 0$ and have no contribution from the scalar field. In general the KK field equations $G_{AB} = 0$ or $R_{AB} = 0$ describe a spin 2 graviton, a spin 1 photon, and a spin 0 boson (Higgs). We omit discussion of supergravity and superstrings.

## 3. INDUCED MATTER THEORY

The idea here is that one extra dimension is enough to explain the phenomenological properties of classical matter. One chooses units so that $c = G = \hbar = 1$ and the metric signature is $(+, +, \cdots, \pm)$. The 5-D field equations for an apparent vacuum are (**3A**) $R_{AB} = 0$ or using a 5-D Ricci scalar (**3B**) $G_{AB} = R_{AB} - Rg_{AB}/2 = 0$ while the 4-D field equations are (**3C**) $G_{\alpha\beta} = 8\pi T_{\alpha\beta}$. The central thesis of Wesson's induced matter theory is that (**3C**) is a subset of (**3B**) with an induced 4-D energy momentum tensor which contains the classical properties of matter. Thus let the 5-D metric have the form

$$dS^2 = e^\nu dt^2 - e^\omega(dr^2 + r^2 d\Omega^2) - e^\mu d\ell^2 \tag{3.1}$$

where $x^0 \sim t$ and $x^{1,2,3} \sim r, \theta, \phi$ with $d\Omega^2 = d\theta^2 + Sin^2(\theta)d\phi^2$ and $x^4 \sim \ell$. The coefficients $\nu$, $\omega\ \mu$ will depend on $t$ and $\ell$ with corresponding partial derivatives denoted via an overdot and an asterisk respectively. The Einstein tensor will have then the form

$$G^0_0 = e^{-\nu}\left(-\frac{3\dot\omega^2}{4} - \frac{3\dot\omega\dot\mu}{4}\right) + e^{-\mu}\left(\frac{\overset{**}{\omega}}{2} + \frac{3\,\overset{*}{\omega}{}^2}{2} - \frac{3\,\overset{**}{\mu}\,\omega}{4}\right); \tag{3.2}$$

$$G^0_4 = e^{-\nu}\left(\frac{3\,\overset{.*}{\omega}}{2} + \frac{3\dot\omega\,\overset{*}{\omega}}{4} - \frac{3\dot\omega\,\overset{*}{\nu}}{4} - \frac{3\,\overset{*}{\omega}\,\dot\mu}{4}\right);$$

$$G^1_1 = G^2_2 = G^3_3 = -e^{-\nu}\left(\ddot\omega + \frac{3\dot\omega^2}{4} + \frac{\ddot\mu}{2} + \frac{\dot\mu^2}{4} + \frac{\dot\omega\dot\mu}{2} - \frac{\dot\nu\dot\omega}{2} - \frac{\dot\nu\dot\mu}{4}\right)$$

$$+e^{\mu}\left(\overset{**}{\omega}+\frac{3\overset{*}{\omega}^{2}}{4}+\frac{\overset{**}{\nu}}{2}+\frac{\overset{*}{\nu}^{2}}{4}+\frac{\overset{*}{\omega}\overset{*}{\nu}}{2}-\frac{\overset{*}{\mu}\overset{*}{\omega}}{2}-\frac{\overset{*}{\nu}\overset{*}{\mu}}{4}\right);$$

$$G_4^4=-e^{-\nu}\left(\frac{3\ddot{\omega}}{2}+\frac{3\dot{\omega}^2}{2}-\frac{3\dot{\nu}\dot{\omega}}{4}\right)+e^{-\mu}\left(\frac{3\overset{*}{\omega}^{2}}{4}+\frac{3\overset{*}{\omega}\overset{*}{\nu}}{4}\right)$$

One wants to match these to $T_{\alpha\beta}=(p+\rho)u_\alpha u_\beta-pg_{\alpha\beta}$ where $u^\alpha=\dot{x}^\alpha$ and $T_0^0=\rho$ with $T_1^1=-p$. This leads to

$$8\pi\rho=-\frac{3}{4}e^{-\nu}\dot{\omega}\dot{\mu}+\frac{3}{2}e^{-\mu}\left(\overset{**}{\omega}+\overset{*}{\omega}^{2}-\frac{\overset{*}{\mu}\overset{*}{\omega}}{2}\right); \tag{3.3}$$

$$8\pi p=e^{-\nu}\left(\frac{\ddot{\mu}}{2}+\frac{\dot{\mu}^2}{4}+\frac{\dot{\omega}\dot{\mu}}{2}-\frac{\dot{\nu}\dot{\mu}}{4}\right)$$

$$-e^{-\mu}\left(\overset{**}{\omega}+\frac{\overset{*}{\omega}^{2}}{4}+\frac{\overset{**}{\nu}}{2}+\frac{\overset{*}{\nu}^{2}}{4}+\frac{\overset{*}{\omega}\overset{*}{\nu}}{2}-\frac{\overset{*}{\mu}\overset{*}{\omega}}{2}-\frac{\overset{*}{\nu}\overset{*}{\mu}}{4}\right)$$

In order to see if this makes sense one combines this with $G_B^A=0$ from (**3B**) which gives

$$G_0^0=-\frac{3}{4}e^{-\nu}\dot{\omega}^2+8\pi\rho=0;\ G_1^1=-e^{-\nu}\left(\ddot{\omega}+\frac{3}{4}\dot{\omega}^2-\frac{\dot{\nu}\dot{\omega}}{2}\right)-8\pi p=0; \tag{3.4}$$

$$G_4^0=e^{-\nu}\left(\frac{3}{2}\dot{\overset{*}{\omega}}+\frac{3}{4}\dot{\omega}\,\overset{*}{\omega}-\frac{3}{4}\dot{\omega}\,\overset{*}{\nu}-\frac{3}{4}\overset{*}{\omega}\,\dot{\mu}\right)=0;$$

$$G_4^4=-e^{-\nu}\left(\frac{3}{2}\ddot{\omega}+\frac{3\dot{\omega}^2}{2}-\frac{3\dot{\nu}\dot{\omega}}{4}\right)+e^{-\mu}\left(\frac{3}{4}\overset{*}{\omega}^{2}+\frac{3\overset{*}{\omega}\overset{*}{\nu}}{4}\right)=0$$

Here $\rho$ must be positive from $G_0^0$ and $p$ could be negative in principle (via $G_1^1$) as is needed in classical descriptions of particle production in QFT. A simple solution of (**3A**)-(**3B**) which does not depend on $\ell$ has $\nu=0,\ \omega=log(t)$, and $\mu=-log(t)$ in (3.1) which becomes

$$dS^2=dt^2-t(dr^2+r^2d\Omega^2)-t^{-1}d\ell^2 \tag{3.5}$$

(with $8\pi\rho=3/4t^2$ and $8\pi p=1/4t^2$). If these are combined to form the gravitational density $\rho+3p$ and the proper radial distance $a=exp(\omega/2)r$ is used then the mass of a porotion of the fluid is $M=4\pi a^3(\rho+3p)/3$ and the field equations ensure that $\ddot{a}=-M/a^2$. Similarly the first law of thermodynamics $dE+pdV=0$ is recovered via $[\rho exp(3\omega/2)]^\bullet+p[exp(3\omega/2)]^\bullet=0$. The equation of state of the fluid described via (3.5) is the radiation type $p=\rho/3$ and the space has a shrinking fifth dimension.

To go beyond radiation one goes to solutions from [**717, 718, 719**] with $exp(\nu)=\ell^2$, $exp(\omega)=t^{2/\alpha}t^{2/(1-\alpha)}$, and $exp(\mu)=\alpha^2(1-\alpha)^{-2}t^2$ in (3.1) leading to

$$dS^2=\ell^2dt^2-t^{2/\alpha}\ell^{2/(1-\alpha)}(dr^2+r^2d\Omega^2)-\alpha^2(1-\alpha)^{-2}t^2d\ell^2 \tag{3.6}$$

This has a growing fifth dimension and (**3D**) $8\pi\rho = 3/\alpha^2t^2\ell^2$ with $8\pi p = (2\alpha - 3)/\alpha^2t^2\ell^2$ (the proper time is $T = \ell t$). For $\alpha = 3/2$ one has a 4-D Einstein-deSitter model for the late universe with dust and for $\alpha = 2$ one has an early universe standard model with radiation or highly relativistic particles (see [**868**]). Further the Ponce de Leon metric (3.6) in standard form becomes (**3E**) $dS^2 = dT^2 - (da^2 + a^2d\Omega^2) - dL^2$ where

$$(3.7)\qquad T = \left(\frac{\alpha}{2}\right)t^{1/\alpha}\ell^{1/(1-\alpha)}\left(1+\frac{r^2}{\alpha^2}\right) - \frac{\alpha}{2(1-2\alpha)}[t^{-1}\ell^{\alpha/(1-\alpha)}]^{(1-2\alpha)/\alpha};$$

$$L = \frac{\alpha}{2}t^{1/\alpha}\ell^{1/(1-\alpha)}\left(1-\frac{r^2}{\alpha^2}\right) + \frac{\alpha}{2(1-2\alpha)}[t^{-1}\ell^{\alpha/(1-\alpha)}]^{((1-2\alpha)/\alpha}$$

with $a = rt^{1/\alpha}\ell^{1/(1-\alpha)}$, which indicates that the universe can either be viewed as a 4D spacetime curved by matter or as an empty 5D space (see Wesson [**868, 869**] for more details - there are here some indications that $\ell$ is related to mass).

KK theory in 5-D has traditionally identified the $g_{4\alpha}$ components with the EM potentials $A_\alpha$ and setting these equal to zero gives in some sense a description of neutral matter. Another natural case to consider is $g_{4\alpha} = 0$ with $g_{44} \neq 0$; in particular taking $g_{\alpha\beta} = g_{\alpha\beta}(x^A)$ with $g_{44} = g_{44}(x^A)$ is not restricted by the cylinder conditions of classical KK theory. Thus one has a natural 5-D interval $dS^2 = g_{AB}dx^Adx^B$ with

$$(3.8)\qquad g_{\alpha\beta} = g_{\alpha\beta}(x^A);\ g_{4\alpha} = 0;\ g_{44} = \epsilon\Phi^2(x^A);\ g_{44} = \epsilon\Phi^2(x^A) = \frac{1}{g_{44}} = \frac{\epsilon}{\Phi^2}$$

where $\epsilon^2 = 1$ (cf. [**871**]). In this regard one has (**3F**) $R_{AB} = \partial_C(\Gamma^C_{AB}) - \partial_B\Gamma^C_{AC} + \Gamma^C_{AB}\Gamma^D_{CD} - \Gamma^C_{AD}\Gamma^D_{BC}$ (see [**868**] for calculations of ${}^5R$, ${}^4R$, the 4-D energy momentum tensor, and the Christoffel symbols. The result is a set of equations for the 4-D Ricci tensor $R_{\alpha\beta}$, a wave equation for $\Phi$ from $R_{44} = 0$, and a set of conservation laws from $R_{4\alpha} = 0$.

## 4. MORE ON ELECTROMAGNETISM

We go now to [**868**], Chapter 5 and begin with $x^4 = \ell$ while (**4A**) $dS^2 = \gamma_{AB}dx^Adx^B$. It is convenient to define a unit 5-D vector $\psi^4$ tangent to the fifth dimension (**4B**) $\psi^A = \delta^A_4/\sqrt{\epsilon\gamma_{44}}$ and a projector (**4C**) $g_{AB} = \gamma_{AB} - \epsilon\psi_A\psi_B$ where (**4D**) $g_{AB} = \gamma_{AB} - [\gamma_{4A}\gamma_{4B}/\gamma_{44}]$ satisfying (**4E**) $g_{44} = g_{4A} = 0$ so the 5-D line element (**4A**) becomes the sum of a 4-D line element and an extra part

$$(4.1)\qquad dS^2 = g_{\mu\nu}dx^\mu dx^\nu + \gamma_{44}\left(dx^4 + \frac{\gamma_{4\alpha}}{\gamma_{44}}dx^\alpha\right)^2;\ ds^2 = g_{\mu\nu}dx^\mu dx^\nu$$

Here $g_{\mu\nu}$, $\gamma_{44}$, $\gamma_{4\alpha}$ can depend on $x^4$ and the signature is open via the choice $\epsilon = \pm 1$. Introduce now a 4-vector and a scalar defined via (**4F**) $A_\mu = \gamma_{\mu 4}/\gamma_{44}$ and $\Phi^2 = \epsilon\gamma_{44}$ so that (4.1) gives

$$(4.2)\qquad dS^2 = ds^2 + \epsilon\Phi^2(dx^4 + A_\mu dx^\mu)$$

Thus (**4D**) with $A = \alpha$ and $B = \beta$ when combined with (**4F**) gives (**4G**) $\gamma_{\alpha\beta} = g_{\alpha\beta} + \epsilon\Phi^2A_\alpha A_\beta$ while for $\gamma^{AC}\gamma_{BC} = \delta^A_B$ with $A = \mu$, $B = 4$, and $C = \lambda, 4$ one obtains $\gamma^{4\mu} = -\gamma^{\mu\lambda}\gamma_{4\lambda}/\gamma_{44}$. Using this with (**4D**) in the expansion of $\gamma^{AC}\gamma_{BC} =$

$\delta^A_B$ for $A = \mu$, $B = \nu$, and $C = \lambda, 4$ gives (**4H**) $\gamma^{\mu\nu} = g^{\mu\nu}$. Further since $\gamma^{\mu 4} = -\gamma^{\mu\lambda}A_\lambda$ one has (**4I**) $\gamma^{4\mu} = -A^\mu$ while from (**4F**) one has (**4J**) $\gamma_{4\mu} = \epsilon\Phi^2 A_\mu$. These relations are useful below.

By minimizing the interval $S$ one obtains the geodesic equation

$$\frac{d^2x^A}{dS^2} + \Gamma^A_{BC}\frac{dx^B}{dS}\frac{dx^C}{dS} = 0 \tag{4.3}$$

where

$$\Gamma^A_{BC} = \frac{\gamma^{AD}}{2}\left(\frac{\partial\gamma_{BD}}{\partial x^C} + \frac{\partial\gamma_{CD}}{\partial x^B} + \frac{\partial\gamma_{BC}}{\partial x^D}\right) \tag{4.4}$$

The $A = 4$ component of (4.3) can be rewritten as

$$\gamma_{4A}\frac{d^2X^A}{dS^2} + \Gamma_{4,CD}\frac{dx^C}{dS}\frac{dx^D}{dS} = 0 \equiv \frac{dn}{dS} = \frac{1}{2}\frac{\partial\gamma_{CD}}{\partial x^4}\frac{dx^C}{dS}\frac{dx^D}{dS} \tag{4.5}$$

where (**4K**) $n = \epsilon\Phi^2[(dx^4/dS) + A_\alpha(dx^\alpha/dS)]$. From (4.5) one sees that if the 4-D metric $\gamma_{CD}$ were to be independent of $x^4$ then $n$ would be a constant of the motion (i.e. $dn/dS = (\partial n/\partial x^A)(dx^A/dS) = 0$ even though $n$ could depend on $x^A$); however this will not be the case in general. Note also from (4.2) and (**4J**) that

$$dS = \frac{ds}{[1 - \epsilon(n^2/\Phi^2)]^{1/2}} \tag{4.6}$$

which gives the 5-D interval in terms of its 4-D part. Explicit equations are derived then for the Christoffel coefficients and geodesic equations which reveal the gravitational and electromagnetic parts (we omit details); note that mass and charge are in practice defined via their equations of motion. In particular one obtains the equations

$$\frac{d^2x^\mu}{ds^2} + \Gamma^\mu_{\alpha\beta}\frac{dx^\alpha}{ds}\frac{dx^\beta}{ds} = \frac{n}{[1 - (\epsilon n^2/\Phi^2]^{1/2}}\left[F^\mu_\nu\frac{dx^\nu}{ds} - \frac{A^\mu}{n}\frac{dn}{ds} - g^{\mu\lambda}\frac{\partial A_\lambda}{\partial x^4}\frac{dx^4}{ds}\right] \tag{4.7}$$

$$+ \frac{\epsilon n^2}{[1 - (\epsilon n^2/\Phi^2)]\Phi^3}\left[\nabla_\mu\Phi + \left(\frac{\Phi}{n}\frac{dn}{ds} - \frac{d\Phi}{ds}\right)\frac{dx^\mu}{ds}\right] - g^{\mu\lambda}\frac{\partial g_{\lambda\nu}}{\partial x^4}\frac{dx^\nu}{ds}\frac{dx^4}{ds}$$

This is the fully general equation of motion in KK theory and it is discussed further in some detail in [**868**]). Calculations show that the mass of a test particle is related to the extra coordinate and the metric while its electric charge is related to its rate of motion in the extra dimension. There is also more material on black holes and induced matter, along with the so called fifth force which we omit.

## 5. 5-D PHYSICS

We go now to the book [**869**] which has fewer equations but a more up to date perspective. One looks first at the Einstein equations

$$R_{\alpha\beta} - \frac{1}{2}Rg_{\alpha\beta} + \Lambda g_{\alpha\beta} = \frac{8\pi G}{c^4}T_{\alpha\beta} \tag{5.1}$$

$$\equiv R_{\alpha\beta} = \frac{8\pi G}{c^4}\left(T_{\alpha\beta} - \frac{1}{2}Tg_{\alpha\beta}\right) + \Lambda g_{\alpha\beta}$$

(using $R = 4\Lambda - (8\pi G/c^4)T$. This shows in particular that $\Lambda$ measures the mean curvature of a 4-D vacuum manifold. The 5-D equations are (**5A**) $R_{AB} = 0$ with coordinates $x^A = (t, xyz, \ell)$ and $dS^2 = g_{AB}dx^A dx^B$ (thus 15 equations for $g_{AB}$ reduced to 10 by covariance in 5 coordinates). The electromagnetic (EM) gauge involves ($A_\mu$, $\Phi$, $\epsilon$) with (**5B**) $dS^2 = ds^2 + \epsilon\Phi^2(dx^4 + A - \mu dx^\mu)$ where $\epsilon = \pm 1$, $g_{44} = \epsilon\Phi^2$, and $\epsilon = (1, -1) \sim$ (wave, particle) behavior. For convenience one works with particle behavior and signature $(+, -, -, -, -)$ where $(c, G)$ have been absorbed by choice of units and dynamics determined via $\delta \int dS = 0$. For neutral matter one has

$$dS^2 = g_{\alpha\beta}(x^\gamma, \ell)dx^\alpha dx^\beta + \epsilon\Phi^2(x^\gamma, \ell) \tag{5.2}$$

The components of the Ricci tensor are

$$^5R_{\alpha\beta} = {}^4R_{\alpha\beta} - \frac{\partial_\alpha \nabla_\beta \Phi}{\Phi} \tag{5.3}$$

$$+\frac{\epsilon}{2\Phi^2}\left(\frac{\partial_4\Phi\partial_4 g_{\alpha\beta}}{\Phi} - \partial_4^2 g_{\alpha\beta} + g^{\lambda\mu}\partial_4 g_{\alpha\lambda}\partial_4 g_{\beta\mu} - \frac{g^{\mu\nu}\partial_4 g_{\mu\nu}\partial_4 g_{\alpha\beta}}{2}\right);\ R_{4\alpha}$$

$$= \Gamma\partial_\beta\left(\frac{g^{\beta\lambda}\partial_4 g_{\lambda\alpha} - \delta^\beta_\alpha g^{\mu\nu}\partial_4 g_{\mu\nu}}{2\Gamma}\right) + \frac{g^{\mu\beta}\partial_\lambda g_{\mu\beta}g^{\lambda\sigma}\partial_4 g_{\sigma\alpha}}{4} - \frac{g^{\lambda\beta}\partial_\alpha g_{\mu\beta}g^{\mu\sigma}\partial_4 g_{\sigma\lambda}}{4};$$

$$R_{44} = -\epsilon\Phi\Box\Phi - \frac{\partial_4 g^{\lambda\beta}\partial_4 g_{\lambda\beta}}{2} - \frac{g^{\lambda\beta}\partial_4^2 g_{\lambda\beta}}{2} + \frac{\partial_4\Phi g^{\lambda\beta}\partial_4 g_{\lambda\beta}}{2\Phi} - \frac{g^{\mu\beta}g^{\lambda\sigma}\partial_4 g_{\lambda\beta}\partial_4 g_{\mu\sigma}}{4}$$

where $\Box\Phi = g^{\mu\nu}\partial_\mu\nabla_\nu\Phi$ and $\Gamma = |\epsilon\Phi^2|^{1/2}$. The tensor components of (5.3) with the 5-D field equations $R_{AB} = 0$ give the 10 Einstein equations (cf. [**871**]). For this one forms the 4-D Ricci tensor and with it construct the 4-D Einstein tensor $G_{\alpha\beta} = {}^4R_{\alpha\beta} - (1/2)^4Rg_{\alpha\beta}$; the remaining terms in $^5R_{\alpha\beta}$ are used to construct an effective or induced energy momentum tensor. In particular one finds

$$^4R = \frac{\epsilon}{4\Phi^2}\left[\partial_4 g^{\mu\nu}\partial_4 g_{\mu\nu} + (g^{\mu\nu}\partial_4 g_{\mu\nu})^2\right] \tag{5.4}$$

This shows

(1) The curvature of 4-D spacetime can be regarded as the result of embedding it in an $x^4$ dependent 5-D manifold.
(2) The sign of the 4-D curvature depends on the signature of the 5-D metric.
(3) The magnitude of the 4-D curvature depends strongly on the scalar field or the size of the extra dimension ($g_{44} = \epsilon\Phi^2$).
(4) The form of the 4-D energy momentum tensor is

$$8\pi T_{\alpha\beta} = \frac{\partial_\alpha\nabla_\beta\Phi}{\Phi} - \frac{\epsilon}{2\Phi^2}\left[\frac{\partial_4\Phi\partial_4 g_{\alpha\beta}}{\Phi} - \partial_4^2 g_{\alpha\beta} + g^{\lambda\mu}\partial_4 g_{\alpha\lambda}\partial_4 g_{\beta\mu}\right. \tag{5.5}$$

$$\left. - \frac{g^{\mu\nu}\partial_4 g_{\mu\nu}\partial_4 g_{\alpha\beta}}{2} + \frac{g_{\alpha\beta}}{4}\left(\partial_4 g^{\mu\nu}\partial_4 g_{\mu\nu} + (g^{\mu\nu}\partial_4 g_{\mu\nu})^2\right)\right]$$

In particular one sees that an apparently empty 5-D manifold contains a 4-D manifold with sources. Note also that the vector components of (5.3) in conjunction with (**4A**) can be considered as a set of conservation equations resembling those of Maxwellian EM; they are (**5C**) $\nabla_\beta P^\beta_\alpha = 0$ where $P^\beta_\alpha = (1/2\Phi)[g^{\beta\sigma}\partial_4 g_{\sigma\alpha} -$

$\delta^\beta_\alpha g^{\mu\nu}\partial_4 g_{\mu\nu}]$. These are an inherent part of the field equations. The scalar or last component of (5.3) when set to zero according to (**5A**) yields a wave-type equation

$$\Box\Phi = -\frac{\epsilon}{2\Phi}\left[\frac{\partial_4 g^{\lambda\beta}\partial_4 g_{\lambda\beta}}{2} + g^{\lambda\beta}\partial_4^2 g_{\lambda\beta} - \frac{\partial_4\Phi g^{\lambda\beta}\partial_4 g_{\lambda\beta}}{\Phi}\right] \quad (5.6)$$

One considers next the canonical (or Einstein) metric of Mashhoon (cf. [**586**]), namely (**5D**) $dS^2 = (\ell^2/L^2)g_{\alpha\beta}(x^\gamma,\ell)dx^\alpha dx^\beta - d\ell^2$ where $x^4 = \ell$ and L is a constant length introduced for consistency purposes. Following [**870**] one summarizes:

(1) (**5D**) is general in that all 5 available coordinate degrees of freedom have been used to set $g_{4\alpha} = 0$ and $g_{44} = -1$. Physically this removes the potentials of EM type and flattens the potential of scalar type.
(2) (**5D**) involves many known solutions for the field equations involving both particles and waves.
(3) When $\partial g_{\alpha\beta}/\partial\ell = 0$ in (**5D**) the 15 field equations $R_{AB} = 0$ give back the Einstein equations in the form $G_{\alpha\beta} = 3g_{\alpha\beta}/L^2$ (i.e. $\Lambda = 3/L^2$).
(4) The factorization in (**5D**) says that the 4-D part of the 5-D interval is $(1/L)ds$ which defines a momentum space if $\ell$ is related to the rest mass $m$. There is some confirmation for this via geodesic equations and energy formulas $E = \ell(1-\nu^2)^{-1/2}$ in 5-D which agrees with 4-D for $\ell = m$.
(5) The 5 components of the geodesic equation for (**5D**) split naturally into 4 spacetime components ($\sim u^\alpha$ for $\partial_\ell g_{\alpha\beta} \neq 0$) and an extra component. For $\partial g_{\alpha\beta}/\partial\ell = 0$ the motion is not only geodesic in 5-D but also in 4-D and the 4-D weak equivalence principle is recovered as a kind of symmetry in 5-D.

**REMARK 5.1.1.** One refer here to the standard 5-D models of Ponce de Leon (cf. [**717, 718, 719, 790, 869**]). These are commonly written in coordinates $x^0 = t$, $x^{1,2,3} = (r,\theta,\phi)$, and $x^4 = \ell$ (c and G are set equal to one). The line element is

$$dS^2 = \ell^2 dt^2 - t^{2/\alpha}\ell^2(1-\alpha)(dr^2 + r^2 d\Omega^2) - \frac{\alpha^2 t^2}{(1-\alpha)^2}d\ell^2 \quad (5.7)$$

where $d\Omega^2 = d\theta^2 + Sin^2(\theta)d\phi^2$. Here the effective or induced energy-momentum tensor is taken as that for a perfect fluid with

$$8\pi\rho = \frac{3}{\alpha^2\tau^2};\ 8\pi p = \frac{2\alpha-3}{\alpha^2\tau^2} \quad (5.8)$$

where $\tau = \ell t$ is the proper time, and the equation of state is $p = [(2\alpha/3) - 1]\rho$. Models with $\alpha < 1$ expand faster than the standard FRW universes and have inflationary equations of state. Physically these cosmologies are flat in 3-D, curved in 4-D, and flat in 5-D; hence one has a 5-D Minkowski space in some other coordinates with (**5E**) $dS^2 = dT^2 - (da^2 + a^2 d\Omega^2) - dL^2$. This does not have a big bang (in 5-D) but the 4-D part does. The coordinate transformation between (5.7) and (**5E**) is given via

$$a(t,r,\ell) = rt^{1/\alpha}\ell^{1/(1-\alpha)};\ T = \frac{\alpha}{2}\left[\left(1+\frac{r^2}{\alpha^2}\right)t^{1/\alpha}\ell^{1/(1-\alpha)} - \frac{t^{\frac{2\alpha-1}{\alpha}}\ell^{\frac{1-2\alpha}{1-\alpha}}}{1-2\alpha}\right] \quad (5.9)$$

$$L(t,r\ell) = \frac{\alpha}{2}\left[\left(1-\frac{r^2}{\alpha^2}\right)t^{1/\alpha}\ell^{\frac{1}{1-\alpha}} + \frac{t^{\frac{2\alpha-1}{\alpha}}\ell^{\frac{1-2\alpha}{1-\alpha}}}{1-2\alpha}\right]$$

(see e.g.[**869**] for an extensive study of these relations). ■

**REMARK 5.1.2.** Here another solution of $R_{ABCD} = 0$ depending on $u = t - \ell$ (i.e. describing a wave). The solution has line element

$$dS^2 = b^2dt^2 - a^2(dr^2 + r^2d\Omega^2) - b^2d\ell^2;\ a = (hu)^{\frac{1}{2+3\alpha}};\ b = (hu)^{\frac{1+3\alpha}{2(2+3\alpha)}} \tag{5.10}$$

There is an associated equation of state (**5F**) $8\pi p = [3h^2/(2+3\alpha)]a^{-3(1+\alpha)}$ so $\alpha = 0$ is the (late) dust universe and $\alpha + 1/3$ is the early (radiation) universe ($h$ is a parameter with dimensions of an inverse length or time, related to the Hubble parameter $H$). Changing coordinates one has for $dT = bdt$

$$T = \frac{2}{3}\left(\frac{2+3\alpha}{1+\alpha}\right)\frac{1}{h}(hu)^{\frac{3(1+\alpha)}{2(2+3\alpha)}} \tag{5.11}$$

The 4-D scale factor which determines the dynamics for (5.10)-(5.11) is

$$a(T) = \left[\frac{3}{2}\left(\frac{1+\alpha}{2+3\alpha}\right)hT\right]^{\frac{2}{3(1+\alpha)}} \tag{5.12}$$

For $\alpha = 0$ one has $a(T) \propto T^{2/3}$ as in the standard Einstein-deSitter dust model. For $\alpha = 1/3$ one has $a(T) \propto T^{1/2}$ as in the standard radiation model. Hubble's paramter is

$$H \equiv \frac{1}{a}\partial_T a = \frac{1}{a}\partial_t a \partial_T t = \frac{h}{2+3\alpha}(hu)^{\frac{-3(1+\alpha)}{2(2+3\alpha)}} = \frac{2}{3(1+\alpha)T} \tag{5.13}$$

This leads also to (**5G**) $8\pi\rho = (4/3)[(1+\alpha)^2T^2]^{-1}$ so for $\alpha = 0$ one has $\rho = 1/6\pi T^2$ and $\alpha = 1/3$ yields $\rho = 3/32\pi T^2$ which are standard FRW values. Hence the 5-D solution (5.10) contains standard 4-D dynamics for the late and early universes (cf. [**552, 553, 717, 718, 719, 868, 869**]). Moreover some insight is obtained via an interpretation of the big bang as either a singularity in 4-D or a hypersurface $t = \ell$ representing a plane wave in 5-D. ■

**REMARK 5.1.3.** Another class of models (cf. [**552, 553**]) extends the FRW ones with line element

$$dS^2 = B^2dt^2 - A^2\left(\frac{dr^2}{1-kr^2} + r^2d\Omega^2\right) - dy^2;\ B = \frac{1}{\mu}\partial_t A = \frac{\dot{A}}{\mu} \tag{5.14}$$

$$A^2 = (\mu^2 + k)y^2 + 2\nu y + \frac{\nu^2 + K}{\mu^2 + K}$$

Here $\mu$, $\nu$ are arbitrary functions of $t$ and $k$ is a 3-D curvature index ($k = \pm 1, 0$); the Kretschmann invariant has the form (**5H**) $I = R_{ABCD}R^{ABCD} = 72K^2/A^8$. The 4-D line element is

$$ds^2 = g_{\alpha\beta}dx^\alpha dx^\beta = B^2dt^2 - A^2\left(\frac{dr^2}{1-kr^2} + r^2d\Omega^2\right) \tag{5.15}$$

and one calculates

$$(5.16)\qquad {}^4R_0^0 = -\frac{3}{B^2}\left(\frac{\ddot{A}}{A} - \frac{\dot{A}\dot{B}}{AB}\right);\ {}^4R_1^1 = {}^4R_2^2 = {}^4R_3^3$$

$$= -\frac{1}{B^2}\left[\frac{\ddot{A}}{A} + \frac{\dot{A}}{A}\left(\frac{2\dot{A}}{A} - \frac{\dot{B}}{B}\right) + 2k\frac{B^2}{A2}\right]$$

and from (5.14) results

$$(5.17)\qquad B = \frac{\dot{A}}{\mu};\ \dot{B} = \frac{\ddot{A}}{\mu} - \frac{\dot{A}}{\mu}\frac{\dot{\mu}}{\mu}$$

Using this in (5.16) one can eliminate $B$ and $\dot{B}$ to get

$$(5.18)\qquad {}^4R_0^0 = -\frac{3\mu\dot{\mu}}{A\dot{A}};\ {}^4R = {}^4R_2^2 = {}^4R_3^3 = -\left(\frac{\mu\dot{\mu}}{A\dot{A}} + \frac{2(\mu^2+k)}{A^2}\right)$$

leading to the Ricci scalar

$$(5.19)\qquad {}^4R = -6\left(\frac{\mu\dot{\mu}}{A\dot{A}} + \frac{\mu^2+k}{A^2}\right)$$

The nonvanishing components of the 4-D Einstein tensor are then

$$(5.20)\qquad G_0^0 = \frac{3(\mu^2+k)}{A^2};\ G_1^1 = G_2^2 = G_3^3 = \frac{2\mu\dot{\mu}}{A\dot{A}} + \frac{\mu^2+k}{A^2}$$

As in the FRW models one can take the matter to be comoving in 3-D so $u^\alpha = (u^0, 0, 0, 0)$ and $u^0 u_0 = 1$. Then (★) $G_{\alpha\beta} = 8\pi[(\rho+p)u_\alpha u_\beta + (\{(\Lambda/8\pi) - p\}g_{\alpha\beta}]$ and (★) $G_{\alpha\beta} = 8\pi[(\rho+p)u_\alpha u_\beta + \{(\Lambda/8\pi) - p\}g_{\alpha\beta}]$ with

$$(5.21)\qquad 8\pi\rho + \Lambda = \frac{3(\mu^2+k)}{A^2};\ 8\pi p - \Lambda = -\frac{2\mu\dot{\mu}}{A\dot{A}} - \frac{\mu^2+k}{A^2}$$

(these are analogous to (5.14) for FRW). One is free to choose an equation of state which is taken here as the isothermal one (**5I**) $p = \gamma\rho$ with $\gamma$ constant. Putting this in (5.21) yields

$$(5.22)\qquad 8\pi\rho = \frac{2}{1+\gamma}\left(\frac{\mu^2+k}{A^2} - \frac{\mu\dot{\mu}}{A\dot{A}}\right);$$

$$\Lambda = \frac{2}{1+\gamma}\left[\left(\frac{1+3\gamma)}{2}\right)\left(\frac{\mu^2+k}{A^2}\right) + \frac{\mu\dot{\mu}}{A\dot{A}}\right]$$

In this spirit one dwells upon bangs and bounces. ■

## 6. SOME QUANTUM CONNECTIONS

One now involves $\hbar$, $c$, $G$, $\Lambda$ and $L$ where $L$ scales the canonical metric in 5-D theory via $\Lambda = 3/L^2$. $L$ also measures the scale of a potential well in which a particle finds itself and may therefore be related to the vacuum or ZPF (zero point fields - cf. [**170**]). Recall the canonical line element is $dS^2 = (\ell/L)^2 ds^2 - d\ell^2$ with coordinates $(x^\alpha, \ell)$ and $ds^2 = g_{\alpha\beta}dx^\alpha dx^\beta$, $(\alpha, \beta = 0, 123)$. Recall also $g_{4\alpha} = 0$ with $g_{44} = -1$ (with no EM connections) and if $g_{\alpha\beta} = g_{\beta\alpha}$ are functions of $x^\gamma$ only one recovers the weak equivalence principle (cf. [**869**]). Then one write $x^4 = \ell = Gm/c^2$ ($m \sim$ rest mass). Since this is the particle's Schwarzschild radius the metric was called canonical and there is a coordinate transformation $\ell_P \to L^2/\ell$ leads to $\ell = \hbar/mc$ (Compton wavelength) and a nomenclature "Planck gauge", specified by

$$dS^2 = (L^2/\ell_P^2)ds^2 - (L^4/\ell_P^4)d\ell^2 \tag{6.1}$$

$(m_P = (\hbar/c)(\Lambda/3)^{1/2})$. The Lagrangian density $L = (dS/ds)^2$ for (6.1) has associated momenta

$$P_\alpha = \frac{\partial L}{\partial(dx^\alpha/ds)} = \frac{2L^2}{\ell^2}g_{\alpha\beta}\frac{dx^\beta}{ds};\ P_\ell = \frac{\partial L}{\partial(d\ell/ds)} = \frac{2L^4}{\ell^4}\frac{d\ell}{ds} \tag{6.2}$$

These define a 5-D scalar

$$\int P_A dx^A = \int (P_\alpha dx^\alpha + P_{|ell}d\ell) = \int \frac{2L^2}{\ell^2}\left[1 - \left(\frac{L}{\ell}\frac{d\ell}{ds}\right)^2\right] \tag{6.3}$$

This is zero for $dS^2 = 0$ since (6.1) gives then (**6A**) $\ell = \ell_0 exp[\pm(s/\ell]$ with $d\ell/ds = \pm(\ell/L)$. Thus in e.g. [**870**] $m$ is related to $\ell$ while in [**900**] m is related to $d\ell/ds$; this is essentially the same and in any case the variation is slow if $s/\ell << 1$. Using $m \sim \ell$ the first part of the line element in (6.1) is essentially $mcds$ with the identification $\ell = \hbar/mc$ leading in 4-D to

$$\int p_\alpha dx^\alpha = \int m u_\alpha dx^\alpha = \int \frac{\hbar ds}{c\ell} = \pm\frac{\hbar L}{c\ell} \tag{6.4}$$

Putting $L/\ell = n$ (possibly rational) (6.4) implies (**6B**) $\int mcds = n\hbar$ leading from a null line element in (6.1) to a conventional action in 4-D. Another scalar quantity of interest here is $dp_\alpha dx^\alpha$ (note $dx^\alpha$ transforms as a tensor - but not $x^\alpha$). Then as above

$$dp_\alpha dx^\alpha = \frac{\hbar}{c}\left(\frac{du_\alpha}{ds}\frac{dx^\alpha}{ds} - \frac{1}{\ell}\frac{d\ell}{ds}\right)\frac{ds^2}{\ell} \tag{6.5}$$

The first term in parenthesis is zero if the acceleration is zero or if the scalar product with the velocity is zero as in conventional 4-D dynamics (cf. [**869**] for details). However there is a contribution from the second term in parenthesis due to the change of mass of the particle, namely

$$-|dp_\alpha dx^\alpha| = \frac{\hbar}{c}\left|\frac{d\ell}{ds}\right|\frac{ds^2}{\ell^2} = \frac{\hbar}{c}\frac{ds^2}{L\ell} = n\frac{h}{c}\left(\frac{d\ell}{\ell}\right)^2 \tag{6.6}$$

(using (**6A**) and $n = (L/\ell)$ which implies $dn/n = -d\ell/\ell = dK_\ell/K_\ell$ where $K_\ell = 1/\ell$ is the wavenumber for the extra dimension. Evidently (6.6) is a Heisenberg type relation which can be written as (**6C**) $|dp_\alpha dx^\alpha| = (\hbar/c)(dn^2/n)$ (see [**869**] for discussion). A particular argument looks at the fundamental mode $n = 1$ where $\ell = \hbar/mc$ with $dS^2 = 0$. One arrives at $|dm| = m(ds/L)$ which with (**6B**) gives $m = [(\int mcds)/cL = h\hbar/cL$ yielding a fundamental mass $m_0 = \hbar/cL$ (quantum mass from null 5-D path - ?). We refer to [**869**] for a discussion of microscopic and macroscopic masses and related matters. The Klein-Gordon and Dirac equations are also discussed.

Some further information on particles and waves involves considering a 5-D canonical analogue of the 4-D Milne model (cf. [**549, 550**]). Thus one desires $R_{AB} = 0$ and $R_{ABCD} = 0$ with an appropriate solution

$$dS^2 = \left(\frac{\ell}{L}\right)^2 dt^2 - \left[\ell Sinh\left(\frac{t}{L}\right)\right]^2 d\sigma^2 - d\ell^2 \tag{6.7}$$

Here the 3-space is given via $d\sigma^2 = (dx^2 + dy^2 + dz^2)[1 + (kr^2/4)]$ with $k = -1$. Note the time dependence is different from an FRW model

$$ds^2 = dt^2 - \frac{a^2(t)}{[1 + (kr^2/4)]^2}(dx^2 + dy^2 + dz^2) \;\; (k = \pm 1, 0) \tag{6.8}$$

due to the dimensional contributions. However the local situation for (6.7) is close to that of (6.8) and to see this note that for laboratory conditions $t/L << 1$ in (6.7) one has

$$dS^2 \simeq \left(\frac{1}{L}\right)^2 dt^2 - \left(\frac{\ell t}{L}\right)^2 d\sigma^2 - d\ell^2 \tag{6.9}$$

Now multiply this by $L^2$, divide by $ds^2$, and take the null-path hypothesis, which for any canonical metric results in a constraint $(d\ell/ds) = \pm(\ell/L)$; then (6.9) gives

$$0 = \ell^2\left(\frac{dt}{ds}\right)^2 - (\ell t)^2\left[\left(\frac{dx}{ds}\right)^2 + \left(\frac{dy}{ds}\right)^2 + \left(\frac{dz}{ds}\right)^2\right] - \ell^2 \tag{6.10}$$

Then putting $\ell = m$ as above and recalling that proper distances are defined via $\int t dx$ one arrives at $0 = E^2 - p^2 - m^2$. To convert (6.9) into a wave change $t \to exp(i\omega t)/i\omega$ and $x \to exp(ik_x x)$, etc. where $\omega$ is a frequency and $k_x$ etc. are wave numbers for the $x, y, z$ directions. After setting the phase velocity to 1 (6.9) becomes

$$dS^2 = \left(\frac{1}{L}\right)^2 e^{2i\omega t} dt^2 - \left(\frac{1}{L}\right)^2 [exp\{2i(\omega t + k_x x)]dx^2 + \cdots] - d\ell^2 \tag{6.11}$$

Then with the null condition the analogue of (6.10) becomes

$$0 = \left[\ell e^{i\omega t}\frac{dt}{ds}\right]^2 - \left[\ell exp\{i(\omega t + k_x x)\}\frac{dx}{ds}\right]^2 - etc. - \ell^2 \tag{6.12}$$

Setting $\ell = m$ and defining (**6D**) $\tilde{E} = \ell exp(i\omega t)(dt/ds)$ with $\tilde{p} = \ell exp[i(\omega t + k_x x)](dx/ds)$ etc. (6.12) is equivalent to (**6E**) $0 = \tilde{E}^2 - \tilde{p}^2 - m^2$ which is the wave analogue of $E^2 = p^2 + m^2$. It should possible to do this for other 5-D flat solutions such as (6.7).

## 7. REMARKS ON COSMOLOGY

We extract here from Chapter 5 of [**869**]. One writes

$$dS^2 = g_{\alpha\beta}(x^\gamma, \ell)dx^\alpha dx^{\beta\epsilon\Phi^2(x^\gamma,\ell} d\ell \tag{7.1}$$

$(\epsilon = \pm 1)$. The field equations are (**7A**) $G_{\alpha\beta} = 8\pi T_{\alpha\beta}$ and one writes

$$8\pi T_{\alpha\beta} = \frac{\nabla_\beta \Phi_\alpha}{\Phi} - \frac{\epsilon}{2\Phi^2}\left[\frac{\overset{*}{\Phi}\overset{*}{g}_{\alpha\beta}}{\Phi} - \overset{**}{g}_{\alpha\beta} + g^{\lambda\mu}\overset{*}{g}_{\alpha\lambda}\overset{*}{g}_{\beta\mu}\right. \tag{7.2}$$

$$-\frac{g^{\mu\nu}\overset{*}{g}_{\mu\nu}\overset{*}{g}_{\alpha\beta}}{2}+\frac{g_{\alpha\beta}}{4}\left\{\overset{*}{g}^{\mu\nu}\overset{*}{g}_{\mu\nu}\left(g^{\mu\nu}\overset{*}{g}_{\mu\nu}\right)^2\right\}\Bigg]$$

(7.3) $$\nabla_\beta P^\beta_\alpha=0;\ P^\alpha_\beta=\frac{1}{2\Phi}\left(g^{\beta\sigma}\overset{*}{g}_{\sigma\alpha}-\delta^\beta_\nu g^{\mu\nu}\overset{*}{g}_{\mu\nu}\right);\ \Box\Phi=g^{\alpha\beta}\nabla_\beta\Phi_\alpha;$$

$$\epsilon\Phi\Box\Phi=-\frac{\overset{*}{g}^{\lambda\beta}\overset{*}{g}_{\lambda\beta}}{4}-\frac{g^{\lambda\beta}\overset{**}{g}_{\lambda\beta}}{2}+\frac{\overset{*}{\Phi}g^{\lambda\beta}\overset{*}{g}_{\lambda\beta}}{2\Phi}$$

(recall an asterisk denotes $\partial/\partial x^4$). The canonical metric is obtained from (7.1) via (**7B**) $dS^2=(\ell^2/L^2)[g_{\alpha\beta}(x^\gamma,\ell)dx^\alpha dx^\beta]-d\ell^2$ which causes the Einstein tensor to be

(7.4) $$G_{\alpha\beta}=\frac{1}{L^2}\left[-\ell^2\overset{**}{g}_{\alpha\beta}+\ell^2g^{\lambda\mu}\overset{*}{g}_{\alpha\lambda}\overset{*}{g}_{\beta\mu}-4\ell^2\overset{*}{g}_{\alpha\beta}-\frac{\ell^2}{2}g^{\mu\nu}\overset{*}{g}_{\mu\nu}\overset{*}{g}_{\alpha\beta}\right]$$

$$-\frac{1}{2L^2}\left[6+2\ell g^{\mu\nu}\overset{*}{g}_{\mu\nu}+\frac{\ell^2}{4}\overset{*}{g}^{\mu\nu}\overset{*}{g}_{\mu\nu}+\frac{\ell^2}{4}\left(g^{\mu\nu}\overset{*}{g}_{\mu\nu}\right)^2\right]g_{\alpha\beta}$$

(dnote $dS^2=(\ell_E/L))^2ds^2-d\ell_E^2$ is the Einstein gauge with $\ell_E=Gm/c^2$ and $m_E=(c^2/G)(3/\Lambda)^{1/2}$). Here $\eta_{\alpha\beta}=diag(1,-1,-1,-1)$ is the metric for flat Minkowski space. To determine the dependence in $\ell$ of $f(x^\gamma,\ell)$ we need to solve the field equations. One could use here (7.2)-(7.3) but since the scalar field is suppressed in (**7B**) it is more convenient to calculate the components of the 5-D Ricci tensor directly. These are

(7.5) $$R_{44}=-\frac{\partial A^\alpha_\alpha}{\partial\ell}-\frac{2}{\ell}A^\alpha_\alpha-A_{\alpha\beta}A^{\alpha\beta};$$

$$R_{\mu4}=\nabla_\alpha A^\alpha_\mu-\frac{\partial\Gamma^\alpha_{\mu\alpha}}{\partial\ell};\ {}^5R_{\mu\nu}={}^4R_{\mu\nu}-S_{\mu\nu}$$

where

(7.6) $$S_{\mu\nu}=\frac{\ell^2}{L^2}\left[\frac{\partial A_{\mu\nu}}{\partial\ell}+\left(\frac{4}{\ell}+A^\alpha_\alpha\right)A_{\mu\nu}-2A^\alpha_\mu A_{\nu\alpha}\right]+\frac{1}{L^2}(3+\ell A^\alpha_\alpha)g_{\mu\nu}$$

Moreover

(7.7) $$A_{\alpha\beta}=\frac{1}{2}\frac{\partial g_{\alpha\beta}}{\partial\ell}$$

One wants to solve (7.5) in the form $R_{\alpha\beta}=0$ subject to putting $g_{\mu\nu}(x^\gamma,\ell)=f(x^\gamma,\ell)\eta_{\mu\nu}$, which ensures conformal flatness. Note that $g^{\mu\nu}=\eta^{\mu\nu}/f$ and $A_{\mu\nu}=\overset{*}{f}\eta_{\mu\nu}/2$ where $\overset{*}{f}=\partial f(x^\gamma,\ell)/\partial\ell$. Also $A^{\alpha\beta}=\overset{*}{f}\eta^{\alpha\beta}/2f^2$ and $A^\alpha_\alpha=2\overset{*}{f}/f$ with $A^\alpha_{gb}=\overset{*}{f}\eta^\alpha_\beta/2f$. Then the scalar component of the field equation (7.5) becomes

(7.8) $$2\frac{\partial}{\partial\ell}\left(\frac{\overset{*}{f}}{f}\right)+\left(\frac{\overset{*}{f}}{f}\right)^2+\frac{4}{\ell}\left(\frac{\overset{*}{f}}{f}\right)=0$$

and one finds a solution

(7.9) $$f(x^\gamma,\ell)=\left[1-\frac{\ell_0(x^\gamma)}{\ell}\right]^2k(x^\gamma)$$

with $k(x^\gamma)$ an arbitrary function. Consequently (after some calculation)

$$S_{\mu\nu} = \frac{3}{L^2} k(x^\gamma)\eta_{\mu\nu} \tag{7.10}$$

and one arrives at

$$^4R_{\mu\nu} = \frac{3}{L^2}\frac{\ell^2}{(\ell-\ell_0)^2} g_{\mu\nu} \tag{7.11}$$

This is equivalent to the Einstein field equation for the deSitter metric tensor $k\eta_{\mu\nu}$ and in any event (7.11) defines an Einstein space $^4R = \Lambda g_{\mu\nu}$ with (**7C**) $\Lambda = (3/L^2)[1/(\ell - \ell_0)]^2$ which reduces for $\ell_0 = 0$ to the standard deSitter value $\Lambda = 3/L^2$ (no $\ell$ dependence - $L\ell \sim L'$). There is more about this and many other things in [**869**].

## 8. EXTENSIONS

We go here to some recent papers [**7, 8, 466, 465, 579, 609, 852, 870**] (cf. also [**239, 478, 696, 790**]). First from [**170, 185, 178, 465, 466, 467, 468**] we sketch how Weyl-Dirac (WD) theory is related to Wesson's induced matter theory (IMT) from Section 3 (as in [**868, 869**]). Many details are developed in [**465, 466, 467, 468**] and we concentrate here on the study of the creation of 4-D neutral and electrically charged classical particles induced by the 5-D bulk in the framework of Weyl-Dirac theory (cf. also Chapter 2). One imagines here (classical) particles of charge 0 or $\pm(1/3)$ characterized by charge, radius, and mass. Spin is expected to arise upon quantization. Generally partial derivatives and denoted by commas, Riemannian covariant 4-D derivatives by a semi-colon, and Riemannian covariant 5-D derivatives by a colon. The 5-D metric tenson is $g_{AB}$ and its 4-D counterpart is $h_{\mu\nu}$; sometimes 5-D quantities have a tilde. One assumes that all gravitational waves have the same speed and therefore in the 5-D bulk the isotropic interval $dS^2 = 0$ is invariant but a line element $g_{AB}dx^A dx^B$ may vary. The situation resembles the 4-D Weyl geometry where the light cone is the principal phenomenon describing the space-time and $ds^2 = 0$ is invariant rather than a line element $ds^2 = h_{\alpha\beta}dy^\alpha dy^\beta$. In the WD version of the Wesson IMT at every point of the 5-D bulk, in addition to the metric tensor $g_{AB}(x^D)$ the existence of a Weylian length connection vector $\tilde{w}^A(x^D)$ and a gauge function $\beta \sim \Omega$ is assumed. In the 5-D (empty) manifold an action integral can be built from the $g_{AB}$, $\tilde{w}_C$, and $\beta$ and 5-D field equations derived (cf. [**466**], 1-3) and it turns out that $\beta$ may be chosen arbitrarily (since its equation follows from equations for $g_{AB}$ and $\tilde{w}_C$ - cf . also [**465**]). There is no matter in the 5-D theory. Now the 5-D quantities $g_{AB}$, $\tilde{w}_C$, and $\tilde{W}_{AB} = \partial_B\tilde{w}_A - \partial_A\tilde{w}_B$ have 4-D counterparts on the brane, namely the 4-D metric tenor $h_{\mu\nu}$, $w_\mu$, and the Maxwell field tensor $W_{\mu\nu} = \partial_\nu w_\mu - \partial_\mu w_\nu$. Thus one has a 4-D geometrically based theory of gravitation and electromagnetism induced by the 5-D bulk.

The 5-D signature is $(+,-,-,-,\epsilon)$ with $\epsilon = \pm 1$ and line element (**8A**) $dS^2 = g_{AB}dx^A dx^B$ $(A, B = 0, 1, 2, 3, 4)$ and one introduces a scalar function $\ell(x^A)$ that defines the foliation of the bulk $M$ by 4-D hypersurfaces $\Sigma_\ell$ at a chosen $\ell = c$ as well as the normal vector $n^A$ to $\Sigma_\ell$. If there is only one time like direction in

$M$ it will be assumed that $n^A$ is spacelike ($\epsilon = -1$); if $M$ possesses 2 timelike direction ($\epsilon = +1$) then $n^A$ will be timelike. In any event the brane $\Sigma_\ell$ contains 3 space-like directions and a time-like one as well; it is mapped by coordinates $y^\mu$ and the metric $h_{\alpha\beta}$ has signature $(+,-,-,-)$. The line element on the brane is (**8B**) $ds^2 = h_{\mu\nu}dy^\mu dy^\nu$ for $\mu,\ \nu = 0,1,2,3$). One assumes that $y^\mu(x^A)$ and $\ell = \ell(X^A)$ as well as $x^A = x^A(y^\mu, \ell)$ are well defined and well behaved. A given 5-D (vector, tensor) in the bulk has a 4-D counterpart on the brane which can be formed via a system of basis vectors orthogonal to $n_A$, namely (**8C**) $e^A_\nu = \partial x^A/\partial y^\nu$ with $n_A e^A_\nu = 0$. In addition there are associated vectors $(e^\nu_A, n^A)$ with $e^\nu_A n^A = 0$ and one has (**8D**) $e^A_\nu e^\mu_A = \delta^\mu_\nu$; $e^A_\sigma e^\sigma_B = \delta^A_B - \epsilon n^A n_B$; $n^A n_A = \epsilon$.

Consider now a 5-D vector $V_A$; $V^A$ in the bulk M. Its 4-D counterpart on the brane $\Sigma_\ell$ is given via (**8E**) $V_\mu = e^A_\mu V_A$; $V^\mu = e^\mu_B V^B$. On the other hand one can write

$$(8.1)\qquad {}_A = e^\mu_A V_\mu + \epsilon(V_S n^S)n_A;\ V^A = e^A_\mu V^\mu + \epsilon(V^S n_S)n^A$$

The 4-D and 5-D metric tensors are related via

$$(8.2)\qquad h_{\mu\nu} = e^A_\mu e^B_\nu g_{AB};\ h^{\mu\nu} = e^\mu_A e^\nu_B g^{AB};$$

$$g_{AB} = e^\mu_A e^\nu_B h_{\mu\nu} + \epsilon n_A n_B;\ g^{AB} = e^A_\mu e^B_\nu h^{\mu\nu} + \epsilon n^A n^B$$

Starting from the 5-D equations for $g_{AB}$ and using the Gauss-Codazzi equations the 4-D equations of gravitation are (cf. [**465, 466**] - we use $\Omega \sim \beta$ to distinguish $\beta$ from subscripts, etc. and define various quantities in an enumertion below - note $\Omega_A = \partial_A \Omega$ etc. with $\Omega^A = g^{AB}\Omega_A$)

$$(8.3)\ G_{\alpha\beta} = -\frac{8\pi}{\Omega^2}M_{\alpha\beta} - \frac{2\epsilon}{\Omega^2}\left(\frac{1}{2}h_{\alpha\beta}B - B_{\alpha\beta}\right) + \frac{6}{\Omega^2}\Omega_\alpha\beta_\beta - \frac{3}{\Omega}(\nabla_\beta\Omega_\alpha - h_{\alpha\beta}\nabla_\sigma\Omega^\sigma)$$

$$+\frac{3\epsilon}{\Omega}(\Omega_S n^S)(h_{\alpha\beta}C - C_{\alpha\beta}) + \epsilon\left[E_{\alpha\beta} - h_{\alpha\beta}E + h^{\mu\nu}C_{\mu[\nu}C_{\lambda]\sigma}\right.$$

$$\left.\times\left(h_{\alpha\beta}h^{\sigma\lambda} - 2\delta^\sigma_\alpha\delta^\lambda_\beta\right)\right] - \frac{1}{2}h_{\alpha\beta}\Omega^2\Lambda$$

Further from the equation for the source-free 5-D Weylian field in the bulk (i.e. $\tilde{\nabla}_B(\Omega\tilde{W}^{AB}) = 0$) the 4-D equation for $W_{\mu\nu}$ on the brane is

$$(8.4)\qquad \nabla_\beta W^{\alpha\beta} = -\frac{\Omega_\beta}{\Omega}W^{\alpha\beta}$$

$$+\epsilon n_S\left[\tilde{W}^{AS}(e^\beta_A h^{\alpha\lambda} - e^\alpha_A h^{\beta\lambda})C_{\beta\lambda} + n^C e^\alpha_A\left(\tilde{\nabla}_C\tilde{W}^{AS} + \tilde{W}^{AS}\frac{\Omega_C}{\Omega}\right)\right]$$

The quantities $M$, $B$, $E$, etc. are defined via

(1) The energy momentum tensor for the 4-D EM field is

$$(8.5)\qquad M_{\alpha\beta} = \frac{1}{4\pi}\left(\frac{1}{4}h_{\alpha\beta}W_{\lambda\sigma}W^{\lambda\sigma} - W_{\alpha\lambda}W^\lambda_\beta\right)$$

(2) The 4-D energy momentum quantities formed from the 5-D field $\tilde{W}_{AB}$ are

$$(8.6)\qquad B_{\alpha\beta} = \tilde{W}_{AS}\tilde{W}_{BL}e^A_\alpha e^B_\beta n^S n^L;\ B = h^{\lambda\sigma}B_{\lambda\sigma} = \tilde{W}_{AS}\tilde{W}_{BL}g^{AB}n^S n^L$$

(3) The extrinsic curvature $C_{\mu\nu}$ of the brane $\Sigma_\ell$ and its contraction are

$$C_{\mu\nu} = e^A_\mu e^B_\nu \tilde{\nabla}_A n_B = e^A_\mu e^B_\nu \left(\partial_A n_B - n_S \tilde{\Gamma}^S_{AB}\right);\ C = h^{\lambda\sigma} C_{\lambda\sigma} \tag{8.7}$$

(4) A quantity $E_{\alpha\beta}$ formed from the 5-D curvature tensor, namely

$$E_{\alpha\beta} = \tilde{R}_{MANB} n^M n^N e^A_\alpha e^B_\beta;\ E = h^{\lambda\sigma} E_{\lambda\sigma} = -\tilde{R}_{MN} n^M n^N \tag{8.8}$$

Note that in the Einstein gauge ($\Omega = 1$) and when $\tilde{w} = 0$ equation (8.4) disappears and one is left with the original equations of Wesson for IMT where (8.3) becomes the gravitational equation (cf. [**465, 466**]).

To describe a particle-like entity in the 4-D brane mapped by the coordinates $y^0 = t$, $y^1 = r$, $y^2 = \theta$, and $y^3 = \phi$ one starts from a spherically symmetric static line element (**8F**) $ds^2 = exp(\nu(r))dt^2 - exp(\lambda(r))dr^2 - r^2(d\theta^2 + Sin^2(\theta)d\phi^2)$ and thinks of the entity restricted by a spherical boundary surface of radius $r = r_b$ with $r \leq r_b$ and filled with a substance induced by the bulk with matter density $\rho$, charge density $\rho_e$, and pressure $P$. For $r > r_b$ there is a vacuum and there is no singularity at $r = 0$. The bulk is to be mapped by $x^0 = exp[-(1/2)N(\ell)]y^0$, $x^1 = exp[-(1/2)L(\ell)]y^1$, $x^2 = y^2$, $x^3 = y^3$, and $x^4 = \ell$ with line element

$$dS^2 = g_{AB} dx^A dx^B \tag{8.9}$$

$$= e^{\tilde{N}(r,\ell)}(dx^0)^2 - e^{\tilde{L}(r,\ell)}(dx^1)^2 - r^2(d\theta^2 + Sin^2(\theta)d\phi^2) + \epsilon e^{\tilde{F}(r,\ell)} d\ell^2$$

One assumes that the $(r, \ell)$ dependence may be separated so that

$$\tilde{N}(r,,\ell) = N(\ell) + \nu(r);\ \tilde{L}(r,\ell) = L(\ell) + \lambda(r);\ \tilde{F}(r,\ell) = F(\ell) + \psi(r) \tag{8.10}$$

One denotes partial derivatives in $r$ with a prime and in $\ell$ with a dot. With no restriction one can write $H(\ell_0) = L(\ell_0) = F(\ell_0) = 0$ for the values on the brane $\ell = \ell_0$ (our 4-D space-time). The basic vectors, along with the metrics and Christoffel symbols for (**8F**), and (8.9)-(8.10) are given in [**465, 466**] and will be recalled as needed. The bulk Weyl vector has the following non-zero components (**8G**) $\tilde{w}_0(x^1, \ell)$ and $\tilde{w}_4(x^1, \ell)$ leading to

$$\tilde{W}_{01} = \partial_1 \tilde{w}_0,\ \tilde{W}^{10} = -e^{-(\tilde{L}+\tilde{N})}\partial_1 \tilde{w}_0;\ \tilde{W}_{1,4} = -\partial_1 \tilde{w}_4; \tag{8.11}$$

$$\tilde{W}^{14} = \epsilon e^{-(\tilde{L}+\tilde{F})}\partial_1 \tilde{w}_4;\ \tilde{W}_{04} = \dot{\tilde{w}}_0;\ \tilde{W}^{04} = \epsilon e^{(\tilde{N}+\tilde{F})}\dot{\tilde{w}}_0$$

Hence, since $N(\ell_0) = L(\ell_0) = F(\ell_0) = 0$ one has for the 4-D Maxwell field on the brane (**8H**) $W_{01} = \tilde{w}_0'(r, \ell_0) = w_0'$. For the Dirac gauge function $\Omega = \beta$ one assumes $\Omega = \Omega(r)$ leading to (**8I**) $\Omega_A = 0$ for $A \neq 1$. The functions $\Omega$, $\tilde{w}_0(x^1, \ell)$, $\tilde{w}_4(x^1, \ell)$, and $\tilde{W}_{AB}$ are parts of the 5-D WD geometric framework while their 4-D counterparts $w_0'$ and $W_{01}$ represent the Maxwell field with sources induced by the bulk. It is convenient to write (8.3) as follows (taking into account (**8I**) and $\Omega_S n^S = 0$ from [**465, 466**]; further one discards terms with $\Lambda$ when considering spatially small regions to obtain

$$G^\beta_\alpha = -\frac{8\pi}{\Omega^2} M^\beta_\alpha - \frac{2\epsilon}{\Omega^2}\left(\frac{1}{2}\delta^\beta_\alpha B - B^\beta_\alpha\right) + \frac{6h^{\beta\lambda}}{\Omega^2}\Omega_\alpha \Omega_\lambda \tag{8.12}$$

$$-\frac{3}{\Omega}(h^{\beta\lambda}\nabla_\lambda \Omega_\alpha - \delta^\beta_\alpha \nabla_\sigma \Omega^\sigma) + \epsilon[E^\beta_\alpha - \delta^\beta_\alpha E + h^{\mu\nu} C_{\mu[\nu} C_{\lambda]\sigma}(\delta^\beta_\alpha h^{\lambda\sigma} - 2\delta^\sigma_\alpha h^{\lambda\beta}]$$

(cf. here [**465, 466**]). Then one rewrites $G_0^0$, $G_1^1$, and $G_2^2$ in terms of an auxiliary gauge function $\omega(r) = log[\Omega(r)]$. For the case under consideration the Maxwell equations (8.4) become then, upon integration

$$w_0' = -\frac{\epsilon}{r^2} e^{(1/2)(\lambda+\nu+\psi-2\omega)} \tag{8.13}$$

$$\times \left[ \int_0^r e^{(1/2)(\lambda-\nu-3\psi+2\omega)} \left\{ \ddot{\tilde{w}}_0 + \frac{1}{2}(\dot{F} + \dot{L} + \dot{N})\dot{\tilde{w}}_0 \right\} r^2 dr + c \right]$$

The three equations for $G_0^0$, $G_1^1$, and $G_2^2$ plus (8.13) give 4 equations for 6 functions $\lambda$, $\nu$, $\psi$, $\omega$, $\tilde{w}_0$, and $\tilde{w}_4$ depending on $r$ (here $\ddot{L}$, $\ddot{N}$, $\dot{L}$, $\dot{N}$, $\dot{F}$ are constants on the brane $\ell = \ell_0$). Thus one can impose 2 conditions and this freedom can be used in order to regard the interior substance of the entity in question as a non-rotating perfect fluid satisfying a special equation of state, $\rho + P = 0$. Rewrite now the $G_i^i$ equations in the form

$$G_0^0 = e^{-\lambda}\left(-\frac{\lambda'}{r} + \frac{1}{r^2}\right) - \frac{1}{r^2} = \frac{\tilde{q}^2}{r^4} - 8\pi\rho; \tag{8.14}$$

$$G_1^1 = e^{-\lambda}\left(\frac{\nu'}{r} + \frac{1}{r^2}\right) - \frac{1}{r^2} = -\frac{\tilde{q}^2}{r^4} + 8\pi P_n;$$

$$G_2^2 = e^{-\lambda}\left[\frac{\nu''}{2} - \frac{\lambda'\nu'}{4} + \frac{(\nu')^2}{4} + \frac{\nu' - \lambda'}{2r}\right] = \frac{\tilde{q}^2}{r^4} + 8\pi P_\tau$$

Here $\tilde{q}(r)$ is the effective charge inside a sphere of radius $r$ and is given via

$$\tilde{q} = -\epsilon e^{(1/2)(\psi-4\omega)} \int_0^r e^{(1/2)(\lambda-\nu-3\psi+2\omega)} \left[ \ddot{\tilde{w}}_0 + \frac{1}{2}(\dot{F} + \dot{L} + \dot{N})\dot{\tilde{w}}_0 \right] r^2 dr \tag{8.15}$$

The constant term in (8.13) which leads to a singular point charge is discarded here. The term in $G_0^0$ and $G_2^2$ of the form $\tilde{q}^2/r^4 = exp[-(\lambda + \nu + 2\omega)](\tilde{w}_0')^2$ is the EM energy inside the sphere of radius $r$. Also $8\pi\rho(r)$ which includes the remaining terms in the right side of the $G_0^0$ equation is the matter density inside the spherically symmetric entity. One is looking here for a non-rotating entity filled with perfect fluid so one imposes the condition (**8J**) $P_\tau = P_n = P$ and a second condition is (**8K**) $\rho + P = 0$ for the prematter equation of state. Then (**8J**) yields

$$2\epsilon e^{-(\lambda+\psi+2\omega)}(\partial_1 \tilde{w}_4)^2 = -\frac{\epsilon}{2}e^{-\psi}\left[\ddot{L} + \frac{1}{2}(\dot{L})^2 - \frac{1}{2}\dot{F}\dot{L} + \frac{1}{2}\dot{L}\dot{N}\right] \tag{8.16}$$

$$+\frac{1}{2}e^{-\lambda}\left[\psi'' + \frac{1}{2}(\psi')^2 - \frac{1}{2}\lambda'\psi' - \frac{\psi'}{r}\right] + +3e^{-\lambda}\left[\omega'' - (\omega')^2 - \frac{1}{2}\lambda'\omega' - \frac{\omega'}{r}\right]$$

while (**8K**) leads to

$$2\epsilon e^{-(\nu+\psi+2\omega)}(\dot{\tilde{w}}_0)^2 = e^{-\lambda}\left(\frac{1}{2}\psi' + 3\omega'\right)\left(\frac{1}{r} - \frac{1}{2}\nu'\right) \tag{8.17}$$

$$+\frac{\epsilon}{2}e^{-\psi}\left[\ddot{N} + \frac{1}{2}(\dot{N})^2 - \frac{1}{2}\dot{F}\dot{N} + \frac{1}{2}\dot{L}\dot{N}\right]$$

There is however a restriction. For the metric in (**8F**) one obtains $G_{01} = 0$ and this leads to $B_{01} = 0$; since $B_{01} = -exp[-(1/2)(L + N + 2\tilde{F})]\dot{\tilde{w}}_0 \partial_1 \tilde{w}_4$ one must have (**8L**) $\dot{\tilde{w}}_0 = 0$ or $\partial_1 \tilde{w}_4 = 0$.

## 8.1. MORE EXAMPLES.

**EXAMPLE** 8.1. Following Israelit [**465, 466**] one constructs a spatially restricted entity in the Einstein gauge. Set then $\Lambda = 0$ and $\Omega = 1 \Rightarrow \omega = 0$ and take coordinates and line element as in (**8F**) but take also $L(\ell) = 0$ so that the bulk is mapped by (**8M**) $x^0 = exp[-(1/2)N(\ell)]t$, $x^{1,2,3} = y^{1,2,3}$, and $x^4 = \ell$ with 5-D metric

$$g_{00} = e^{\tilde{N}(r,\ell)} = e^{N(\ell)+\nu(r)};\ g_{11} = h_{11};\ g_{22} = h_{22};\ g_{33} = h_{33};\ g_{44} = \epsilon e^{\tilde{F}} \tag{8.18}$$

The corresponding basis and normal vectors are as before but now $L(\ell) = 0$ and the metric functions are chosen so that (**8N**) $\dot{F}(\ell_0) = \dot{N}(\ell_0) = 0$. Guided by symmetry and the restriction (**8L**) one takes for the 5-D Weyl vector $\tilde{w}_A$ only one non-zero component (**8O**) $\tilde{w}_0 \neq 0$ and on the brane the Weyl vector is (**8P**) $w_0(r) = \tilde{w}_0(r, \ell_0)$ with $w_1 = w_2 = w_3 = 0$. Consequently one obtains gravitational equations on the brane

(8.19)

$$G_0^0 = -e^{-(\lambda+\nu)}(w_0')^2 + \epsilon e^{-(\nu+\psi)}(\dot{\tilde{w}}_0)^2 - \frac{1}{2}e^{-\lambda}\left[\psi'' + \frac{1}{2}(\psi')^2 - \frac{1}{2}\psi'\lambda' + \frac{2\psi'}{r}\right];$$

$$G_1^1 = -e^{-(\lambda+\nu)}(w_0')^2 - \epsilon e^{-(\nu+\psi)}(\dot{\tilde{w}}_0)^2 - \frac{1}{2}\left[\frac{1}{2}\nu'\psi' + \frac{2\psi'}{r}\right] + \frac{\epsilon}{2}e^{-\psi}\ddot{N};$$

$$G_2^2 = e^{-(\lambda+\nu)}(w_0')^2 - \epsilon e^{-(\nu+\psi)}(\dot{\tilde{w}}_0)^2$$
$$-\frac{1}{2}e^{-\lambda}\left[\psi'' + \frac{1}{2}(\psi')^2 + \frac{1}{2}\psi'(\nu' - \lambda') + \frac{\psi'}{r}\right] + \frac{\epsilon}{2}e^{-\psi}\ddot{N}$$

In order to have a non-rotating fluid ($P_\tau = P_n = P$) one imposes (8.16). Then with $\tilde{w}_4 = 0$, $L(\ell) = 0$, and $\omega' = 0$ one can take $\psi' = 0$ so $\psi = c$. Further since the multiplier $exp[\psi = c]$ can cause only rescaling of $\ddot{\tilde{w}}_0$ and $\ddot{N}$ one sets $\psi = 0$ and the $G_i^i$ equations become

$$e^{-\lambda}\left(-\frac{\lambda'}{r} + \frac{1}{r^2}\right) - \frac{1}{r^2} = e^{-(\lambda+\nu)}(w_0')^2 + \epsilon e^{-\nu}(\dot{\tilde{w}}_0)^2; \tag{8.20}$$

$$e^{-\lambda}\left(\frac{\nu'}{r} + \frac{1}{r^2}\right) - \frac{1}{r^2} = -e^{-(\lambda+\nu)}(w_0')^2 - \epsilon e^{-\nu}(\dot{\tilde{w}}_0)^2 + \frac{\epsilon}{2}\ddot{N};$$

$$e^{-\lambda}\left[\frac{\nu''}{2} - \frac{\lambda'\nu'}{4} + \frac{(\nu')^2}{4} + \frac{\nu' - \lambda'}{2r}\right] = e^{-(\lambda+\nu)}(w_0')^2 - \epsilon e^{-\nu}(\dot{\tilde{w}}_0)^2 + \frac{\epsilon}{2}\ddot{N}$$

The 4-D Maxwell equations (8.13) become then (cf. (**8M**) and (**8N**))

$$w_0' = -\epsilon\frac{e^{(1/2)(\lambda+\nu)}}{r^2}\int_0^r \ddot{\tilde{w}}_0 e^{(1/2)(\lambda-\nu)} r^2 dr + \epsilon\frac{ce^{(1/2)(\lambda+\nu)}}{r^2} \tag{8.21}$$

In order to avoid a singularity at $r = 0$ one takes $c = 0$ and writes

$$w_0' = -\epsilon\frac{e^{(1/2)(\lambda+\nu)}}{r^2}\int_0^r \ddot{\tilde{w}}_0 e^{(1/2)(\lambda-\nu)} r^2 dr \tag{8.22}$$

One can compare this with the expression following from the Maxwell equations in GR where $w' = -[exp\{(1/2)(\lambda+\nu)\}q]/r^2$ with $q$ being the charge within a sphere of radius $r$ given via $q = 4\pi\int_0^r exp[(1/2)\lambda]\rho_e r^2 dr$. One sees that in the present case the charge is $q = \epsilon\int_0^r \ddot{\tilde{w}}_0 exp[(1/2)(\lambda-\nu)]r^2 dr$ with charge density $4\pi\rho_e = \epsilon ex[-(1/2)\nu]\ddot{\tilde{w}}_0$. These equations describe a spherically symmetric distribution of charged matter but choosing a suitable expression for $\tilde{w}_0(r,\ell)$ one can obtain a model of a neutral spatially closed entity (a particle).

Indeed choose (**8Q**) $\tilde{w}_0(\ell,r) = Sin[\kappa(\ell-\ell_0)e^{\nu/2}$ where $\kappa$ is arbitrary and $\nu = \nu(r)$. By (**8Q**) one has on the brane $\Sigma_{\ell_0}$ (**8R**) $\tilde{w}_0(\ell_0) = w_0'(\ell_0) = \ddot{\tilde{w}}_0(\ell_0) = 0$ but $\dot{\tilde{w}}_0 = \kappa exp(\nu/2)$. Thus (8.22) is satisfied identically and one is left with

$$e^{-\lambda}\left(-\frac{\lambda'}{r}+\frac{1}{r^2}\right)-\frac{1}{r^2}=\epsilon\kappa^2; \tag{8.23}$$

$$e^{-\lambda}\left(\frac{\nu'}{r}+\frac{1}{r^2}\right)-\frac{1}{r^2}=-\epsilon\kappa^2+\frac{\epsilon}{2}\ddot{N};$$

$$e^{-\lambda}\left[\frac{\nu''}{2}-\frac{\lambda'\nu'}{4}+\frac{(\nu')^2}{4}+\frac{\nu'-\lambda'}{2r}\right]=-\epsilon\kappa^2+\frac{\epsilon}{2}\ddot{N}$$

From (8.23) one has for the matter density and pressure (**8S**) $8\pi\rho = -\epsilon\kappa^2$ and $8\pi P = -\epsilon\kappa^2 + (1/2)\epsilon\ddot{N}$ and one notes that $\ddot{N}$ is constant on the brane and $\kappa$ is an arbitrary constant which will be chosen so that (**8T**) $\kappa^2 = (1/4)\ddot{N}$. Then from (**8S**) one has (**8U**) $\rho = -P = -(1/8\pi)\epsilon\kappa^2$. For $\epsilon = -1$ the matter density is positive and the pressure negative and according to (**8U**) the prematter equation of state is $\rho + P = 0$ (cf. (**8K**)). Now going back to (8.23) one can make use of the equilibrium equation $8\pi P' + 8\pi(\nu/2)(\rho+P) = 2qq'/r^4 = -8\pi exp(-\nu/2)\rho_e w_0'$. But this is satisfied identically by (**8R**) and (**8U**) so one is left with the first two equations in (8.23) which via (**8T**) and (**8U**) take the form

$$e^{-\lambda}\left(-\frac{\lambda'}{r}+\frac{1}{r^2}\right)-\frac{1}{r^2}=-8\pi\rho;\ e^{-\lambda}\left(\frac{\nu'}{r}+\frac{1}{r^2}\right)-\frac{1}{r^2}=8\pi P \tag{8.24}$$

Via (**8U**) one has $\lambda+\nu = 0$ so one can write for the solution of (8.23) (■) $exp(-\lambda) = exp(\nu) = 1-(r^2/a^2)$ with $a^2 = (3/8\pi\rho) = (3/\kappa^2)$. One is looking for a spatially restricted spherically symmetric entity having a boundary at $r_b$ where $P = 0$ must hold. This is impossible however since the pressure is constant via (**8U**). This obstacle can be overcome by taking $r_b = a$ and the metric inside the entity is

$$ds^2=\left(1-\frac{r^2}{r_b^2}\right)dt^2-\left(1-\frac{r^2}{r_b^2}\right)dr^2-r^2(d\theta^2+Sin^2(\theta)d\phi^2)\ \ (r\le r_b) \tag{8.25}$$

This is the metric of a deSitter universe with a positive cosmological constant. If one introduces $r = f_b Sin(\chi)$ $(0 \le \chi \le \pi/2)$ the line element (8.25) can be rewritten as

$$ds^2=Cos^2(\chi)dt^2-a^2(d\chi^2+Sin^2(\chi)d\Omega^2)\ \ (d\Omega^2=d\theta^2+Sin^2(\theta)d\phi^2) \tag{8.26}$$

which describes a closed universe - there is no boundary and hence no boundary condition on $P$. Outside of the entity ($r \geq r_b$) one has the Schwarzschild solution

$$ds^2 = \left(1 - \frac{2M}{r}\right) dt^2 - \left(1 - \frac{2M}{r}\right)^{-1} dr^2 - r^2 d\Omega^2 \tag{8.27}$$

with mass $M$ given by

$$M = \frac{4\pi}{3}\rho r_b^3 = \frac{1}{2}r_b = \frac{1}{2}a = \frac{\sqrt{3}}{2\kappa} \tag{8.28}$$

Recall that the mass density is given via $8\pi\rho = -\epsilon exp(-\nu)(\dot{\tilde{w}}_0)^2$ showing how matter arises from the fifth dimension. The entity just described could be considered a classical model of a neutral particle induced by the bulk. ■

**EXAMPLE** 8.2. Next in [**465, 466**] one constructs a neutral particle in a general gauge and a charged particle. First (neglecting $\Lambda$) one adopts (**8F**)-(8.10) and (**8I**) but now takes $L(\ell) = 0$. In order to have no Maxwell field on the 4-D brane it is assumed that the Weyl vector in the bulk $\tilde{w}_A$ has only one non-zero component $\tilde{w}_4(r, \ell)$ so that the 5-D Weyl field is given by $\tilde{W}_{14} = -\tilde{w}_4'$ and on the brane $w_\nu = 0$ with $W_{\mu\nu} = 0$. Since $L(\ell) = 0$ and $\Omega = \Omega(r)$ one has from [**465, 466**] (**8W**) $C_{\mu[\nu}C_{\lambda]\sigma} = 0$ and $\Omega_S n^S = 0$ so that the gravitational equations take the simple form

$$G_\alpha^\beta = -\frac{2\epsilon}{\Omega^2}\left(\frac{1}{2}\delta_\alpha^\beta B - B_\alpha^\beta\right) + \frac{6}{\Omega^2}\Omega_\alpha\Omega_\lambda h^{gl\beta} \tag{8.29}$$

$$- \frac{3}{\Omega}(\nabla_\lambda\Omega_\alpha h^{\lambda\beta} - \delta_\alpha^\beta\nabla_\sigma\Omega^\sigma) + \epsilon[E_\alpha^\beta - \delta_\alpha^\beta E]$$

From the $G_i^i$ equations in [**465, 466**] one obtains somewhat more complicated equations than in (8.19) where it is noted that $\psi$, $\omega$, and $\tilde{w}_4'$ are arbitrary functions and on the brane the constant $C_N = [\ddot{N} + (1/2)(\dot{N})^2 - (1/2)\dot{F}\dot{N}]$ is also arbitrary. For a spherically symmetric non-rotating entity one obtains the following conditions (cf. (**8J**) and (8.16))

$$-2\epsilon e^{-(\lambda+\psi+2\omega)}(\tilde{w}_4')^2 - 3e^{-\lambda}\left[(\omega')^2 + \frac{\omega'}{r} - \omega'' + \frac{1}{2}\lambda'\omega'\right] \tag{8.30}$$

$$= -\frac{1}{2}e^{-\lambda}\left[\psi'' + \frac{1}{2}(\psi')^2 - \frac{1}{2}\lambda'\psi' - \frac{\psi'}{r}\right]$$

Equation (8.30) can be regarded as a condition imposed on 3 functions, $\psi$, $\omega$, $\tilde{w}_4'$ and in order to get prematter $\rho + P = 0$ one obtains from (**8K**) and (8.17) another condition

$$-e^{-\lambda}\left[3\omega' + \frac{1}{2}\psi'\right]\left(\frac{1}{r} - \frac{1}{2}\nu'\right) = \epsilon e^{-\psi}C_N \tag{8.31}$$

One can choose $N(\ell)$ and $F(\ell)$, so that $C_N = [\ddot{N} + (1/2)(\dot{N})^2 - (1/2)\dot{F}\dot{N}] = 0$ on the brane $\Sigma_{\ell_0}$. Then one has a simple gauge condition (**8X**) $\omega' = -(1/6)\psi'$ which, inserted into (8.30) yields the result (**8Y**) $exp(2\omega)(\omega')^2 = (\epsilon/3)exp(-\psi)(\tilde{w}_4')^2$. Finally using (**8X**)-(**8Y**) and the Einstein tensor one arrives at

$$e^{-\lambda}\left(-\frac{\lambda'}{r}+\frac{1}{r^2}\right)-\frac{1}{r^2} = -3\epsilon e^{-(\lambda+\psi+2\omega)}(\tilde{w}_4')^2 \doteq -9e^{-\lambda}(\omega')^2; \tag{8.32}$$

$$e^{-\lambda}\left(\frac{\nu'}{r}+\frac{1}{r^2}\right)-\frac{1}{r^2} = -3\epsilon e^{-(\lambda+\psi+2\omega)}(\tilde{w}_4')^2 \doteq -9e^{-\lambda}(\omega')^2;$$

$$e^{-\lambda}\left(\frac{\nu''}{2}-\frac{\lambda'\nu'}{4}+\frac{(\nu')^2}{4}+\frac{\nu'-\lambda'}{2r}\right) = -3\epsilon e^{-(\lambda+\psi+2\omega)}(\tilde{w}_4')^2 \doteq -9e^{-\lambda}(\omega')^2$$

Instead of solving the last equation one makes use of the equilibrium equation $P' + (\nu'/2)(\rho + P) = 0$ which via $P = -\rho$ gives $P' = 0$ so that

$$8\pi\rho = -8\pi P = 3\epsilon e^{-(\omega+\psi+2\omega)}(\tilde{w}_4') = c = 8\pi\rho_0 \tag{8.33}$$

Thus the entity is filled with prematter having constant density and pressure (and for positive matter density one takes $\epsilon = 1$).

Now from (8.32) one has $\lambda + \nu = 0$ so the solution is

$$e^{-\lambda} = e^{\nu} = 1 - \frac{r^2}{r_b^2}\ with\ f_b^2 = \frac{3}{8\pi\rho_0} \tag{8.34}$$

$$ds^2 = \left(1-\frac{r^2}{r_b^2}\right)dt^2 - \left(1-\frac{r^2}{r_b^2}\right)^{-1}dr^2 - r^2(d\theta^2 + Sin^2(\theta)d\phi^2);\ (r \le r_b) \tag{8.35}$$

This is formally identical with the previous model (8.25). There is a deSitter universe with a positive cosmological constant and if one writes $r = r_b Sin(\chi)$, $(0 \le \chi \le (\pi/2))$ (8.26) comes up again (as a description of a closed universe with no boundaries and hence no boundary conditions on the pressure at $r = r_b$). Outside of the entity one has again the Schwarzschild solution (8.27) with mass as in (8.28). ∎

**EXAMPLE** 8.3. Again following [**465, 466**], in order to create a charged particle entity one uses the spherically symmetric 4-D line element (**8F**) and for the metric function (8.10) one chooses (•) $N(\ell) = L(\ell) = 0$. Thus the 5-D line element is $dS^2 = exp(\nu)(dt^2) - exp(\lambda)(dr)^2 - r^2(d\theta^2 + Sin^2(\theta)d\phi^2) + \epsilon exp(\tilde{F}(r,\ell))d\ell^2$ (cf. (8.9)) with $\tilde{F} = F(\ell) + \psi(r)$. Recalling $B_{01} = -exp[-\tilde{F})]\dot{\tilde{w}}_0\partial_1\tilde{w}_4 = 0$ (cf. (**8L**)) one chooses the possibility $\dot{\tilde{w}}_0 = 0$ in (**8L**). Further imposing (**8K**) and using (**8L**) one obtains from (8.17)

$$\left(\frac{1}{2}\psi' + 3\omega'\right)\left(\frac{1}{r} - \frac{1}{2}\nu'\right) = 0 \tag{8.36}$$

This results in the simple gauge condition (cf. (**8X**)) (♦) $\omega' = -(1/6)\psi'$; $\omega = -(1/6)\psi$ (discarding a possible constant). Since one is looking for a non-rotating entity filled with perfect fluid one takes $P_\tau = P_n = P$ (cf. (**8J**)) and imposes (8.16). Inserting (•) and (♦) into (8.16) yields then

$$2\epsilon e^{-(\lambda+\psi+2\omega)}(\partial_1\tilde{w}_4)^2 = \frac{1}{6}e^{-\lambda}(\psi')^2 \tag{8.37}$$

Discarding the cosmological term as irrelevant for a spatially restricted entity one obtains from the $G_i^i$ equations

(8.38) $$G_0^0 = -e^{-(\lambda+\nu-(\psi/3))}(\tilde{w}_0')^2 - \frac{e^{-\lambda}(\psi')^2}{4};$$

$$G_1^1 = -e^{-(\lambda+\nu-(\psi/3))}(\tilde{w}_0')^2 - \frac{e^{-\lambda}(\psi')^2}{4};\ G_2^2 = e^{-(\lambda+\nu-(\psi/3))}(\tilde{w}_0')^2 - \frac{e^{-\lambda}(\psi')^2}{4}$$

leading to the conclusion $\lambda + \nu = 0$.

One goes back now to the Maxwell equations (8.13) and taking into account (**8L**) and (♦♦) as well as $\lambda + \nu = 0$ one obtains the Maxwell equations for the present model

(8.39) $$w_0' = -\frac{\epsilon}{r^2}e^{(2/3)\psi}\int_0^r e^{(\lambda-(5/3)\psi)}\ddot{\tilde{w}}_0 r^2 dr$$

According to (8.38) and (8.39) we can introduce the effective charge inside the sphere of radius $r$ (cf. (8.15))

(8.40) $$\tilde{q}(r) = e^{(5/6)\pi}\int_0^r e^{(\lambda-(5/3)\psi)}\ddot{\tilde{w}}_0 r^2 dr$$

With (8.40) one can write $e^{-(\lambda+\nu-(\psi/3))}(\tilde{w}_0')^2 = \tilde{q}^2/r^4$ for the EM energy inside the sphere of radius $r$. Further from (8.38) it follows that inside the entity

(8.41) $$8\pi\rho = -8\pi P = \frac{1}{4}e^{-\lambda}(\psi')^2$$

(the substance is in the state of prematter). With (8.39)-(8.41) one rewrites (8.38) as

(8.42)
$$G_0^0 = e^{-\lambda}\left(-\frac{\lambda'}{r} + \frac{1}{r^2}\right) - \frac{1}{r^2} = -8\pi\rho - \frac{\tilde{q}^2}{r^4};\ G_1^1 = e^{-\lambda}\left(\frac{\nu'}{r} + \frac{1}{r^2}\right) - \frac{1}{r^2} = 8\pi P$$
$$-\frac{\tilde{q}^2}{r^4};\ G_2^2 = \frac{e^{-\lambda}}{2}\left(\nu'' - \frac{\lambda'\nu'}{2} + \frac{(\nu')^2}{2} + \frac{\nu' - \lambda'}{r}\right) = 8\pi P + \frac{\tilde{q}^2}{r^4}$$

Thus inside the sphere of radius $r_b$ there is a prematter substance and outside a vacuum. Introducing the function $y(r) = exp(-\lambda) = exp(\nu)$ one obtains the following solution of (8.38)

(8.43) $$y(r) = e^{\nu} = e^{-\lambda} = 1 - \frac{8\pi}{r}\int_0^r \rho r^2 dr - \frac{1}{r}\int_0^r \frac{\tilde{q}^2}{r^2}dr\ (r \leq r_b);$$

(8.44) $$y(r) = 1 - \frac{2M}{r} + \frac{Q^2}{r^2}\ (r \geq r_b);\ Q = \tilde{q}(r_b)$$

In (8.44) $M$ stands for the mass of the whole entity while via (8.40) the total charge $Q$ is given via

(8.45) $$Q = \tilde{q}(r_b) = e^{(5/6)\pi}\int_0^{r_b} e^{(\lambda-(5/3)\psi)}\ddot{\tilde{w}}_0 r^2 dr$$

From (8.43)-(8.44) we obtain for the mass seen by an external observer

$$M = \frac{Q^2}{2r_b} + 4\pi \int_0^{r_b} \rho r^2 dr + \frac{1}{2} \int_0^{r_b} \frac{\tilde{q}^2}{r^2} dr \tag{8.46}$$

Consider now the last equation in (8.38) and, instead of solving it, make use of the equilibrium relation $8\pi\rho' + 8\pi(\rho + P) = -[2\tilde{q}\tilde{q}'/r^4] = -(\tilde{q}^2)'/r^4$ arising from the Bianchi identity. For prematter this relation gives (★) $8\pi\rho' = -(\tilde{q}^2)'/r^4$ leading to

$$(\tilde{q})^2 = -8\pi r^4 \rho + 32\pi \int_0^r \rho r^2 dr \tag{8.47}$$

However as noted above, $\rho(r_b) = 0$ so the total charge of the entity is (••) $Q^2 = 32\pi \int_0^{r_b} \rho r^3 dr$. Now going back to (8.42) substitute $y(r) = exp(-\lambda) = exp(\nu)$ as well as (8.47) to obtain

$$y' r^3 + y r^2 - r^2 = -32\pi \int_0^r \rho r^3 dr \tag{8.48}$$

Thus the entity is described by the equation

$$y'' + \frac{4}{r} y' + \frac{2}{r^2} y - \frac{2}{r^2} = -32\pi\rho \tag{8.49}$$

Assume now that one has a known expression for $y(r)$; then one can deduce from (8.49) the matter density $\rho(r)$, from (8.47) the charges $\tilde{q}(r)$ and $Q$, and from (8.46) the mass $M$. A suitable representation for $0 \leq r \leq r_b$ is in fact the bell like function (♦♦) $y = [1/k^2r^2]Sin^2(kr)$ with $k = \pi/r_b$ and $|k|$ measured in $cm^{-1}$. Inserting (♦♦) into (8.49) gives

$$8\pi\rho = \frac{1 - Cos(2kr)}{2r^2} = \frac{Sin^2(kr)}{r^2} \equiv k^2 y \tag{8.50}$$

Thus for the mass density $\rho(r) \geq 0$, $8\pi\rho(0) = k^2$ with $\rho(r_b) = 0$. Further putting $\rho$ into (8.47) and choosing a suitable constant of integration one obtains for the effective charge inside a sphere of radius $r \leq r_b$

$$\tilde{q}^2(r) = \left[ rCos(kr) - \frac{1}{k} Sin(kr) \right]^2 \tag{8.51}$$

Consequently via (8.51) (★★) $\tilde{q}(0) = 0$ and $|Q| = |\tilde{q}(r_b)| = r_b$. To obtain $\psi$ one can equate (8.40) and (8.50) to get

$$(\psi')^2 = 4k^2 \Rightarrow \psi' = \pm 2k;\ \psi = \pm 2kr + c \tag{8.52}$$

Choosing $c = \mp 2\pi$ gives then (• • •) $\psi = \pm(2kr - 2\pi)$. To accound for the external mass $M$ one starts from (8.46) and uses (8.50)-(8.51) to get (♦♦♦) $M = (1/2)r_b + (Q^2/2r_b)$ so via (★★) we get (★★★) $M = Q = r_b$. Recall for neutral particles $M = (1/2)r_b$ so (♦♦♦) consists of two parts representing proper gravitational mass and EM mass. For the charge density $\rho_e$ inside the entity recall that for a spherically symmetric distribution of matter the charge is given by $q = 4\pi \int_0^r exp(\lambda/2)\rho_e r^2 dr$. Making use of (♦♦) and (8.51) yields then

$$4\pi|\rho_e| = \frac{Sin^2(kr)}{r^2} \Rightarrow |\rho_e| = 2\rho \tag{8.53}$$

There is much more in [**465, 466, 467, 468**] and the other references cited in this chapter.

CHAPTER 5

# REMARKS ON THERMODYNAMICS AND GRAVITY

## 1. INTRODUCTION

Since early work of Beckenstein and Hawking on black holes, entropy, radiation, and temperature there has been enormous interest in connections between thermodynamics and gravity. This was especially enhanced by work in [**474**] connecting entropy, horizons, and the Einstein equations and subsequently there has been a deluge of information connecting gravity with the laws of thermodynamics (see e.g. [**15, 16, 38, 149, 178, 219, 286, 318, 319, 367, 471, 524, 570, 607, 620, 661, 662, 663, 664, 665, 666, 668, 669, 676, 702, 703, 704, 834, 860, 904, 905, 906**]). We sketched some of this material in [**170, 178**] based on the geometric apparatus related to geometrical information theory in [**154, 155, 194, 196, 175**] and refer to [**29, 93, 146, 147, 154, 155, 170, 194, 195, 196, 197, 199, 318, 319, 320, 321, 348, 349, 350, 383, 480, 490, 724, 725, 756**] for information theory in general. In particular Fisher information is developed at great length in [**170, 318, 319**]. We envision here a rather long range project designed to relate gravitational thermodynamics to a geometry of information theory; as a background we begin with some material from [**196**] supporting the idea of a statistical geometrodynamics (SGD) related to information theory (this follows [**170**] and see [**197**] for more recent developments). Here one builds a model of SGD based on (i) Positing that the geometry of space is of statistical origin and is explained in terms of the distinguishability Fisher-Rao (FR) metric and (ii) Assuming the dynamics of the geometry is derived solely from principles of inference. There is no external time but an intrinsic one à la [**58**]. A scale factor $\sigma(x)$ is required to assign a Riemannian geometry and it is conjectured that it can be chosen so that the evolving geometry of space sweeps out a 4-D spacetime. The procedure defines only a conformal geometry but that is entirely appropriate d'après [**899**]. One uses the FR metric in two ways, one to distinguish neighboring points and the other to distinguish successive states. Consider then a "cloud" of dust with coordinate values $y^i$ ( $i = 1, 2, 3$) and estimates $x^i$ with $p(y|x)dy$ the probability that the particle labeled $x^i$ should have been labeled $y^i$ (the FR metric encodes the use of probability distributions - instead of structureless points). Thus

$$\frac{p(y|x+dx)-p(y|x)}{p(y|x)} = \frac{\partial log[p(y|x)]}{\partial x^i}dx^i \tag{1.1}$$

$$(1.2)\qquad d\lambda^2 = \int d^4yp(y|x)\frac{\partial log[p(y|x)]}{\partial x^i}\frac{\partial log[p(y|x)]}{\partial x^j}dx^i dx^j = \gamma_{ij}dx^i dx^j$$

and $d\lambda^2 = 0 \iff dx^i = 0$. The FR metric $\gamma_{ij}$ is the only local Riemannian metric reflecting the underlying statistical nature of the manifold of distributions $p(y|x)$ and a scale factor $\sigma$ giving a metric $g_{ij}(x) = \sigma(x)\gamma_{ij}(x)$ is needed for a Riemannian metric (cf. [**194, 196**]). Also the metric $d\lambda^2$ is related to the entropy of $p(y|x+dx)$ relative to $p(y|x)$, namely

$$(1.3)\qquad S[p(y|x+dx)|p(y|x)] = -\int d^3yp(y|x+dx)log\frac{p(y|x+dx)}{p(y|x)} = -\frac{1}{2}d\lambda^2$$

and maximizing the relative entropy $S$ is equivalent to minimizing $d\lambda^2$. One thinks of $d\lambda$ as a spatial distance in specifying that the reason that particles at $x$ and $x+dx$ are considered close is because they are difficult to distinguish. To assign an explicit $p(y|x)$ one assumes the relevant information is given via $<y^i>=x^i$ and the covariance matrix $<(y^i-x^i)(y^j-x^j)>=C^{ij}(x)$; this leads to

$$(1.4)\qquad p(y|x) = \frac{C^{1/2}}{(2\pi)^{3/2}}exp\left[-\frac{1}{2}C_{ij}(y^i-x^i)(y^j-x^j)\right]$$

where $C^{ik}C_{kj} = \delta^i_j$ and $C = det(c_{ij})$. Subsequently to each $x$ one associates a probability distribution

$$(1.5)\qquad p(y|x,\gamma) = \frac{\gamma^{1/2}(x)}{(2\pi)^{3/2}}exp\left[-\frac{1}{2}\gamma_{ij}(x)(y^i-x^i)(y^j-x^j)\right]$$

where $\gamma_{ij}(x) = C_{ij}(x)$ (extreme curvature situations are avoided here). One deals with a conformal geometry described via $\gamma_{ij}$ and a scale factor $\sigma(x)$ will be needed to compare uncertainties at different points; the choice of $\sigma$ should then be based on making motion "simple".

Thus define a macrostate via

$$(1.6)\qquad P[y|\gamma] = \prod_x p(y(x)|x,\gamma_{ij}(x))$$

$$= \left[\prod_x \frac{\gamma^{1/2}(x)}{(2\pi)^{3/2}}\right]exp\left[-\frac{1}{2}\sum_x \gamma_{ij}(x)(y^i-x^i)(y^j-x^j)\right]$$

Once a dust particle in an earlier state $\gamma$ is identified with the label $x$ one assumes that this particle can be assigned the same label $x$ as it evolves into the later state $\gamma+\Delta\gamma$ (equilocal comoving coordinates). Then the change between $P[y|\gamma+\Delta\gamma]$ and $P[y|\gamma]$ is denoted by $\Delta\ell$ and is measured via their relative entropy (this is a form of Kullback-Liebler entropy - cf. [**194, 196**])

$$(1.7)\quad S[\gamma+\Delta\gamma|\gamma] = -\int\left(\prod_x dy(x)\right)P[y|\gamma+\Delta\gamma]log\frac{P[y|\gamma+\Delta\gamma]}{P[y|\gamma]} = -\frac{1}{2}\Delta\ell^2$$

Since $P[y|\gamma]$ and $P[y|\gamma+\Delta\gamma]$ are products one can write

$$(1.8)\qquad S[\gamma+\Delta\gamma,\gamma] = \sum_x S[\gamma(x)+\Delta\gamma(x),\gamma(x)]$$

$$= -\frac{1}{2}\sum_x \Delta\ell^2(x);\ \Delta\ell^2(x) = g^{ijk\ell}\Delta\gamma_{ij}(x)\Delta\gamma_{k\ell}(x)$$

where, using (1.5)

$$g^{ijk\ell} = \int d^3yp(y|x,\gamma)\frac{\partial log[p(y|x,\gamma)]}{\partial\gamma_{ij}}\frac{\partial log[p(y|x,\gamma)]}{\partial\gamma_{k\ell}} \tag{1.9}$$

$$= \frac{1}{4}\left(\gamma^{ik}\gamma^{ji} + \gamma^{i\ell}\gamma^{jk}\right)$$

Then $\Delta L^2 = \sum_x \Delta\ell^2(x)$ can be written as an integral if we note that the density of distinguishable distributions is $\gamma^{1/2}$. Thus the number of distinguishable distributions, or distinguishable points in the interval $dx$ is $dx\gamma^{1/2}$ $(dx \sim d^3x)$ and one has

$$\Delta L^2 = \int dx\gamma^{1/2}\Delta\ell^2 = \int dx\gamma^{1/2}g^{ijk\ell}\Delta\gamma_{ij}\Delta\gamma_{k\ell} \tag{1.10}$$

Thus the effective number of distinguishable points in the interval $dx$ is finite (due to the intrinsic fuzziness of space). Now to describe the change $\Delta\gamma_{ij}(x)$ one introduces an arbitrary time parameter $t$ along a trajectory

$$\Delta\gamma_{ij} = \gamma_{ij}(t+\Delta t, x) - \gamma_{ij}(t,x) = \partial_t\gamma_{ij}\Delta t \tag{1.11}$$

Thus $\partial_t\gamma_{ij}$ is the "velocity" of the metric and (1.10) becomes

$$\Delta L^2 = \int dx\gamma^{1/2}g^{ijk\ell}\partial_t\gamma_{ij}\partial_t\gamma_{k\ell}\Delta t^2 \tag{1.12}$$

Now go to an arbitrary coordinate frame where equilocal points at $t$ and $t+\Delta t$ have coordinates $x^i$ and $\tilde{x}^i = x^i - \beta^i(x)\Delta t$. Then the metric at $t+\Delta t$ transforms into $\tilde{\gamma}_{ij}$ with

$$\gamma_{ij}(t+\Delta t, x) = \tilde{\gamma}_{ij}(t+\Delta t, x) - (\nabla_i\beta_j + \nabla_j\beta_i)\Delta t \tag{1.13}$$

where $\nabla_i\beta_j = \partial_i\beta_j - \Gamma^k_{ij}\beta_k$ is the covariant derivative associated to the metric $\gamma_{ij}$. In the new frame, setting $\tilde{\gamma}_{ij}(t+\Delta t, x) - \gamma_{ij}(t,x) = \Delta\gamma_{ij}$ one has

$$\Delta_\beta\gamma_{ij} = \Delta\gamma_{ij} - (\nabla_i\beta_j + \nabla_j\beta_i)\Delta t \sim \Delta_\beta\gamma_{ij} = \dot{\gamma}_{ij}\Delta t \tag{1.14}$$

$$\dot{\gamma}_{ij} = \partial_t\gamma_{ij} - \nabla_i\beta_j - \nabla_j\beta_i$$

leading to

$$\Delta_\beta L^2 = \int dx\gamma^{1/2}g^{ijk\ell}\dot{\gamma}_{ij}\dot{\gamma}_{k\ell}\Delta t^2 \tag{1.15}$$

Next one addresses the problem of specifying the best matching criterion, i.e. what choice of $\beta^i$ provides the best equilocality match. This is treated as a problem in inference and asks for minimum $\Delta_\beta L^2$ over $\beta$. Hence one gets

$$\delta(\Delta_\beta L^2) = 2\int dx\gamma^{1/2}g^{ijk\ell}\dot{\gamma}_{ij}\dot{\gamma}_{k\ell}\Delta t^2 = 0 \tag{1.16}$$

$$\Rightarrow \nabla_\ell(2g^{ijk\ell}\dot{\gamma}_{ij}) = 0 \equiv \nabla_\ell\dot{\gamma}^{k\ell} = 0$$

(using (1.9) and $\dot{\gamma}^{k\ell} = \partial_t\gamma^{k\ell} + \nabla^k\beta^\ell + \nabla^\ell\beta^k$). These equations determine the shifts $\beta^i$ giving the best matching and equilocality for the geometry $\gamma_{ij}$ and alternatively they could be considered as constraints on the allowed change $\Delta\gamma_{ij} = \partial_t\gamma_{ij}\Delta t$ for given shifts $\beta^i$. In describing a putative entropic dynamics one assumes now e.g. continuous trajectories with each factor in $P[y|\gamma]$ evolving continuously through intermediate states labeled via $\omega(x) = \omega\zeta(x)$ where $\zeta(x)$ is a fixed positive function and $0 < \omega < \infty$ is a variable parameter (some kind of many fingered time à la Schwinger, Tomonaga, Wheeler, et al). It is suggested that they dynamics be determined by an action

$$J = \int_{t_i}^{t_f} dt \int dx \gamma^{1/2} [g^{ijk\ell}\dot{\gamma}_{ij}\dot{\gamma}_{k\ell}]^{1/2} \tag{1.17}$$

The similarities to "standard" geometrodynamics are striking.

**1.1. INFORMATION DYNAMICS.** We go here to [**172, 173**] and consider the idea of introducing some kind of dynamics in a reasoning process. One looks at the Fisher metric defined by

$$g_{\mu\nu} = \int_X d^4x p_\theta(x) \left(\frac{1}{p_\theta(x)}\frac{\partial p_\theta(x)}{\partial\theta^\mu}\right)\left(\frac{1}{p_\theta(x)}\right)\left(\frac{\partial p_\theta(x)}{\partial\theta^\nu}\right) \tag{1.18}$$

and constructs a Riemannian geometry via

$$\Gamma^\sigma_{\lambda\nu} = \frac{1}{2}g^{\nu\sigma}\left(\frac{\partial g_{\mu\nu}}{\partial\theta^\lambda} + \frac{\partial g_{\lambda\nu}}{\partial\theta^\mu} - \frac{\partial g_{\mu\lambda}}{\partial\theta^\nu}\right); \tag{1.19}$$

$$R^\lambda_{\mu\nu\kappa} = \frac{\partial\Gamma^\lambda_{\mu\nu}}{\partial\theta^\kappa} - \frac{\partial\Gamma^\lambda_{\mu\kappa}}{\partial\theta^\nu} + \Gamma^\eta_{\mu\nu}\Gamma^\lambda_{\kappa\eta} - \Gamma^\eta_{\mu\kappa}\Gamma^\lambda_{\nu\eta}$$

Then the Ricci tensor is $R_{\mu\kappa} = R^\lambda_{\mu\lambda\kappa}$ and the curvature scalar is $R = g^{\mu\kappa}R_{\mu\kappa}$. The dynamics associated with this metric can then be described via functionals

$$J[g_{\mu\nu}] = -\frac{1}{16\pi}\int \sqrt{g(\theta)}R(\theta)d^4\theta \tag{1.20}$$

leading upon variation in $g_{\mu\nu}$ to equations

$$R^{\mu\nu}(\theta) - \frac{1}{2}g^{\mu\nu}(\theta)R(\theta) = 0 \tag{1.21}$$

Contracting with $g_{\mu\nu}$ gives then the Einstein equations $R^{\mu\nu}(\theta) = 0$ (since $R = 0$). $J$ is also invariant under $\theta \to \theta + \epsilon(\theta)$ and variation here plus contraction leads to a contracted Bianchi identity. Constraints can be built in by adding terms $(1/2)\int \sqrt{g}T^{\mu\nu}g_{\mu\nu}d^4\theta$ to $J[g_{\mu\nu}]$. If one is fixed on a given probability distribution $p(x)$ with variable $\theta^\mu$ attached to give $p_\theta(x)$ then this could conceivably describe some gravitational metric based on quantum fluctuations for example. As examples a Euclidean metric is produced in 3-space via Gaussian $p(x)$ and complex Gaussians will give a Lorentz metric in 4-space.

**1.2. ENTROPY AND THE EINSTEIN EQUATIONS.** In [**662, 664**] (cf. also [**170, 178, 670, 671, 673, 770**]) one takes an entropy functional ($u^a = \bar{x}^a - x^a$ is a perturbation)

$$S = \frac{1}{8\pi}\int d^4x\sqrt{g}\left[M^{abcd}\nabla_a u_b \nabla_c u_d + N_{ab}u^a u^b\right] \tag{1.22}$$

Extremizing with respect to $u_b$ leads to ($N_{ab}u^a u^b = N^{ab}u_a u_b$)

$$\nabla_a \left(M^{abcd}\nabla_c\right) u_d = N^{bd}u_d \tag{1.23}$$

Note $\int d^4x\sqrt{-g}f\nabla_a u_b = -\int d^4x\sqrt{-g}u_b\nabla_a f$ since $\delta\sqrt{-g} = -(1/2)\sqrt{-g}g_{\mu\nu}\delta g^{\mu\nu}$ and $\nabla_a g^{\mu\nu} = 0$. Choosing $M$ and $N$ such that (1.23) (for all $u_d$) implies the Einstein equations entails

$$M^{abcd} = g^{ad}g^{bc} - g^{ab}g^{cd};\ N_{ab} = 8\pi\left(T_{ab} - \frac{1}{2}g_{ab}T\right) \tag{1.24}$$

Consequently $S$ becomes

$$S = \frac{1}{8\pi}\int d^4x\sqrt{-g}\left[(\nabla_a u^b)(\nabla_b u^a) - (\nabla_b u^b)^2 + N_{ab}u^a u^b\right] \tag{1.25}$$

$$= \frac{1}{8\pi}\int d^4x\sqrt{-g}\left[Tr(J^2) - (Tr(J))^2 + 8\pi\left(T_{ab} - \frac{1}{2}g_{ab}T\right)u^a u^b\right]$$

where $J_a^b = \nabla_a u^b$. Note here

$$\int d^4x\sqrt{-g}g^{ad}g^{bc}\nabla_a u_b\nabla_c u_d = \int d^4x\sqrt{-g}(\nabla^a u^b)(\nabla^c u^d) \tag{1.26}$$

and also

$$\nabla_a \left(M^{abcd}\nabla_c\right) u_d = \nabla_a\left[g^{ad}g^{bc} - g^{ab}g^{cd}\right]\nabla_c u_d \tag{1.27}$$

$$= \nabla_a g^{ad}g^{bc}\nabla_c u_d - \nabla_a g^{ab}g^{cd}\nabla_c u_d = \nabla_a\nabla^b u^a - \nabla^b\nabla_c u^c \sim (\nabla_a\nabla^b - \nabla^b\nabla_a)u^a$$

Further (as in (1.27))

$$M^{abcd}\nabla_a u_b\nabla_c u_d = g^{ad}g^{bc}\nabla_a u_b\nabla_c u_d - g^{ab}g^{cd}\nabla_a u_b\nabla_c u_d \tag{1.28}$$

$$= \nabla^d u_b\nabla^b u_d - \nabla_a u^a\nabla_c u^c$$

which confirms (1.25). We record also from [**646**] that

$$(\nabla_\mu\nabla_\nu - \nabla_\nu\nabla_\mu)\alpha(w) = R(\alpha, \partial_\mu, \partial_\nu, w) \tag{1.29}$$

which identifies $\nabla_\mu\nabla_\nu - \nabla_\nu\nabla_\mu$ with $R_{\mu\nu}$ and allows us to imagine (1.27) as $R_a^b u^a$ with Einstein equations

$$R_a^b u^a = N_a^b u^a\ (= N^{bc}g_{ca}g^{ca}u_c) \tag{1.30}$$

for example, which is of course equivalent to $R_{ab} = N_{ab}$ (cf. also [**662, 665, 676**]). Note also $G_{ab} = R_{ab} - (1/2)Rg_{ab} = kT_{ab}$ implies that $R^\mu_\nu - (1/2)R\delta^\mu_\nu = kT^\mu_\nu$ which upon contraction gives $R = -kT$ (since $\delta^\mu_\mu = 4$) and hence $R_{ab} = k(T_{ab} - (1/2)Tg_{ab})$.

For completeness we sketch here a derivation of the Einstein equations from an action principle (cf. [**225, 610, 860**] and see Chapter 6, Section 5, for updated and expanded coverage of the entropy field equations connections). The Einstein-Hilbert action is $A = \int_\Omega[\mathfrak{L}_G + \mathfrak{L}_M]d^4x$ where $\mathfrak{L}_G = (1/2\chi)\sqrt{-g}{}^4R$ ($\chi = 8\pi$ and ${}^4R$ is the Ricci scalar). Following [**225**] we list a few useful facts first (generally we will write if necessary $g_{ab}T^{cb} = T^{c}_{\cdot a}$ and $g_{ab}T^{bc} = T_a^{\cdot c}$).

(1) $\nabla_\gamma g^{\alpha\beta} = 0$ (by definitions of covariant derivative and Christoffel symbols).

(2) $\delta\sqrt{-g} = (1/2)\sqrt{-g}g^{\alpha\beta}\delta g_{\alpha\beta}$ and $(\delta g_{\alpha\beta})g^{\alpha\beta} = -(\delta g^{\alpha\beta})g_{\alpha\beta}$ (see e.g. **[860]** for the calculation).

(3) For a vector field $v^a$ one has $\nabla_a v^a = \partial_a(\sqrt{-g}v^a)(1/\sqrt{-g})$ and in addition $\nabla_\beta T^{\alpha\beta} = \partial_\beta(\sqrt{-g}T^{\alpha\beta})(1/\sqrt{-g}) + \Gamma^\alpha_{\sigma\beta}T^{\sigma\beta}$ ($\Gamma^\sigma_{\sigma\alpha} = (1/2)(\partial_\alpha g_{\mu\nu})g^{\mu\nu}$ and $\partial_\alpha(log(\sqrt{-g}) = \Gamma^\sigma_{\sigma\alpha})$.

(4) For two metrics $g$, $g^*$ one shows that $\delta\Gamma^\alpha_{\beta\gamma} = \Gamma^{*\alpha}_{\beta\gamma} - \Gamma^\alpha_{\beta\gamma}$ is a tensor.

(5) $\delta R_{\alpha\beta} = \nabla_\sigma(\delta\Gamma^\sigma_{\alpha\beta} - \nabla_\beta(\delta\Gamma^\sigma_{\alpha\sigma})$ (see **[225]** for the calculations).

(6) Recall also Stokes theorem $\int_\Omega \nabla_\sigma v^\sigma\sqrt{-g}d^4x = \int_\Omega \partial_\sigma(v^\sigma\sqrt{-g})d^4x = \int_{\partial\Omega}\sqrt{-g}v^\sigma d^3\Sigma_\sigma$.

Now requiring a stationary action for arbitrary $\delta g^{ab}$ (with certain derivatives of the $g^{ab}$ fixed on the boundary of $\Omega$ one obtains ($\mathfrak{L}_M$ is the matter Lagrangian)

$$\delta I = \frac{1}{2\chi}\int_\Omega\left(R_{\alpha\beta} - \frac{1}{2}g_{\alpha\beta}R\right)\sqrt{-g}\delta g^{\alpha\beta}d^4x \tag{1.31}$$
$$+\frac{1}{2\chi}\int_\Omega g^{\alpha\beta}\sqrt{-g}\delta R_{\alpha\beta}d^4x + \int_\Omega\frac{\delta\mathfrak{L}_M}{\delta g^{\alpha\beta}}\delta g^{\alpha\beta}d^4x = 0$$

The second term can be written

$$\frac{1}{2\chi}\int_\Omega g^{\alpha\beta}\sqrt{-g}\delta R_{\alpha\beta}d^4x = \frac{1}{2\chi}\int g^{\alpha\beta}\sqrt{-g}[\nabla_\sigma(\delta\Gamma^\sigma_{\alpha\beta}) - \nabla_\beta(\delta\Gamma^\sigma_{\alpha\sigma}]d^4x \tag{1.32}$$
$$= \frac{1}{2\chi}\int_\Omega\sqrt{-g}[\nabla_\sigma(g^{\alpha\beta}\delta\Gamma^\sigma_{\alpha\beta}) - \nabla_\beta(g^{\alpha\beta}\delta\Gamma^\sigma_{\alpha\sigma})]d^4x$$
$$= \frac{1}{2\chi}\int_\Omega\partial_\sigma[(\sqrt{-g}g^{\alpha\beta}\delta\Gamma^\sigma_{\alpha\beta}) - (\sqrt{-g}g^{\alpha\sigma}\delta\Gamma^\rho_{\alpha\rho})]d^4x$$

where $\delta\Gamma^\alpha_{\beta\gamma} = (1/2)[\nabla_\gamma(\delta g_{\beta\sigma}) + \nabla_\beta(\delta g_{\sigma\gamma}) - \nabla_\sigma(\delta g_{\gamma\beta})]$. This can be transformed into an integral over the boundary $\partial\Omega$ where it vanishes if certain derivatives of $g_{\alpha\beta}$ are fixed on the boundary. In fact the integral over the boundary $\partial\Omega = \sum S_i$ can be written as $\sum_i(\epsilon_I/2\chi)\int_{S_i}\gamma_{\alpha\beta}\delta\tilde{N}^{\alpha\beta}d^3x$ where $\epsilon_i = \mathbf{n}_i\cdot\mathbf{n}_i = \pm 1$ ($\mathbf{n}_i$ normal to $S_i$) and $\gamma_{\alpha\beta} = g_{\alpha\beta} - \epsilon_i\mathbf{n}_\alpha\cdot\mathbf{n}_\beta$ is the 3-metric on the hypersurface $S_i$ (cf. **[899]**). Further

$$\tilde{N}^{\alpha\beta} = \sqrt{|\gamma|}(K\gamma^{\alpha\beta} - K^{\alpha\beta}) = -\frac{1}{2}g\gamma^{\alpha\mu}\gamma^{\beta\nu}\mathcal{L}_{\mathbf{n}}(g^{-1}\gamma_{\mu\nu}) \tag{1.33}$$

where $K_{\alpha\beta} = -(1/2)\mathcal{L}_{\mathbf{n}}\gamma_{\alpha\beta}$ is the extrinsic curvature of each $S_i$ and $\mathcal{L}_{\mathbf{n}}$ is the Lie derivative. Consequently if the quantities $\tilde{N}^{\alpha\beta}$ are fixed on the boundary for an arbitrary $\delta g_{\alpha\beta}$ one gets from the first and last equations in (1.31) the Einstein field equations

$$G_{\alpha\beta} = R_{\alpha\beta} - \frac{1}{2}Rg_{\alpha\beta} = \chi T_{\alpha\beta};\ T_{\alpha\beta} = -2\frac{\delta\mathfrak{L}_M}{\delta g^{ab}} + \mathfrak{L}_M g_{\alpha\beta} \tag{1.34}$$

We note here that

$$\delta\int\mathfrak{L}_m\sqrt{-g}d^4x = \int\frac{\delta\mathfrak{L}_m}{\delta g^{ab}}\sqrt{-g}d^4x + \int\mathfrak{L}_m\delta(\sqrt{-g})d^4x \tag{1.35}$$
$$= \int\frac{\delta\mathfrak{L}_m}{\delta g^{ab}}\sqrt{-g}d^4x - \frac{1}{2}\int\mathfrak{L}_m g_{ab}(\delta g^{ab})\sqrt{-g}d^4x$$

A factor of 2 arises from the $2\chi$ in (1.31).

**REMARK 1.1.1.** Let us rephrase some of this following [**860**] for clarity. Thus e.g. think of functionals $F(\psi)$ with $\psi = \psi_\lambda$ a one parameter family and set $\delta\psi = (d\psi_\lambda/d\lambda)|_{\lambda=0}$. For $F(\psi)$ one writes then $dF/d\lambda = \int \phi\delta\psi$ and sets $\phi = (\delta F/\delta\psi)|_{\psi_0}$. Then (assuming all functional derivatives are symmetric with no loss of generality) one has for $\mathfrak{L}_G = \sqrt{-g}R$ and $S_G = \int \mathfrak{L}_G d^4x$

$$\frac{d\mathfrak{L}_G}{d\lambda} = \sqrt{-g}(\delta R_{ab})g^{ab} + \sqrt{-g}R_{ab}\delta g^{ab} + R\delta(\sqrt{-g}) \tag{1.36}$$

But $g^{ab}\delta R_{ab} = \nabla^a v_a$ for $v_a = \nabla^b(\delta g_{ab}) - g^{cd}\nabla_a(\delta g_{cd})$. Further there results $\delta\sqrt{-g} = -(1/2)\sqrt{-g}g_{ab}\delta g^{ab}$ so one has

$$\frac{dS_G}{d\lambda} = \int \frac{d\mathfrak{L}_G}{d\lambda}d^4x = \int \nabla^a v_a\sqrt{-g}d^4x + \int \left(R_{ab} - \frac{1}{2}Rg_{ab}\right)(\delta g^{ab})\sqrt{-g}d^4x \tag{1.37}$$

Discarding the first term as a boundary integral we get the first term in (1.31).■

**REMARK 1.1.2.** From [**664**] we see that the entropy in S in (1.22) reduces to a 4-divergence when the Einstein equations are satisfied "on shell" making S a surface term

$$S = \frac{1}{8\pi}\int_V d^4x\sqrt{-g}\nabla_i(u^b\nabla_b u^i - u^i\nabla_b u^b) \tag{1.38}$$

$$= \frac{1}{8\pi}\int_{\partial V} d^3x\sqrt{h}n_i(v^b\nabla_b u^i - u^i\nabla_b u^b)$$

Thus the entropy of a bulk region $V$ of spacetime resides in its boundary $\partial V$ when the Einstein equations are satisfied. In varying (1.22) to obtain (1.23) one keeps the surface contribution to be a constant. Thus in a semiclassical limit when the Einstein equations hold to the lowest order the entropy is contributed only by the boundary term and the system is holographic. ■

## 2. HORIZONS, ENTROPY, AND ALL THAT

There is abundant information about horizons in e.g. [**47, 256, 413, 471, 661, 662, 663, 664, 668, 669, 710, 797, 856, 860**] and we will not try to survey this. The basic horizons seem to be the event and particle horizons (cf. [**388, 710**]). Thus e.g. in an FRW universe the particle horizon is a surface dividing space into a region that can be seen at time $t$ by an observer at the origin and one that cannot yet be seen. The event horizon divides space into two regions, from one of which light can reach the observer in a finite time and from the other where this is not possible. The radii are given by (starting at $t = 0$)

$$\int_0^{r_{PH}} \frac{dr}{(1-kr^2)^{1/2}} = \int_0^t \frac{dt}{a(t)}; \quad \int_0^{r_{EH}} \frac{dr}{(1-kr^2)^{1/2}} = \int_t^\infty \frac{dt}{a(t)} \tag{2.1}$$

where the FRW universe has the form

$$ds^2 = -dt^2 + a^2(t)\left[\frac{dr^2}{1-kr^2} + r^2(d\theta^2 + Sin^2(\theta)d\phi^2)\right] \tag{2.2}$$

**EXAMPLE** 2.1. The Schwarzschild solution for empty space is

$$ds^2 = -\left(1 - \frac{2M}{r}\right)dt^2 + \frac{dr^2}{1-(2M/r)} + r^2(d\theta^2 + Sin^2(\theta)d\theta^2) \tag{2.3}$$

($M$ is an arbitrary constant at this stage). For large $r$ this is essentially Minkowski space and there are singularities at $r = 0$ (singularity) and $r = 2M$ (horizon). For a gravitating mass $m$ one has $M \sim Gm/c^2$ and $R_s = (2Gm/c^2)$ is called the Schwarzschild radius (note $R_s$ is well within the radius of the earth or sun for these bodies). An observer at infinity perceives a horizon at $R_s$; nothing can escape this horizon (not even light) and we have a black hole with radius $R_s$. ∎

**EXAMPLE** 2.2. We consider next the Kerr black hole following [**388**]. A general axisymmetric metric has a canonical form

$$ds^2 = -V(dt - hd\phi)^2 + V^{-1}[e^{2\gamma}(d\rho^2 + dz^2) + \rho^2 d\phi^2] \tag{2.4}$$

where $V = [(\Delta - a^2Sin^2(\theta))/\Sigma]$ (see below) and using the formula (**2A**) $\rho = k\sqrt{(x^2-1)(1-y^2)}$ with $z = kxy$ and (**2B**) $exp(2\gamma') = exp(2\gamma)(x^2 - y^2)$ one arrives at

$$ds^2 = -\frac{\Delta - a^2Sin^2\theta}{\Sigma}dt^2 - \frac{4MarSin^2\theta}{\Sigma}dtd\theta \tag{2.5}$$

$$+\frac{\Sigma}{\Delta}dr^2 + \left[\frac{(r^2+a^2) - \Delta a^2 Sin^2(\theta)}{\Sigma}\right]Sin^2(\theta)d\phi^2 + \Sigma d\theta^2$$

where (**2C**) $\Sigma = r^2 + a^2Cos^2(\theta)$ and $\Delta = r^2 + a^2 - 2Mr$ with (**2D**) $r = (M^2 - a^2)^{1/2}x + M$ and $\theta = Cos^{-1}(y)$. The Kerr metric (2.4) describes the spacetime outside a rotating mass distribution $M$ with angular momentum $J \sim Ma$. The equations (**2E**) $\Delta = r^2 + a^2 - 2Mr = 0$ describe a horizon and is no real singularity but points where (**2F**) $\Sigma = r^2 + a^2Cos^2(\theta) = 0$ represent physical singularities. The exterior solution of $\Delta = 0$ is $r_+ = M + \sqrt{M^2 - a^2}$ which is the radius of the horizon and (**2G**) $r_s = M + \sqrt{M^2 - a^2Cos^2(\theta)}$ is called the stationary limit. Particles in the region $r_+ < r < r_s$ lie in the so called ergosphere and are dragged along by the black hole. ∎

**EXAMPLE** 2.3. The horizon of a black hole is a null surface (i.e. any normal vector is a null vector). To see this consider the Kruskal-Szekares (KS) coordinate system which arises as follows (cf. [**388**]). Write the Schwarzschild metric as

$$ds^2 = -\left(1 - \frac{2M}{r}\right)du^2 - 2dudr + r^2(d\theta^2 + Sin^2(\theta)d\phi^2) \tag{2.6}$$

with (**2H**) $u = r^* - t$ with $r^* = r + 2Mlog[(r/2M) - 1]$ and $v = r^* + t$. Then putting (**2I**) $U = exp(-u/4\lambda t)$ and $V = exp(v/4\lambda t)$ one arrives at the KS coordinates

$$ds^2 = -\frac{32M^3}{r}e^{-(r/2M)}dUdV + r^2(d\theta^2 + Sin^2(\theta)d\phi^2) \tag{2.7}$$

These coordinates represent a maximaly expanded Schwarzschild solution. There are no coordinate singularities except for $r = 0$ which is in fact a physical singularity. Some calculations based on Killing vectors shows that the surface gravitation will be $\kappa = 1/4M$ (see below for some comments on Killing vectors).

**EXAMPLE** 2.4. One has now four laws of black hole thermodynamics (cf. [**388**] again) and this applies to Schwarzschild and Kerr black holes. Thus

(1) $\kappa$ is constant over the horizon of a black hole.
(2) $dM = (\kappa/8\pi)dA + \Omega dJ$ ($A \sim$ surface area) which is equivalent to $d(mc^2) = (\kappa c^2/8\pi G)dA + \Omega dJ$ where $\Omega dJ$ is the work performed on a black hole when its spin changes by $dJ$ (see [**388**] for a derivation). This can be compared to the first law of thermodynamics $dU = TdS + dW$.
(3) No classical process can make the horizon of a black hole decrease.
(4) No naked singularity with $J > M^2$ can exist.

We add to this the Hawking-Beckenstein rule that the a black hole emits electromagnetic radiation with temperature (**2J**) $T = \hbar\kappa/2\pi k_B c$ where $k_B$ is the Boltzman constant. For a Schwarzschild black hole this reduces to $T = \hbar c^3/8\pi G k_B m$. Generally the temperature in (**2J**) is associated with an entropy of the form $S_{BH} = (1/4)(k_B c^3/G\hbar)A$ (see [**388**] for details).

**REMARK 2.1.1.** We recall that Killing vectors arise from symmetries of the metric. Thus if $g \sim g_{ab}dx^a \otimes dx^b$ and if $\phi_t(x) = (\hat{x}^\alpha)$ is a 1-parameter group of diffeomorphisms one can find a new metric via (**2K**) $\hat{g}_{\alpha\beta} = g_{ab}(dx^a/d\hat{x}^\alpha)(dx^b/d\hat{x}^\beta)$. If then $\hat{g} = g$ one speaks of an isometry generated by a vector $\xi$ where (**2L**) $\mathfrak{L}_\xi = lim_{t\to 0}(1/t)(\phi_t^* g - g)$ ($\mathfrak{L}_\xi \sim$ Lie derivative). Since however (**2M**) $(\mathfrak{L}_\xi g)_{ab} = \nabla_\alpha g_{ab}\xi^\alpha + g_{a\alpha}\nabla_b\xi^\alpha + g_{\alpha b}\nabla_a\xi^\alpha$ and $\nabla_\alpha g_{ab} = 0$ we have (**2N**) $\nabla_b\xi_a + \nabla_a\xi_b = 0 \equiv \partial_b\xi_a + \partial_a\xi_b = 2\xi_\alpha\Gamma^\alpha_{ab}$.

## 3. THERMODYNAMICS AND FRW FORMATS

We will consider several sources of material which we sketch, namely (**1**) [**662, 664, 673, 797, 770**], (2) [**62, 263, 653**], (3) [**524, 634, 676, 797**], and (4) [**254, 372, 875, 884, 874**], (5) [**53, 16, 151, 149, 178, 498**], (6) [**26, 518, 729, 730**], (7) [**54, 480, 490, 626, 756**], (8) [**319, 466, 799**], and (**9**) [**471, 299, 860**]. (**1**) has been treated in part in Section 1.2 for general actions and its continuation (with general actions) involves many subtleties involving covariance, foliation dependence, etc.; we will try to pick this up later and concentrate here on FRW type situations.

**3.1. METRICS FROM (1).** We extract from [**662, 797**] and think of $d\Omega^2 = d\theta^2 + Sin^2(\theta)d\phi^2$ with (**3A**) $ds^2 = f(r)dt^2 - (1/f(r))dr^2 - r^2 d\Omega^2$. If $f(r)$ has a simple zero at $r = a$ with $f'(a) = B$ finite then there is a horizon at $r = a$. One now wants to rewrite the Einstein equations in a form analogous to $TdS - dE = PdV$. The metric (**3A**) will satisfy the Einstein equations provided the source stress tensor has the form

$$T_t^t = T_r^r = \frac{\epsilon(r)}{8\pi};\ T_\theta^\theta = T_\phi^\phi = \frac{\mu(r)}{8\pi} \tag{3.1}$$

(note $g_{00} = (-1/g_{11}$ - more general situations can be treated) and the Einstein equations reduce to

$$\frac{1}{r^2}(1-f) - \frac{f'}{r} = \epsilon;\ \nabla^2 f = -2\mu \tag{3.2}$$

Given any $\epsilon(r)$ the solution becomes

$$f(r) = 1 - \frac{a}{r} - \frac{1}{r}\int_a^r \epsilon(\rho)\rho^2 d\rho;\ \mu(r) = \epsilon + \frac{1}{2}r\epsilon'(r) \tag{3.3}$$

Evidently $\epsilon = 0 \Rightarrow \mu = 0$ and yields the Schwarzschild spacetime and other choices lead to a deSitter spacetime or a Reissner-Nordstrom-deSitter metric (cf. [**662**] for details and also for multiple horizons).

There is also some interesting development in [**62, 797**] regarding dual thermodynamic systems. Thus the Bekenstein-Hawking (BH) entropy is $S = (1/4)(4\pi a^2) = \pi a^2$ (assume $f'(a) < 0$ and see [**662**]). The duality transformation of [**62**] involves

$$S \to E;\ E \to S;\ T \to \frac{1}{T};\ \frac{\mu}{T} \to -\mu \tag{3.4}$$

where $\mu$ is the chemical potential. Assuming this is valid and mapping a general spherically symmetric horizon to its dual system one arrives at

$$S_d = E;\ E_d = \pi a^2;\ T_d = \frac{4\pi}{f'(a)} \tag{3.5}$$

Taking the dual system to be a 1-D Bose gas one has from [**62, 797**]

$$S_d = E;\ E_d = \pi a^2;\ T_d = \frac{4\pi}{f'(a)};\ E = \frac{a^2 f'(a)}{2};\ T = \frac{f'(a)}{4\pi};\ S = \pi a^2 \tag{3.6}$$

For a Schwarzschild spacetime one has $\pi a^2 = 2E\pi/f'(a)$ leading to $E = a^2 f'(a)/2$ as the energy associated with the horizon; this is correct via $a = 2M$ and $f'(a) = 1/2M$ with $E = M$ provided $af'(a) = 1$.

**3.2. METRICS FROM (3).** Following [**524, 676**] one finds a spherically symmetric black hole spacetime with the Einstein equations equivalent to the thermodynamic law (★) $TdS = dE + PdV$. Indeed consider from [**676**] the metric

$$ds^2 = -f(r)c^2dt^2 + \frac{1}{f(r)}dr^2 + r^2 d\Omega^2 \tag{3.7}$$

where $f(a) = 0$ with $f'(a) \neq 0$ finite and $\kappa = f'(a)/2$. Periodicity in Euclidean time (Wick transformed $t$) permits one to associate a temperature with the horizon of the form $k_B T = \hbar c\kappa/2\pi = \hbar c f'(a)/4\pi$ (this will hold at each horizon for spacetimes with multiple horizons). The Einstein equations for this metric have the form (**3B**) $rf'(r) - (1-f) = (8\pi G/c^4)Pr^2$ (cf. (3.2)) where $P$ is the radial pressure; evaluation at $r = a$ gives then

$$\frac{c^4}{G}\left[\frac{1}{2}f'(a)a - \frac{1}{2}\right] = 4\pi P a^2 \tag{3.8}$$

Consider two solutions with radii $a$ and $a + da$ for horizons; multiply (3.8) by $da$ and introduce a factor $\hbar$ by hand into an otherwise classical equation to obtain

$$\left(\frac{\hbar c f'(a)}{4\pi}\right)\frac{c^3}{G\hbar}d\left(\frac{1}{4}4\pi a^2\right) - \frac{1}{2}\frac{c^4 da}{G} = Pd\left(\frac{4\pi}{3}a^3\right) \tag{3.9}$$

This can be interpreted as (**3C**) $k_BTdS - dE = PdV$ via

$$S = \frac{1}{4L_P^2}(4\pi a^2) = \frac{1}{4}\frac{A_H}{L_P^2};\ E = \frac{c^4}{2G}a = \frac{c^4}{G}\left(\frac{A_H}{16\pi}\right) \tag{3.10}$$

where $A_H$ is the horizon area and $L_P^2 = G\hbar/c^3$ ($k_B$ is of course the Boltzman constant and $G$ the gravitational constant). This exhibits the Einstein equations as a thermodynamic identity.

(1) The combination $TdS$ is completely classical and independent of $\hbar$ given that $T \propto \hbar$ and $S \propto (1/\hbar)$. This is analogous to comparing classical thermodynamics and statistical mechanics where $TdS$ is independent of $k_B$ but $S \propto k_B$ and $T \propto 1/k_B$.

(2) Equation (3.9) is really different from the conventional first law of black hole thermodynamics because of the $PdV$ term, which is closer to the membrane paradigm for black holes. For example one needs the $PdV$ term to get $TdS = dM$ when a chargeless particle is dropped into a Reissner-Nordstrom black hole.

(3) See [**524, 662, 676**] for further discussion - the above discussion is extended to Gauss-Bonnet and and Lanczos-Lovelock gravity (see also [**634**] for evolving horizons). In [**634**] one considers time dependent spherically symmetric situations (cf. also [**524**])

$$ds^2 = e^{-2\Phi(r,t)}\left[1 - \frac{2m(r,t)}{r}\right]dt^2 + \frac{dr^2}{1-(2m(r,t)/r)} + r^2d\Omega^2 \tag{3.11}$$

In order to avoid non-symmetric slicings of space time and other problems in [**634**] one goes from (3.11), where $2m(r_H(t),t) \sim r_H(t)$ gives the horizon radius, to a new time variable $t \to \tilde{t}(t,r)$ (Painlevé-Gullstrand (PG) time) so that (**3C**) $d\tilde{t} = (\partial\tilde{t}/\partial t)dt + (\partial\tilde{t}/\partial r)dr = \dot{\tilde{t}}dt + \tilde{t}'dr$. Putting this in (3.11) gives

$$ds^2 = -e^{-2\Phi(r,t)}\left(1 - \frac{2m(r,t)}{r}\right)\left(\frac{1}{\dot{\tilde{t}}}d\tilde{t} - \frac{\tilde{t}'}{\dot{\tilde{t}}}\right)^2 + \frac{dr^2}{1-(2m(r,t)/r)} + r^2d\Omega^2 \tag{3.12}$$

Expanding and requiring that $g_{rr} = 1$ yields the condition

$$\tilde{t}' = \pm\frac{\sqrt{2m(r,t)/r}}{1-2m(r,t)/r}e^{\Phi(r,t)}\dot{\tilde{t}} \tag{3.13}$$

This differential equation can be shown to have a unique solution via characteristic curve methods and, imposing (3.13), leads to

$$g_{\tilde{t}r} = \pm\frac{e^{-\phi(r,t)}}{\dot{\tilde{t}}}\sqrt{2m(r,t)/r} = \pm c(r,\tilde{t})\sqrt{2m(r,\tilde{t})/r} = v(r,\tilde{t}) \tag{3.14}$$

This defines $c(r,\tilde{t})$ and implies $g_{\tilde{t}\tilde{t}} = -(c^2 - v^2)$ producing the P-G metric

$$ds^2 = -c^2(r,\tilde{t})^2d\tilde{t}^2 + [dr + v(r,\tilde{t})d\tilde{t}]^2 + r^2d\Omega^2 \tag{3.15}$$
$$\equiv ds^2 = -[c(r,\tilde{t})^2 - v(r,\tilde{t})^2]d\tilde{t}^2 + 2v(r,\tilde{t})drd\tilde{t} + dr^2 + r^2d\Omega^2$$

Now surfaces of constant $\tilde{t}$ are spatially flat and there is a natural idea of out and in associated with increasing or decreasing the $r$ coordinate. Thus one can define the location of an "evolving" horizon via

**(3D)** $c(r,\tilde{t}) = v(r,\tilde{t})$ which determines a function $v(r_H(\tilde{t}),\tilde{t}) = c(r_H(\tilde{t}),\tilde{t})$ which is equivalent to **(3E)** $2m(r_H(\tilde{t}),\tilde{t}) = r_H(\tilde{t})$. In [**524**] one uses the PG time starting from

$$ds^2 = -f(t,r)dt^2 + \frac{1}{g(t,r)}dr^2 + r^2 d\omega^2 \tag{3.16}$$

where the evolving horizon appears at $r = r_H$ via the condition $f(t, r_H) = 0$. Regularity of the Einstein equations and Ricci scalar on the horizon involves $f(t, r_H) = g(t, r_H) = 0$ and $f'(t, r_H) = g'(t, r_H)$. To determine the surface gravity $\kappa$ one writes the PG form of the metric as

$$ds^2 = [c^2(\tau,r) - v^2(\tau,r)]d\tau^2 + 2(r,\tau)drd\tau + dr^2 + r^2 d\Omega^2 \tag{3.17}$$

$$c(\tau,r) = \frac{1}{\dot{\tau}}\sqrt{\frac{f(\tau,r)}{g(\tau,r)}};\ v(\tau,r) = c(\tau,r)\sqrt{1-g(\tau,r)} \tag{3.18}$$

(where $\dot{\tau}$ is the time derivative). One defines an outward radial null vector

$$\ell^a = \frac{(1, c(\tau,r) - v(\tau,r), 0, 0)}{c(\tau,r)} \tag{3.19}$$

and checks that $g_{ab}\ell^a\ell^b = 0$. Then because of spherical symmetry **(3F)** $\ell^a\nabla_a\ell^b = \kappa_\ell\ell^b$ where $\kappa_\ell$ is defined everywhere and should reduce to surface gravity $\kappa$ on the horizon; this gives **(3G)** $\kappa = g'(\tau, r_H)/2$. Although $\kappa$ does not involve a time derivative it is still dynamic via $r_H(\tau)$ and this result seems to be in accord with that of [**634**]. The corresponding temperature associated with the evolving horizon is then $T = g'(t, r_H(t))/4\pi$ and the entropy is one fourth of the instantaneous area. The Einstein equation for (3.16) evaluated on the horizon is

$$G^t_t = G^r_r = \frac{g'(r_H(t),t)}{r_H(t)} - \frac{1}{r_H^2(t)} = 8\pi T^r_r \tag{3.20}$$

Via consideration of two solutions with horizon radii $r_H$ and $r_H + dr_H$ one obtains from (3.20)

$$TdS[r_H(t)] - dE[t, r_H(t)] = PdV \tag{3.21}$$

where $P = T^r_r$ is the radial pressure, $S[r_H(t)] = \pi r_H^2$ is the entropy, and $dV = 4\pi r_H^2(t)dr_H$ is the areal volume. This implies $r_H(t)/2$ is the instantaneous Misner-Sharp energy of the horizon and one concludes that the near horizon structure of the Einstein equations can be interpreted as a thermodynamic identity $TdS = dE + PdV$ where the differentials arise from the evolution of the horizon with time.

From the discussion in [**524**] it is suggested that the presence of a causal horizon is a unique feature of gravity because only gravity can effect the causal structure of the space time described by light cones. The dynamics of gravity then allow a thermodynamic description that makes gravity essentially holographic. Inverting the logic one might then argue that the thermodynamic interpretation is generic and may be a consequence of the underlying microscopic statistical mechanical quantum theory. This seems in accord with the approach of [**62, 797**] and perhaps also [**263**].

**3.3. METRICS FROM (2).** In the second paper of [**653**] one looks first at a gravitating system (spherically symmetric perfect fluid in equilibrium) which does not have a horizon and discovers an identity between the thermodynamics of this system and general relativity. By extremizing the total entropy of the matter while keeping the total particle number and energy fixed one finds that the most probable configuration is one which obeys a linear combination of the two independent Einstein equations (the other 8 components being trivially satisfied). However there does not seem to be any reason a priori why the entropy should only be an extremum for the Einstein equations and this suggests that one can derive one of the Einstein equations from thermodynamics. In this case one need only assume one of the Einstein equations and the remaining ones will be automatically satisfied due to thermodynamical considerations. Thus one of the two independent Einstein equations has no physical content - it is just a restatement of the laws of equilibrium, statistical mechanics in curved space (and hence would not be involved in any subsequent quantization). In this direction the author refers to [**474**] for support.

Thus one begins with no assumption about the Einstein equations and assumes only a metric theory of gravity which for a spherically symmetric system gives a metric (**3H**) $ds^2 = -exp(2\Phi)dt^2 + exp(2\Lambda)dr^2 + r^2 d\Omega^2$ (independent of time). Assume also the stress energy of a perfect fluid (**3I**) $T^{ab} = (\rho+p)u^a(r)u^b(r)+p(r)g^{ab}$ where $u^a(r)$ is the 4-velocity of the fluid (note that spherical symmetry would follow automatically for a perfect fluid in equilibrium if one assumes asymptotic flatness). One assumes the fluid is "extensive" which means that the entropy, particle number, and energy scale as the size of the system when the intensive variables (temperature, pressure, and chemical potential) are held fixed. This assumption is only assumed to hold in the absence of gravity and then extensivity plus the first law imply the Gibbs-Duhem relation (**3J**) $\rho = Ts - p + \mu n$ where $s$, $\rho$, $n$ are the entropy, energy, and particle number densities (cf. [**211**]). A gravitating perfect fluid then obeys the relations

$$\frac{dp}{dr} = -(\rho+p)\frac{d\Phi}{dr};\ e^{-2\Lambda} = 1 - \frac{2m(r)}{r};\ \frac{d\Phi}{dr} = \frac{m+4\pi r^2 p}{r(r-2m)} \tag{3.22}$$

The first equation follows from the vanishing of $\nabla_b T^{ab}$ while the second and third items are Einstein's equations written in the orthonormal reference frame with tetrads

$$\mathbf{e}_{\hat{t}} = \frac{\partial}{e^{\Phi}\partial t};\ \mathbf{e}_{\hat{r}} = \frac{\partial}{e^{\Lambda}\partial r};\ \mathbf{e}_{\hat{\theta}} = \frac{\partial}{r\partial\theta};\ \mathbf{e}_{\hat{\phi}} = \frac{\partial}{rSin(\theta)\partial\phi} \tag{3.23}$$

Note that (3.22B) is equivalent to the constraint on initial data (**3K**) ${}^3R = 16\pi\rho$ where ${}^3R$ is the scalar curvature of the intrinsic geometry defined by the tetrads above. One observes that while the system obeys (**3J**) locally it does not necessarily obey it globally (see paper 1 of [**653**]). Given the system is in thermal and chemical equilibrium one can use a relation of Tolman asserting that the temperature and chemical potential at any two points are related by the redshift factor in the form

$$T(r) = T_0 e^{-\Phi(r)};\ \mu(r) = \mu_0 e^{-\Phi(r)} \tag{3.24}$$

One can now derive (3.22A) from the first law of thermodynamics using the relations (**3J**) and (3.24) without using any other machinery of differential geometry (thus only local extinsivity and the Tolman relation are used). Thus combining (**3J**) and (3.24) one has (**3L**) $s = \beta_0 exp(\Phi)(\rho + p) - \mu_0\beta_0 n$ where $\beta_0$ is the inverse temperature at infinity. Taking the derivative in $r$ gives then

$$\frac{ds}{dr} = \beta_0 e^{\Phi}\left(\frac{d\rho}{dr} + \frac{dp}{dr}\right) + \beta_0 \frac{d\Phi}{dr} e^{\Phi}(\rho + p) - \beta_0\mu_0 \frac{dn}{dr} \tag{3.25}$$

Then using the first law one obtains

$$d\rho = Tds - \mu dn \Rightarrow \frac{dp}{dr} = -(\rho + p)\frac{d\Phi}{dr} \tag{3.26}$$

as required in (3.22). Next one derives one of the Einstein equations by extremizing the total entropy of the matter, using a variation at fixed particle number $N$ and fixed energy at infinity, where the energy at infinity is given via the ADM mass (**3M**) $M = \int_0^\infty 4\pi r^2 \rho dr$). $N$ is given via (**3N**) $N = \int_0^\infty n 4\pi r^2 exp(\Lambda) dr$ and one changes variables via (**3O**) $(dm/dr) = 4\pi r^2 \rho$. Then it is asserted that

**THEOREM 3.1.1.** Assuming $\Lambda$ an arbitrary function of $(m, r)$ one has

$$4\pi r^2(\rho + p)\frac{\partial\Lambda}{\partial m} = \frac{d\Lambda}{dr} + \frac{d\Phi}{dr} \tag{3.27}$$

and this equation is a linear combination of the Einstein equations (3.22B) and (3.22C).

As corollaries one has

(1) If (3.22B) and (3.22C) are satisfied then the entropy is an extremum.
(2) If the constraint (**3K**) holds and the entropy is an extremum then the rest of the Einstein equations hold (i.e. if one assumes the $tt$ component of the Einstein equations, which gives $\Lambda$, then the $rr$ component which gives $\Phi$ can be derived).

To prove this one writes total entropy as (**3P**) $S = \int_0^\infty s 4\pi r^2 exp(\Lambda) dr$; then adding (**3N**) one defines (**3Q**) $L = S + \lambda(\int_0^\infty 4\pi r^2 exp(\Lambda) n dr - N)$. Since $\Lambda$ depends on $(m, r)$ one can vary it with respect to m, keeping the variation fixed at the end points; the vanishing of $\delta m$ at the endpoints is equivalent to varying the entroy at fixed total energy. Thus treating $\rho(r)$ and $n(r)$ as the independent thermodynamical variables one writes

$$\delta L = \int_0^R 4\pi r^2 dr e^{\Lambda}\left[\beta_0 e^{\Phi}\delta\rho + (\lambda - \mu_0\beta_0)\delta n + (s(\rho, n) + \lambda n)\delta\Lambda\right] \tag{3.28}$$

where

$$\beta = \beta_0 e^{\Phi} = \left(\frac{\partial s}{\partial \rho}\right)_n;\ \mu\beta = \mu_0\beta_0 = -\left(\frac{\partial s}{\partial n}\right)_\rho \tag{3.29}$$

Write now, via (**3O**), (**3R**) $\delta\rho = \delta\dot{m}/(4\pi r^2)$ where $\dot{} = d/dr$, and put this and (**3L**) into (3.28) to get after integration by parts

$$\delta S = \int_0^R dr\beta_0 e^{\Lambda+\Phi}(4\pi r^2(\rho+p)\delta\Lambda - (\dot{\Lambda}+\dot{\Phi})\delta m) \tag{3.30}$$

$$+\int_0^R 4\pi r^2 dr e^{\Lambda}(\lambda - \mu_o\beta_o)(\delta n + n\delta\Lambda)$$

The vanishing of $\delta n(R) = \delta M$ is equivalent to performing the variation at fixed energy while the vanishing of $\delta n(0)$ is a necessary boundary condition to keep $\rho(0)$ finite. Then put in the relation (**3S**) $\delta\Lambda(m,r) = (\partial\Lambda/\partial m)\delta m$ into (3.30) and since the system is in equilibrium the variation of total entropy must vanish and the vanishing of $\delta L$ implies (**3T**) $\beta_0\mu_0 = \lambda$ (consistent with the Tolman relation) and

$$4\pi r^2(\rho+p)\frac{\partial\Lambda}{\partial m} = \frac{d\Lambda}{dr} + \frac{d\Phi}{dr} \tag{3.31}$$

which is (3.27). One sees also that (3.31) is related to a linear combination of Einstein equations given via

$$G_{\hat{t}\hat{t}} - 8\pi T_{\hat{t}\hat{t}} + G_{\hat{r}\hat{r}} - 8\pi T_{\hat{r}\hat{r}} = 0 \tag{3.32}$$

The first corollary is proved by verifying that $\Phi$ and $\Lambda$ as obtained from (3.22B) and (3.22C) satisfy (3.31); the entropy is then an extremum. The second corollary is verified by substituting the solution from the $G_{\hat{t}\hat{t}}$ component of (3.22B) into (3.31) yielding the correct result for the $G_{\hat{r}\hat{r}}$ component in (3.22C).

Thus one can recast the Einstein equations (3.22) into (3.31), which can be classified as thermodynamical in nature, and an additional equation representing the true physical theory. The latter equation is somewhat arbitrary; one could pick (**3K**) for example, which determines $\Lambda$ in terms of the matter fields, and $\Phi$ is then determined by extremizing the entropy to find the most probable configuration. Quantization of the full system is straightforward since the quantization of $\Lambda$ follows by reducing the phase space via (**3K**) and the full Lagrangian will be

$$L = 4\pi\int \mathfrak{L} r^2 e^{\Phi(r)}\sqrt{1-(2\hat{m}/r)}dr \tag{3.33}$$

where $\hat{m}$ is the operator version of $m$ in (**3O**). $\Phi$ acts as a potential as far as the quantization of matter fields is concerned. One could also choose for the physical theory the trace of the Einstein equations (**3U**) $R = [16/(2-d)]\pi T$ and the physical theory would then give rise to (3.31) via the theorem stated.

## 4. THE GEOMETRY OF THERMODYNAMICS

We go first to [**480, 490**] and look at metrics from (**7**). Recall first (cf. [**480**]) from parametric statistics that, given a probability distribution $p(x,\theta)$ with $\theta$ a parameter (to be estimated) and samples $x_i$, the loglikelihood function is defined as (**4A**) $logL(\theta) = \sum_1^n log[p(x_i|\theta)]$. The "score" function is (**4B**) $U(\theta) =$

$d[log(L(\theta))]/d\theta$ and **(4C)** $Var[U(\theta)] = E(-\partial_\theta^2 log[L(\theta)]$ is called the expected information. Expanding one has

$$log[L(\theta + d\theta)] - log[L(\theta)] = \delta\theta\partial_\theta L(\theta) + \frac{(\delta\theta)^2}{2}\frac{\partial^2 log[L(\theta)]}{\partial\theta^2} \tag{4.1}$$

The first term $\partial[log[L(\theta)]]/\partial\theta = 0$ at the true $\theta$ value and the closeness of two probability distributions characterized by $\theta$ and $\theta + d(\theta)$ is estimated by the second term (called Fisher information - cf. [**170, 318, 319**]). For higher dimensional parameter space with parameters $\theta_i$ one defines Fisher information via

$$G_{ij}(\theta) = -E\left[\frac{\partial^2 log(p(x|\theta))}{\partial\theta_i\partial\theta_j}\right] = -\int p(x|\theta)\frac{\partial^2 log[p(x|\theta)]}{\partial\theta_i\partial\theta_j} \tag{4.2}$$

and in fact this is a metric in parametric statistics called the Fisher-Rao metric (cf. here Sections 1 and 1.1 along with [**170, 194, 196, 197, 318, 319**]). In generic statistical physics models one often has two parameters $\beta_i$ (e.g. inverse temperature $\beta$ and external field $h$) and a simple Fisher-Rao metric **(4D)** $G_{ij} = \partial_i\partial_j f$ where $f$ is a reduced free energy. In that case the scalar curvature is

$$\mathfrak{R} = -\frac{1}{2G^2}\begin{vmatrix} \partial_\beta^2 f & \partial_{|gb}\partial_h f & \partial_h^2 f \\ \partial^3 f & \partial_\beta^2\partial_h f & \partial_\beta\partial_h^2 f \\ \partial_\beta^2\partial_h f & \partial_\beta\partial_h^2 f & \partial_h^3 f \end{vmatrix} \tag{4.3}$$

where $G = det(G_{ij})$ and this plays an important role in studying phase transitions (where the curvature diverges - cf. [**490, 480**]).

A related Ruppeiner metric (cf. [**54, 626, 756**]) arises in various contexts. The idea here is that the Hessian matrix of the second derivatives of energy or entropy can be used as a Riemannian metric in the space of thermodynamic states. In general we have then an idea about making thermodynamics geometrical (see e.g. [**26, 54, 28, 93, 97, 170, 220, 490, 480, 724, 725, 756**]) and this is clearly related to information metrics as in Sections 1 and 1.1 based on [**154, 155, 194, 196**]. We want to sketch some of the common features here and begin with thermodynamics. Thus go first to [**54**] where the Ruppeiner metric is taken to be of the form **(4E)** $g_{ij} = -\partial_i\partial_j f(X)$ where $X^i$ corresponds to extensive variables of the system (extensive variables are proportional to the size of the system - note pressure P and temperature T are intrinsic and extensive variables give rise to intensive ones when divided by a size factor - cf. [**211**]). In a black hole physics one uses entropy S or energy $E \sim M$ as dependent quantities $f$ with e.g. mass (M) and spin (Q) as variables X.

**REMARK 4.1.1.** We refer here to [**170, 736**] and take $P(y^i)$ to be a probability density with a cross entropy defined via

$$J = \int P(y^i + \Delta y^i) log\frac{P(y^i + \Delta y^i)}{P(y^i)} d^m y \tag{4.4}$$

$$\simeq \left[\frac{1}{2}\int \frac{1}{P(y^i)}\partial_i P\partial_j P d^m y\right] \Delta y^i \Delta y^j = I_{ij}\Delta y^i \Delta y^j$$

where $I_{ij}$ is the Fisher information matrix which can also be written somewhat more generally as

$$I_{jk}(\theta^i) = \frac{1}{2}\int \frac{1}{P(x^i|\theta^i)}\frac{\partial P(x^i|\theta^i)}{\partial\theta^j}\frac{\partial P(x^i|\theta^i)}{\partial\theta^k}d^m x \tag{4.5}$$

corresponding to (4.2) (up to a factor of (1/2)). Note here $\partial_i\partial_j log(p) \sim \partial_i[\partial_j p/p] \sim [\partial_i\partial_j p/p] - [\partial_j p\partial_i p/p^2]$ and $p\partial_i\partial_j log(p) \sim \partial_i\partial_j p - [\partial_i p\partial_j p/p]$ with the first term vanishing upon integration. It is often associated with a differential (or Nash) entropy (**4F**) $\mathfrak{S}(\rho) = -\int \rho(x)log[\rho(x)]dx$ (cf. [**170**]) which gives rise to a FI term upon differentiation in various ways (cf. [**170, 172**]). Indeed from [**170**] given a continuity equation $\partial_t\rho = -\nabla(v\rho)$ in a quantum mechanical problem with e.g. $\rho = |\psi|^2$ and $v = D\nabla(log\rho)$ for $D \sim \hbar/2m$ (Brownian motion) one obtains in e.g. $\mathbf{R}^3$

$$\partial_t\mathfrak{S}(\rho) = -\int \rho_t(1+log(\rho))d^3x = -\int \nabla(v\rho)(1+log(\rho))d^3x \tag{4.6}$$

$$= \int v\nabla(\rho)d^3x = D\int \frac{(\nabla\rho)^2}{\rho}d^3x$$

On the other hand from [**170, 172, 621, 836**] one recalls the Perelman entropy functional $mfF = \int_M (R + |\nabla f|^2)exp(-f)dV$ with $dV = \sqrt{|g|}d^3x$ and a so-called Nash entropy $N(u) = \int_M ulog(u))dV$ for $u = exp(-f)$ where $f_t + \Delta f = |\nabla f|^2 - R$ and $\partial_t g = -2Ric$. It follows then that $\partial_t N = \mathfrak{F}$. This shows two important roles for Fisher information in mathematics and physics - aside from its use in purely statistical matters. ■

Returning to [**54**] one looks at $f(x) = \sum_1^N x^i log(x^i)$ which for $f = -S(M,Q)$ is the Ruppeiner metric while for $f = M(S,Q)$ it is called the Weinhold metric. They are related by a conformal transformation via

$$ds^2_W = Tds^2_R;\ T = \left(\frac{\partial M}{\partial S}\right)_Q \tag{4.7}$$

For the Reissner-Nordström (RN) black hole $Q \sim$ electric charge while for Kerr (K) it is angular momentum. One shows e.g. that if $f(x,y) = x^a F(y/x)$ then for $\sigma = y/x$

$$ds^2 = \left(\frac{a-1}{a}\right)\frac{df^2}{f} + f\left(\frac{F''}{F} - \frac{a-1}{a}\frac{(F')^2}{F^2}\right)d\sigma^2 \tag{4.8}$$

is diagonal and flat (for $a \neq 1$). This leads to the statement that if $S = M^a F(M^b Q)$ then for $b = -1$ the $R$-metric is flat while if $a+b = 0$ the $W$-metric is flat. For black holes one sets $S = kA/4$ ($k \sim k_B =$ Boltzman constant) and looks at dimensions

$$[S] = L^{d-2};\ [M] = L^{d-3};\ [Q] = L^{d-3};\ [J] = L^{d-2} \tag{4.9}$$

where $d$ is the spacetime dimension. The previous analysis then leads to

$$S = M^{\frac{d-2}{d-3}} f\left(\frac{Q}{M}\right);\ S = M^{\frac{d-2}{d-3}} f\left(\frac{J}{M^{\frac{d-2}{d-3}}}\right) \tag{4.10}$$

The RN black hole has the form

$$S = M^c \left(1 + \sqrt{1 - \frac{c}{2}\frac{Q^2}{M^2}}\right); \; c = \frac{d-2}{d-3} \tag{4.11}$$

and for the Kerr black hole

$$M = \frac{d-2}{4} S^{\frac{d-3}{d-2}} \left(1 + \frac{4J^2}{S^2}\right)^{1/(d-2)} \tag{4.12}$$

In [28] one produces a flat $R$-metric for the RN black hole in Rindler coordinates

$$(\bigstar\bigstar)\; ds_R^2 = \frac{1}{T} ds_W^2 = -\frac{dS^2}{2S} + \frac{4S du^2}{1-u^2} \sim d\tau^2 + \tau^2 d\chi^2$$

where $u = Q/\sqrt{2}$, $\tau = \sqrt{2S}$, and $Sin(\chi/\sqrt{2}) = u$ while for the Kerr black hole (4.12) is confirmed and a flat $W$-metric is found in the form

$$(\bigstar\bigstar\bigstar)\; ds_W^2 = -d\tau^2 + \frac{2(d-3)}{d-2}\frac{\left(1 - 4\frac{d-5}{d-3}u^2\right)}{(1+4u^2)^2}\tau^2 du^2$$

where $u = (1/2)Sinh(2\sigma)$ gives Rindler coordinates for $d = 4$ with $ds_W^2 = -d\tau^2 + \tau^2 d\sigma^2$.

**4.1. UNIFIED LEGENDRE FORMS.** We go here to [26] and note first as background that the Kerr-Newman (KN) black hole satisfies (**4G**) $dM = TdS + \phi dQ + \Omega_H dJ$ where $T$ is the Hawking temperature (proportional to the surface gravity on the horizon), $S = A/4$ is the entropy, $\Omega_H$ is the angular velocity on the horizon, and $\phi$ is the electric potential. All of the thermodynamic information is contained in Smarr's equation

$$M = \left[\frac{\pi J^2}{S} + \frac{S}{4\pi}\left(1 + \frac{\pi Q^2}{S}\right)^2\right] \tag{4.13}$$

$$\equiv S = \pi\left(2M^2 - Q^2 + 2\sqrt{M^4 - M^2Q^2 - J^2}\right)$$

One looks now for a description of phase transitions in terms of curvature singularities and one follows the approach of [**724, 725**] in which Legendre invariance is incorporated in a natural way to derive metrics. Since $R$ and $W$ metrics are not Legendre invariant one provides a generalization which reproduces the thermodynamic behavior of black holes, including the KN one. Thus the starting point of geometrothermodynamics (GTD) is a phase space $\mathcal{T}$ which in the case of Einstein-Maxwell black holes is a 7-dimensional space with coordinates $Z^A = (M, S, Q, J, T, \phi, \Omega_H)$ for $A = 0, \cdots, 6$. In the cotangent space $\mathcal{T}^*$ one defines the 1-form (**4H**) $\Theta_M = dM - TdS - \phi dQ - \Omega_H dJ$ which satisfies the condition $\Theta_M \wedge (d\Theta_M)^3 \neq 0$. A nondegenerate metric G is introduced so that $(\mathcal{T}, \Theta_M, G)$ is a Riemannian contact manifold. Thus let $\mathcal{E}$ be a 3-D subspace of $\mathcal{T}$ with coordinates $E^a = (S, Q, J)$, $a = 1, 2, 3$, defined via a smooth mapping $\phi : \mathcal{E} \to \mathcal{T}$. $\mathcal{E}$ is called the space of equilibrium states if $\phi_M^*(\Theta_M) = 0$ and a metric structure is naturally induced on $\mathcal{E}$ via $g = \phi_M^*(G)$. Evidently the condition $\phi_M^*(\Theta_M) = 0$ leads immediately to the first law (**4G**). It also implies the existence of the fundamental equation $M = M(S, Q, J)$ and the conditions of thermodynamic equilibrium

(**4I**) $T = \partial M/\partial S$, $\phi = \partial M/\partial Q$, and $\Omega_H = \partial M/\partial J$. If we define the intensive variables as $I^a = (T, \phi, \Omega_H)$ then a Legendre transformation is defined via

$$(4.14)\quad (M, E^a, I^a) \to (\tilde{M}, \tilde{E}^a, \tilde{I}^a);\ M = \tilde{M} - \delta_{ab}\tilde{E}^a\tilde{I}^b;\ E^a = -\tilde{I}^a;\ I^a = \tilde{E}^a$$

One sees that $\Theta_M$ is invariant under Legendre transformations and via [**724**] if $G$ is Legendre invariant so is $g = \phi_M^*(G)$. For the $S$-representation, $S = S(M, Q, J)$ one considers the fundamental 1-form

$$(4.15)\quad \Theta_S = dS - \frac{1}{T}dM + \frac{\phi}{T}dQ + \frac{\Omega_H}{T}dJ$$

so the coordinates of $\mathcal{T}$ are $Z^A = (S, E^a, I^a) = (S, M, Q, J, 1/T, -\phi, T, -\Omega_H/T)$ and the space of equilibrium states of $\mathcal{E}$ can be introduced with a smooth mapping of the form (**4J**) $\phi_S : (M, Q, J) \to (M, S(M, Q, J), Q, J, I^a(M, Q, J))$ which from the condition $\phi_S^*(\Theta_S) = 0$ generates the first law of thermodynamics (**4G**) and the conditions of thermodynamic equilibrium conditions

$$(4.16)\quad \frac{1}{T} = \frac{\partial S}{\partial M};\ \frac{\phi}{T} = -\frac{\partial S}{\partial Q};\ \frac{\Omega_H}{T} = -\frac{\partial S}{\partial J}$$

Consider now the following metric on $\mathcal{T}$

$$(4.17)\quad G = \left(dS - \frac{1}{T}dM + \frac{\phi}{T}dQ + \frac{|gO_H}{T}dJ\right)^2$$

$$+\left(\frac{M}{T} - \frac{Q\phi}{T} - \frac{J\Omega_H}{T}\right)\left[dMd\left(\frac{1}{T}\right) + dQd\left(\frac{\phi}{T}\right) + dJd\left(\frac{\Omega_H}{T}\right)\right]$$

One can show that this metric is invariant with respect to Legendre transformations (4.14). The first term can be written in the form $\Theta_S \otimes \Theta_S$ so that its projection on $\mathcal{E}$ vanishes due to the condition $\phi_S^*(\Theta_S) = 0$; however this term is need for non-degeneracy. One then calculates
(4.18)

$$g = (MdS_M + QdS_Q + JdS_J)\left(S_{MM}dM^2 - S_{QQ}dQ^2 - S_{JJ}dJ^2 - 2S_{QJ}dQdJ\right)$$

This gives rise to a Legendre invariant Riemannian metric in the space of equilibrium states $\mathcal{E}$. We give a few examples and refer to [**26, 724, 725**] for more related information.

**EXAMPLE** 4.1. The RN metric involves $J = 0$ and describes a static spherically symmetric black hole with two horizons (**4K**) $r_\pm = M \pm \sqrt{M^2 - Q^2}$ (assume $Q < M$ to avoid naked singularities). One has (**4L**) $S = \pi(M + \sqrt{M^2 - Q^2})^2$ and "traditionally" the phase transition can be derived from the heat capacity

$$(4.19)\quad C_Q = \frac{4TM^3S^3}{-2M^6 + 3M^4Q^2 - 2(M^4 - M^2Q^2)^{3/2}} = -\frac{2\pi^2 r_+^2(r_+ - r_-)}{r_+ - 3r_-}$$

For the geometric approach all that is needed here is the fundamental equation as given in (**4L**) from which one obtains

$$(4.20)\quad g_{ab}^{RN} = (MS_M + QS_M)\begin{pmatrix} S_{MM} & 0 \\ 0 & -S_{QQ} \end{pmatrix}$$

$$= \frac{8\pi^2 r_+^3}{(r_+ - r_-)^3}\begin{pmatrix} 2r_+(r_+ - 3r_-) & 0 \\ 0 & r_+^2 + 3r_-^2 \end{pmatrix}$$

Note this metric is singular in the extremal limit $r_+ = r_-$. This could indicate a breakdown of the geometrical approach but the analysis of the scalar curvature

$$R^{RN} = \frac{(r_+^2 - 3r_-r_+ + 6r_-^2)(r_+ + 3r_-)(r_+ - r_-)^2}{\pi^2 r_+^3 (r_+^2 + 3r_-^2)^2 (r_+ - 3r_-)^2} \tag{4.21}$$

shows that in the extremal limit the space of equilibrium states becomes flat. In any event one sees from the scalar curvature that the only singular point corresponds to the point $r_+ = 3r_-$ which is exactly the point of a phase transition in the heat capacity (4.19).

**EXAMPLE** 4.2. The Kerr metric corresponds to the limit $Q = 0$ of the KN metric and it describes the gravitational field of a stationary axially symmetric rotating black hole with two horizons at radial distances (**4M**) $r_\pm M \pm \sqrt{M^2 - (J^2/M^2)}$. The corresponding fundamental thermodynamic equation in the entropy representation is (**4N**) $S = 2\pi(M^2 + \sqrt{M^4 - J^2})$ and second order phase transitions occur at points where the heat capacity $C_J$ diverges, where

$$C_J = \frac{4TM^3S^3}{6M^2J^2 - 2M^6 - 2(M^4 - J^2)^{3/2}} = \frac{2\pi^2 r_+ (r_+ + r_-)^2 (r_+ - r_-)}{r_+^2 - 6r_+r_- - 3r_-^2} \tag{4.22}$$

(assume $M^2 \geq J$). The Legendre invariant metric is then

$$g_{ab}^K = (MS_M + JS_J) \begin{pmatrix} S_{MM} & 0 \\ 0 & -S_{JJ} \end{pmatrix} \tag{4.23}$$

$$= \frac{16\pi^2 r_+^2 (r_+ + r_-)}{(r_+ - r_-)^4} \begin{pmatrix} r_+(r_+^2 - 6r_+r_- - 3r_-^2) & 0 \\ 0 & r_+ + r_- \end{pmatrix}$$

The scalar curvature is then

$$R^K = \frac{(3r_+^3 + 3r_+^2 r_- + 17r_+r_-^2 + 9r_-^3)(r_+ - r_-)^3}{2\pi^2 r_+^2 (r_+ + r_-)^4 (r_+^2 - 6r_+r_- - 3r_-^2)^2} \tag{4.24}$$

Thus the metric singularity at $r_+ = r_-$ is only a coordinate singularity and the curvature singularities are situated at the roots of the equation $r_+^2 - 6r_+r_- - 3r_-^2 = 0$ - which are exactly the points of phase transition via (4.21).

**EXAMPLE** 4.3. The KN black hole has outer and inner horizons given via (**4O**) $r_\pm = M \pm \sqrt{M^2 - a^2 - Q^2}$ where $a = J/M$ and one has

$$g_{ab}^{KN} = (MS_M + QS_Q + JS_J) \begin{pmatrix} S_{MM} & 0 & 0 \\ 0 & -S_{QQ} & -S_{QJ} \\ 0 & -S_{QJ} & -S_{JJ} \end{pmatrix} \tag{4.25}$$

The metric becomes cumbersome but the scalar curvature is

$$R^{KN} = \frac{N}{D};\ D = 4(MS_M + QS_Q + JS_J)^3 (S_{QJ}^2 - S_{QQ}S_{JJ})^3 S_{MM}^2 \tag{4.26}$$

and it can be shown that the curvature singularities of $g^{KN}$ are situated at those points where phase transitions can occur.

## 5. FRW FOLLOWING (4) AND (5)

First from [**388**] (cf. also [**188, 710**]) let $d\Omega_3^2$ be the metric on the unit 3-sphere and define R via $R^2 = x^2 + y^2 + z^2 + w^2$. The deSitter hyperboloid is then (**5A**) $-T^2 + R^2 = R_0^2$ leading to (**5B**) $ds^2 = R_0^2(-dt^2 + cosh^2(t)d\Omega_3^2)$. By rescaling this is the same as the deSitter solution of the Einstein-Friedman equations

$$3\frac{\dot{a}^2 + k}{a^2} - \Lambda = 0;\ -2\frac{\ddot{a}}{a} - \frac{\dot{a}^2 + k}{a^2} + \Lambda = 0 \tag{5.1}$$

in the form $\dot{a}^2 - \omega^2 a^2 = -k$ with

$$a(t) = \begin{cases} \frac{\sqrt{k}}{\omega}Cosh(\omega t) & k > 0 \\ exp(\omega t) & k = 0 \\ \frac{\sqrt{k}}{\omega}Sinh(\omega t) & k < 0 \end{cases} \tag{5.2}$$

To compare with (**5B**) set $k > 0$ and $R_0 = 1/\omega$.

Anti-deSitter (AdS) space has a different flavor. It can be considered as the hyperboloid (**5C**) $-v^2 - u^2 + x_1^2 + \cdots + x_{n-1}^2 = -R^2$ embedded in a flat $(n+1)$-dimensional space with metric (**5D**) $ds^2 = -dv^2 - du^2 + \sum_1^{n-1} dx_i^2$. This would involve closed time like curves with periodic time variable so one goes to the universal covering $\widetilde{AdS}$ with $-\infty < t < \infty$. Given a global parametrization of (**5C**) via

$$U = \sqrt{r^2 + k^2}Sin(t/k);\ V = \sqrt{r^2 + k^2}Cos(t/k);\ r^2 = \sum_1^{n-1} x_i^2 \tag{5.3}$$

one can write

$$ds^2 = -\left(1 + \frac{r^2}{R^2}\right)dt^2 + \frac{dr^2}{1 + (r^2/R^2)} + r^2 d\Omega_{n-2}^2 \tag{5.4}$$

and this is usually taken for the metric on $\widetilde{AdS}$. One has another static version of AdS via

$$ds^2 = -\left(\frac{r^2}{R^2} - 1\right)dt^2 + \frac{dr^2}{(r^2/R^2) - 1} + r^2 dH_{n-1}^2 \tag{5.5}$$

with $H^{n-2}$ a hyperbolic space (i.e. a hyperboloid in a flat $(n-1)$-dimensional space of the form $-x_{n-1}^2 + \sum_1^{n-2} x_i^2 = -1$ with $x_{n-1} > 0$). The deSitter metric can also be written as

$$(\bullet\bullet)\quad ds^2 = -dt^2 + Cosh^2(t)\left[\frac{dr^2}{1 - r^2} + f^2(d\theta^2 + Sin^2(\theta)d\phi^2)\right]$$

to compare with (5.4) for AdS. We refer to [**388**] for other metrics, black holes, and Penrose diagrams.

We go now to [**149**] involving the Friedman equations and the first law of thermodynamics. One thinks of a 4-D deSitter space with radius $\ell$ and a cosmological event horizon associated with a Hawking temperature $T = 1/4\pi\ell$ and entropy $S = A/4G$ where $A = 4\pi\ell^2$. For an asymptotic deSitter space such as Schwarzschild-deSitter space there still exists a cosmological horizon for which the area law $S = A/4G$ holds with $T = \kappa/2\pi$ ($\kappa \sim$ surface gravity of the horizon). If some matter with energy $dE$ passes through the horizon one has then $-dE = TdS$ (first law of thermodynamics). The paper is concerned with deriving the Friedman equations from thermodynamics (cf. [**151, 152, 254, 319, 421, 422**]). Thus consider an $(n+1)$ dimensional FRW universe with metric

$$ds^2 = -dt^2 = a^2(t)\left(\frac{dr^2}{1-kr^2} + r^2 d\Omega_{n-1}^2\right) \tag{5.6}$$

where $d\Omega_{n-1}^2$ refers to an $(n-1)$ dimensional sphere and $k = 1, 0, -1$ corresponds to a closed, flat, and open universe. This metric can be rewritten as (**5E**) $ds^2 = h_{ab}dx^a dx^b + \tilde{r}^2 d\Omega_{n-1}^2$ where $\tilde{r} = a(t)r$, $x^0 = t$, $x^1 = r$, and $h_{ab} \sim diag(-1, a^2/(1-kr^2))$. The dynamical apparent horizon (a marginally trapped surface with vanishing expansion) is determined via ${}_{ab}\partial_a\tilde{r}\partial_b\tilde{r}$ and one finds for the apparent horizon (**5F**) $\tilde{r}_A = 1/\sqrt{H^2 + (k/a^2)}$ where $H$ is the Hubble parameter $H = \dot{a}/a$. Evidently from (**5F**) when $k = 0$ one has $\tilde{r}_A = \tilde{r}_H = 1/H$ (Hubble horizon) whereas the cosmological event horizon, defined via (•) $\tilde{r}_E = a(t)\int_t^\infty (dt/a(t))$, exists only for an accelerated expanding universe. Now following [**421, 422**] one defines the work density by (**5G**) $W = -(1/2)T^{ab}h_{ab}$ and the energy supply vector via (**5H**) $\psi_a = T_a^b\partial_b\tilde{r} + W\partial_a\tilde{r}$ where $T^{ab}$ is the projection of the $(n+1)$ dimensional energy-momentum tensor $T^{\mu\nu}$ of a perfect fluid matter in the FRW universe in the normal direction of the $(n-1)$ sphere. The work density should be regarded as the work done by a change of the apparent horizon while the energy supply at the horizon is the total energy flow through the apparent horizon. Then from [**86, 421, 422**] one has (**5I**) $\nabla EA\psi + W\nabla V$ where $A = n\Omega_n\tilde{r}^{n-1}$ and $V = \Omega_n\tilde{r}^n$ are the area and volume of an $n$-dimensional space with radius $\tilde{r}$ where $\Omega_n = \pi^{n/2}/\Gamma((n/2)+1)$ is the volume of an $n$-dimensional unit ball; the total energy inside the space with radius $\tilde{r}$ is defined by (**5J**) $E = [n(n-1)\Omega_n/16\pi G]\tilde{r}^{n-2}(1 - h^{ab}\partial_a\tilde{r}\partial_b\tilde{r}$. The equation (**5I**) is called the unified first law and the entropy is associated with heat flow $\delta Q = TdS$ related to the change of energy of the system. Hence the entropy is associated with the energy supply term and one can write

$$A\psi = \frac{\kappa}{8\pi G}\nabla A + \tilde{r}^{n-2}\nabla\left(\frac{E}{\tilde{r}^{n-2}}\right);\ \kappa = \frac{1}{2\sqrt{-h}}\partial_a(\sqrt{-h}h^{ab}\partial_b\tilde{r}) \tag{5.7}$$

On the apparent horizon the last term in (5.7A) vanishes and one can then assign an entropy $S = A/4G$ to the apparent horizon.

Now $\delta Q = -dE$ is the change of energy inside the apparent horizon and if $T_{\mu\nu} = (\rho+p)U_\mu U_\nu + pg_{\mu\nu}$ then

$$\psi_a = \left(-\frac{1}{2}(\rho+p)H\tilde{r}, \frac{1}{2}(\rho+p)a\right) \tag{5.8}$$

During the interval $dt$ one has for the amount of energy crossing the apparent horizon (**5K**) $-dE = -A\psi = A(\rho+p)H\tilde{r}_A dt$ where $A = n\Omega_n\tilde{r}_A^{n-1}$. Then for horizon associated S and T of the form (**5L**) $S = A/4G$ and $T = 1/2\pi\tilde{r}_A$ one uses $-dE = TdS$ to obtain (**5M**) $\dot{H} - (k/a^2) = -[8\pi G/(n-1)](\rho+p)$ which is one of the Friedman equations. Note that one has used (**5N**) $\dot{\tilde{r}}_A = -H\tilde{r}_A^3[\dot{H} - (k/a^2)]$. Once the continuity (conservation) equation of the perfect fluid is given via e.g.

(**5O**) $\dot{\rho}+nH(\rho+p)=0$ one can put $H(\rho+p)$ into (**5M**) and integrate to obtain (**5P**) $H^2+(k/a^2)=[16\pi G\rho/n(n-1)]$ which is another Friedman equation; an integration constant (cosmological constant) has also been dropped here. Thus one can obtain the Friedman equations from the first law of thermodynamics. This can also be applied to an inflationary model with a homogeneous scalar field (inflaton) $\phi(t)$. Indeed the field equation is (**5Q**) $\ddot{\phi}+nH\dot{\phi}V'(\phi)=0$ and putting

$$\rho=\frac{1}{2}\dot{\phi}^2+V(\phi);\ p=\frac{1}{2}\dot{\phi}^2-V(\phi) \tag{5.9}$$

into (**5M**) and integrating one obtains the Friedman equation for the inflationary model

$$H^2+\frac{k}{a^2}=\frac{16\pi G}{n(n-1)}\left(\frac{1}{2}\dot{\phi}^2+V(\phi)\right) \tag{5.10}$$

**5.1. HIGHER DERIVATIVE GRAVITIES.** The area formula of black hole entropy no longer holds in higher derivative gravity theories so one examines now in [**149**] whether the correct Friedman equations can be derived from a suitable formula connecting area and entropy. The action of the Gauss-Bonnet gravity can be written as

$$\mathfrak{A}=\frac{1}{16\pi G}\int d^{m+1}x\sqrt{-g}(R+\alpha R_{GB})+\mathfrak{A}_M \tag{5.11}$$

where $\alpha$ is a constant with dimension *length*$^2$ and $R_{GB}=R^2-4R_{\mu\nu}R^{\mu\nu}+R_{\mu\nu\gamma\delta}R^{\mu\nu\gamma\delta}$. This term appears naturally in the low energy effective action of heterotic string theory and although it includes higher derivative curvature terms there are no more than second order derivative terms of metrics in the equations of motion. Varying the action gives

$$8\pi GT_{\mu\nu}=R_{\mu\nu}-\frac{1}{2}g_{\mu\nu}R \tag{5.12}$$

$$-\alpha\left(\frac{1}{2}g_{\mu\nu}R_{GB}-2RR_{\mu\nu}+4R_{\mu\gamma}R^{\gamma}_{\nu}+4R_{\gamma\delta}R^{\gamma\ \delta}_{\ \mu\ \nu}-2R_{\mu\gamma\delta\lambda}R_{\nu}^{\ \gamma\delta\lambda}\right)$$

Static black hole solutions are known (see e.g. [**151, 152**]) and one form involves (**5R**) $ds^2=-exp(\lambda(r))dt^2+exp(\nu(r))dr^2+r^2d\Omega^2_{n-1}$ with

$$e^{\lambda(r)}=e^{-\nu(r)}=1+\frac{r^2}{2\tilde{\alpha}}\left(1-\sqrt{1+\frac{64\pi G\tilde{\alpha}M}{(n-1)\Omega_n r^n}}\right) \tag{5.13}$$

where $\tilde{\alpha}=(n-2)(n-3)\alpha$ and M is the black hole mass. The entropy has the form (**5S**) $S=(A/4G)[1+\frac{n-1}{n-3}\frac{2\tilde{\alpha}}{r_+^2}]$ where $A=n\Omega_n r_+^{n-1}$ is the horizon area and $r_+$ the horizon radius. Now apply (**5S**) to the apparent horizon (assuming the same expression but replacing $r_+$ by $r_A$); this yields (**5T**) $S=(A/4G)[1+\frac{n-1}{n-3}\frac{2\tilde{\alpha}}{\tilde{r}_A^2}]$ with $A=n\Omega_n\tilde{r}_A^{n-1}$ the apparent horizon area (assume also the horizon has temperature $T=1/(2\pi\tilde{r}_A)$). Using (**5K**) and applying the first law $-dE=TdS$ leads to

$$\left[1+2\tilde{\alpha}\left(H^2+\frac{k}{a^2}\right)\right]\left(\dot{H}-\frac{k}{a^2}\right)=-\frac{8\pi G}{n-1}(\rho+p) \tag{5.14}$$

Inserting (**5O**) into (5.14) one obtains finally

$$H^2 + \frac{k}{a^2} + \tilde{\alpha}\left(H^2 + \frac{k}{a^2}\right)^2 = \frac{16\pi G}{n(n-1)}\rho \tag{5.15}$$

Equations (5.14) and (5.15) are of course the Friedman equations for a FRW universe in the Gauss-Bonnet gravity given in [**152**] but derived in a different manner.

Next one looks at Lovelock gravity with Lagrangian $\mathfrak{L} = \sum_0^m c_i \mathfrak{L}_i$ where $c_i$ is arbitrary and $\mathfrak{L}_i$ is the Euler density of a $(2i)$-dimensional manifold (**5U**) $\mathfrak{L}_i = 2^{-i}\delta^{a_1 b_1 \cdots a_i b_i}_{c_1 d_1 \cdots c_i d_i} R^{c_1 d_1}_{a_1 b_1} \cdots R^{c_i d_i}_{a_i b_i}$. Here $\mathfrak{L}_0 = 1$ with $c_0 = \Lambda$, $c_1 = 1$, $\mathfrak{L}_2$ is the Gauss-Bonnet term, and although there are higher order derivative curvature terms there are no terms with more that second order derivatives of the metric. For an $(n+1)$-dimensional static, spherically symmetric black hole with metric (**5V**) $ds^2 = -f(r)dt^2 + f^{-1}(r)dr^2 + r^2 d\Omega^2_{n-1}$ the metric function is given by $f(r) = 1 - r^2 F(r)$ where $F(r)$ is determined by solving for the real roots of (**5W**) $\sum_0^m \hat{c}_i F^i(r) = [16\pi G M/n(n-1)\Omega_n r^n]$ where $M$ is an integration constant (black hole mass). The coefficients are given recursively by

$$(\bullet\bullet\bullet)\quad \hat{c}_0 = \frac{c_0}{n(n-1)};\ \hat{c}_1 = 1;\ \hat{c}_i = c_i \prod_{j=3}^{2m}(n+1-j)\ (i > 1)$$

In terms of the horizon radius $r_+$ the black hole entropy is given via (**5X**) $S = (A/4G)\sum_1^m [i(n-1)/(n-2i+1)]\hat{c}_i r_+^{2-2i}$ where $A = n\Omega_n r_+^{n-1}$ is the horizon area. Again one assumes that the apparent horizon of the FRW universe has an entropy of the form (**5X**) with $r_+$ replaced by the apparent horizon radius $\tilde{r}_A$. The temperature of the apparent horizon is $T = 1/(2\pi\tilde{r}_A)$ and the first law $-dE = TdS$ leads to

$$\sum_1^m i\hat{c}_i \left(H^2 + \frac{k}{a^2}\right)^{i-1}\left(\dot{H} - \frac{k}{a^2}\right) = -\frac{8\pi G}{n-1}(\rho + p); \tag{5.16}$$

$$\sum_1^m \hat{c}_i \left(H^2 + \frac{k}{a^2}\right)^i = \frac{16\pi G}{n(n-1)}\rho$$

These two equations are the Friedman equations for an FRW universe in the Lovelock gravity. When $m = 2$ they reduce to the equations for Gauss-Bonnet gravity. In [**16**] the derivation of Friedman equations from thermodynamics is carried out for scalar-tensor and $f(R)$ gravity. Further elaboration and extensions appear in [**16, 53, 151, 152, 178, 372, 421, 422, 476, 498, 884, 874, 875**].

# 6. THERMODYNAMICS AND QUANTUM GRAVITY - I

In Sections 5.6-5.8 we deal with some fundamental articles by L. Glinka (see [**353, 354, 355, 357**]) for which some general theory is also motivated by theoretical material involving second quantization and Bogoliubov transformations as spelled out in [**74, 75, 76, 110, 136, 356, 358, 503, 683, 697, 698, 796**] (see also Sections 7 and 8). There have been updated and polished versions of papers in [**353, 354, 355, 357**] and we suggest printing out the latest versions and working from them - we can only give a sketch here and remark that some of the work has a visionary nature which is valuable in itself. One goes first to the Friedman-Robertson-Walker-Lemaitre (sometimes FRW, sometimes FL) model ([**353**]) and recalls the general Einstein-Hilbert action

$$A_{EH} = \int d^4x\sqrt{-g}\left(-\frac{1}{2\kappa}R + L_{fields}\right) \tag{6.1}$$

where $\kappa = 8\pi G$ and

$$R_{\mu\nu} = \partial_\alpha\Gamma^\alpha_{\mu\nu} - \partial_\nu\Gamma^\alpha_{\mu\alpha} + \Gamma^\alpha_{\beta\alpha}\Gamma^\beta_{\mu\nu} - \Gamma^\alpha_{\beta\nu}\Gamma^\beta_{\mu\alpha} \tag{6.2}$$

$$\Gamma^\rho_{\mu\nu} = \frac{1}{2}g^{\rho\sigma}(\partial_\nu g_{\mu\sigma} + \partial_\mu g_{\sigma\nu} - \partial_\sigma g_{\mu\nu})$$

Variation of the action (6.1) gives the equations of motion for geometry

$$R_{\mu\nu} - \frac{1}{2}Rg_{\mu\nu} = \kappa T_{\mu\nu};\ T_{\mu\nu} = -2\frac{\delta L}{\delta g^{\mu\nu}} + Lg_{\mu\nu} \tag{6.3}$$

($L = L_{fields}$). One considers the FRWL model for homogeneous and isotropic spacetime

$$g_{\mu\nu} = \begin{pmatrix} N_d^2(t) & 0 \\ 0 & -a^2(t)\delta_{ij} \end{pmatrix} \tag{6.4}$$

Nonvanishing Christoffel symbols for (6.4) are

$$\Gamma^0_{00} = \frac{\dot N_d}{N_d};\ \Gamma^0_{ii} = \frac{a\dot a}{N_d^2};\ \Gamma^i_{i0} = \frac{\dot a}{a} \tag{6.5}$$

Using (6.5) one obtains the Ricci tensor

$$R_{\mu\nu} = \begin{pmatrix} -3\left[\frac{\ddot a}{a} - \frac{\dot a\dot N_d}{aN_d}\right] & 0^T \\ 0 & \left[\frac{a\ddot a}{N_d^2} - 2\frac{\dot a^2}{N_d^2} - \frac{a\dot a\dot N_d}{N_d^2}\delta_{ij}\right] \end{pmatrix} \tag{6.6}$$

The curvature scalar is then

$$R = -6\left(\frac{\ddot a}{aN_d^2} - \frac{\dot a^2}{a^2N_d^2} - \frac{\dot a\dot N_d}{aN_d^3}\right) \tag{6.7}$$

and one introduces the conformal time $\tau$ and mass scale $\phi$ via

$$\tau(t) = \int_0^t \frac{N_d(t')}{a(t')}dt';\ \phi(\tau) = \sqrt{\frac{3}{8\pi}}M_P a(\tau) \tag{6.8}$$

($M_P = 1/\sqrt{G}$ is the Planck mass). The space time interval is then (**6A**) $ds^2 = a^2(\tau)[d\tau^2 - (dx^2 + dy^2 + dz^2)]$ and the action in conformal time is

$$A[\phi] = \int d\tau\left[-V\frac{d\phi}{d\tau}\frac{d\phi}{d\tau} - \left(\frac{8\pi}{3M_P^2}\right)^2\phi^4 H_{fields}(\tau)\right] \tag{6.9}$$

where $V = \int_{\mathbf{R}^3} d^3x$ and (**6B**) $H = H_{fields} = \int_{\mathbf{R}^3} d^3x\mathcal{H}_{fields}(x,\tau)$. The canonical momentum for (6.9) is

$$P_\phi = \frac{\delta A[\phi]}{\delta(d\phi/d\tau)} = -2V\frac{d\phi}{d\tau} \tag{6.10}$$

and the Hamiltonian for (6.9) is
(6.11)

$$A[\phi] = -\int d\tau \left[P_\phi \frac{d\phi}{d\tau} - H(\tau)\right];\ H(\tau) = \frac{1}{4V^2}\left[P_\phi^2 - \left(\frac{16\pi V}{3M_P^2}\right)^2 \phi^4 H(\tau)\right] = 0$$

The shell energy constraint equation defines values of the canonical momentum (**6C**) $P_\phi = \pm E_\phi$ where (**6D**) $E_\phi = [16\pi V/3M_P^2]\phi^2\sqrt{H(\tau)}$. Equations (6.10) and (**6D**) give the equation for $\phi(\tau)$

$$\frac{d\phi(\tau)}{d\tau} = \pm\frac{8\pi}{3M_P^2}\sqrt{H(\tau)}\phi^2(\tau) \tag{6.12}$$

leading to

$$\frac{1}{\phi(\tau)} - \frac{1}{\phi(\tau_0)} = \pm\frac{8\pi}{3M_P^2}\int_{\tau_0}^{\tau}\sqrt{H(\tau')}d\tau' \tag{6.13}$$

Upon introducing the redshift $z$ via

$$z(\tau_0,\tau) = \pm\frac{8\pi\phi(\tau_0)}{3M_P^2}\int_{\tau_0}^{\tau}\sqrt{H(\tau')}d\tau' \tag{6.14}$$

(6.13) takes the form (**6E**) $[\phi(\tau)/\phi(\tau_0)] = [1/(1+z(\tau_0,\tau))]$.

Now first quantization starts with the shell-energy constraints of the classical theory (**6F**) $P_\phi^2 - E_\phi^2 = 0$ with $E_\phi$ given by (**6D**) and quantizes this via a canonical commutation relation (CCR) (**6G**) $i[P_\phi,\phi] = 1$ while assuming there is a wave function $\psi(\phi)$ of a quantum theory which should satisfy (**6H**) $[(\partial^2/\partial\phi^2) + E_\phi^2]\psi(\phi) = 0$. This is referred to here as a WDW equation (as a Klein-Gordon equation of a massive scalar field in 0-D space - (cf. [**110**])). The next step (second quantization) involves building the Fock space and considering the action functional needed to obtain (**6H**) as an Euler-Lagrange (EL) equation. One finds

$$A[\psi] = \int d\phi\left[\frac{1}{2}\partial_\phi\psi\partial_\phi\psi - \frac{1}{2}E_\phi^2\psi^2\right] \tag{6.15}$$

for which a canonical momentum is (**6I**) $\Pi_\psi = \delta A[\psi]/\delta(\partial_\phi\psi)] = \partial_\phi\psi$ leading to an action

$$A[\psi] = \int d\phi\left[\Pi_\psi\partial_\phi\psi - \frac{1}{2}(\Pi_\psi^2 + E_\phi^2\psi^2)\right] \tag{6.16}$$

and a Hamiltonian (**6J**) $H(\Pi_\psi\psi) = (1/2)[\Pi_\psi^2 + E_\phi^2\psi^2]$ (corresponding to the form for a harmonic oscillator with frequency $\omega = E_\phi$

Thus the wave equation is of Klein-Gordon type and the Fock space should be of boson type; hence one builds from operators $\mathfrak{A}$ and $\mathfrak{A}^\dagger$ to be called universe creation and annihilation operators. The boson type CCR's are (**6K**) $[\mathfrak{A},\mathfrak{A}^\dagger] = 1$, $[\mathfrak{A},\mathfrak{A}] = [\mathfrak{A}^\dagger,\mathfrak{A}^\dagger] = 0$. In this basis the canonical fields $\psi$ and $\Pi_\psi$ have representations

$$\psi = \frac{1}{\sqrt{E_\phi}}\frac{\mathfrak{A}+\mathfrak{A}^\dagger}{\sqrt{2}};\ \Pi_\psi = i\sqrt{E_\phi}\frac{\mathfrak{A}^\dagger - \mathfrak{A}}{\sqrt{2}};\ i[\Pi_\psi,\psi] = 1 \tag{6.17}$$

Note the normalization coefficients depend on $\phi$ - by analogy with the harmonic oscillator. In the Fock space (**6K**) the reduced WDW action (6.16) takes the form

$$A(\mathfrak{A},\mathfrak{A}^\dagger)=\int d\phi\left[i\frac{\mathfrak{A}^\dagger\partial_\phi\mathfrak{A}-\mathfrak{A}\partial_\phi\mathfrak{A}^\dagger}{2}+i\frac{\mathfrak{A}\mathfrak{A}-\mathfrak{A}^\dagger\mathfrak{A}^\dagger}{2}\frac{\partial_\phi E_\phi}{2E_\phi}-\mathfrak{H}\right] \tag{6.18}$$

where $\mathfrak{H}=[\mathfrak{A}^\dagger\mathfrak{A}+(1/2)]=E_\phi$. A variational principle applied to the action (6.18) gives then the EL equations

$$i\partial_\phi\begin{pmatrix}\mathfrak{A}\\ \mathfrak{A}^\dagger\end{pmatrix}=\begin{pmatrix}E_\phi & i\frac{\partial_\phi E_\phi}{2E_\phi}\\ i\frac{\partial_\phi E_\phi}{2E_\phi} & -E\phi\end{pmatrix}\begin{pmatrix}\mathfrak{A}\\ \mathfrak{A}^\dagger\end{pmatrix} \tag{6.19}$$

In order to deal with the nondiagonal situation one introduces a "diagonalization ansatz" in the form (we just spell out the necessary equations)

(1) The $\mathfrak{A}$ operators arise via a Bogoliubov transformation from operators $U$

$$\begin{pmatrix}\mathfrak{A}^\dagger\\ \mathfrak{A}\end{pmatrix}=\begin{pmatrix}A(\phi) & B^*(\phi)\\ B(\phi) & A^*(\phi)\end{pmatrix}\begin{pmatrix}U^\dagger\\ U\end{pmatrix} \tag{6.20}$$

where $det(A,A^*,B,B^*)=1$

(2) The $\phi$-evolution of $U$, $U^\dagger$ is described by Heisenberg equations

$$i\partial_\phi\begin{pmatrix}U\\ U^\dagger\end{pmatrix}=\begin{pmatrix}\tilde{E}_\phi & 0\\ 0 & -\tilde{E}_\phi\end{pmatrix}\begin{pmatrix}U\\ U^\dagger\end{pmatrix} \tag{6.21}$$

where $\tilde{E}_\phi$ is the diagonalization energy.

(3) In the $(U,U^\dagger)$ basis the Lagrangian is diagonal $L(U^\dagger,U)=H(U^\dagger,U)=E_U U^\dagger U$ where $E_U$ is the energy of elementary excitations.

(4) The operator $N=U^\dagger U$ is the density of excitations number operator and is an integral of motion, i.e. $\partial_\phi N=0$, and hence there is a stable vacuum state $|0>$ with $U|0>=0$ and $<0|U^\dagger=0$

The A and B equations of motion arise from the diagonalization ansatz in the Bogoliubov transformation (6.20), i.e.

$$\partial_\phi\begin{pmatrix}A(\phi)\\ B(\phi)\end{pmatrix}=\begin{pmatrix}iE_\phi & \frac{\partial_\phi E_\phi}{2E_\phi}\\ \frac{\partial_\phi E_\phi}{2E_\phi} & iE_\phi\end{pmatrix}\begin{pmatrix}A(\phi)\\ B(\phi)\end{pmatrix} \tag{6.22}$$

with initial values $A(\phi(\tau_0))=1$ and $B(\phi(\tau_0))=0$ and this is solved explicitly via

$$A(x)=\frac{1}{2x}e^{i\lambda(x^3-1)}\left[-x^2-1+2i\lambda\frac{x^3-1}{log(x)}(x^2-1)\right]; \tag{6.23}$$

$$B(x)=\frac{1}{2x}e^{i\lambda(x^3-1)}(x^2-1)$$

where $x=[\phi(\tau)/\phi(\tau_0)]$ is dimensionless and is expressed via the equation (**6L**) $\lambda=[16\pi V/9M_P^2]\sqrt{H(\tau)}\phi^3(\tau_0)$. The diagonalization energy $\tilde{E}_\phi$ has the form

$$\tilde{E}_\phi=-\frac{\lambda}{\phi(\tau_0)}\left[3x^4-4x^2+\frac{x^3-1}{x^3log(x)}(x^2-1)^2+3\right]-i\frac{x^4-1}{2x^3} \tag{6.24}$$

whille the energy of elementary excitations is $E_U=-[3\lambda/2\phi(\tau_0)](x^4+1)$.

From elementary thermodynamics one has (**6N**) $T^{-1}=(\partial S/\partial E)_V$ and the von Neumann entropy is a vacuum expectation value (**6O**) $S = k_B < \mathfrak{N}log(\mathfrak{N} >$ where $\mathfrak{N} = \mathfrak{A}^\dagger\mathfrak{A}$ is the density of universes number operator and $k_B$ is the Boltzmann constant. The energy of the system is (**6P**) $E =< \mathfrak{N}\mathfrak{H} >$ and from this follows

$$\left(\frac{\partial S}{\partial E}\right)_V = k_B\frac{\partial < \mathfrak{N}log(\mathfrak{N}) >}{\partial < \mathfrak{N}\mathfrak{H} >} = k_B\frac{\partial_\phi < \mathfrak{N}log(\mathfrak{N}) >}{\partial_\phi < \mathfrak{N}\mathfrak{H} >} \tag{6.25}$$

$$= k_B\frac{< \partial_\phi\mathfrak{N}log(\mathfrak{N}) + \partial_\phi\mathfrak{N} >}{< \partial_\phi\mathfrak{N}\mathfrak{H} + \mathfrak{N}\partial_\phi\mathfrak{H} >}$$

Now calculating the vacuum expectation values in (6.25) one arrives at

$$\left(\frac{\partial S}{\partial E}\right)_V = \frac{2k_B}{E_\phi}\frac{1+ < log(\mathfrak{N}) >}{1 + 4 < \mathfrak{N} > + [2 < \mathfrak{N}^2 > + < \mathfrak{N} >]\frac{1}{E_\phi}\frac{\partial_\phi E_\phi}{\partial_\phi<\mathfrak{N}>}} \tag{6.26}$$

leading to

$$T = \frac{3\lambda}{8k_B\phi(\tau_0)}\left[x^5 - x^3 + 15x^2 - 6 + \frac{3}{x^2} - \frac{15x^2}{x^3+4} - \frac{63}{4x^3(x^3+4)}\right] \tag{6.27}$$

Next from [**355**], one works again with the classical Friedmann-Lemaitre (FL) spacetime characterized by the interval $ds^2 = dt^2 - a^2(t)dx^idx^i$ and we summarize some of this in Remarks.

**REMARK 6.1.1.** Making a change of variables $\eta = \int_{t_0}^t [dt'/a(t')]$ one arrives at a conformal flat form and the interval becomes (cf. (**6A**)) $ds^2 = a^2(\eta)(d\eta^2 - dx^idx^i)$. The lapse function is defined via $d\eta = N_d(x^0)dx^0$ (cf. (6.8)) while $x^0 \to \tilde{x}^0 = \tilde{x}(x^0)$. The action is again $A = \int dx^4\sqrt{-g}[-(1/6)R + L_M]$ which in FL form is

$$A[a] = -V\int dx^0\left[\frac{1}{N_d}\left(\frac{da}{dx^0}\right)^2 + N_da^4 < \mathcal{H}(x^0) >\right] \tag{6.28}$$

$$< \mathcal{H}(x^0) > = \frac{1}{V}\int d^3x\mathcal{H}_M(x^i, x^0);\ V = \int d^3x < \infty$$

The latter formulas represent the zeroth Fourier harmonic of the matter Hamiltonian and the spatial volume. One now calculates (**6Q**) $p_a = -(2V/N_d)(da/dx^0)$ so that the action becomes

$$A[a] = -V\int dx^0\left[\frac{p_a^2}{4V^2} + a^4 < \mathcal{H}(x^0) >\right] \tag{6.29}$$

leading to

$$A[a] = \int dx^0\left[p_a\frac{da}{dx^0} - H(p_a, a)\right]; \tag{6.30}$$

$$H(p_a, a) = N_d\left[-\frac{p_a^2}{4V} + V < \mathcal{H}(x^0) > a^4\right]$$

Via the Dirac approach the action principle with respect to the lapse function $N_d$ applied to (6.30) produces the Hamiltonian constraint equation; hence the constraint is

$$\frac{\delta A[a]}{\delta N_d} = 0 = -\frac{p_a^2}{4V} + V < \mathcal{H}(x^0) > a^4 \tag{6.31}$$

which can be solved via

$$\frac{a(t)}{a(t_0)} = exp\left[sgn(t-t_0)\int_{t_0}^{t} N_d(x^0)dx^0\sqrt{< \mathcal{H}(x^0) >}\right] \tag{6.32}$$

(Hubble law). From the other side the constraint equation (6.31) expressed in Dirac conformal time has a solution **(6R)** $p_a = -2V(da/d\eta) = \pm\omega_a$ which defines the values of the canonical conjugate momentum **(6Q)**. In **(6R)** one has a time diffeomorphism variable **(6S)** $\omega_a = 2V\sqrt{< \mathcal{H}(\eta) >} a^2(\eta)$ and **(6R)** gives an equation for $a(\eta)$, namely **(6T)** $-(da/d\eta) = \pm\sqrt{< \mathcal{H}(\eta) >} a^2(\eta)$ leading to

$$a(\eta) = \frac{a(\eta_0)}{1+z(\eta_0,\eta)};\ z(\eta_0,\eta) = a(\eta_0)sgn(\eta-\eta_0)\int_{\eta_0}^{\eta} d\eta'\sqrt{< \mathcal{H}(\eta') >} \tag{6.33}$$

Note that $z(\eta_0,\eta)$ is the redshift, $H_0$ is the Hubble parameter, and $q_0$ is the deceleration parameter, all related via $z(\eta,\eta) = H_0(\eta-\eta_0) + ([1+(q_0/2)]H_0^2(\eta-\eta_0)^2 + \cdots$ with

$$H_0 = \sqrt{< \mathcal{H}(\eta_0) >} a(\eta_0);\ q_0 = \frac{2}{H_0}\frac{< \dot{\mathcal{H}}(\eta_0) >}{< \mathcal{H}(\eta_0) >} - 2 \tag{6.34}$$

The Hubble parameter and the deceleration parameter are diffeomorphism invariants. ■

**REMARK 6.1.2.** Quintessence is presumed to be a kind of matter characterized by constant energy, namely the cosmological constant $\Lambda$, which is supposed to be the zero mode of the matter Hamiltonian. It's properties are $< \mathcal{H}(\eta) > = < \mathcal{H}(\eta_0) > \Lambda$ and $< \dot{\mathcal{H}}(\eta) > = < \dot{\mathcal{H}}(\eta_0) = 0$. In this approximation the Hubble constant and deceleration parameter have a simple form $H_0 = \Lambda^{1/2}a(\eta_0)$ and $q_0 = -2$ while the redshift is $z(\eta_0,\eta) = H_0|\eta-\eta_0|$. The solution of classical constraints **(6S)** for the quintessence has a form

$$p_a = \pm\omega_a(\eta) = \pm 2V\Lambda^{1/2}a^2 = \pm\omega_a(\eta_0)\left(\frac{a(\eta)}{a(\eta_0)}\right)^2 \tag{6.35}$$

where $\omega_a(\eta_0) = 2V\frac{H_0^2}{\sqrt{\Lambda}}$. ■

The program now for quantizing the FL spacetime with quintessence involves using a procedure based on the Hamiltonian equations of motion. Thus (cf. [**110, 111, 112, 723**] for background information on Fock space, Bogolubov transformations, quantization, etc.)

(1) First quantization gives the WDW equation for the wave function of the universe $\psi$.
(2) Treating $\psi$ as a classical field one constructs classical field theory for canonical Hamiltonian equations.

(3) One quantizes the canonical Hamiltonian equations by non-Fockian distributions in the Fock space of annihilation and creation operators.
(4) One applies the Bogoliubov transformation and by diagonalization of the quantized canonical Hamiltonian equations in the Fock space one carries evolution to the Bogoliubov coefficients.
(5) One finds the field operator $\Psi$ of the universe and the conjugate momentum $\Pi_\Psi$.

Thus, recall first that the FL universe Hamiltonian constraint has a form (cf. (**6R**) $p_a^2 - \omega_a^2 = 0$ with $\omega$ given by (**6S**). A classical solution of this is given by the Hubble law (6.33) with redshift as indicated. The first quantization of this constraint equation is given via the CRR ($\bullet$) $i[\hat{p}_a, a] = 1$ where $\hat{p}_a = -i(\partial/\partial a)$ with result (**6U**) $(\partial_a\partial_a + \omega_a^2)\psi(a) = 0$ (WDW equation). This equation (**6U**) looks like the KG equation for the boson with mass $\omega_a$ so look at the classical action producing this equation, namely

$$S[\psi] = \frac{1}{2}\int_{a(\eta_0)}^{a(\eta)} da\,\left[(\partial_a\psi)^2 - \omega_a^2\psi^2\right] \tag{6.36}$$

To see that this produces the WDW equation one has

$$\delta S[\psi] = \frac{\delta S[\psi]}{\delta\psi}\delta\psi + \frac{\delta S[\psi]}{\delta\partial_a\psi}\delta\partial_a\psi = \frac{\delta S[\psi]}{\delta\psi}\delta\psi + \frac{\delta S[\psi]}{\delta\partial_a\psi}\partial_a\delta\psi \tag{6.37}$$
$$= \left[\frac{\delta S[\psi]}{\delta\psi} - \partial_a\frac{\delta S[\psi]}{\delta\partial_a\psi}\right]\delta\psi + \partial_a\left[\frac{\delta S[\psi]}{\delta\partial_a\psi}\delta\psi\right]$$

The second term vanishes on boundaries leading to

$$\frac{\delta S[\psi]}{\delta\psi} - \partial_a\frac{\delta S]\psi]}{\delta\partial_a\psi} = 0 \Rightarrow \tag{6.38}$$
$$\int da\,\left[\omega_a^2\psi + \partial_a\partial_a\psi\right] = 0 \Rightarrow (\partial_a\partial_a + \omega_a^2)\psi = 0$$

(upon using (6.36). Thus this is the WDW equation (**6U**); it is an equation of motion for the classical field $\psi$ and the heuristic action (6.36) is correct. Now one has a canonical conjugate momentum ($\blacklozenge$) $\Pi_\psi = \delta S[\psi]/\delta(\partial_a\psi) = \partial_a\psi$ and with this the action (6.36) reduces to the form

$$S[\psi] = \int da\,[\pi_\psi\partial_a\psi - H(\Pi_\psi,\psi)]\,;\; H(\Pi_\psi,\psi) = (1/2)(\Pi_\psi^2 + \omega_a^2\psi^2) \tag{6.39}$$

as the Hamiltonian describing the evolution of the classical field $\psi$. The canonical Hamilton equations for this classical field theory are

$$\frac{\partial H(\Pi_\psi,\psi)}{\partial\Pi_\psi} = \partial_a\psi;\; -\frac{\partial H(\Pi_\psi,\psi)}{\partial\psi} = \partial_a\Pi_\psi \tag{6.40}$$
$$\equiv \partial_a\begin{pmatrix}\psi\\ \Pi_\psi\end{pmatrix} = \begin{pmatrix}0 & 1\\ -\omega_a^2 & 0\end{pmatrix}\begin{pmatrix}\psi\\ \Pi_\psi\end{pmatrix}$$

Thus the equation (6.40A) leads to the relation ($\blacklozenge$) and (6.40B) is equivalent to the WDW equation (**6U**) after using (6.40A). Standard approaches to quantization of FL spacetime and in general to quantization problems for gravity are based on testing of solutions or on the second quantization of the WDW equation. The present method of Glinka is different and a quantum theory of gravity is based on the canonical Hamiltonian equations of motion.

In this spirit now given a system described by a boson field one creates a boson Fock space of creation and annihilation operators ($\mathfrak{G}^{\dagger}$ and $\mathfrak{G}$ respectively) satisfying

$$[\mathfrak{G}(a(\eta)), \mathfrak{G}^{\dagger} a(\eta'))] = \delta(a(\eta) - a(\eta')); \; [\mathfrak{G}(a(\eta)), \mathfrak{G}(a(\eta'))] = 0 \tag{6.41}$$

By analogy with the KG theory one proposes a second quantization by the non-Fockian type distributions in the Fock space

$$\psi(a) = \frac{1}{\sqrt{2\omega_a}}(\mathfrak{G}(a) + \mathfrak{G}^{\dagger}(a)); \; \Pi_{\psi}(a) = -i\sqrt{\frac{\omega_a}{2}}(\mathfrak{G}(a) - \mathfrak{G}^{\dagger}(a)) \tag{6.42}$$

or equivalently

**REMARK 6.1.3.** We now write $\mathbf{G}$ instead of $\mathfrak{G}$ because the Latex software treats $\mathfrak{G}$ strangely in array patterns. ■

$$\begin{pmatrix} \psi \\ \Pi_{\psi} \end{pmatrix} = \begin{pmatrix} \frac{1}{\sqrt{2\omega_a}} & \frac{1}{\sqrt{2\omega_a}} \\ -i\sqrt{\frac{\omega_a}{2}} & i\sqrt{\frac{\omega_a}{2}} \end{pmatrix} \begin{pmatrix} \mathbf{G} \\ \mathbf{G}^{\dagger} \end{pmatrix} \tag{6.43}$$

The correct CCR (**6V**) $[\psi(a(\eta'), \Pi_{\psi}(a(\eta))] = i\delta(a(\eta) - a(\eta'))$ is preserved automatically. Now using the non-fockian distributions (6.42)-(6.43) one can translate the WDW action (6.39) into the Fock space language

$$S(\mathbf{G}, \mathbf{G}^{\dagger}) = \int \mathcal{D}\left[i\frac{\mathbf{G}^{\dagger}\partial_a\mathbf{G} - \mathbf{G}\partial_a\mathbf{G}^{\dagger}}{2} - \mathcal{H}\right] \tag{6.44}$$

($\mathcal{D}$ is a Feynman type measure) and

$$\mathcal{H} = \left(\mathbf{G}^{\dagger}\mathbf{G} + \frac{1}{2}\right)\omega_a + \frac{i}{2}(\mathbf{G}^{\dagger}\mathbf{G}^{\dagger} - \mathbf{G}\mathbf{G})\Delta \tag{6.45}$$

where $\Delta = [\partial_a\omega_a/2\omega_a]$ has a meaning of coupling. This Hamiltonian is known in many particle theories as the boson superfluidity phenomenon and $\Delta$ manifests collective phenomena (cf. Section 8 and [**353**]-2).

Thus by quantization of (6.40) one obtains equations of motion for the creation and annihilation operators in the Fock space

$$i\partial_a \begin{pmatrix} \mathbf{G} \\ \mathbf{G}^{\dagger} \end{pmatrix} = \begin{pmatrix} -\omega_a & 2i\Delta \\ 2i\Delta & \omega_a \end{pmatrix} \begin{pmatrix} \mathbf{G} \\ \mathbf{G}^{\dagger} \end{pmatrix} \tag{6.46}$$

These are the Heisenberg equations for $\mathbf{G}$, $\mathbf{G}^{\dagger}$ and the evolution matrix in (6.46) is not diagonal; hence to diagonalize this one changes the $\mathbf{G}$ basis to a $\mathbf{W}$ basis via

$$\begin{pmatrix} \mathbf{W}(a) \\ \mathbf{W}^{\dagger}(a) \end{pmatrix} = \begin{pmatrix} u(a) & v(a) \\ v^*(a) & u^*(a) \end{pmatrix} \begin{pmatrix} \mathbf{G}(a) \\ \mathbf{G}^{\dagger}(a) \end{pmatrix} \tag{6.47}$$

If one wishes to preserve the CCR in the basis $(\mathbf{W}, \mathbf{W}^{\dagger})$, namely

$$[\mathbf{W}(a(\eta)), \mathbf{W}^{\dagger}(a(\eta'))] = \delta(a(\eta) - a(\eta')); \; [\mathbf{W}(a(\eta)), \mathbf{W}(a(\eta'))] = 0 \tag{6.48}$$

then there is a required relation (★) $|u(a)|^2 - |v(a)|^2 = 1$. In these circumstances one now arrives at an evolution

$$i\partial_a \begin{pmatrix} \mathbf{W} \\ \mathbf{W}^\dagger \end{pmatrix} = \begin{pmatrix} \omega_1 & 0 \\ 0 & \omega_2 \end{pmatrix} \begin{pmatrix} \mathbf{W} \\ \mathbf{W}^\dagger \end{pmatrix} \quad (6.49)$$

This procedure produces equations for the coefficients $u$, $v$ in the form

$$i\partial_a \begin{pmatrix} v \\ u \end{pmatrix} = \begin{pmatrix} -\omega_a & -2i\Delta \\ -2i\Delta & \omega_a \end{pmatrix} \begin{pmatrix} v \\ u \end{pmatrix} \quad (6.50)$$

and the values of the diagonalization energies are $\omega_1 = \omega_2 = 0$. Thus one has solutions of (6.49) (••) $\mathbf{W}(a) = \mathbf{W}(a_0)$ and $\mathbf{W}^\dagger(a) = \mathbf{W}^\dagger(a_0)$ and one sees that the operator $\mathfrak{N}_W = \mathbf{W}^\dagger\mathbf{W} = \mathbf{W}^\dagger(a_0)\mathbf{W}(a_0)$ is an integral of motion (♦♦) $\partial_a\mathfrak{N}_W = 0$. Hence the stable Bogoliubov vacuum state $|0>$ exists with (★★) $\mathbf{W}|0> = 0$ and $<0|\mathbf{W}^\dagger = 0$. The hyperbolic identity (★) can be parametrized via (**6W**) $v(a) = exp(i\theta(a))Sinh(\phi(a))$ and $u(a) = exp(i\theta(a))Cosh(\phi(a))$ so (6.50) is equivalent to

$$\partial_a\theta(a) = \pm\omega_a = p_a;\ \partial_a\phi(a) = -2\Delta = -\frac{\partial_a\omega}{\omega} = -\partial_a log(|\omega_a|) \quad (6.51)$$

with solutions

$$\theta(a) = \int_{a_0}^a p_a da;\ \phi(a) = -log\left|\frac{\omega_a(\eta)}{\omega_a(\eta_0)}\right|; \quad (6.52)$$

$$v(a) = \frac{1}{2}exp\left[i\int_{a_0}^a p_a da\right]\left(\frac{\omega_a(\eta_0)}{\omega_a(\eta)} - \frac{\omega(\eta)}{\omega(\eta_0)}\right);$$

$$u(a) = \frac{1}{2}exp\left[i\int_{a_0}^a p_a da\right]\left(\frac{\omega_a(\eta_0)}{\omega_a(\eta)} + \frac{\omega_a(\eta)}{\omega_a(\eta_0)}\right)$$

**REMARK 6.1.4.** The explanation here is that spacetime is described in the language of collective phenomena, which take place in a "graviton-matter" gas mixture of quanta of gravity and quintessence which produces the boson fields involved (see Sections 7-8 for some elaboration). ∎

In the same spirit one defines now a field operator for the universe by expressing the $\mathbf{G} \equiv \mathfrak{G}$ operators in terms of $\mathbf{W}$ and $\mathbf{W}^\dagger$ and thence to

(6.53)

$$\begin{pmatrix} \psi(a) \\ \Pi_\psi(a) \end{pmatrix} = \begin{pmatrix} \frac{1}{\omega_a(\eta_0)}\sqrt{\frac{|\partial_a\theta|}{2}}e^{-i\theta} & \frac{1}{\omega_a(\eta_0)}\sqrt{\frac{|\partial_a\theta|}{2}}e^{i\theta} \\ -i\omega_a(\eta_0)\sqrt{\frac{1}{2|\partial_a\theta|}}e^{-i\theta} & i\omega_a(\eta_0\sqrt{\frac{1}{2|\partial_a\theta|}}e^{i\theta} \end{pmatrix} \begin{pmatrix} \mathbf{W}(a_0) \\ \mathbf{W}^\dagger(a_0) \end{pmatrix}$$

where (**6X**) $\partial_a\theta(a0 = \pm\omega_a(\eta_0)[a(\eta)/a(\eta_0)]^2$ and $\theta(a) = \int_{a_0}^a p_a da$. Here we have the field operator of the universe and momentum field

$$\psi(a) = \frac{1}{\omega_a(\eta_0)}\sqrt{\frac{|\partial_a\theta|}{2}}(e^{i\theta}\mathbf{W}^\dagger(a_0) + e^{-i\theta}\mathbf{W}(a_0)); \quad (6.54)$$

$$\Pi_\psi(a) = i\omega_a(\eta_0)\sqrt{\frac{1}{2|\partial_a\theta|}}[e^{i\theta}\mathbf{W}^\dagger(a_0) - e^{-i\theta}\mathbf{W}(a_0)]$$

**6.1. THERMODYNAMICS OF THE UNIVERSE.** In connection with the universe one should see also Sections 6.2-6.5. Glinka looks at a graviton-matter gas as an open quantum system which should be described by nonequilibrium quantum statistical mechanics. In such a situation the one particle density operator is the particle number operator and in the graviton-matter gas a role of particles is played by elements of this gas with density operator (**6Y**) $\rho_{\mathbf{G}} = \mathbf{G}^\dagger\mathbf{G}$ which becomes (**6Z**) $\rho_{\mathbf{G}} = \mathcal{W}^\dagger \rho \mathcal{W}$ where

$$\mathcal{W} = \begin{pmatrix} \mathbf{W} \\ \mathbf{W}^\dagger \end{pmatrix}; \ \rho = \begin{pmatrix} |u|^2 & -uv \\ -u^*v^* & |v|^2 \end{pmatrix} \tag{6.55}$$

The physical entropy is played by the quantum information theory formula

$$S = -\frac{Tr(\rho log(\rho)}{Tr(\rho)} = log(\Omega); \ \Omega = \frac{1}{2|u|^2 - 1} \tag{6.56}$$

There will be thermodynamical nonequilibrium since particles of the gas to out from the system. As a result one has diagonalized equations of motion and since one has found a basis where the particle number operator is an integral of motion one has thermodynamical equilibrium in this basis. If one identifies the partition function of the graviton - quintessence gas (6.56) with the Bose-Einstein type partition function there results

$$\Omega = \frac{1}{2|u|^2 - 1} = \frac{1}{exp(E/T) - 1} \Rightarrow T = \frac{E}{log(2|u|^2)} \tag{6.57}$$

(Gibbs state type). This type of identification has meaning if and only if one identifies (•••) $E = U - \mu N$ where U is internal energy, $\mu$ is the chemical potential, and N is the number of particles. One has seen that the Hamiltonian is

$$\mathcal{H} = \left(\mathfrak{G}^\dagger\mathfrak{G} + \frac{1}{2}\right)\omega_a + \frac{i}{2}\left(\mathfrak{G}^\dagger\mathfrak{G}^\dagger - \mathfrak{G}\mathfrak{G}\right)\Delta \tag{6.58}$$

which in a diagonalized basis becomes an effective Hamiltonian $\mathcal{H} = \mathcal{W}^\dagger \mathbf{H} \mathcal{W}$ where (6.59)

$$\mathbf{H} = \begin{pmatrix} \frac{1}{2}(|u|^2 + |v|^2)\omega_a + \frac{i}{2}(u^*v - uv^*) & -uv\omega_a + \frac{i}{2}(uu - vv)\Delta \\ -u^*v^*\omega_a + \frac{i}{2}(v^*v^* - u^*u^*)\Delta & \frac{1}{2}(|u|^2 + |v|^2)\omega_a + \frac{i}{2}(u^*v - uv^*)\Delta \end{pmatrix}$$

In quantum statistical mechanics the internal energy U is defined by the quantum mechanical average Hamiltonian of the thermodynamical system so

$$U = <\mathbf{H}> = \frac{Tr(\rho\mathbf{H})}{Tr(\rho)}; \tag{6.60}$$

$$U = \left(\frac{1}{2} + 2N + \frac{N}{2N+1}\right)(\sqrt{N+1} - \sqrt{N})\omega_a(\eta_0); \ N = |v|^2$$

Hence the chemical potential for the gas is (6.61)

$$\mu = \left(2 + \frac{1}{(2N+1)^2} - \frac{(1/2) + 2N + [N/(2N+1)]}{2\sqrt{N(N+1)}}\right)(\sqrt{N+1} - \sqrt{N})\omega_a(\eta_0)$$

and the temperature is

$$T = \frac{\sqrt{N+1}-\sqrt{N}}{log(2N+2)}\left[\left(\frac{1}{2}+2N+\frac{N}{2N+1}\right)\left(1+\frac{1}{2}\sqrt{\frac{N}{N+1}}\right) - 2N - \frac{N}{(2N+1)^2}\right]\omega_a(\eta_0) \tag{6.62}$$

For $N=0$ one sees a finite contribution from the gas temperature, namely $T[0] = \omega_a(\eta_0)/log(4)$ and one concludes that the equation of state for the graviton-matter gas with quintessence is

$$\frac{U}{T} = \frac{log(2N+2)}{1+\frac{1}{2}\sqrt{\frac{N}{N+1}} - \frac{N}{2N+1}\frac{1+2(2N+1)^2}{N-1+3(2N+1)^2}} \tag{6.63}$$

The paper concludes with some graphs and philosophy, the main point of which is that one obtains a nontrivial formulation of cosmology in terms of collective phenomena.

## 7. THERMODYNAMICS AND QUANTUM GRAVITY - II

We sketch here some ideas from [**93, 354, 353, 355**] (cf. also [**171, 509, 750, 831, 860, 899**]). The main theme is that the gravitational space time of general relativity (GR) can be described as an effect of the thermodynamics of quantum states of the spacetime. One eliminates the superspace (deWitt) metric of the (quantum) Wheeler-deWitt (WDW) equation and refers all variations to the determinant $h$ of the 3-D $h_{ij}$. The WDW wave function $\psi$ is then treated as a 1-D Bose field associated with a 3-D Riemannian manifold. The first part (transformation of the WDW equation) is fairly straightforward but there are some delicate and "visionary" aspects as one goes along which we can only partially capture here - there are updated versions of a number of the articles in [**353, 354, 355**], up to versions 5 and 6, which we recommend for detailed study. Thus first write the 4-D GR action in the form

$$S[g] = -\frac{1}{3}\int_{\partial M} d^3x\sqrt{h}K[h] + \int_M d^4x\sqrt{-g}\left[-\frac{1}{6}R[g]+\frac{\Lambda}{3}+L\right] \tag{7.1}$$

where $\partial M$ is the boundary of a pseudo-Riemannian manifold M and $K[h] = Tr(K_{ij})$ is the extrinsic curvature (cf. [**509, 610**] where one defines $K_{\mu\nu} = (1/2)\mathfrak{L}_{\mathbf{n}}h_{\mu\nu} = h^{\rho}_{\mu}\nabla_{\rho}n_{\nu}$ with $\mathbf{n}$ is the unit normal vector and $\mathfrak{L}$ the Lie derivative). Then via (**7A**) $\delta S[g]/\delta g_{\mu\nu} = 0$ with $\delta S[g]|_{\partial M} = 0$ one arrives at the Einstein equations for GR (**7B**) $G_{\mu\nu} + \Lambda g_{\mu\nu} = 3T_{\mu\nu}$ where $G_{\mu\nu} = R_{\mu\nu} - (R/2)g_{\mu\nu}$ (one is using units where $8\pi G/3 = c = \hbar = k_B = 1$) and (**7C**) $T_{\mu\nu} = (2/\sqrt{-g})[\delta(\sqrt{-g}\mathfrak{L})/\delta g^{\mu\nu}$. One can also write the Einstein equations as

$$R_{\mu\nu} = \Lambda g_{\mu\nu} + 3\left(T_{\mu\nu} - \frac{1}{2}Tg_{\mu\nu}\right) \tag{7.2}$$

where $T = g^{\mu\nu}T_{\mu\nu}$.

Supposing that the boundary surface is a constant time surface one uses an ADM form

$$ds^2 = g_{\mu\nu}dx^\mu dx^\nu = -N^2dt^2 + h_{ij}(dx^i + N^i dt)(dx^j + N^j dt) \tag{7.3}$$
$$= -(N^2 - N_iN^i)dt^2 + 2N_i dx^i dt + h_{ij}dx^i dx^j \quad (N_i = h_{ij}N^j)$$

where

$$h_{ij} = g_{ij};\ N = \frac{1}{\sqrt{-g^{00}}};\ N_i = g_{0i};\ K_{ij} = \frac{1}{N}\left(-\frac{1}{2}\partial_t h_{ij} + \nabla_j N_i\right) \tag{7.4}$$

This can be written as

$$g_{\mu\nu} = \begin{bmatrix} -N^2 + N_iN_i & N_i \\ N_i & h_{ij} \end{bmatrix} \tag{7.5}$$

The Hamiltonian constraint for GR in the ADM decomposition is

$$H = \frac{\delta S}{\delta N} = \sqrt{h}[K^2 - K_{ij}K^{ij} + R[h] - 2\Lambda - 6T_{nn}] = 0 \tag{7.6}$$

where $\rho = T_{nn} = n^\mu n^\nu T_{\mu\nu}$ is the normal component of the stress-energy tensor. The canonical conjugate momenta conjugate are (**7E**) $\hat{\pi}^{ij} = -i(\delta/\delta h_{ij})$, $\pi^j = -i(\delta/\delta N_j)$, and $\hat{\pi} = -i(\delta/\delta N)$ and the classical geometro-dynamics is given via the ADM Hamiltonian constraint (7.6) and is quantized by the Dirac conditions (**7F**) $i[\hat{\pi}^{ik}, h_{kj}] = \delta^i_j$; this leads to the Wheeler-deWitt (WDW) equation (cf. [**170, 171, 509, 750, 831**])

$$\left[-G_{ijk\ell}\frac{\delta}{\delta h_{ij}}\frac{\delta}{\delta h_{k\ell}} + \sqrt{h}(R[h] - 2\Lambda - 6T_{nn}\right]\psi[h] = 0 \tag{7.7}$$

where $G_{ijk\ell} = (1/2\sqrt{h})(h_{ik}h_{j\ell} + h_{i\ell}h_{jk} - h_{ij}h_{k\ell})$ is the so called supermetric. A lovely way to eliminate the supermetric is contrived as follows. Write (**7G**) $\delta h = hh^{ij}\delta h_{ij}$ (**7H**) $(\delta/\delta h_{ij}) = hh^{ij}(\delta/\delta h)$ leading to

$$\frac{\delta}{\delta h_{ij}}\frac{\delta}{\delta h_{k\ell}}\psi[h] = hh^{ij}\frac{\delta}{\delta h}\left(hh^{k\ell}\frac{\delta}{\delta h}\right)\psi[h] \tag{7.8}$$
$$= hh^{ij}\left(h^{k\ell}\frac{\delta}{\delta h} + h\frac{\delta h^{k\ell}}{\delta h}\frac{\delta}{\delta h} + hh^{k\ell}\frac{\delta^2}{\delta h^2}\right)\psi[h]$$

Note here that (**7G**)-(**7H**) follow from calculations in [**269, 388**] involving determinants and cofactors. Then using (**7I**) $\delta h^{ij} = \delta\left(\frac{1}{h_{ij}}\right) = -\frac{\delta h_{ij}}{(h_{ij})^2}$ one obtains (**7J**) $\delta h^{ij} = -(h^{ij}/h)\delta h \Rightarrow (\delta h^{ij}/\delta h) = -(h^{ij}/h)$ leading to

$$\frac{\delta}{\delta h_{ij}}\frac{\delta}{\delta h_{k\ell}}\psi[h] = h^2h^{ij}h^{k\ell}\frac{\delta^2\psi[h]}{\delta h^2} \tag{7.9}$$

This allows us to eliminate the supermetric via

$$G_{ijk\ell}h^{ij}h^{k\ell} = \frac{1}{2\sqrt{h}}(h_{ik}h_{j\ell} + h_{i\ell}h_{jk} - h_{ij}h_{k\ell})h^{ij}h^{k\ell} = -\frac{3}{2\sqrt{h}} \tag{7.10}$$

and for the WDW equation one can utilize the relation

$$-G_{ijk\ell}\frac{\delta^2\psi[h]}{\delta h_{ij}\delta h_{k\ell}} = \frac{3h^2}{2\sqrt{h}}\frac{\delta^2\psi[h]}{\delta h^2} \tag{7.11}$$

leading to the WDW equation in the form

$$\left[\frac{3h^2}{2\sqrt{h}}\frac{\delta^2}{\delta h^2}+\sqrt{h}(R[h]-2\Lambda-6T_{nn})\right]\psi[h]=0 \tag{7.12}$$

This can be rewritten in the form of a functional 2-D Klein-Gordon-Fock equation for the classical massive Bose field $\psi[h]$, namely (**7K**) $[(\delta^2/\delta h^2)+m^2[h]]\psi[h]=0$ where

$$\frac{2}{3h}(R[h]-2\Lambda-6T_{nn})\equiv m^2[h] \tag{7.13}$$

is understood as a mass squared term for the classical bosonic field $\psi[h]$.

**7.1. BOSON FIELDS AND QUANTIZATION.** One begins with (**7K**) $[(\delta^2/\delta h^2)+m^2[h]]\psi[h]=0$ as the Euler-Lagrange (EL) equation of motion from some field theory Lagrangian $L[\psi[h],\delta\psi[h]/\delta h]$ according to the equations

$$\frac{\delta\Pi_\psi[h]}{\delta h}-\frac{\partial}{\partial\psi[h]}L\left[\psi[h],\frac{\delta\pi[h]}{\delta h}\right]=0; \tag{7.14}$$

$$\Pi_\psi[h]-\frac{\partial}{\partial(\delta\psi[h]/\delta h)}L\left[\psi[h],\frac{\delta\psi[h]}{\delta h}\right]=0$$

One can construct the appropriate Lagrangian $S[\psi]$ directly by using (**7K**) in the form

$$S[\psi]=-\frac{1}{2}\int\delta h\psi[h]\left[\frac{\delta^2}{\delta h^2}+m^2[h]\right]\psi([h]) \tag{7.15}$$

$$=-\frac{1}{2}\int\delta h\left[\frac{\delta}{\delta h}\left(\psi[h]\frac{\delta\psi[h]}{\delta h}\right)-\left(\frac{\delta\psi[h]}{\delta h}\right)^2+m^2[h]\psi^2[h]\right]$$

$$=\frac{1}{2}\int\delta h\left[\left(\frac{\delta\pi[h]}{\delta h}\right)^2-m^2[h]\psi^2[h]\right]=\int\delta hL\left[\psi[h],\frac{\delta\psi[h]}{\delta h}\right]$$

where one has integrated a divergence and $\psi^\dagger[h]=\psi[h]$. This yields a classical system of EL equations

$$\frac{\delta\Pi_\psi[h]}{\delta h}+m^2[h]\psi[h]=0;\ \Pi_\psi[h]-\frac{\delta\psi[h]}{\delta h}=0 \tag{7.16}$$

The classical field theory Hamiltonian can be constructed immediately as

$$H[\Pi_\psi,\psi]=\Pi_\psi[h]\frac{\delta\psi[h]}{\delta h}-L\left[\psi[h],\frac{\delta\psi[h]}{\delta h}\right]=\frac{1}{2}(\Pi_\psi^2[h]+m^2[h]\psi^2[h]) \tag{7.17}$$

One can rewrite (7.17) as a definition of mass

$$m^2[h]\psi^2[h]=2H[h]-\left(\frac{\delta\psi[h]}{\delta h}\right)^2 \tag{7.18}$$

and after elimination of the mass squared by using (**7K**) this leads to the functional differential equation for $\psi[h]$, namely

$$\frac{\delta^2\psi[h]}{\delta h^2}\psi[h]=2H[h]-\left(\frac{\delta\psi[h]}{\delta h}\right)^2 \tag{7.19}$$

After collecting terms this leads to

$$\frac{\delta}{\delta h}\left(\frac{\delta\psi[h]}{\delta h}\psi[h]\right)=2H[h] \tag{7.20}$$

Integration gives then formally

$$\psi^2[h]=4\int_{h_I}^{h}\delta h'\int_{h_I}^{h'}\delta h''h[h''] \tag{7.21}$$

and this gives a solution of the classical wave equation (**7K**) in the form $\psi[h]=(\psi^2[h])^{1/2}$. From the other side of (7.19) integration gives

$$\Pi_\psi[h]\psi[h]=2\int_{h_I}^{h}\delta h'H[h'] \tag{7.22}$$

which, combined with the solution of (7.21) fixes values of the canonical conjugate momentum $\Pi_\psi[h]$ with respect to values of $H[h]$ via

$$\Pi_\psi[h]=\frac{\int_{h_I}^{h}\delta h'H[h']}{(\int_{h_I}^{h}\delta h'\int_{h_I}^{h'}\delta h''H[h''])^{1/2}} \tag{7.23}$$

We refer to [**353, 354, 355, 357**] for further calculations in the classical context but mention here that one can arrive at a reduced form of (7.16)-(7.17)

$$\frac{\delta}{\delta h}\begin{pmatrix}\psi\\ \Pi_\psi\end{pmatrix}=\begin{pmatrix}0 & 1\\ -m^2[h] & 0\end{pmatrix}\begin{pmatrix}\psi\\ \Pi_\psi\end{pmatrix};\ \Phi_\mu=\begin{pmatrix}\psi\\ \Pi_\psi\end{pmatrix};\ \partial_\nu=\begin{pmatrix}\delta/\delta h\\ 0\end{pmatrix} \tag{7.24}$$

which can be presented in a form similar to the Dirac equation, namely (**7L**) $(i\Gamma^\mu\partial_\nu-M^\mu_\nu)\Phi_\mu=0$ $(\Gamma^\mu=[-iI_1,O_2]$ with mass matrix

$$M^\mu_\nu=\begin{pmatrix}0 & -1\\ -m^2 & 0\end{pmatrix}\geq 0 \tag{7.25}$$

Note also (**7M**) $[\Gamma^\mu,\Gamma^\nu]=2\eta^{\mu\nu}I_2$ where $I_2$ is the 2-D unit matrix and $O_2$ is the 2-D null matrix with $\eta^{\mu\nu}=diag(-1,0)$. Note that the classical field theory Hamiltonian given by (7.17) can be written in matrix form as

$$H[h]=[\psi,\Pi_\psi]\begin{pmatrix}\alpha & \beta\\ \gamma & \delta\end{pmatrix}\begin{pmatrix}\psi\\ \Pi_\psi\end{pmatrix}=\Phi^\dagger_\mu H^{\mu\nu}\Phi_\nu \tag{7.26}$$

where $\alpha$, $\beta$, $\gamma$, $\delta$ are functionals of $h$. From a classical point of view (7.26) can be written as

$$H[h]=\alpha\pi^2+\delta\Pi^2_\psi+\gamma\Pi_\psi\psi+\beta\psi\Pi_\psi;\ \alpha=\frac{1}{2},\ \beta=\gamma=0;\ \delta=\frac{1}{2}m^2[h] \tag{7.27}$$

**7.2. FOCK SPACE CONSTRUCTIONS.** One now quantizes the reduced Klein-Gordon-Fock field equation (**7L**) in a standard manner - we refer here to [**74, 110, 354, 353, 355, 697**] for second quantization and Fock spaces and will follow at first the development of the second paper in [**354**]. Thus one looks for the field quantization of the equation (**7L**), namely

$$\Phi[h]\to\mathbf{\Phi}[h]\Rightarrow(i\mathbf{\Gamma}\vec{\partial}-\mathbf{M})\mathbf{\Phi}[h]=0;\ \psi\to\mathbf{\Psi};\ \Pi_\psi\to\mathbf{\Pi_\Psi} \tag{7.28}$$

($\mathbf{M} \sim (\mathbf{M}^{\mu}_{\nu})$) where canonical commutation rules appropriate for Bose statistics will hold, namely

$$[\mathbf{\Pi}_{\Psi}[h'], \mathbf{\Psi}[h]] = -i\delta(h'-h);\ [\mathbf{\Pi}_{\Psi}[h'], \mathbf{\Pi}_{\Psi}[h]] = 0;\ [\mathbf{\Psi}[h'], \mathbf{\Psi}[h]] = 0 \tag{7.29}$$

Via [**110**] for example one knows that the field operator $\mathbf{\Phi}[h]$ in (7.28) can be represented in the Fock space of annihilation and creation functional operators as (**7M**) $\mathbf{\Phi}[h] = \mathbf{Q}[h]\mathbf{B}[h]$ where $\mathbf{B}[h]$ is a dynamical basis in the Fock space

$$\mathbf{B}[h] = \left\{ \left[ \begin{array}{c} \mathbf{G}[h] \\ \mathbf{G}^{\dagger}[h] \end{array} \right] : [\mathbf{G}[h], \mathbf{G}^{\dagger}[h]] = \delta(h'-h);\ [\mathbf{G}[h'], \mathbf{G}[h]] = 0 \right\} \tag{7.30}$$

where

$$\mathbf{Q}[h] = \begin{pmatrix} \frac{1}{\sqrt{2|m[h]|}} & \frac{1}{\sqrt{2|m[h]|}} \\ -i\sqrt{|m[h]|/2} & i\sqrt{|m[h]|/2} \end{pmatrix} \tag{7.31}$$

In this way the operator equation (7.28) becomes the equation for a basis $\mathbf{B}[h]$

$$\frac{\delta \mathbf{B}[h]}{\delta h} = \begin{pmatrix} -im[h] & \frac{1}{2m[h]}\frac{\delta m[h]}{\delta h} \\ \frac{1}{2m[h]}\frac{\delta m[h]}{\delta h} & im[h] \end{pmatrix} \mathbf{B}[h] \tag{7.32}$$

Here there is a nonlinearity given via coupling between annihilation and creation operators present as nondiagonal terms in (7.32) so in order to solve such equations one imagines a new basis $\mathbf{B}'[h]$ in the Fock space

$$\mathbf{B}'[h] = \left\{ \begin{pmatrix} \mathbf{G}'[h] \\ \mathbf{G}'^{\dagger}[h] \end{pmatrix} : [\mathbf{G}'[h'], \mathbf{G}'^{\dagger}[h]] = \delta(h'-h);\ [\mathbf{G}'[h'], \mathbf{G}'[h]] = 0 \right\} \tag{7.33}$$

for which the Bogoliubov transformation

$$\mathbf{B}'[h] = \begin{pmatrix} u(h) & v[h] \\ v^*[h] & u^*[h] \end{pmatrix} \mathbf{B}[h];\ |u[h]|^2 - |v[h]|^2 = 1 \tag{7.34}$$

and the Heisenberg evolution

$$\frac{\delta \mathbf{B}'[h]}{\delta h} = \begin{pmatrix} -i\lambda[h] & 0 \\ 0 & i\lambda[h] \end{pmatrix} \mathbf{B}'[h] \tag{7.35}$$

both hold. This will be called the Fock-Bogoliubov-Heisenberg basis (FBH) and the diagonalization procedure (7.34)-(7.35) converts the operator basis evolution (7.32) into the Bogoliubov coefficients basis evolution

$$\frac{\delta}{\delta h}\begin{pmatrix} u[h] \\ v[h] \end{pmatrix} = \begin{pmatrix} -im[h] & \frac{1}{2m[h]}\frac{\delta m[h]}{\delta h} \\ \frac{1}{2m[h]}\frac{\delta m[h]}{\delta h} & im[h] \end{pmatrix} \begin{pmatrix} u[h] \\ v[h] \end{pmatrix} \tag{7.36}$$

and the basis $\mathbf{B}'[h]$ takes the meaning of a static operator basis associated with initial data

$$\mathbf{B}'[h] = \mathbf{B}_I = \left\{ \begin{pmatrix} \mathbf{G}_I \\ \mathbf{G}_I^{\dagger} \end{pmatrix} : [\mathbf{G}_I, \mathbf{G}_I^{\dagger}] = 1;\ [\mathbf{G}_I, \mathbf{G}_I] = 0 \right\} \tag{7.37}$$

within which a vacuum state can be correctly defined via (**7N**) $|0>_I = \{|0>_I :$ $\mathbf{G}_I|p>_I = 0;\ 0 = {}_I<0|\mathbf{G}_I^{\dagger}\}$. Thus the integrability problem involves the equations (7.36). However the Bogoliubov coefficients are additionally constrained by the

hyperbolic rotation conditions (**7O**) $|u[h]|^2-|v[h]|^2=1$ from (7.34). It is therefore useful to apply the superfield parametrization with solutions
(7.38)

$$u[h]=\frac{1+\mu[h]}{2\sqrt{\mu[h]}}exp\left(im_I\int_{h_I}^{h}\frac{\delta h'}{\mu[h']}\right);\ v[h]=\frac{1-\mu[h]}{2\sqrt{\mu[h]}}exp\left(-im_I\int_{h_I}^{h}\frac{\delta h'}{\mu[h']}\right)$$

where $\mu[h]$ is a mass scale (**7P**) $\mu[h]=(m_I/m[h])$. This establishes the relation between a dynamical basis $\mathbf{B}[h]$ and the initial data FBH basis $\mathbf{B}_I$ via (**7Q**) $\mathbf{B}[h]=\mathcal{G}[h]\mathbf{B}_I$ where

$$\mathcal{G}[h]=\begin{pmatrix}\frac{\mu[h]+1}{2\sqrt{\mu[h]}}exp(-i\theta[h] & \frac{\mu[h]-1}{2\sqrt{\mu[h]}}exp(i\theta[h]\\ \frac{\mu[h]-1}{2\sqrt{\mu[h]}}exp(-i\theta[h]) & \frac{\mu[h]+1}{2\sqrt{\mu[h]}}exp(i\theta[h])\end{pmatrix} \tag{7.39}$$

where $i\theta[h]$ is given by a phase of (7.38). Hence the solution of (7.32) can be expressed in the initial data basis as (**7R**) $\mathbf{\Phi}[h]=\mathbf{Q}[h]\mathcal{G}[h]\mathbf{B}_I=\mathbf{G}[h]\mathbf{B}_I$. Thus for $m_I=m[h_I]$ with

$$\theta[h]=\pm m_I\int_{h_I}^{h}\sqrt{\left|\frac{m^2[h]}{m_I^2}\right|}\delta h;\ \phi[h]=log\left[\left|\frac{m^2[h]}{m_I^2}\right|\right]^{1/4} \tag{7.40}$$

and (7.37) can be solved for the Bogoliubov coefficients via Cayley-Hamilton, etc. One concludes that the quantum theory of gravitation is completely determined by the correct choice of the monodromy matrix between dynamic and static bases in the Fock space of creation and annihilation functional operators. This means that the Fock formulation of quantum gravitation is determined by the space geometry of the Riemannian manifold given by a solution of the Einstein equations since this is what determines the monodromy matrix (cf. [**354**] for more on this and calculation of fields etc.).

## 8. THERMODYNAMICS AND QUANTUM GRAVITY - III

We consider here [**353, 355**] for more on quantum cosmology and bosonic strings. There is some repetition from Sections 6 and 7 and in particular the Hamiltonian in (6.39) is used here for simplicity (no superfluidity term); this leads to

$$H=\begin{pmatrix}\frac{1}{2}(|u|^2+|v|^2)\omega & -uv\omega\\ -u^*v^*\omega & \frac{1}{2}(|u|^2+|v|^2)\omega\end{pmatrix} \tag{8.1}$$

Then, understanding internal energy as the average $U=[Tr(\rho H)/Tr(\rho)]$, one has

$$U=\left[\frac{1}{2}+\frac{4n+3}{2n+1}n\right]\omega(a) \tag{8.2}$$

where

$$n=<0|\mathbf{G}^{\dagger}\mathbf{G}|0>=\frac{1}{4}\left|\sqrt{\left|\frac{\omega(a_I)}{\omega(a)}\right|}-\sqrt{\left|\frac{\omega(a)}{\omega(a_I)}\right|}\right|^2 \tag{8.3}$$

which is the number of particles produced from the vacuum. The averaged number of particles is (**8A**) $<n>=2n+1$ and the chemical potential $\mu=\partial U/\partial n$ is then

$$\mu=\left(1+\frac{1}{(2n+1)^2}-\frac{1}{2}\frac{4n+1}{4n^2+2n}\sqrt{\frac{n}{n+1}}\right)\omega(a) \tag{8.4}$$

The system is characteried by CCR's and leads to an entropy

$$S=-\frac{Tr(\rho log(\rho)}{Tr(\rho)}=log\left[\frac{1}{2|u|^2-1}\right]=\left[exp\left(\frac{U-\mu n}{T}\right)-1\right]^{-1} \tag{8.5}$$

This leads to a temperature formula (cf. (6.62))

$$T=\frac{1+\left(\frac{2n}{2n+1}\right)^2+\frac{8n^2+8n+1}{4n+2}\sqrt{\frac{n}{n+1}}}{2log(2n+2)}\omega(a) \tag{8.6}$$

**8.1. QUANTUM COSMOLOGY.** We go now to [**355**] where the theory is looked at from the viewpoint of bosonic strings (cf. [**716, 912**]) with the fundamental mass groundstate excitation a hypothetical particle with negative mass squared and velocity greater than light speed $c$ (a tachyon). This leads to a so-called extremal tachyon mass model for description of the universe. One takes the standard Einstein-Hilbert action (6.28) and the Friedmann metric (**6A**). Using the ADM framework the universe is then unambiguously characterized by the constraints (**8A**) $p_a^2-\omega^2=0$ where (**8C**) $p_a=-2V_0(da/d\eta)$ and (**8D**) $\omega=2V_0\sqrt{\rho(a)}$ with $\rho(a)=(a^4/V_0)\int d^3x\mathcal{H}(x)$ as before. One sees that the Dirac Hamiltonian primary constraints (**8B**) with the formal correspondence $m^2\to-\omega^2$ can be rewritten in the form (**8E**) $p_a^2+m^2=0$ that describes the primary constraints for the free bosonic string with mass $m$. Since the mass squared is negative one is dealing with a tachyon as above. Recall also $d\eta=dt/a(t)$ with $t=\tau+x^0$ ($\tau=constant$) and $H(a)=(1/a)(da/dt)$. One looks now at the identification (**8F**) $H^2(a)=\rho(a)/a^4=(1/V_0)\int d^3x\mathcal{H}(x)$ leading to (**8G**) $m=m(a)=2iV_0a^2H(a)$ and recall the Hubble law which is a nontrivial additional condition for the tachyon mass, namely

$$\frac{1}{i(t-t_I)}\int_{a_I^2}^{a^2}\frac{dy}{m(y)}=\frac{1}{V_0} \tag{8.7}$$

where $y\sim a^2$ and $I$ refers to initial data. One recalls the equal time commutation relation $i[p_a,a]=1\Rightarrow p_a=(1/i)(\partial/\partial a)$ and this leaves the classical field equations in the form

$$\partial_a\begin{pmatrix}\mathbf{\Psi}\\ \mathbf{\Pi}_\psi\end{pmatrix}=\begin{pmatrix}0 & 1\\ m^2 & 0\end{pmatrix}\begin{pmatrix}\mathbf{\Psi}\\ \mathbf{\Pi}_\psi\end{pmatrix} \tag{8.8}$$

where $\mathbf{\Psi}$ is the WDW wave function (cf. [**407, 408**]).

Now one must realize CRR (**8H**) $[\mathbf{\Pi}_\mathbf{\Psi}[a],\mathbf{\Psi}[a']]=-i\delta_{aa'}$ where $a=a(\eta)$ and $a'=a(\eta')$ and this leads as in (6.43) to second quantization via

$$\begin{pmatrix}\mathbf{\Psi}[a]\\ \mathbf{\Pi}_\mathbf{\Psi}[a]\end{pmatrix}=\begin{pmatrix}\frac{1}{\sqrt{2\omega}} & \frac{1}{\sqrt{2\omega}}\\ -i\sqrt{\frac{\omega}{2}} & i\sqrt{\frac{\omega}{2}}\end{pmatrix}\begin{pmatrix}\mathbf{G}[a]\\ \mathbf{G}^\dagger[a]\end{pmatrix} \tag{8.9}$$

which realizes (**8H**) if and only if $\mathbf{G}$ and $\mathbf{G}^\dagger$ create a bosonic type dynamical operator basis as in (7.30) (with h replaced by a), i.e.

$$\mathbf{B}_a = \left[\begin{pmatrix} \mathbf{G}[a] \\ \mathbf{G}^\dagger[a] \end{pmatrix} : [\mathbf{G}[a], \mathbf{G}^\dagger[a']] = \delta_{aa'};\ [\mathbf{G}[a], \mathbf{G}[a']] = 0\right] \tag{8.10}$$

Evolution in this basis is governed by quantized classical field theory

$$\frac{\partial}{\partial a}\begin{pmatrix} \mathbf{G}[a] \\ \mathbf{G}^\dagger[a] \end{pmatrix} = \begin{pmatrix} -m & \frac{1}{2m}\frac{\partial m}{\partial a} \\ \frac{1}{2m}\frac{\partial m}{\partial a} & m \end{pmatrix}\begin{pmatrix} \mathbf{G}[a] \\ \mathbf{G}^\dagger[a] \end{pmatrix} \tag{8.11}$$

As shown before in Sections 6-7 this is fully integrable under certain conditions. Namely one must use the Bogoliubov-Heisenberg basis, diagonalization, etc. Thus for

$$\mathbf{B}_0 = \left[\begin{pmatrix} w \\ w^\dagger \end{pmatrix} : [w, w^\dagger] = 1;\ [w, w] = 0\right] \tag{8.12}$$

one has (**8I**) $\mathbf{B}_a = \mathcal{M}(a)\mathbf{B}_0$ with monodromy matrix

$$\mathcal{M}(a) = \begin{bmatrix} \left(\sqrt{\left|\frac{m}{m_I}\right|} + \sqrt{\left|\frac{m_I}{m}\right|}\right)\frac{e^{\lambda}}{2} & \left(\sqrt{\left|\frac{m}{m_I}\right|} - \sqrt{\left|\frac{m_I}{m}\right|}\right)\frac{e^{-\lambda}}{2} \\ \left(\sqrt{\left|\frac{m}{m_I}\right|} - \sqrt{\left|\frac{m_I}{m}\right|}\right)\frac{e^{\lambda}}{2} & \left(\sqrt{\left|\frac{m}{m_I}\right|} + \sqrt{\left|\frac{m_I}{m}\right|}\right)\frac{e^{-\lambda}}{2} \end{bmatrix} \tag{8.13}$$

where (**8J**) $\lambda = \lambda(a) = \pm\int_{a_I}^{a} m da$ is the integrated mass of the free bosonic string.

As for the tachyon thermodynamics one obtains from [**355**]

(1) Occupation number $n = <0|\mathbf{G}^\dagger\mathbf{G}|0> = (1/4)|\sqrt{|m/m_I|} - \sqrt{|m_I/m|}|^2$ with $<n> = 2n+1$
(2) Entropy $S = S(a) = -log(2n+1)$
(3) Internal energy $U = U(a) = [(1/2) + n(4n+3)/(2n+1)]|m|$ (cf. (**8A**))
(4) Chemical potential (cf. (8.4) with $\omega(a)$ replaced by $|m|$.
(5) Temperature (cf. (8.6) with $\omega(a)$ replaced by $|m|$.

Next the extremal tachyon mass model as a special case of the above involves (**8K**) $\lambda = \lambda(a) = \int_{a_I}^{a} m(a)da$ and via (**8I**) $\lambda$ can be understood as the Weyl characteristic scale for the universe, which means that the scale must fulfill the d'Alembert equation (**8L**) $\Delta\lambda = 0$ (presumably $\Box\lambda = 0$) which in the 1-D situation considered means (**8M**) $\partial_a^2\lambda(a) = 0$ so in terms of the tachyon mass $\partial_a m(a) = 0 \Rightarrow m(a) = m_I(a) = m_I$. Using (**8G**) then produces the conclusion that the Hubble evolution parameter for the tachyon must have the form (**8N**) $H(a) = Q/a^2$ where $Q = m_I/2iV_0 = a_I^2 H(a_I)$ is a constant. This Hubble parameter will be called extremal. We omit a brief discussion of dark matter and go to the so-called extremal thermodynamics. Thus the solution of the Dirac constraints for the model (**8N**) has a form ($\Omega_R$ refers to radiation density)

$$a(t) = a_I\sqrt{1 + 2H_0\sqrt{\Omega_R}|t - t_I|} \tag{8.14}$$

leading to

$$\lambda(t) = m_I a_I\left|\sqrt{1 + 2H_0\sqrt{\Omega_R}|t - t_I|} - 1\right| \tag{8.15}$$

Here there is a simple form of operator evolution

$$\left(\begin{array}{c}\mathbf{G}(t)\\ \mathbf{G}^{\dagger}(t)\end{array}\right)=\left(\begin{array}{cc}exp(-\lambda t) & 0\\ 0 & exp(\lambda t)\end{array}\right)\left(\begin{array}{c}w\\ w^{\dagger}\end{array}\right) \tag{8.16}$$

and via (**8N**) (which implies $n=0$) one obtains

$$T=\frac{|m_I|}{2log(2)}=V_0\frac{a_I^2H_0\sqrt{\Omega_R}}{log(2)} \tag{8.17}$$

Thus in spite of a finite constant value of $T$ the value of the Boltzman entropy is $S=0$ but $U=(1/2)|m_I|$ and (**8O**) $\mu(|m|=|m_I|)=3|m_I|$. This apparently means that the extremal tachyon mass model describes the universe as an open quantum system with a stabilized number of quantum states produced from the vacuum state. This model is claimed to be the minimal model for the theory given by the monodromy matrix $\mathcal{M}$. It is suggested that the approach here provides the correct definition of many-particle quantum cosmology (perhaps modulo bosonic string dynamics and second quantization).

**8.2. PARTIAL SUMMARY.** Going to [**355**] now for a partial summing up (note this is superceded by some of the material in Section 7 but it gives a limited summation of material in Sections 6 and 8 - with obvious connections to Section 7). Thus first one looks at GR in the form

$$S_{EH}=\int d^4x\sqrt{-g}\left(-\frac{1}{6}R+L\right);\ \frac{\delta S_{EH}}{\delta g^{\mu\nu}}=0\Rightarrow R_{\mu\nu}-\frac{1}{2}Rg_{\mu\nu}=3T_{\mu\nu} \tag{8.18}$$

One looks at a Riemannian manifold with FLRW metric or the form (**8P**) $ds^2=g_{\mu\nu}dx^{\mu}dx^{\nu}=(dx^0)^2-a^2(t)(dx^i)^2$ which can also be written in ADM form with (**8Q**) $d\eta=N(x^0)dx^0=dt/a(t)$ with $t=\tau+x^0$ and $\tau$ constant ($t$ is called cosmological time and $\eta$ is conformal time). The ADM metrics in (**8O**) lead to Hamiltonian constraints (**8R**) $H=p_a^2-4V_0^2a^4H^2(a)=0$ where

$$p_a=-2V_0\frac{da}{d\eta};\ H^2(a)=\left(\frac{\dot{a}}{a}\right)^2=\left(\frac{1}{a^2}\frac{da}{d\eta}\right)^2=\frac{1}{V_0}\int d^3x\mathcal{H}(x) \tag{8.19}$$

The Hubble law results from (**8R**) in the form (**8S**) $\int_{a_I}^{a}[da'/a'H(a')]=t_I-t$ where $I$ refers to initial data. The goal here is find a road from GR to the generalized thermodynamics of quantum states for a Riemannian manifold satisfying GR - such a result could be considered a a form of quantum gravity.

Consider (**8R**) and apply to this the identification (**8T**) $m^2(a)=-4V_0^2a^4H^2(a)$ connecting the geometrical object (**8P**) with a free bosonic string of negative squared mass (called a tachyon) defined via (**8U**) $p_a^2+m^2=E^2;\ E=0,\ m^2=m^2(a)\leq 0$. The Hubble law (**8S**) can be treated for the tachyon as a definition of spatial volume $V_0$ via (**8V**) $(\int_{a_I^2}^{a^2}[dy/|m(y)|])^{-1}\delta t=V_0$ where $\Delta t=t-t_I$. First quantization of (**8U**) gives the WDW equation for (**8P**)

$$i[p_a,a]=1\Rightarrow\left[-\frac{\partial^2}{\partial a^2}+m^2(a)\right]\Psi=0;\ \Psi=\Psi^* \tag{8.20}$$

One can consider this WDW equation as a 0+1 Klein-Gordon-Fock equation which treats a Riemannian manifold as a Bose field. Separation of variables gives then

$$(8.21)\qquad (i\Gamma^{\mu}\partial_{\nu} - M^{\mu}_{\nu})\Phi_{\mu} = 0;\ M^{\mu}_{\nu} = \begin{pmatrix} 0 & -1 \\ -m^2 & 0 \end{pmatrix} \geq 0$$

where the matrices $\Gamma^{\mu} = [\mathbf{0}_2, i\mathbf{I}_2]$ create a Clifford algebra $[\Gamma^{\mu}, \Gamma^{\nu}] = 2\eta^{\mu\nu}\mathbf{I}_2$ and $\Phi_{\mu} = (\mathbf{\Psi}, \mathbf{\Pi_{\Psi}})^T$ with $\partial_{\nu} = (0, -\partial_a)^T$. Quantization of (8.21) via (8.20) has a proposed form

$$(8.22)\qquad \begin{pmatrix} \hat{\mathbf{\Psi}}[a] \\ \hat{\mathbf{\Pi}}_{\mathbf{\Psi}}[a] \end{pmatrix} = \begin{pmatrix} \frac{1}{\sqrt{2|m(a)|}} & \frac{1}{\sqrt{2|m(a)|}} \\ -\frac{i}{\sqrt{|m(a)|/2}} & \frac{i}{\sqrt{|m(a)|/2}} \end{pmatrix} \begin{pmatrix} \mathbf{G}[a] \\ \mathbf{G}^{\dagger}[a] \end{pmatrix}$$

which automatically fulfills the CCR (**8W**) $[\hat{\mathbf{\Pi}}_{\mathbf{\Psi}}[a], \hat{\mathbf{\Psi}}[a']] = -i\delta(a-a')$ $(a = a(\eta),\ a' = a(\eta'))$. The Bose system is described by the dynamical basis in a Fock space (8.10) with evolution equation (8.11) and can be diagonalized to the Heisenberg form via $(|u|^2 - |v|^2 = 1)$

$$(8.23)\qquad \mathbf{B}'_a = \begin{pmatrix} u & v \\ v^* & u^* \end{pmatrix}\mathbf{B}_a;\ \partial_a\mathbf{B}'_a = \begin{pmatrix} -i\lambda & 0 \\ 0 & i\lambda \end{pmatrix}\mathbf{B}'_a$$

$$\mathbf{B}'_a = \left[\begin{pmatrix} \mathbf{G}'[a] \\ \mathbf{G}'^{\dagger}[a] \end{pmatrix} :\ [\mathbf{G}'[a], \mathbf{G}'^{\dagger}[a']] = \delta(a-a');\ [\mathbf{G}'[a], \mathbf{G}'[a']] = 0\right]$$

is some new basis; the Bogoliubov coefficients $u$, $v$ are then functions of $a$. This procedure simply gives $\lambda = 0$ and thus $\mathbf{G}'[a] = w_I = constant$ defines a stable vacuum $|0>_I, \{|0>_I\!: w_I|0>_I = 0,\ 0 = {}_I<0|w_I^{\dagger}\}$ in the static Fock space basis $\mathbf{B}'_a = \mathbf{B}_I$, where

$$(8.24)\qquad \mathbf{B}_I = \left[\begin{pmatrix} w_I \\ w_I^{\dagger} \end{pmatrix} :\ [w_I, w_I^{\dagger}] = 1;\ [w_I, w_I] = 0\right]$$

As a result one obtains (cf. (8.11))

$$(8.25)\qquad \partial_a \begin{pmatrix} u \\ v \end{pmatrix} = \begin{pmatrix} -im & \frac{1}{2m}\partial_a m \\ \frac{1}{2m}\partial_a m & im \end{pmatrix} \begin{pmatrix} u \\ v \end{pmatrix}$$

Then applying the hyperbolic parametrization this system can be solved in the form

$$(8.26)\qquad u(a) = \frac{1}{2}exp\left[\pm i\int_{a_I}^{a} mda\right]\left(\sqrt{\left|\frac{m}{m_I}\right|} + \sqrt{\left|\frac{m_I}{m}\right|}\right);$$

$$v(a) = \frac{1}{2}exp\left[\pm i\int_{a_I}^{a} mda\right]\left(\sqrt{\left|\frac{m}{m_I}\right|} - \sqrt{\left|\frac{m_I}{m}\right|}\right)$$

where $m = m(a)$ and $m_I = m(a_I)$. In this approach now quantum gravity is defined by the monodromy between bases in the Fock space, thus (cf. (8.13))

$$(8.27)\quad \mathbf{B}_a = \begin{bmatrix} \left(\sqrt{\left|\frac{m_I}{m(a)}\right|} + \sqrt{\left|\frac{m(a)}{m_I}\right|}\right)\frac{e^{-i\lambda(a)}}{2} & \left(\sqrt{\left|\frac{m_I}{m(a)}\right|} - \sqrt{\left|\frac{(a)}{m_I}\right|}\right)\frac{e^{i\lambda(a)}}{2} \\ \left(\sqrt{\left|\frac{m_I}{m(a)}\right|} - \sqrt{\left|\frac{m(a)}{m_I}\right|}\right)\frac{e^{-i\lambda(a)}}{2} & \left(\sqrt{\left|\frac{m_I}{m(a)}\right|} + \sqrt{\left|\frac{m(a)}{m_I}\right|}\right)\frac{e^{i\lambda(a)}}{2} \end{bmatrix}$$

where $\lambda(a) = \pm \int_{a_I}^{a} m(a')da'$. The field operator $\hat{\Psi}$ that represents the spacetime (**8P**) as a boson and defines quantum states of the manifold is

$$\hat{\Psi}[a] = \frac{1}{2m(a)}\sqrt{\frac{m_I}{2}}\left(e^{-i\lambda(a)}w_I + e^{i\lambda(a)}w_I^{\dagger}\right) \tag{8.28}$$

The initial data given by the basis (8.24) defines a thermal equilibrium state for the ensemble of quantum states of the manifold. One considers now one particle approximations defined by the density functional $\rho_G$ in the initial data basis

$$\rho_G = \mathbf{G}^{\dagger}[a]\mathbf{G}[a] = \mathbf{B}_a^{\dagger}\begin{pmatrix} 1 & 0 \\ 0 & 0 \end{pmatrix}\mathbf{B}_a \tag{8.29}$$

$$= \mathbf{B}_I^{\dagger}\begin{pmatrix} |u|^2 & -uv \\ -u^*v^* & |v|^2 \end{pmatrix}\mathbf{B}_I \equiv \mathbf{B}_I^{\dagger}\rho_{eq}\mathbf{B}_I$$

The equilibrium entropy $S$ and the partition function $\Omega_{eq}$ of the system in the initial data basis are

$$S = -\frac{Tr(\rho_{eq}log(\rho_{eq}))}{Tr(\rho_{eq})} \equiv log(\Omega_{eq});\ \Omega_{eq} = \frac{1}{2|u|^2 - 1} \tag{8.30}$$

Physical quantum states of the manifold under consideration are assumed to be described in the Gibbs ensemble. One can compute directly the entropy via (**8X**) $S = -log < n >$, $< n >= 2n + 1$ where $< n >$ is the averaged occupation number n of quantum states determined via

$$n = < 0|\mathbf{G}^{\dagger}[a]\mathbf{G}[a]|0 > = \frac{1}{4}\left(\left|\frac{m}{m_I}\right| + \left|\frac{m_I}{m}\right|\right) - \frac{1}{2} \tag{8.31}$$

which together with natural conditions $n \geq 0$ and $|m| \geq |m_I|$ leads to the energy spectrum

$$\left|\frac{m}{m_I}\right| = \left|\frac{mc^2}{m_I c^2}\right| = < n > + \sqrt{< n >^2 - 1} \tag{8.32}$$

The internal energy U and chemical potential $\mu$ are computed via proper averaging as

$$U = \left[1 + \left(2 + \frac{1}{< n >}\right)(< n > - 1)\right]\left(< n > + \sqrt{< n >^2 - 1}\right)\frac{|m_I|}{2}; \tag{8.33}$$

$$\mu = \left[1 + \frac{1}{< n >^2} - \left(2 - \frac{1}{< n >}\right)\sqrt{\frac{1}{< n >^2 - 1}}\right]\left(< n > + \sqrt{< n >^2 - 1}\right)|m_I|$$

From bosonic properties one can assume for equilibrium in Bose-Einstein statistics

$$\Omega_{eq} = \left(exp\left[\frac{U - \mu n}{T}\right] - 1\right)^{-1} \tag{8.34}$$

(see [**355**] for $T$). Then, maximal entropy quantum states (MEQS) of the Riemannian manifold are defined via (**8Y**) $S_{max} = 0 \Rightarrow n = 0$, $< n >= 1$, $m = m_I$, $U_{max} = (1/2)|m_I|$, $\mu_{max} = -\infty$, $T_{max} = [U_{max}/log(2)]$. This permits the conclusion that for the metrics the MEQS are fully determined by initial data. The last equation in (**8Y**) is an equation of state for MEQS that describes a classical ideal Bose gas. The infinite value of $\mu$ means that MEQS create closed systems and the point $n = 0$ is a phase transition point for MEQS. The initial data point can be interpreted as a birth-point of the Riemannian manifold. Measurement of MEQS formally defines the initial data point and the initial data reference frame. Furthermore

$$m_I = m = 2iV_0 a^2 H(a) \Rightarrow H(a) = \frac{Q}{a^2};\ Q = \frac{|m_I|}{2V_0} = \frac{U_{max}}{V_0} = a_I^2 H(a_I) \tag{8.35}$$

and for $U_{max} = p_{max} V_0$ one has a pressure $p_{max} = a_I^2 H(a_I)$. The form of the Hubble parameter in (8.35) means that the MEQS for the manifold describe primordial radiation and the initial data basis $\mathbf{B}_I$ is related to this light (CMB radiation). Further one sees that $\lambda$ in (8.27) is linear for MEQS and fulfills the d'Alembert equation

(8.36)

$$\lambda(a \int_{a_I}^{a} m(a')da'|_{m=m_I} = m_I(a - a_I) \Rightarrow \Delta\lambda = 0;\ a(t) = a_I\sqrt{1 + 2Q(t - t_I)}$$

and hence can be treated as the Weyl scale of the ensemble.

CHAPTER 6

# GEOMETRY AND MECHANICS

## 1. MORE ON EMERGENCE

In [**444, 445**] one provides a number of interesting examples concerning the emergence of quantum mechanics and relations to deterministic systems (cf. also [**105, 107, 106, 290**]). In particular Hooft sketches a result (in 0707.4568) suggesting that for any quantum system there exists at least one deterministic model which reproduces all the dynamics after "prequantization". The example deals with a finite dimensional system to simplify the argument and starts with a SE

$$\frac{d\psi}{dt} = iH\psi;\ H = \begin{pmatrix} H_{11} & \cdots & H_{1N} \\ \cdot & \cdot & \cdot \\ H_{N1} & \cdots & H_{NN} \end{pmatrix} \tag{1.1}$$

Now the dynamics can be reproduced by using a deterministic system with two degrees of freedom $\omega$ and $\phi$ with $\phi$ periodic ($\phi \in [0, 2\pi)$ or $\psi(\omega, \phi) = \psi(\omega, \phi + 2\pi)$) and satisfying the classical equations

$$\frac{d\phi(t)}{dt} = \omega(t);\ \frac{d\omega(t)}{dt} = -\kappa f(\omega) f'(\omega);\ f(\omega) = det(H - \omega) \tag{1.2}$$

The function $f(\omega)$ has zeros at the eigenvalues of H and $f'(\omega)$ has zeros between the zeros of $f$. A little picture is helpful in seeing that all eigenvalues $\omega_i$, $i = 1, \cdots, N$ are attractive zeros and, depending on the initial configuration, one of the eigenvalues of the matrix $H$ is rapidly approached. If $\omega_{i+(1/2)}$ is defined to be the zero of $f'(\omega)$ between $\omega_i$ and $\omega_{i+1}$ then the regions $\omega_{i-(1/2)} < \omega < \omega_{i+(1/2)}$ are the equivalence classes associated to the final state $\omega = \omega_i$. Since the convergence towards $\omega_i$ is exponential in time one can say that for all practical purposes the limit value $\omega_i$ is soon reached, at which point the system enters into a limit cycle wherein $\phi$ is periodic in time with period $T = 2\pi/\omega$. The prequantum states can be Fourier transformed in $\phi$ via (**1A**) $\psi(\omega, \phi) = \sum_n e^{in\phi}\psi_n(\omega)$ and as $t \to \infty$ one has (**1B**) $\psi(\omega, \phi, t) \to \sum_n \psi_n(\omega_i, 0) e^{in(\phi - \omega_i t)}$. In fact the quantum number $n$ is absolutely conserved so take $n = 1$ and the states $\psi_1(\omega_i)$ can be identified with the energy eigenstates of the original quantum system. Here interference is a formality since one is always free to consider probabilistic superpositions of different $\psi_n$.

To deal with information loss which seems essential in generating quantum mechanics one can look at a simple automaton having 4 states $\{1\} \to \{2\} \to \{3\} \to \{1\}$ and $\{4\} \to \{2\}$. Here $\{1\}$ and $\{4\}$ form one equivalence class, say $|1>$ and one obtains a unitary model in terms of (gauge) equivalence classes $|1> \to |2> \to |3> \to |1>$ (see [**444, 445**] for further discussion). Omitting here the need for a lower bound on the Hamiltonian one can argue that the energy eigenstates of a quantum system correspond to the limit cycles of the deterministic model. This seems to be a very deep connection in view of the Gutzwiller trace formula connecting energy eigenvalues to periodic orbits with underlying relations to number theory.

To continue this line we go to Elze (0710.2765) where one notes that the initial conditions for $\omega$ in (1.2) determine which eigenvalue $E_i$ is approached, resulting in a limit cycle for $\phi$ of period $T = 2\pi\omega_i^{-1} = 2\pi E_i^{-1}$. Introduce now two auxiliary operators (**1C**) $\hat{p}_\phi = -i(\partial/\partial\phi)$ and $\hat{p}_\omega = -i(\partial/\partial\omega)$ (which do not correspond to classically observable quantities). Define also the operator (**1D**) $\hat{h} = \omega\hat{p}_\phi - (\kappa/2)\{f(\omega)f'(\omega), \hat{p}_\omega\}$ where $\{x,y\} = xy + yx$. This leads to

$$\frac{d\phi(t)}{dt} = -i[\phi(t), \hat{h}];\ \frac{d\omega(t)}{dt} = -i[\omega(t), \hat{h}] \tag{1.3}$$

where $[x,y] = xy - yx$. Remarkably the generator $\hat{h}$ is Hermitian despite the dissipative character of the motion equations. The Hilbert space on which these operators act will have elements called prequantum states and one considers (cf. (**1A**)-(**1B**))

$$\psi(\phi,\omega,t) = \sum e^{in\phi}\psi_n(\omega,t) \to \sum e^{in(\phi-\omega_i t)}\psi_n(\omega_i,0)\ \ (t\to\infty) \tag{1.4}$$

where $\omega_i$ is the particular fixed point to which $\omega(t)$ is attracted, depending on initial conditions (the Fourier transform takes periodicity in the angular variable into account). One sees that in a superselection sector, where the quantum number $n$ is fixed to be a particular value $n'$, the prequantum states are related to the energy eigenstates via (**1E**) $exp(-iE_it')\psi(E_i) = exp(-in'\omega_i t)\psi_{n'}(\omega_i,0)$ (where $t' = n't$). Thus characteristic features of quantum systems described by (1.1) emerge from the dissipative evolution of deterministic systems. The prequantum approach here is designed in part to elucidate the appearance of the set of eigenvalues $\{E_i\}$ of the Hamiltonian in $det(H-\omega) = \prod(E_i - \omega)$ $(H \sim \hat{H})$ in the classical context of (1.2).

A complete characterization of the state of a quantum mechanical object with a finite number of degrees of freedom generally requires a set of simultaneous eigenvalues of a number of linearly independent and mutually commuting Hermitian operators $\hat{A}_n,\ n = 1,\cdots,N$ called "beables". A specific set of beables $(\hat{A}_1,\cdots,\hat{A}_N)^t$ can be interpreted as a set of coordinates of a point in an $N$-dimensional vector space. There is no absolute meaning attached to a given set of beables $\vec{A}$ compared to a second set $\vec{A}'$ obtained from $\vec{A}$ by the action of a group of symmetry transformations, given compatibility with Hilbert space properties. Since operators with zero eigenvalues are not excluded some of the $\hat{A}'s$ may not have inverses so one restricts symmetry transformations of beables into linear transformations $\hat{A}'_n = \sum_1^N M_{nm}\hat{A}_m,\ M \in GL(N,\mathbf{R})$. The quantum states evolve according to the SE (1.1) and one goes now to equations (**1F**) $(d\vec{\phi}(t)/dt) = \vec{\omega}$ for periodic $\vec{\phi} = (\phi_1,\cdots,\phi_n)^T$ with $\phi_n \in [0,2\pi]$ and $\vec{\omega} = (\omega_1,\cdots,\omega_N)^T$. First however one introduces an auxiliary real field $F$ on the space of matrices $M$ representing the symmetry group of beables, namely $M \in GL(N, \mathbf{R})$, with a wave equation

$$(\partial_t + \dot{\vec{\omega}} \cdot \partial_{M \cdot \vec{A}})(\partial_t - \dot{\vec{\omega}} \cdot \partial_{M \cdot \vec{A}})F(M,t) = 0 \tag{1.5}$$

where $\dot{\vec{\omega}} = d\vec{\omega}/dt$. The field depends on real numbers $A_n^j$, $n = 1, \cdots, N$ and $j = 1, \cdots, d$ which define the sets of simultaneous eigenvalues of $N$ commuting Hermitian operators denoted via $\vec{A} = (\hat{A}_1, \cdots, \hat{A}_N)^T$. This kind of dependence is visible in the general solution of the wave equation

$$F(M,t) = f(\vec{\omega}(t) - M \cdot \vec{A}) + g(\vec{\omega}(t) + M \cdot \vec{A}) \tag{1.6}$$

where $f$ and $g$ are two real functions of (respective combinations of Hilbert space operators) with suitable differentiability properties. A property of beables here is that related eigenvalues are invariant under unitary transformations in Hilbert space and hence one requires $F$ to be a scalar under such transformations. Furthermore the arguments $\vec{\omega}(t) \pm M \cdot \vec{A}$ in (1.6) transform covariantly with respect to beable symmetry, i.e. under transformations

$$\vec{A} \to \vec{A}' = S\vec{A};\ \vec{\omega} \to \vec{\omega}' = S\vec{\omega};\ M \to M' = SMS^{-1} \tag{1.7}$$

for $S \in GL(N, \mathbf{R})$. Next one wants an initial condition that breaks the $GL(N, \mathbf{R})$ symmetry (in order to describe a quantum mechanical object with a fixed set of beables - related specifically to $\vec{A}$) in the deterministic model. Thus one would like the solution F to be left invariant only by the subgroup of $GL(N, \mathbf{R})$ which effects permutations of the $A_n$'s - while remaining invariant under unitary transformations in Hilbert space. This leads one to the initial condition

$$F(M, t_0) = f(\vec{\omega}(t_0) - M \cdot \vec{A}) = \sum_{n=1}^{N} \prod_{j=1}^{d} [\sum_{m=1}^{N} M_{nm} A_m^j - \omega_n(t_0)]^2 \tag{1.8}$$

$$= \sum_{1}^{N} det^2 [M \cdot \vec{A} - \vec{\omega}(t_0)]_n$$

(where the determinant refers to a $d$-dimensional Hilbert space on which the operators act that are collected in $\vec{A}$). Note also that the sum of squares of determinants in (1.8) is zero if and only if the $N$-dimensional vector $\omega$ corresponds to one of the points of the $N$-dimensional finite lattice defined by the $d \times N$ numbers $A_n^j$ or, rather, by the $d$ eigenvalues of each one of the $N$ operators $M \cdot \vec{A}$. This initial condition represents a generalization of the function $f(\omega)$ in (1.2). Finally a boundary condition on the solution for the field $F$ is imposed via

$$\frac{d\omega}{dt} = -\kappa \frac{\partial}{\partial \vec{\omega}} F^2(M^*, t);\ \kappa > 0;\ M^* \in GL(N, \mathbf{R}) \tag{1.9}$$

($M^*$ fixed but arbitrary). This situation then determines $\vec{\omega}$ once its initial value is supplied and generalizes (1.2). It turns out that the zeros of $F$ (corresponding to points on the $d \times N$ lattice above) are attractive and hence $\vec{\omega}$ is attracted to a fixed point as $t \to \infty$

$$\omega_n(t) \to \sum_{1}^{N} M_{nm}^* A_m^{j(m)} = \omega_n^* \ \ (t \to \infty) \tag{1.10}$$

Further $F$ decays to zero on the boundary ($M = M^*$) and approaches a constant value $F(M, t \to \infty) = f(\vec{\omega}^* - M \cdot \vec{A})$ elsewhere. Through the zeros of $F$ and depending on $\omega(t_0)$, asymptotically $\vec{\omega} \approx \vec{\omega}^*$ also defines a vector of eigenvalues for the $N$ operators $M \cdot \vec{A}$, given via $M \cdot (M^*)^{-1}\vec{\omega}^*$ for any choice of $M \in GL(N, \mathbf{R})$. This all seems sort of complicated at first sight but is actually rather "nifty". Note for the role of the auxiliary field $F$ that via Ockham's razor one should omit (1.5), since only (**1F**) and (1.8)-(1.10) are needed for what follows. However only a field equation of motion allows one to separately interpret (1.8) and (1.9) as a symmetry breaking initial condition and a boundary condition respectively. The boundary condition introduces dissipation which leads to the attractive periodic orbits of $\vec{\phi}$ that are essential for the quantum mechanical features.

Now for emergent quantum mechanics consider prequantum states $\psi$ which describe the trajectory of the deterministic system for an arbitrary but fixed initial condition

$$\psi(\vec{\phi}, \vec{\omega}, t) = \sum_{\vec{n}} e^{i\vec{n}\cdot\vec{\phi}}\psi_{\vec{n}}(\vec{\omega}, t) \to \sum_{\vec{n}} e^{i\vec{n}\cdot(\vec{\phi}-\vec{\omega}^* t)}\psi_{\vec{n}}(\vec{\omega}^*, 0) \;\; (t \to \infty) \tag{1.11}$$

where $\vec{\omega}^*$ is the fixed point to which $\vec{\omega}(t)$ is attracted. Here the states fall into superselection sectors that can be classified by the absolutely conserved vector $\vec{n}$ and the various possibilities are enumerated and explored in [**290**] (0710.2765).

The whole matter is further generalized and expanded in [**290**] (0806.3408) and we sketch a few points. Thus consider a situation with conserved forces and (**1G**) $H(x,p) = (1/2)p^2 + V(x)$. An ensemble of such objects with distribution function $f(x,p,t)dxdp$ evolves via a Liouville equation

$$-\partial_t f = \frac{\partial H}{\partial x}\frac{\partial f}{\partial x} - \frac{\partial H}{\partial x}\frac{\partial f}{\partial p} = \{p\partial_x - V'(x)\partial_p\}f \tag{1.12}$$

A Fourier transformation $f(x,p,t) = \int dyexp[-ipy]f(x,y,t)$ replaces the Liouville equation by (**1H**) $i\partial_t f = \{-\partial_y\partial_x + yV'(x)\}f$ (note the symbol $f$ is unchanged here when changing variables). Thus momentum is eliminated by doubling the number of coordinates. Finally write (**1I**) $Q = x + (y/2)$; $q = x - (y/2)$ to obtain the Liouville equation in the form

$$i\partial_t f = \{\hat{H}_Q - \hat{H}_q + \Delta(Q,q)\}f; \; H_\chi = -\frac{1}{2}\partial_\chi^2 + V(\chi) \tag{1.13}$$

for $\chi = Q, q$ and $\Delta(Q,q) = (Q-q)V'[(Q+q)/2] - V(Q) + V(q) = -\Delta(q,Q)$. There are several relevant comments:

(1) The reformulation here of classical dynamics is independent of the number of degrees of freedom and applies to matrix valued as well as Grassman valued variables. Field theories require the classical functional formalism as in [**290**] (0512016, 0510267, etc.). Gauge theories or theories with constraints have to be examined carefully.

(2) The first equation in (1.13) appears as the vonNeumann equation for a density operator $\hat{f}(t)$ considering $f(Q,q,t)$ as its matrix elements. However a crucial difference is found in the interaction $\Delta$ between the bra and ket states which couples the Hilbert space and its dual.
(3) Alternatively (1.13)-1 might be read as the SE for two identical (sets of) degrees of freedom which respective Hamiltonians $\hat{H}_{Q,q}$ contributing with opposite signs. Since their interaction $\Delta$ is antisymmetric under $Q \leftrightarrow q$, the complete (Liouville) operator on the right side of (1.13)-1 has a symmetric spectrum with respect to zero and, generically, will not be bounded below. This Kaplan-Sundrum energy parity symmetry (cf. [**290, 501**]) has been invoked previously as a protection for a (near) zero cosmological constant which is otherwise threatened by many orders of magnitude too large zeropoint energies. $\Delta$ vanishes only for a free particle or harmonic oscillator and generally, with a coupling of the Hilbert space and its dual, or without a stable ground state, the reformulation of Hamiltonian dynamics above does not qualify as a quantum theory (cf. also [**105, 107, 290**])

A dissipative process has been of importance for deterministic models of quantum processes (cf. [**290, 444, 445**]) and here an attractor mechanism is proposed referring to two assumptions:

(1) The emergence of quantum states originates from a microscopic process applying to all physical objects.
(2) The statistical interpretation of quantum states (Born's rule) orginates from the classical ensemble theory.

One begins with the normalization of the classical probability distribution

$$1 = \int \frac{dxdp}{2\pi} f(x,p,t) = \int dQdq\delta(Q-q)f(Q,q,t) = Tr[\hat{f}(t)] \tag{1.14}$$

incorporating the transformations indicated above. Consider a complete set of orthonormal eigenfunctions of $\hat{H}_\chi$ of (1.13)-3, defined by $g_j(\chi,t) = exp(-iE_jt)g_j(\chi)$ and $\hat{H}_\chi g_j(\chi) = E_j g_j(\chi)$ respectively, with a discrete spectrum for simplicity. Then one can expand (**1J**) $f(Q,q,t) = \sum_{j,k} f_{jk}(t)g_j(Q,t)g_k^*(q,t)$ with normalization (1.14) in the form

$$1 = \sum_{j,k} f_{jk}(t)e^{-i(E_j-E_k)t} \int dQg_j(Q)g_k(Q) = \sum f_{jj}(t) \tag{1.15}$$

Since the classical phase space is real one has $f_{ij} = f_{ji}^*$ which is denoted also by $\hat{f}$. The classical expectation values are

(1.16)
$$<x>= \int \frac{dxdp}{2\pi} xf(x,p,t) = \int dQdq\delta(Q-q)\frac{Q+q}{2}f(Q,q,t) = Tr(\hat{X}\hat{f}(t))$$

(1.17)
$$<p>= \int \frac{dxdp}{2\pi} pf(x,p,t \int dQdq\delta(Q-q)(-i)\frac{\partial_Q - \partial_q}{2}f(Q,q,t) = Tr(\hat{P}\hat{f}(t))$$

where $X(q,Q) = \delta(Q-q)(Q+q)/2$ and $P(q,Q) = -i[\delta(Q-q)\overrightarrow{\partial}_Q - \overleftarrow{\partial}_q\delta(Q-q)]$. Eliminating one of the two integrations via partial integrations and delta functions one recognizes the coordinate and momentum operators. Further one finds that

$$(1.18) \qquad Tr((\hat{X}\hat{P}+\hat{P}\hat{X})\hat{f}(t)) = \frac{i}{2} + 2\int \frac{dxdp}{2\pi} xpf(x,p,t) - \frac{i}{2}$$

The operators appear here strictly by rewriting classical statistical formulae and not by a quantization rule, such as replacing $x$ and $p$ by operators $\hat{X}$ and $\hat{P}$ with $[\hat{X},\hat{P}] = i$ acting on some Hilbert space (not necessarily related to phase space). Further (1.14)-(1.18) are in accordance with the interpretation of $f(Q,q,t)$ as matrix elements of a density operator $\hat{f}(t)$.

**REMARK 1.1.1.** There is an important caveat: The eigenvalues of normalized quantum mechanical density operators are usually constrained to lie between zero and one (as probabilities). This is not necessarily the case with the operator $\hat{f}$ obtained from a classical probability distribution. Similarly the Wigner distribution, obtained from the matrix elements of a quantum mechanical density operator by applying the transformation $f(x,p) \to f(Q,q)$ in reverse, is not generally positive semi-definite on phase space and does not necessarily qualify as a classical probability density. ∎

**1.1. FLUCTUATIONS AND DISSIPATION.** One considers now a generalization of (1.13) that incorporates dissipation as well as diffusion. For sufficiently long times the evolving density operator should be attracted to solutions of the QM von Neumann equation (**1K**) $i\partial_t\hat{f} = [\hat{H}_\chi, \hat{f}]$. Equivalently the expansion coefficients $f_{ij}$ in (**1J**) should become constants. Then (1.13)-1 can be rewritten as a matrix equation for the coefficients (**1L**) $i\partial_t f_{jk}(t) = \sum_{\ell,m}\Delta_{jk\ell m}f_{\ell m}(t)$ with

$$(1.19) \qquad \Delta_{jk\ell m} = \int dQdqg_j(Q)g_k(q)\Delta(Q,q)g_\ell(Q)g_m(q) = -\Delta_{kjm\ell}$$

(employing the antisymmetry of $\Delta$ from (1.13)). Consequently $i\Delta$ maps a Hermitian matrix, such as $\hat{f}$, to a Hermitian matrix and this map produces then zero when taking traces, i.e. $Tr(\hat{\Delta}\hat{M}) = 0$ for any matrix $\hat{M}$ since (**1M**) $\sum_j \Delta_{jj\ell m} = 0$ by (1.19), completeness, and (1.13). For example the solution of (**1L**), namely $\hat{f}(y) = exp(-i\hat{\Delta}t)\hat{f}(0)$ conserves the normalization of $\hat{f}$, namely (1.15). The interaction $\Delta$ presents the unfamiliar coupling between Hilbert space and its dual which prevented one from considering (1.13)-1 as a truly quantum mechanical equation.

There is thens some interesting discussion of causal sets, fluctuations, dissipation, etc. for which we refer to [**290**] (0806.3408 - cf. also [**501**]) leading to a minimalist model

$$(1.20) \qquad i\partial_t\hat{f}(t) = (\hat{\Delta} + \delta H)(\hat{f}(t) - \hat{g}(t))$$

which generalizes (**1L**). The dissipative term comes from $\delta Hf(Q,q)$ entering the right side of (1.13)-1 which means adding $\delta H\sum_{\ell,m}\delta_{j\ell}\delta_{km}f_{\ell m}(t) = \delta Hf_{jk}(t)$ in

(**1L**). Together with its prefactors the matrix $\hat{g}$ enters here as a source term. The solution of the linear first order equation (1.20) is

$$\hat{f}(t) = e^{-i(\hat{\Delta}+\delta H)t}[\hat{f}(0) + i(\hat{\Delta} + \delta H)\int_0^t dse^{i(\hat{\Delta}+\delta H)s}\hat{g}(s)] \tag{1.21}$$

Averaging over the Gaussian fluctuations $\delta H$ gives

$$\hat{f}(t) = e^{-i(\hat{\Delta}-i\epsilon)t}[\hat{f}(0) + i(\hat{\Delta} - i\epsilon)\int_0^t dse^{i(\hat{\Delta}-i\epsilon)s}\hat{g}(s)] \tag{1.22}$$

$$= \hat{g}(t) + e^{-i(\hat{\Delta}-i\epsilon)t}[\hat{f}(0) - \hat{g}(0)) - \int_0^t dse^{-i(\hat{\Delta}-i\epsilon)(t-s)}\partial_s\hat{g}(s)$$

which shows the nonunitary dissipative decay caused by the fluctuations. Taking the trace of (1.21) or (1.22), with the help of (**1M**) and with $Tr\hat{f}(0) = 1$ however, one finds that probability is conserved, i.e. the normalization $Tr\hat{f} = 1$, provided that $Tr\hat{g}(t) = 1$. Thus the source term compensates the dissipative loss. Further if $\hat{g}(t)$ becomes constant sufficiently fast, then for large $t >> 1/\epsilon$, $\hat{f}(t) \approx \hat{g}(t) \to \hat{g}(\infty)$. In this limit dissipation effectively eliminates the coupling $\Delta$ and the above simplistic account of dissipation/diffusion leads to constant matrix elements $f_{ij}(t) \to g_{ij}(\infty)$. Via (**1J**) this implies that the von Neumann equation (**1K**) becomes a valid approximation for sufficiently large $t$, i.e. in this limit quantum theory will be recovered (one still needs a relation between $\hat{g}(t)$ and $\hat{f}(t)$ however).

The minimalist model of (1.20) is now completed by a nonlinear equation for the source matrix (**1N**) $\partial_t\hat{g}(t) = \tau^{-1}(\hat{f}(0) - < \hat{f}(0) >_{\hat{g}(t)})\hat{g}(t)$ with $\tau$ a time scale and $< \hat{f} >_{\hat{g}} = Tr(\hat{f}\hat{g})/Tr(\hat{g})$ (some motivation is indicated in [**290**]). The solution of (**1N**) is given by (**1O**) $\hat{g}(t) = exp[\hat{f}(0)t/\tau]/Tr[exp(\hat{f}(0)t/\tau)]$ (which implies $Tr\hat{g}(t) = 1$ as well). There is considerable discussion of features and possibilities and generally one arrives at stationary states or superpositions of stationary states. The quantum states emerge asymptotically and are represented by properly normalized density matrices with eigenvalues in $[0, 1]$. There are also some outstanding issues and one asks, motivated by an assumption of an atomistic (e.g. causal set) structure of spacetime, whether quantum mechanics originates, as a phenomena of coarse graining, from fluctuations induced by the growth of a discrete space-time.

**1.2. PREQUANTIZATION AND BORN'S RULE.** We go now to [**107**] (cf. also [**105**] with a path integral approach in [**106**]). A basic premise here is that quantum mechanics (QM) is actually not a complete ontological system but represents a very accurate low-energy approximation to a deeper level of dynamics. What this deeper level is remains however unspecified but it is thought by many that at a very high energy scale the fundamental rules should be deterministic. First one notes that Bell's inequalities are not a result about QM but they simply state that usual determinisitic theories obeying the usual (Kolmogorovian) probability theory inevitably satisfy certain inequalities between the mean values of certain (suitably chosen) observed quantities. This means theories that are "realistic" in that the system has an intrinsic existence independent of observation and local in that sufficiently separated measurements should not influence each other. However the prequantization ideas mentioned above involving enormous loss of information in forming equivalence classes etc. just do not fit into this kind of scheme. In [**107**] one looks at Born's rule as it arises out of prequantization ideas. Thus start with dynamics at the primordial deterministic level as described by a Hamiltonian (**1P**) $H = \sum p_i f_i(q) + g(q)$ where $q \sim (q_1, \cdots, q_n)$. Note that the equation of motion (**1Q**) $\dot{q}_i = f_i(q)$ is autonomous since the $p_i$ variables are decoupled. The system is obviously deterministic but the Hamiltonian is not bounded below. Note also that

$$(1.23)\quad q_i(t+\Delta t) = q_i(t) + f_i(q)\Delta t + \frac{1}{2} f_k(q) \frac{\partial^2 H}{\partial p_i \partial q_k} (\Delta t)^2 + \cdots = F_i(q(t), \Delta t)$$

where $F_i$ is some function of $q(t)$ and $\Delta t$ but not of $p$. Since (1.23) holds for any $\Delta t$ one has a Poisson bracket (**1R**) $\{q_i(t'), q_k(t)\} = 0$ and because of (**1Q**) one can define a formal Hilbert space $\mathcal{H}$ spanned by the states $\{|q>\}$; then associate with $p_i$ the operator $\hat{p}_i = -i(\partial/\partial q_i)$. Subsequently the generator of time translations (i.e. the Hamiltonian operator) of the form $\hat{H} = \sum \hat{p}_i f_i(\hat{q}) + g(\hat{q})$ generates precisely the deterministic evolution (**1Q**). Indeed since $\hat{H}$ generates time translations one has in the Heisenberg picture (**1S**) $\hat{q}_i(t+\Delta t) = exp[i\Delta t\hat{H}]\hat{q}_i(t)exp[-i\Delta t\hat{H}]$ which for infinitesimal $\Delta t$ means (**1T**) $\hat{q}_i(t+\Delta t) - \hat{q}(t) = i\Delta t[\hat{H}, q_i(t)] \Rightarrow \dot{q}_i = f_i(\hat{q})$. On the other hand for arbitrary finite $\Delta t$ one has from (**1S**)

$$(1.24)\qquad \hat{q}_i(t+\Delta t) = \sum_0^\infty \frac{1}{n!}[\hat{H}, [\hat{H}, [\cdots [\hat{H}, \hat{q}_i(t)]] \cdots ]] = \tilde{F}_i(\hat{q}(t), \Delta t)$$

(here $\hat{H}$ appears in the generic term n times). On the other hand $\tilde{F}_i$ is some function of $\hat{q}(t)$ and $\Delta t$, but not of $\hat{p}$, which means that (**1U**) $[\hat{q}_i(t), \hat{q}_j(t)] = 0$ and hence $F_i = \tilde{F}_i$ (i.e. everything can be simultaneously diagonalized - leading back to (**1Q**)). From the Schrödinger point of view (where base vectors are time independent and fixed) this means that the state vector evolves smoothly from one base vector to another. Hence there is no non-trivial linear superposition of the state vector in terms of base vectors and hence no interference phenomenon shows up when measurement of the $q$-variable is performed. This led to the idea of beables by Bell. Now going back to (**1P**), although the Hamiltonian is not bounded below it has another useful function in formulating a classical statistical mechanics for beables; the emergent Hamiltonians will however be bounded from below (as a consequence of coarse graining the beable degrees of freedom down to the observational ones). Thus consider $\rho(\hat{q})$ to be some positive function of the $\hat{q}_i$ (but not of $\hat{p}$) with $[\hat{\rho}, \hat{H}] = 0$ and define a splitting

$$(1.25)\qquad \hat{H} = \hat{H}_+ - \hat{H}_-;\ \hat{H}_+ = (\hat{\rho} + \hat{H})^2 \frac{\hat{\rho}^{-1}}{4};\ \hat{H}_- = (\hat{\rho} - \hat{H})^2 \frac{\hat{\rho}^{-1}}{4}$$

where $\hat{H}_+$ and $\hat{H}_-$ are positive definite operators satisfying (**1V**) $[\hat{H}_+, \hat{H}_-] = [\hat{\rho}, \hat{H}_\pm] = 0$. Now one introduces the coarse graining operator $\hat{\Phi}$ that describes the loss of information during the passage from the beable to the observational scale. One possible choice is (**1W**) $\hat{\Phi}_E = (1 - exp[-(E_P - E)/E])\hat{H}_-$ where $E$ refers to the observer's energy scale and $E_P$ is the beable energy scale (Planck scale). The operator $\hat{\Phi}_E$ is then implemented as a constraint on the Hilbert space $\mathcal{H}$ with observed physical states $|\psi>_{phys}$ given by the condition (**1X**) $\hat{\Phi}_E|\psi>_{phys} = 0$. This identifies the states that are not affected by the coarse graining, i.e. those that are still distinguishable at the observational scale $E$. The number factor $(1 - exp[-(E_P - E)/E])$ is not irrelevant and one cannot use directly $\hat{H}_-$ instead of $\hat{\Phi}_E$. In fact the constraint (**1X**) is a Dirac first class constraint which means that it generates gauge transformations; this restricts the Hilbert space (from $\mathcal{H}$ to $\mathcal{H}_c$) and produces equivalence classes (generally non-local). Two states are in the same class if they can be transformed into each other by a gauge transformation with generator $\hat{\Phi}_E$ and one denotes by $\mathcal{G}_E$ the one parameter group of such gauge transformations. Then the space of observables can be denoted by (**1Y**) $\mathcal{O}_E = \mathcal{H}_c/\mathcal{G}_E$. When $E << E_P$ one expects $\Phi_E = \hat{H}_-$ and the energy eigenvalues are bounded below since (**1Z**) $\hat{H}|\psi>_{phys} = \hat{H}_+|\psi>_{phys} = \hat{\rho}|\psi>_{phys}$. Hence in the Schrödinger picture

$$\frac{d}{dt}|\psi_t>_{phys} = -i\hat{H}_+|\psi_t>_{phys} \tag{1.26}$$

and this involves only positive frequencies on physical states. The constraining procedure can also be done via the Dirac-Bergman algorithm directly on the phase space (cf. [**107**] for details).

One goes back now to (**1P**) and develops the theory toward an understanding of the Born rule. Thus the dynamics is driven by (**1Q**) and a term $g(\hat{q})$ is added to the Hamiltonian to make it Hermitian; this should be done so that $g^\dagger(\hat{q}) - g(\hat{q}) = \sum_i[\hat{p}_i, f_i(\hat{q})]$, ensuring that $\hat{H}^\dagger = \hat{H}$. When $g(\hat{q})$ is taken to be anti-Hermitian the Hamiltonian will become not only Hermitian but also Weyl ordered. Indeed in this case

$$\hat{H} = \sum_i \hat{p}_i f_i(\hat{q}) + g(\hat{q}) = \frac{1}{2}\sum[\hat{p}_i f_i(\hat{q}) + f_i(\hat{q})\hat{p}_i] + \frac{1}{2}\sum_i[\hat{p}_i, f_i(\hat{q})] + g(\hat{q}) \tag{1.27}$$

$$= \frac{1}{2}\sum_i[\hat{p}_i f_i(\hat{q}) + f_i(\hat{q})\hat{p}_i] = W\left(\sum_i \hat{p}_i f_i(\hat{q})\right)$$

One now shows that the Born rule is closely related to the Koopman-von Neumann operator formulation of classical physics. One asks how an observer living in the beable world would do statistical physics on systems described via (**1Q**). To answer this one goes via (**1P**) and there is a simple recipe for defining the probability function. Thus define a "wave function" $\psi(p, q, t)$ that evolves in time with the Liouville operator, i.e. (•) $i\partial_t\psi = \hat{\mathfrak{H}}\psi$ where

$$\hat{\mathfrak{H}} = -i\partial_{p_i}H(p,q)\partial_{q_i} + i\partial_{q_i}H(p,q)\partial_{p_i} \tag{1.28}$$

The vectors $\psi$ are complex wave functions on the phase space $\Gamma = (p, q)$ with normalization $\int dpdq|\psi(p,q)|^2 = 1$. Conjugation gives (♦) $i\partial_t\psi^* = \hat{\mathfrak{H}}\psi^*$ and together with (•) this leads to

$$i\partial_t\rho = \hat{\mathfrak{H}}\rho \iff \partial_t\rho = (-\partial_{p_i}H\partial_{q_i} + \partial_{q_i}H\partial_{p_i})\rho \tag{1.29}$$

where $\rho = \psi^*\psi$. The last equation is the well known Liouville equation for a classical statistical system and the Liouvillian Hamiltonian $\mathfrak{H}$ is selfadjoint yielding constant state norms consistent with $\rho = \psi^*\psi$ being a probability density on $\Gamma$. However the $p$ variables are only dummy variables in the Hooft theory (true degrees of freedom are beables, i.e. variables $q$) and this means that the reduced density matrix $\tilde{\rho}(q) = \int dp\rho(p,q)$ is the relevant density matrix. Using then (**1P**) and integrating (1.29) over $p$ one arrives at

$$\partial_t\tilde{\rho}(q) = -f_i(q)\partial_{q_i}\tilde{\rho}(q) - [\partial_{q_i}f_i(q)]\tilde{\rho}(q) = -\partial_{q_i}(f_i(q)\tilde{\rho}(q)) \tag{1.30}$$
$$= -i\hat{p}_i f_i(q)\tilde{\rho}(q) = -i\hat{H}(p,q)\tilde{\rho}(q)$$

Define then ($\bigstar$) $\tilde{\psi}(q) = \int_{-\infty}^{\infty} dp\psi(p,q)$ and one can use ($\bullet$) (resp. ($\blacklozenge$)) to compute the evolution of $\tilde{\psi}(q)$ (resp. $\tilde{\psi}^*(q)$) under the Hooft Hamiltonian $\hat{H}$ leading to

$$i\partial_t\tilde{\psi}(q) = \hat{H}(p,q)\tilde{\psi}(q);\ i\partial_t\tilde{\psi}^*(q) = \hat{H}(p,q)\tilde{\psi}^*(q) \tag{1.31}$$

Taking advantage of the particular form of the Hooft Hamiltonian one obtains directly then

$$\partial_t(\tilde{\psi}^*(q)\tilde{\psi}(q)) = -i\hat{H}(p,q)(\tilde{\psi}^*(q)\tilde{\psi}(q) \tag{1.32}$$

which, in view of (1.30) leads to the conclusion $\tilde{\rho}(q) = \tilde{\psi}^*(q)\tilde{\psi}(q)$ and hence the Born rule. Actually the Hooft Hamiltonian has a privileged role and it is shown in [**106**] that there are no other Hamiltonian systems with the peculiar property that their full quantum evolution coincides with the classical one. Thus combining the Koopman-von Neumann operator formulation of classical physics with the Hooft Hamiltonian for beables one arrives at Born's rule. This substantiates Hooft's pre-quantization scheme and answers affirmatively whether QM can be a hidden variable theory (the beable dynamics is fully local). The non-locality that provides the loophole in the Bell inequalities is realized in Hooft's prequantization by identifying the physical space of observables with a quotient space that has a non-local structure (see [**107, 106**] for more discussion of such matters).

## 2. QUANTUM MECHANICS, GRAVITY, AND RICCI FLOW

In Chapter 2 we discussed some aspects of Weyl geometry and relativity in connection with the quantum potential and Ricci flow. This was related to the work of many people in the last 20-25 years and we mention in particular [**53, 54, 55, 97, 117, 171, 173, 177, 178, 179, 185, 191, 193, 297, 319, 401, 402, 440, 465, 466, 467, 468, 506, 507, 621, 642, 643, 702, 703, 770, 797, 836**]. We review here the recent papers in [**2, 456, 517**] where one theme involves Ricci flow and emergent quantum mechanics and there are a number of connections to material in Chapter 2. The first paper in [**456**] deals with 2-D closed Riemannian manifolds $M$ with isothermal metric (**2A**) $g_{ij} = exp(-f)\delta_{ij}$ where $f(x,y)$ is called a conformal factor and the volume element on $M$ is (**2B**) $\sqrt{g}dxdy = exp[-f]dxdy$. Here $g = |det(g_{ij})|$ with Ricci tensor

$$R_{im} = g^{-1/2}\partial_n(\Gamma^n_{im}g^{1/2}) - \partial_i\partial_m(log[g^{1/2}]) - \Gamma^r_{is}\Gamma^s_{mr}; \tag{2.1}$$
$$\Gamma^m_{ij} = \frac{1}{2}g^{mh}(\partial_i g_{jh} - \partial_j g_{hi} - \partial_h g_{ij})$$

Given an arbitrary function $\phi(x, y)$ on $M$ one has

$$\nabla^2\phi = \frac{1}{\sqrt{g}}\partial_m(\sqrt{g}g^{mn}\partial_n\phi) = e^f(\partial_x^2\phi + \partial_y^2\phi) = e^f D^2\phi; \tag{2.2}$$

$$(\nabla\phi)^2 = g^{mn}\partial_m\phi\partial_n\phi = e^f\left[(\partial_x\phi)^2 + (\partial_y\phi)^2\right] = e^f(D\phi)^2$$

The Ricci tensor is given via

$$R_{ij} = \frac{1}{2}D^2 f\delta_{ij} = \frac{1}{2}e^{-f}\nabla^2 f\delta_{ij};\ R = e^f D^2 f = \nabla^2 f \tag{2.3}$$

Thus one is dealing with a compact Riemann surface without boundary.

Referring to [**170, 172, 506, 507, 621, 693, 836**] Perelman's functional $\mathfrak{F}[\phi, g_{ij}]$ on $M$ is defined by

$$\mathfrak{F}[\phi, g_{ij}] = \int_M dxdy\sqrt{g}e^{\phi}\left[(\nabla\phi)^2 + R(g_{ij})\right] \tag{2.4}$$

where $\phi$ is a real function. It can be regarded as providing an action functional for the independent fields $g_{ij}$ and $\phi$. One notes that for $\phi = 0$ one has the Einstein-Hilbert functional for gravity on $M$ (which, being a boundary term in 2-D, is trivial here). Computing the Euler-Lagrange extremals one obtains two evolution equations

$$\partial_t g_{ij} = -2\left(R_{ij} + \nabla_i\nabla_j\phi\right);\ \partial_t\phi = -\nabla^2\phi - R \tag{2.5}$$

Note that the right sides of (2.5), set equal to zero, are the Euler-Lagrange equations corresponding to (2.4), while the time derivatives have been put in by hand; time is here an external parameter and (2.5) is referred to as gradient flow. Via a time-dependent diffeomorphism one can show that (2.5) is equivalent to

$$\partial_t g_{ij} = -2R_{ij};\ \partial_t\phi = -\nabla^2\phi + (\nabla\phi)^2 - R \tag{2.6}$$

Noting that all metrics are conformal for 2-D manifolds $M$ one can substitute (**2A**) throughout and by (**2B**) and (2.3) this leads to

$$\mathfrak{F}[\phi, f] = \mathfrak{F}[\phi, g_{ij}(f)] = \int_M dxdye^{-\phi-f}\left[(\nabla\phi)^2 + \nabla^2\phi\right] \tag{2.7}$$

In order to understand the physical meaning of the flow equations (2.5) one proceeds by using (**2A**) and (2.3) to show that for the first flow equation

$$\partial_t g_{ij} = -2R_{ij} \equiv \partial_t f = \nabla^2 f \tag{2.8}$$

This is the standard heat equation with $\nabla^2$ given via (2.2); note $M$ is not flat but only Conformally flat so conformal metrics on $M$ evolve in time according to the heat equation with respect to the "curved" Laplacian. Setting now $\phi = f$ in (2.7) one obtains

$$\mathfrak{F}[f] = \mathfrak{F}[\phi = f, f] = \int_M dxdye^{-2f}[(\nabla f)^2 + \nabla^2 f] \tag{2.9}$$

There might seem to be a contradiction since for $\phi = f$ there are two different flow equations in (2.6); the procedure here implies reducing this to one flow equation simply by substituting one of the equations in (2.6) into the other. By (2.3) and (2.8) this leads to $R = \partial_t f$ which when put into (2.6) gives

$$\partial_t f = \frac{1}{2}(\nabla f)^2 - \frac{1}{2}\nabla^2 f \tag{2.10}$$

In order to distinguish notationally between the time-independent and the time-dependent conformal factors (in (2.9) and (2.10)) one rewrites (2.10) as (**2C**) $\partial_t \tilde{f} = (1/2)(\nabla \tilde{f})^2 - (1/2)\nabla^2 \tilde{f}$ (the tilde denoting time dependence).

The idea now is to regard the manifold as the configuration space of a mechanical system and to establish a 1-1 correspondence between conformally flat metrics on 2-D $M$ and (classical or quantum) mechanical systems on $M$. First consider classical systems with Hamilton-Jacobi (HJ) equations

$$\partial_t \tilde{S} + \frac{1}{2m}(\nabla \tilde{S})^2 + U = 0;\ \tilde{S} = S - Et \tag{2.11}$$

This produces (**2D**) $(1/2m)(\nabla S)^2 + U = E$ and suggests separating variables in (**2C**) via (**2E**) $\tilde{f} = f + Et$ leading to (**2F**) $(1/2)(\nabla f)^2 - (1/2)\nabla^2 f = E$ (note time is reversed here relative to $\tilde{S} = S - Et$ - cf. [**836**]). Picking $m = 1$ and comparing (**2F**) with (**2D**) one can make the identifications

$$S = f;\ U = -\frac{1}{2}\nabla^2 f = -\frac{1}{2}R \tag{2.12}$$

Hence the potential $U$ is proportional to $R$ with $S \sim f$ and one has constructed a classical mechanics starting from a conformal metric. Conversely given a classical mechanical system determined by $U$ one can produce a conformal metric with $f$ satisfying $-2U = \nabla^2 f$. There will also be a corresponding (stationary) SE involving $U$ and hence a QM.

One can be more explicit regarding the SE. Thus one can exchange a conformally flat metric for a wave function satisfying the SE. First look at

$$S[\psi, \psi^*] = \int_M dxdy\sqrt{g}\left(i\psi^*\partial_t\psi - \frac{1}{2m}\nabla\psi^*\nabla\psi - U\psi^*\psi\right) \tag{2.13}$$

Pick $m = 1$ and write $\psi = exp(i\tilde{f})$; putting this in (2.13) gives (**2G**) $-\partial_t \tilde{f} - (1/2)(\nabla \tilde{f})^2 + (1/2)\nabla^2 \tilde{f}$ in the integrand. Consider the stationary case (as before) so $\tilde{f} \to f$ and (2.13) becomes

$$S[f] = S[\psi = exp(if)] = \frac{1}{2}\int_M dxdy\, e^{-f}[-(\nabla f)^2 + \nabla^2 f] \tag{2.14}$$

Comparing (2.14) and (2.9) one arrives at

$$F[f/2] = S[f] + \frac{3}{2}\mathfrak{K}[f];\ \mathfrak{K}[f] = \frac{1}{2}\int_M dxdy\, e^{-f}(\nabla f)^2 \tag{2.15}$$

Here the Laplacian $\nabla^2 f$ and the gradient in $(\nabla f)^2$ are computed with respect to the conformal factor $f$, even though the functional is evaluated on $f/2$ so $S[f]$ is closely related to $F[f/2]$ and one has produced a Schrödinger QM on a 2-D compact configuration space via the Perelman functional.

Some features of emergent QM are present here, most noticeable being the presence of dissipation or information loss, which underlies the passage from a classical description to a quantum description. Here the classical description rests on the conformal factor $f$ of the metric $g_{ij}$ while the quantum description is given via the wave function $\psi = exp(if)$. The latter contains less information that the former since there can be different conformal factors $f$ giving rise to the same quantum wave function $\psi$. Here, beyond the trivial case of $f_1$ and $f_2$ differing by $2\pi$ times an integer there can be different $f_i$ satisfying $\nabla_1^2 f_1 = \nabla_2^2 f_2$. Another feature of emergence is to establish that to every quantum system there is at least one deterministic system which upon prequantization gives back the original system. Here one takes a quantum system with potential $V$ be given on $M$, satisfying the requirements noted above. Then consider the Poisson equation on $M$, $\nabla_V^2 f_V = -2V$ where $f_V$ is some unknown conformal factor to be determined as the solution of this Poisson equation. Then one can claim that the deterministic system, whose prequantization gives back the original quantum system with potential $V$, is described by the following data: Configuration space $M$ with classical states being conformal factors $f_V$ and mechanics described by the functional (2.9). The lock-in mechanism (following Hooft) is the choice of one particular conformal factor, with respect to which the Laplacian is computed, out of all possible solutions to the Poisson equation $\nabla_V^2 f_V = -2V$. The lock-in mechanism has been translated into a problem concerning the geometry and topology of the configuration space $M$, namely whether or not the Poisson equation possesses solutions on $M$ and how many.

We go now to [**2**]-2 and [**456**]-2 by Isidro et al where these ideas are developed much further. First from [**456**] one considers a $d$-dimensional Hilbert space of quantum states identified as $\mathbf{C}^d$. Let $\mathbf{C}$ be the phase space of a corresponding classical modification. Unitary transformations on Hilbert space correspond to canonical transformations on a classical phase space so one can imagine $SU(d)$ over $\mathbf{C}^d$ corresponding to classical canonical transformations (using $\mathbf{C}^d$ as "carrier" space and restricting attention to canonical transformations represented by unitary matrices with determinant one). Now quantum states are unit rays rather than vectors so in fact the true space of inequivalent quantum states is the complex projective space $\mathbf{CP}^{d-1}$. This can be regarded as a homogeneous manifold (cf. [**93, 433, 516**])

$$\mathbf{CP}^{d-1} = \frac{SU(d)}{SU(d-1) \times U(1)} \tag{2.16}$$

One could take here $SU(d)$ as the total space of a fiber bundle with typical fiber $SU(d-1) \times U(1)$ over a base manifold $\mathbf{CP}^{d-1}$. The projection map (**2H**) $\pi : SU(d) \to \mathbf{CP}^{d-1}$; $\pi(w) = [w]$ arranges points $w \in SU(d)$ into $SU(d-1) \times U(1)$- equivalence classes $[w]$. Classical canonical transformations as represented by $SU(d)$ act on $\mathbf{C}^d$ and this descends to an action $\alpha$ of $SU(d)$ on $\mathbf{CP}^{d-1}$ via

$$\alpha : SU(d) \times \mathbf{CP}^{d-1} \to \mathbf{CP}^{d-1};\ \alpha(u, [v]) = [uv] \tag{2.17}$$

($uv$ denotes $d \times d$ matrix multiplication - and check that the action is well defined on the equivalence classes under right multiplication by elements of the stabilizer subgroup $SU(d-1)\times U(1)$). This enables one to regard quantum states as equivalence classes of classical canonical transformations on $\mathbf{C}$. Physically $u$ in (2.17) denotes (the representative matrix of) a canonical transformation on $\mathbf{C}$ and $[v]$ denotes the equivalence of (representative matrices of) the canonical transformation $v$ or equivalently of the quantum state $|v>$. Thus two canonical transformations are equivalent whenever they differ by a canonical transformation belonging to $SU(d-1)$ and/or whenever they differ by a $U(1)$ transformation. Modding out by $U(1)$ has a clear physical meaning via a standard freedom in the choice of the phase of the wavefunction corresponding to $v\in SU(d)$. Modding out by $SU(d-1)$ also has a physical meaning via thinking of canonical transformations on the $(d-1)$-dimensional $\mathbf{C}^d-1\subset\mathbf{C}^d$ involving a symmetry of $v$. Hence the true quantum state $|v>$ is obtained from $v\in SU(d)$ after modding out by the stabilizer subgroup $SU(d-1)\times U(1)$.

One concludes that this picture contains some of the elements identified for the passage from a classical world (canonical transformations) to a quantum world (equivalence classes of canonical transformations, or unit rays in a Hilbert space). Some kind of dissipative mechanism is at work through the emergence of orbits (or equivalence classes). However the projection (**2H**) is an on-off mechanism and one would prefer to see dissipation arising from a flow with some continuous parameter. In fact the Ricci flow (**2I**) $\partial_t g_{ij}=-2R_{ij}$ is the desired flow. To see this note that the Lie group $SU(d-1)\times U(1)$ is compact but is not semisimple due to the Abelian factor $U(1)$. However $SU(d-1)$ is compact and semisimple and it qualifies as an Einstein space with positive scalar curvature with respect to the Killing-Cartan metric (cf. [**93, 433, 516**]). Now (**2I**) ensures that $SU(d-1)$ contracts to a point under Ricci flow. However the $U(1)$ component renders $SU(d-1)\times U(1)$ nonsemisimple and the Killing-Cartan metric has a vanishing determinant (cf. [**433**]). The Ricci flow can still cancel the $SU(d-1)$ within $SU(d)$ but not the $U(1)$ factor. However after contracting $SU(d-1)$ to a point one is left with $U(1)\times\mathbf{CP}^{d-1}$, or more generally, with a $U(1)$ bundle over the base $\mathbf{CP}^{d-1}$. This $U(1)$ bundle is the Hopf bundle where the total space is the sphere $S^{2d-1}$ in $2d-1$ real dimensions (cf. [**516**]). This sphere falls short of being the true space of quantum states by the unwanted $U(1)$ fiber (that cannot be removed by the Ricci flow). However it can be removed by projection P from the total space to its base; the combination PR (Ricci flow followed by projection) acts on the stabilizer subgroup $SU(d-1)\times U(1)$ and leaves one with $\mathbf{CP}^{d-1}$ as desired. Thus it acts in the same manner as $\pi$ in (**2H**) and in addition provides a differential equation implementing dissipation along a continuous parameter. The Ricci flow is especially important here since the Ricci tensor can be identified as its infinitesimal generator or Hamiltonian. In particular it involves a positive definite Hamiltonian and hence it is bounded below and there will be a stable ground state.

We embellish this further now with some text from [**2**], paper 2. The main result is formally stated first in the form

**THEOREM 2.1.1.** Let $M$ be a smooth $n$-dimensional, compact manifold without boundary endowed with a conformally flat Riemannian metric at some initial time $t_0$ (time is a parameter here). Then a classical mechanical system having $M$ as its configuration space can be defined such that its time-independent mechanical action (Hamilton's principal function) equals the conformal factor of the initial metric. Moreover the corresponding gradient Ricci flow comes from the time-dependent Hamilton-Jacobi (HJ) equation of the mechanical system thus defined.

**COROLLARY 2.1.1.** On $M$ the action functional for Einstein-Hilbert gravity equals the sum of the action functional for Schrödinger quantum mechanics plus Perelman's functional, plus the Coulomb functional.

We will build this up following [**2**] using some of the previous material from [**456**]. For $n > 2$ a metric need not be conformal but when it is it is uniquelly determined by the knowledge of just one function, called the conformal factor, on $M$. First let $M$ have a Riemannian metric $g_{ij}$, $i, j = 1, \cdots, n$ with volume element (**2J**) $dV = \sqrt{g}dx^1 \wedge \cdots \wedge dx^n = \sqrt{g}d^n x$ and let the Christoffel symbols of the corresponding Levi-Civita connection be given as (**2K**) $\Gamma^m_{ij} = (1/2)g^{mh}(\partial_i g_{jh} + \partial_j g_{hi} - \partial_h g_{ij})$ with Ricci tensor (**2L**) $R_{ik} = \partial_\ell \Gamma^\ell_{ik} - \partial_k \Gamma^\ell_{i\ell} + \Gamma^\ell_{ik}\Gamma^m_{\ell m} - \Gamma^m_{i\ell}\Gamma^\ell_{km}$. For a smooth function on $M$ one has (**2M**) $\nabla^2\phi = (1/\sqrt{g})\partial_r(\sqrt{g}g^{rs}\partial_s\phi)$ and $(\nabla\phi)^2 = g^{mn}\partial_m\phi\partial_n\phi$. For $n = 3$ a necessary and sufficient condition for conformality is the vanishing of the Cotton tensor and for $n \geq 4$ a necessary and sufficient for conformality is the vanishing of the Weyl tensor (cf. [**359**]). One assumes the appropriate conformality condition holds as an initial condition for the Ricci flow equations and the $x^i$ will be taken as isothermal coordinates. Let the initial metric be (**2N**) $g_{ij}(t_0) = exp(-f)\delta_{ij}$ where the conformal factor $f$ is smooth and real. It will be seen that the Ricci flow of the metric (**2N**) will lead to a metric (**2O**) $g_{ij}(t) = exp(-Et)g_{ij}(t_0)$ with $E$ a constant (energy of the mechanical system). For the Christoffel symbols leading to (**2N**) one finds (**2P**) $\Gamma^m_{ij} = (1/2)(\delta_{ij}\delta^{mh}\partial_h f - \delta^m_j\partial_i f - \delta^m_i\partial_j f)$ while the volume element will be (**2Q**) $dV = exp(-nf/2)d^n x$. Further a computation gives

$$R_{im} = \frac{1}{2}e^{-f}\left(\frac{2-n}{2}\partial_j f\partial^j f + \partial_j\partial^j f\right)\delta_{im} + \frac{n-2}{2}\left(\partial_i\partial_m f + \frac{1}{2}\partial_i f\partial_m f\right) \tag{2.18}$$

with Ricci scalar

$$R = (n-1)\left(\frac{2-n}{4}\partial_j f\partial^j f + \partial_j\partial^j f\right) \tag{2.19}$$

Moreover for any smooth function one has

$$\nabla^2\phi = \left(1 - \frac{n}{2}\right)\partial_j\partial^j\phi + \partial_j\partial^j\phi;\ (\nabla\phi)^2 = (\partial_j\phi)(\partial^j\phi) \tag{2.20}$$

In particular for $\phi = f$ one has

$$\partial_j\partial^j f = \nabla^2 f + \left(\frac{n}{2} - 1\right)(\nabla f)^2 \tag{2.21}$$

Using (2.21) in (2.18) and (2.19) one has

$$R_{im} = \frac{1}{2}e^{-f}\nabla^2 f\delta_{im} + \frac{n-2}{2}\left(\partial_i\partial_m f + \frac{1}{2}\partial_i f\partial_m f\right); \tag{2.22}$$

$$R = (n-1)\left[\frac{n-2}{4}(\nabla f)^2 + \nabla^2 f\right]$$

Next, following [**693, 836**] one writes (**2R**) $\mathfrak{F}[\phi, g_{ij}] = \int_M exp(-\phi)[(\nabla\phi)^2 + R(g_{ij})]dV$ and the gradient flow is given via

$$\partial_t g_{ij} = -2(R_{ij} + \nabla_i\nabla_j\phi);\ \partial_t\phi = -\nabla^2\phi - R \tag{2.23}$$

Via a time-dependent diffeomorphism these equations are equivalent to

$$\partial_t g_{ij} = -2R_{ij}\ \partial_t\phi = -\nabla^2\phi + (\nabla\phi)^2 - R \tag{2.24}$$

and one will use (2.24) rather than (2.23). Then via (**2Q**) and (2.22) when the metric is conformal as in (**2N**) there results

$$\mathfrak{F}[\phi, f] = \mathfrak{F}[\phi, g_{ij} = e^{-f}\delta_{ij}] \tag{2.25}$$

$$= \int_M e^{-\phi-(nf/2)}\left[(\nabla\phi)^2 + (n-1)\left(\frac{n-2}{4}(\nabla f)^2 + \nabla^2 f\right)\right] d^n x$$

Setting then $\phi = f$ this simplifies to

$$\mathfrak{F}[f] = \mathfrak{F}[\phi = f, f] = \int_M e^{-[1+(n/2)]f}\left[a_n(\nabla f)^2 + b_n\nabla^2 f\right] d^n x \tag{2.26}$$

where (**2S**) $b_n = n-1$ and $a_n = 1 + (1/4)(n-1)(n-2)$. Here $a_n > 0$ and setting $\phi = f$ again one substitutes one of the flow equations in (2.24) into the other to obtain after contraction with $g_{ij}$ (**2T**) $(n/2)\partial_t f = R$ which inserted back into (2.24) leads to

$$\partial_t f + \frac{2}{n+2}\nabla^2 f - \frac{2}{n+2}(\nabla f)^2 = 0 \tag{2.27}$$

One can verify that (2.27) is the gradient flow equation of $\mathfrak{F}[\phi]$ with $\mathfrak{F}$ as in (2.26) and $f = [(6-n)/4]\phi$.

Now for the proof of Theorem 2.1.1 one thinks of $M$ as the configuration space of a mechanical system (to be identified) and recalls that for a particle of mass $m$ the HJ equation for $\tilde{S}$ involving potential $U$ is (**2U**) $\partial_t\tilde{S} + (1/2m)(\nabla\tilde{S})^2 + U = 0$. As before writing (**2V**) $\tilde{S} = S - Et$ one arrives at (**2W**) $(1/2m)(\nabla S)^2 + U = E$. This suggests separating variables in (2.27) via (**2X**) $\tilde{f} = f + Et$ and putting this in (2.27) to get

$$\frac{2}{n+2}(\nabla f)^2 - \frac{2}{n+2}\nabla^2 f = E \tag{2.28}$$

Using (2.22) this becomes

$$\frac{1}{2}(\nabla f)^2 + \frac{2R}{(n+2)(1-n)} = E \tag{2.29}$$

Comparing with (**2W**) and taking $m = 1$ then gives (**2Y**) $S = f$; $U = 2R/(n+2)(1-n)$ so the potential $U$ is proportional to the Ricci curvature $R$ of $M$ while the reduced action $S = f$. This identifies a mechanical system in terms of the metric such that the gradient Ricci flow of latter is the time-dependent HJ equation of the former.

To prove Corollary 2.1.1 we proceed from the time independent SE with the Ricci curvature as potential function as in [**836**], namely

$$i\hbar\partial_t\psi = -\frac{\hbar^2}{2m}\nabla^2\psi + U\psi \tag{2.30}$$

This reduces to (**2U**) for $\psi = exp[i\tilde{S}/\hbar]$ with $\hbar \to 0$. Take $\hbar = 1$ now for convenience and (2.30) can be obtained as an extremal for the action functional

$$\mathfrak{S}[\psi,\psi^*] = \int_M e^{-nf/2}\left(i\psi^*\partial_t\psi - \frac{1}{2m}\nabla\psi^*\nabla\psi - U\psi^*\psi\right)d^nx \tag{2.31}$$

Putting $\psi = exp(i\tilde{f})$ and $m = 1$ and using (2.22) and (2.24) with $\partial_t\tilde{f} = 0$ (stationary case where $\tilde{f} \to f$) (2.31) becomes

$$\mathfrak{S}[f] = \mathfrak{S}[\psi = e^{if}] = \frac{2}{n+2}\int_M e^{-nf/2}[-(\nabla f)^2 + \nabla^2 f]d^nx \tag{2.32}$$

Rescale the conformal factor $f$ now as (**2Z**) $f_n = [n/(n+2)]f$ and then (2.26) can be expressed via

$$\mathfrak{F}[f_n] = \frac{n}{n+2}\int_M e^{-hf/2}\left[\frac{na_n}{n+2}(\nabla f)^2 + b_n\nabla^2 f\right]d^nx \tag{2.33}$$

On the other hand the Einstein-Hilbert gravitational action functional $\mathcal{G}$ on $M$ is ($\bullet$) $\mathcal{G}[g_{ij}] = \int R(g_{ij})dV$ and this acting on the conformal metric (**2N**) becomes by (**2Q**) and (2.22)

$$\mathcal{G}[f] = \mathcal{G}[g_{ij}(f)] = (n-1)\int_M e^{-nf/2}\left[\frac{n-2}{4}(\nabla f)^2 + \nabla^2 f\right]d^nx \tag{2.34}$$

Some calculation shows then that

$$\mathcal{G}[f] = \mathfrak{F}[f_n] + \mathfrak{S}[f] + \mathfrak{C}[f];\ \mathfrak{C}[f] = 2c_n\int_M e^{-hf/2}\left[\frac{1}{2}(\nabla f)^2 + d_n\nabla^2 f\right]d^nx \tag{2.35}$$

is the action functional for a free scalar field $f$ coupled to a Coulomb charge placed at $\infty$ (cf. [**315**]) and

$$c_n = \frac{n^3 - 3n^2 + n + 6}{(n+2)^2};\ d_n = \frac{(n-2)(n+2)}{n^3 - 3n^2 + n + 6} \tag{2.36}$$

which proves the Corollary.

The Einstein-Hilbert functional $\mathcal{G}$, the Coulomb functional $\mathfrak{C}$, and the Perelman entropy functional $\mathfrak{F}$ have all been dimensionless. Introducing Newton's constant $G$, Planck's constant $\hbar$, and Boltzmann's constant $k$ restores the usual dimensions. One sees that on a compact conformally flat Riemannian configuration space without boundary, Einstein-Hilbert gravity arises from Schrödinger quantum mechanics, from Perelman's Ricci flow functional, and from the Coulomb functional. Reading this in reverse one can state that on a conformally flat manifold Perelman's functional contains Einstein-Hilbert gravity and Schrödinger quantum mechanics.

## 3. THERMODYNAMICS, GRAVITY, AND QUANTUM THEORY

There is an emerging "field" of literature on connections between and among thermodynamics, quantum mechanics, and general relativity (some of which has already been discussed). In particular Ricci flow, conformal geometry, and quantum mechanics were linked in Chapter 2 following [**97, 117, 170, 171, 172, 179, 182, 185, 191, 192, 193, 297, 770, 797**] (and the treatment in Chapter 6, Section 2 based on [**2, 456**] just concluded focuses on the emergence theme of quantum mechanics from classical mechanics via Ricci flow). In all this it may be however that the main theme has been lost, namely that the Einstein-Hilbert action functional for general relativity has a quantum mechanical nature! One can search for theories of quantum gravity ad infinitum but it has already been given to us via the gravitational action functional (the form of which may ultimately be optimized). This also indicates the importance of the entropy idea (and hence information theory) and of Ricci flow in gravity theory - and in studying gravity we are automatically studying quantum mechanics. Another way to look at this involves the identification of conformal mass with Bohmian quantum mass (developed in Chapter 2). In [**118**] for example one studies the effect of conformal fluctuations in gravitational theory without recognizing the implicit quantum nature of the conformal factor. In this section however we want to dwell on some aspects of Ricci flow developed in [**769**].

We recall an important remark in [**836**] that the Perelman entropy functional (**3A**) $\mathfrak{F} = \int_M (R + |\nabla f|^2)exp(-f)dV$ can be considered as a Fisher information. Further the $L^2$ gradient flow of $\mathfrak{F}$ is determined by evolution equations

$$\partial_t g_{ij} = -2(R_{ij} + \nabla_i \nabla_j f);\ \partial_t f = -\Delta f - R \tag{3.1}$$

or equivalently by the decoupled family

$$\partial_t g_{ij} = -2R_{ij};\ \partial_t f = -\Delta f + |\nabla f|^2 - R \tag{3.2}$$

via a time dependent diffeomorphism (cf. also [**170, 172, 218, 506, 507, 621, 693, 836**]). This fact allowed us to relate Ricci flow for $M \subset \mathbf{R}^3$ to quantum theory via Weyl geometry, Fisher information, and a quantum potential (cf. [**170, 172, 191, 193, 770, 797**]). Further since $\mathfrak{F}$ also represents the action for conformal general relativity (GR), with no ordinary mass term, it was possible to envision Ricci flow gravity as in [**378**] (cf. also [**172, 185**]). A variation on this was given in [**172**] where we considered FRW metrics of the form

$$ds^2 = -dt^2 + a^2(t)\left[\frac{dr^2}{1-\kappa r^2} + f^2 d\Omega^2\right] = -dt^2 + \gamma_{ij}dx^i dx^j \tag{3.3}$$

with Ricci flow for the $\gamma$ metric (cf. also Chapter 2).

The connection to quantum mechanics developed in [**170, 172**] suggested connections of gravity to entropy and information theory (cf. [**170, 179**]) and in fact $\mathfrak{F}$ is referred to as an entropy functional in [**693**] with $\partial_t N \sim \mathfrak{F}$ where $N$ is the Nash (or differential) entropy with connections to quantum mechanics, Schrödinger equations, statistical mechanics, and information theory (see e.g. [**170, 171, 318, 319, 332, 333, 348, 401, 402, 419, 481, 506, 507, 693, 736, 742**]). Of course today there are many stronger connections between entropy, information, and gravity which have been developed for example in [**16, 15, 150, 149, 170, 179, 171, 196, 194, 197, 286, 339, 367, 474, 523, 524, 659, 660, 661, 662, 663, 664, 665, 666, 667, 668, 669, 670, 671, 672, 673, 674, 675, 676, 721, 724, 725, 756, 769, 776, 818, 820, 888**]; we refer here also to various aspects of Ricci flows in [**60, 163, 303, 317, 652, 769, 886**].

**3.1. GEOMETRIC FLOWS AND ENTROPY.** We follow here [**769**] for some background ideas. In the paper 0711.0428 a thermodynamic perspective is taken and one applies Ricci flow ideas to isolated gravitational situations. 3-D slices $\Sigma$ with induced metrics $h_{ij}$ are considered and for time symmetric initial data one applies Ricci type flows to $h_{ij}$. The dominant energy condition implies $R > 0$ and this is preserved under Ricci flow for example (cf. [**836**]). In GR one is mainly interested in situations where space has an asymptotic region and the ADM mass is defined using a fixed metric at infinity (either flat or AdS). This suggests dealing with non-compact $\Sigma$ (unlike the Perelman development) and in pursuing thermodynamic analogues one restricts attention to static 4-D Lorentzian asymptotically flat spacetime subject to a finite energy condition. After some argument the conclusion is reached that Perelman entropy is not connected to Beckenstein-Hawking entropy and in particular the Perelman entropy is not a so called geometric entropy. However a slight modification of the Ricci-Perelman flow does have relations to geometric entropy. Indeed consider modified Ricci flow (MRF)

$$\partial_t h_{ab} = -2fR_{ab} + 2\nabla_a\nabla_b f;\ \partial_t f = \Delta f \tag{3.4}$$

with $f \geq \epsilon > 0$. Then fixed points of the flow characterized via

$$fR_{ab} = \nabla_a\nabla_b f;\ \Delta f = 0 \tag{3.5}$$

with $f = [1-(2M/r)]^{1/2}$ lead to the Schwartzschild exterior space

$$ds^2 = \left(1 - \frac{2M}{r}\right)^{-1} dr^2 + r^2(d\theta^2 + Sin^2(\theta)d\phi^2);\ (r > 2M) \tag{3.6}$$

as a fixed point of the MRF. Recall here that static spherically symmetric black holes involve maximal geometric entropy. Note that $f \to 0$ on the boundary $r = 2M$ and (**3B**) $(h_{ab}, f) \sim ds^2 = -f^2(x)dt^2 + h_{ab}x^a dx^b$. One can also replace asymptotic flatness by AdS and use MRF of the form

$$\partial_t h_{ab} = -2f(R_{ab} - \lambda h_{ab}) + 2\nabla_a\nabla_b f;\ \partial_t f = \Delta f + \lambda f \tag{3.7}$$

and the AdS Schwartzschild black hole is a "soliton" of this MRF with $f = [1-(2M/r)-(\lambda/3r^2)]^{1/2}$.

Now going to paper 0711.0430 one looks at the unmodified Ricci flow to see how some interesting quantities evolve with the flow. Recall here that the conjectured relation between the ADM mass of an initial data set for GR and the area or its outermost apparent horizon is expressed via $M_{ADM} \geq \sqrt{16\pi A}$ and this is saturated by the Schwarzschild space which means that Schwarzschild spacetime maximizes geometric entropy for a given energy. This (Penrose inequality) is entirely consistent with ascribing thermal character to Hawking radiation from a black hole and seemingly captures something deep about GR with a thermodynamic statement expressed in geometric terms (cf. also the area theorem in [**860**]). One motivation in the paper under discussion is to explore whether the Ricci flow would lead to a new proof of the time symmetric Penrose inequality. Thus let $(\Sigma, h_{ab})$ be an asymptotically flat 3-D Riemannian manifold (cf. [**651**]) and therefore in particular standard definitions of energy, entropy, and black holes apply; the total energy or ADM mass of an initial data set is only well defined if an asymptotic structure (either flat of AdS) is fixed. Hence one assumes $h_{ab} \to \delta_{ab} + O(1/r)$ at infinity with (**3C**) $\partial_{tau} h_{ab} = -2R_{ab}$. In the neighborhood of a point $p \in \Sigma$ one can introduce Riemann normal coordinate leading to (**3D**) $\partial_\tau h_{ab} = \nabla^2 h_{ab}$ (but note (**3C**) is a degenerate parabolic equation because of diffeomorphism invariance. More generally one can consider a modified Ricci flow (**3D**) $\partial_\tau h_{ab} = -2R_{ab} + \nabla_a \xi_b + \nabla_b \xi_a$ where $\xi^a$ is any vector field on $\Sigma$ which vanishes at infinity. It is convenient to consider the flows separately as a pure Ricci flow (**3C**) and a pure diffeomorphism (**3E**) $\partial_\tau h_{ab} = \nabla_a \xi_b + \nabla_b \xi_a$. It is not permissible here to add on term $\lambda h_{ab}$ since this would rescale the metric at infinity. In the standard initial value formulation of GR (cf. [**715**]) the basic variables are $h_{ab}$ on $\Sigma$ and the extrinsic curvature $k_{ab}$ of $\Sigma$. Setting $k = h^{ab} k_{ab}$ one has constraints

$$\nabla_b (k^{ab} - h^{ab} k) = 8\pi j^b; \; R + k^2 - k_{ab} k^{ab} = 16\pi\rho \tag{3.8}$$

where $j^b$ is the matter current and $\rho$ the matter density (the matter being required to satisfy the some energy condition (dominant, weak, or strong). One can view $(\Sigma, h_{ab})$ as a time symmetric initial data set for the Einstein equations (in particular the extrinsic curvature $k_{ab}$ of $\Sigma$ is set equal to zero so that $(\Sigma, h_{ab})$ is totally geodesic). With $k_{ab} = 0$ and $j^b = 0$ the diffeomorphism constraint in (3.5) is automatically satisfied and the Hamiltonian constraint reduces to $R = 16\pi\rho$. The energy conditions all imply the local energy condition ($\geq 0$) which implies that the scalar curvature $R$ of $h_{ab} \geq 0$; this is preserved under Ricci flow

$$\partial_\tau R = \nabla^2 R + 2R^{ab} R_{ab} + L_\xi R \tag{3.9}$$

which implies that $\partial_\tau R$ is positive at a minimum of $R$. In the present situation equation (**3E**) provides a flow on the space of initial data to the Einstein equations which is physically allowed (non-negative scalar curvature data).

Now let $S$ be a closed surface in $\Sigma$, $\gamma_{ij}$ $(i,j = 1,2)$ the induced metric on $S$, $\mathfrak{R}$ the scalar curvature for $(S, \gamma_{ij})$, and $K$ the trace of its extrinsic curvature. Let (**3F**) $\mathfrak{A}(S) = \int_S dA = \int_S d^2x\sqrt{\gamma}$ be the area of $S$ and (**3G**) $\mathfrak{M}_H(S) = (\sqrt{\mathfrak{A}(S)}/64\pi^{3/2}) \int dA(2\mathfrak{R} - K^2)$ the Hawking mass. The area of apparent horizons is related to the entropy of black holes and the Hawking mass is related to the energy. The latter vanishes in the limit that $S$ shrinks to a point and becomes the ADM energy for a round sphere at infinity. Unlike the ADM mass which is only well defined for asymptotic spheres the Hawking mass is defined for any closed surface. It is actually more convenient however to deal with the related dimensionless quantity (**3H**) $\mathfrak{C}(S) = \int_S dA(2\mathfrak{R} - K^2)$ called the compactness of $S$ which tends to zero as $S$ tends to a round sphere of infinitesimal radius and also as $S$ tends to an asymptotic round sphere.

Now one looks for spherically symmetric spaces which are nontrivially asymptotically flat in order to provide guidelines. There are two forms:

(1) The **a**-form with (**3I**) $ds^2 = a(r)dr^2 + r^2(d\theta^2 + Sin^2(\theta)d\phi^2)$ with

$$(3.10)\quad R_{\tau\tau} = \frac{a'}{ra};\ R_{\theta\theta} = \frac{a'r}{2a^2} + 1 - \frac{1}{a};\ R_{\phi\phi} = Sin^2(\theta)(R_{\theta\theta};\ R = \frac{2}{r^2} + \frac{2a'}{ra^2} - \frac{2}{ar^2}$$

(2) The **b**-form (**3J**) $ds^2 = dr^2 + b(r)(d\theta^2 + Sin^2(\theta)d\phi^2$ with

$$(3.11)\quad R_{\tau\tau} = \frac{(b')^2 - 2bb'}{2b^2};\ R_{\theta\theta} = 1 - \frac{b''}{2};\ R_{\phi\phi} = Sin^2(\theta)(R_{\theta\theta});$$

$$R = \frac{(b')^2 - 4b(b'' - 1)}{2b^2}$$

One shows that the **a**-form is unsuitable for treating apparent horizons and goes to the **b**-form with unit normal (**3K**) $h^{ab}\hat{n}_a\hat{n}_b = \hat{n}_r\hat{n}_r = 1$ and $\hat{n} = (1,0,0)$. The area of $S$ is (**3L**) $\mathfrak{A}(r) = \int_S \sqrt{\gamma}d\theta d\phi = 4\pi b(r)$ where $\gamma = b^2 Sin^2(\theta)$ is the determinant of $\gamma_{ij}$ on $S$. The trace of the extrinsic curvature is (**3M**) $K = \nabla_a\hat{n}^a = (b'/b)$ (where a prime indicates differentiation in $r$). The compactness formula reduces to (**3N**) $\mathfrak{C}(r) = 16\pi - \int_S \sqrt{\gamma}d\theta d\phi K^2 = 16\pi - (4\pi(b')^2/b)$ and the Hawking mass is (**3O**) $\mathfrak{M}_H(r) = (\sqrt{\mathfrak{A}(r)}/64\pi^{3/2})\mathfrak{C}(r) = (\sqrt{b}/2)[1 - ((b')^2/4b)]$. Under a pure Ricci flow the **b**-form may not be preserved but viewing S as a fixed surface in $\Sigma$ so that $dr/d\tau = 0$ and from the Ricci flow one has (**3P**) $(\partial h_{\theta\theta}/\partial\tau) = -2R_{\theta\theta}$ leading to (**3Q**) $(d\mathfrak{A}/d\tau) = 4\pi(\partial b/\partial\tau) = -4\pi(2 - b'')$ for the instantaneous rate of change of the area of S. Using (3.8) for the curvature one has in spherical symmetry

$$(3.12)\quad \frac{d\mathfrak{A}}{d\tau} = -\frac{1}{2}\int_S \sqrt{\gamma}d\theta d\phi R - \frac{1}{4}\mathfrak{C} \Rightarrow \frac{d\mathfrak{A}}{d\tau} \leq -\frac{1}{4}\mathfrak{C}$$

For the Schwarzschild space $R = 0$ and the first integral in (2.9) vanishes leading to (**3R**) $(d\mathfrak{A}/d\tau) = -(1/4)\mathfrak{C}(S)$ so this space saturates the inequality in (3.9) just as it saturates the Penrose inequality.

Now let $S$ be a minimal surface in $\Sigma$ (or apparent horizon - which coincides in the case of time symmetric data); thus $S$ is a closed 2-manifold embedded in $\Sigma$ with vanishing trace of the extrinsic curvature. To see how the area of $S$ varies under Ricci flow use the **b**-form of the metric with apparent horizon at $r = r_0$. From (**3M**) one has $K = (b'/b)|_{r=r_0} = 0$ and for $r = r_0$ to be an apparent horizon one requires (**3S**) $(b')|_{r=r_0} = 0$. Then one shows that a minimal surface cannot spontaneously appear if none was present initially (cf. [**769**]). Regions where $b' < 0$ are called trapped regions and since the Ricci flow is continuous such regions evolve continuously and they can shrink to zero and disappear. Setting aside such situations one checks how the area of minimal surfaces evolves under the pure Ricci flow. During the flow the metric changes and the location of the horizon may change so $r_0 = r_0(\tau)$; further the geometry of $S$ can change and both effects could lead to changes in area $\mathfrak{A}(r) = 4\pi b(r)$. Hence (**3T**) $4\pi(db/d\tau) = 4\pi[(\partial b/\partial r)|_{r=r_0}(dr_0/d\tau) + (\partial b/\partial\tau)|_{r=r_0}]$. The first term vanishes because of the apparent horizon condition (**3S**) and the second is evaluated by specializing (3.9) to an apparent horizon so the area $\mathfrak{A} = 4\pi b$ satisfies (**3U**) $(\partial\mathfrak{A}/\partial\tau) \leq -4\pi$. This implies that the area of the horizon is decreasing at least linearly with $\tau$. Since the area was finite to begin with one sees that if the horizon persists the function $b$ evaluated at the horizon goes to zero in a finite "time" $\tau$. Next one shows that as $b \to 0$ one approaches a singularity in $R$ (cf. [**769**]). One shows also that the rate of change of $\mathfrak{C}$ with $\tau$ is not monotonic and the Hawking mass is also not monotonic along the pure Ricci flow.

Now assume no minimal surface is initially present, and, knowing that none can develop under Ricci flow, one can use the **a**-form of the metric (via choice of a diffeomorphism). One has seen that $\mathfrak{C}(r)$ starts from 0 at $r = 0$, reaches a maximum (perhaps several local maxima) and then decays to zero as $r \to \infty$. $\mathfrak{C}_{max}$ is not affected by a diffeomorphism and moving the surface does not affect the value $\mathfrak{C}_{max}$. Choosing a diffeomorphism to preserve the **a**-form leads to

$$\frac{\partial a(r)}{\partial \tau} = \frac{a''(r)}{a(r)} - \frac{3(a')^2}{2a^2} - \frac{2(a-1) + (ra'(1-a)/a)}{r^2} \tag{3.13}$$

Focusing on the maximum value of $a(r)$ one recalls that $a(r) \geq 1$ and $a'(r)|_{max} = 0$ with $a'''(r)|_{max} \leq 0$; hence the maximum value of $a(r)$ is monotone non-increasing as the flow parameter increases, i.e.

$$\frac{\partial a(r)_{max}}{\partial \tau} = \frac{a''(r)}{a(r)} - \frac{2(a(r)|_{max} - 1)}{r^2} \leq 0 \tag{3.14}$$

Via an equation (**3V**) $\mathfrak{C}(r) = 16\pi[1 - (1/a)]$ from [**769**] a max of $\mathfrak{C}$ corresponds to max of $a(r)$ and a maximum principle for $a(r)$ implies a maximum principle for compactness $\mathfrak{C}(r)$ with (**3W**) $[d\mathfrak{C}(r)_{max}/d\tau] \leq 0$. One can use this maximum principle to comment on the long time existence of the spherically symmetric asymptotically flat Ricci flow. Suppose no minimal surface is initially present; then for the left side of (3.11) to vanish one must have $a(r)_{max} = 1$ which implies $a(r) = 1$ identically and describes flat space. If the initial metric is not flat space its $a(r)_{max}$ must decrease with the flow and finally attain the flat space fixed point $a(r) = 1$. This shows that in spherical symmetry the only asymptotically flat fixed point of the flow is flat space (cf. also [**651**]). In summary the maximum principle for compactness leads to a criterion for the existence of the Ricci flow in the asymptotically flat case. If there are no apparent horizons the flow exists for all $\tau$ and converges to flat space; if there are apparent horizons the Ricci flow either terminates in a finite time singularity or removes the horizons by mergers (cf. [**769**] for more detail).

Now one drops the assumption of spherical symmetry and deals with a general asymptotically flat manifold $\Sigma$ with one end at infinity and a fixed closed orientable surface $S$ of arbitrary topology embedded in $\Sigma$. The induced metric of $S$ is $\gamma_{ij}$ with extrinsic curvature $K_{ij}$ where $(i,j)$ denotes two dimensional indices in the tangent space to $S$ (sometimes written as projected indices $(a,b)$. The trace of $K_{ij}$ is $K = \gamma^{ij}K_{ij}$ and one asks how the area changes with $\tau$ (recall that the unit normal to $S$ is changing since the metric of $\Sigma$ is changing). One defines $S$ as the level set of a function $\eta$ on $\Sigma$ which is strictly increasing outward from $S$. Independently of any metric the normal $\eta_a = \nabla_a \eta$ is a well defined covector (not zero since $\eta$ is assumed to be non locally constant). The unit normal

$\hat{n}_a = \eta_a/(\eta\cdot\eta)^{1/2}$ depends on the metric and it can only change by a multiple of itself $d\hat{n}_a/dt = \alpha\hat{n}_a$. Differentiating $\hat{n}\cdot\hat{n} = 1$ gives (**3X**) $\alpha = (1/2)(dh^{ab}/d\tau)\hat{n}_a\hat{n}_a$ where $dh^{ab}/d\tau$ is defined as $dh_{ab}/d\tau$ with indices raised using $h^{ab}$. Starting from (**3F**) one computes now (**3Y**) $(d\mathfrak{A}/d\tau) = \int_S d^2x(d\sqrt{\gamma}/d\tau)$ and one sees that (**3Z**) $(d\sqrt{\gamma}/d\tau) = (1/2)\sqrt{\gamma}\gamma^{ab}(dh_{ab}/d\tau)$ leading to

$$\frac{d\mathfrak{A}}{d\tau} = \frac{1}{2}\int_S \sqrt{\gamma}d^2x(h^{ab} - \hat{n}^a\hat{n}^b)\frac{dh_{ab}}{d\tau} \tag{3.15}$$

Then using Ricci flow as in (**3C**) one obtains

$$\frac{d\mathfrak{A}}{d\tau} = \int_S \sqrt{\gamma}d^2x[\hat{n}^a\hat{n}^b R_{ab} - R] \tag{3.16}$$

From the Gauss-Codazzi formula this gives

$$\hat{n}^a\hat{n}^b R_{ab} - R = -\frac{1}{2}[R + \mathfrak{R} + (K_{ij}K^{ij} - K^2)] \tag{3.17}$$

which can be rearranged to

$$\frac{d\mathfrak{A}}{d\tau} = -\frac{1}{2}\int_S d^2x\sqrt{\gamma}\left[R + \left(K^{ij} - \frac{1}{2}K\gamma^{ij}\right)\left(K_{ij} - \frac{1}{2}K\gamma_{ij}\right)\right] - \frac{1}{4}\iint_S d^2x\sqrt{\gamma}(2\mathfrak{R} - K^2) \tag{3.18}$$

The second integral in (2.15) is $-\mathfrak{C}(S)/4$ and the first integral (of definite sign) can be dropped to yield (•) $(d\mathfrak{A}/d\tau) \leq -(\mathfrak{C}(S)/4)$ which in turn can be reexpressed in terms of the Hawking mass via (★) $(d\mathfrak{A}^{3/2}/d\tau) \leq -24\pi^{3/2}\mathfrak{M}_H(S)$. Thus the rate of decrease of area of a closed 2-surface under Ricci flow is bounded by the Hawking mass. One recalls that (•) is saturated by the spheres of Schwarzschild space where $a(r) = [1-(2M/r)]^{-1}$; in this situation $R = 0$ with shear free spheres ($K_{ij} = (1/2)K\gamma_{ij}$) so the first integral in (2.15) vanishes. One notes that for a flat space $(d\mathfrak{A}/d\tau) = 0$ so the left side of (★) vanishes leading to the conclusion that for all spheres in flat space the Hawking mass is non-positive (cf. also [**420**]). In fact the converse is also true, namely given positiive curvature flat space is the only one for which the Hawking mass is non-positive. To see this note that the supremum over $S$ of the Hawking mass is the ADM mass and if this vanishes the positive mass theorem implies that the space must be flat.

To study the area under diffeomorphism note that the metric changes as in (**3Z**) and hence from (3.12)

$$\frac{d\mathfrak{A}}{d\tau} = \frac{1}{2}\int \sqrt{\gamma}d^2x(h^{ab} - \hat{n}^a\hat{n}^b)2\nabla_a\xi_b = \int \sqrt{\gamma}d^2x[\nabla_a\xi^a - n^a n^b\nabla_a\xi_b] \tag{3.19}$$

Supposing $\xi^a$ is tangent to $S$ with $\tilde{\nabla}_a$ denoting the intrinsic covariant derivative to $(\gamma_{ab}, S)$ one has

$$\tilde{\nabla}_a\xi_b = \gamma_a^{a'}\gamma_b^{b'}\nabla_{a'}\xi_{b'} \Rightarrow \tilde{\nabla}_a\xi^a = \gamma^{ab}\tilde{\nabla}_a\xi_b = (h^{ab} - \hat{n}^a\hat{n}^b)\nabla_a\xi_b \tag{3.20}$$

Consequently

$$\frac{d\mathfrak{A}}{d\tau} = \int \sqrt{\gamma}d^2x[\tilde{\nabla}_a\xi_b] = 0 \tag{3.21}$$

since this is a divergence over a closed surface. Thus it is enough to consider the component of $\xi$ normal to S, namely consider $\xi^a = u\hat{n}^a$ to obtain

$$\frac{d\mathfrak{A}}{d\tau} = \int \sqrt{\gamma}d^2x(h^{ab} - \hat{n}^a\hat{n}^b)\nabla_a(u\hat{n}_b) \tag{3.22}$$

which yields

$$\frac{d\mathfrak{A}}{d\tau} = \int \sqrt{\gamma}d^2xu\gamma^{ab}\nabla_a n_b = \int \sqrt{\gamma}d^2xuK \tag{3.23}$$

As for the area of horizons under Ricci flow one finds from (**2H**) that since $K = 0$ the area changes according to

$$\frac{d\mathfrak{A}}{d\tau} \leq \frac{1}{4}\mathfrak{C}(S) = -\frac{8\pi\chi(S)}{4} \tag{3.24}$$

where $\chi(S)$ is the Euler characteristic of S. For a minimal surface of spherical topology one has (★★) $(d\mathfrak{A}/d\tau) \leq -4\pi$. This is unaffected by adding a diffeomorphism to the Ricci flow because of (2.20) and the minimal surface condition $K = 0$.

Now to study the evolution of Hawking mass it is enough to understand the evolution of compactness and from (**3H**) one sees that the first term $\int d^2x\sqrt{\gamma}\mathfrak{R}$ drops out on differentiation since it is a topological invariant by Gauss-Bonnet. The second term gives

$$\frac{d\mathfrak{C}(S)}{d\tau} = -\frac{d}{d\tau}\int K^2\sqrt{\gamma}d^2x = -\int 2K\sqrt{\gamma}\frac{dK}{d\tau}d^2x - \int K^2\frac{d\sqrt{\gamma}}{d\tau}d^2x \tag{3.25}$$

Using (**3Z**) for the second term in (2.22) and (••) $(dK/d\tau) = (d/d\tau)(\nabla_a\hat{n}^a) = (d\Gamma^a_{am}/d\tau)\hat{n}^m + \nabla_a(d\hat{n}^a/d\tau)$ one arrives at

$$\frac{d\mathfrak{C}}{d\tau} = -\int dA\,K\left[h^{ab}\hat{n}^c\nabla_c\frac{dh_{ab}}{d\tau} - 2\nabla_a\left(\frac{dh^{ab}}{d\tau}\hat{n}^b\right)\right. \tag{3.26}$$
$$\left.+\hat{n}^a\nabla_a\left(\frac{dh^{cd}}{d\tau}\hat{n}_c\hat{n}_d\right)\right] - \int dA\frac{K^2}{2}\left[\frac{dh^{cd}}{d\tau}\hat{n}_c\hat{n}_d + h^{ab}\frac{dh^{ab}}{d\tau}\right]$$

This gives the general evolution of the compactness for any one parameter family $h_{ab}(\tau)$ of metrics. Specializing to $(dh_{ab}/d\tau) = -2R_{ab}$ gives for Ricci flow (via Bianchi)

$$\frac{d\mathfrak{C}}{d\tau} = \int_S dA\{K^2(R + \hat{n}^a\hat{n}^bR_{ab}) - 2K[2R^{ab}\nabla_a\hat{n}_b - \hat{n}^a\nabla_a(R^{cd}\hat{n}_c\hat{n}_d)]\} \tag{3.27}$$

and for diffeomorphisms generated by $\xi^a$ the tangential component of $\xi^a$ does not cause any change in the integral (as for the area) so one writes $\xi^a = u\hat{n}^a$ and putting this into (3.23), with simplifications due to Gauss-Codazzi in the form (♦) $-2n^an^bR_{ab} + R = \mathfrak{R} + (k^{ij}k_{ij} - k^2)$, leads to

$$\frac{d\mathfrak{C}}{d\tau} = \int_S \left[2K\tilde{\nabla}^a\tilde{\nabla}_a u + uK\sigma^{ij}\sigma_{ij} + ukR - \frac{1}{2}uK(2\mathfrak{R} - K^2)\right]\sqrt{\gamma}d^2x \tag{3.28}$$

where $\tilde{\nabla}_a$ denotes the intrinsic covariant derivative within the surface and $\sigma^{ij} = K^{ij} - (1/2)\gamma^{ij}K$. If there exists a diffeomorphism such that $uK = 1$ (inverse mean curvature flow - IMC) one can obtain the Geroch inequality (•••) $(d\mathfrak{C}/d\tau) \geq -C/2$

(this is also saturated by the Schwarzschild space and was used to prove the Penrose inequality in [**479**]). For the Hawking mass (♦♦) $\mathfrak{M}_H(S) = (\sqrt{\mathfrak{A}(S)}/64\pi^{3/2})\mathfrak{C}(S)$ one has

$$\frac{d}{d\tau}\mathfrak{M}_H = \left(\frac{1}{64\pi^{3/2}}\right)\left(\frac{1}{2\sqrt{\mathfrak{A}}}\frac{d\mathfrak{A}}{d\tau}\mathfrak{C} + \sqrt{\mathfrak{A}}\frac{d\mathfrak{C}}{d\tau}\right) \quad (3.29)$$

Using (2.20) and (•••) one sees that the Hawking mass is monotonic under IMC. We refer to [**769**] for more details (see also [**818**] for results on the monotonicity of entaglement entropy along the Ricci flow).

## 4. MORE ON GRAVITATION

In Chapter 2, Section 1 we discussed a derivation of the Einstein equations and subsequently we have discussed in various places some relations between gravitation and thermodynamics (see e.g. Chapter 5 and earlier sections in Chapter 6). We go now to a survey of some fundamental ideas in gravitation and thermodynamics following [**15, 16, 139, 286, 380, 443, 471, 474, 490, 659, 660, 661, 662, 663, 664, 665, 666, 667, 668, 669, 670, 671, 672, 673, 674, 675, 676, 677, 860, 861, 889**]. In order to deal with this we need definitions at least of Killing vectors, horizons, Rindler frames, etc. Thus we recall (cf. [**715, 839, 860**]) that a covariant derivative is defined via a connection $\Gamma^{\alpha}_{\beta\gamma}$ as (**4A**) $A^{\alpha}_{\ ;\beta} = \nabla_{\beta}A^{\alpha} = \partial_{\beta} + \Gamma^{\alpha}_{\mu\beta}A^{\mu}$ and given a symmetric metric $g_{ab}$ the connection symbols (called then Christoffel symbols) are (**4B**) $\Gamma^{\alpha}_{\beta\gamma} = (1/2)g^{\alpha\mu}(\partial_{\gamma}g_{\mu\beta} + \partial_{\beta}g_{\mu\gamma} - \partial_{\mu}g_{\beta\gamma})$. Given a curve $\gamma \sim x^{\alpha}(\lambda)$ with tangent vector $u^{\alpha} = dx^{\alpha}/d\lambda$ one defines the Lie derivative of $A^{\alpha}$ along $\gamma$ via

$$L_uA^{\alpha} = \nabla_{\beta}A^{\alpha}u^{\beta} - (\nabla_{\beta}u^{\alpha})A^{\beta} \equiv L_uA = \partial_{\beta}A^{\alpha}u^{\beta} - \partial_{\beta}(u^{\alpha})A^{\beta} \quad (4.1)$$

$$L_uf = \partial_{\alpha}fu^{\alpha};\ L_up_{\alpha} = (\partial_{\beta}p_{\alpha})u^{\beta} + (\partial_{\alpha}u^{\beta})p_{\beta} = (\nabla_{\beta}p_{\alpha})u^{\beta} + (\nabla_{\alpha}u^{\beta})p_{\beta} \quad (4.2)$$

If $\phi_t : M \to M$ is a one parameter group of isometries $\phi_t g_{ab} = g_{ab}$ then the vector field $\xi^a$ generating $\phi_t$ is called a Killing vector field and a necessary and sufficient for $\phi_t$ to be a group of isometries is $L_{\xi}g_{ab} = 0$ which translates here into (**4C**) $\nabla_a\xi_b + \nabla_b\xi_a = 0$ since $L_{\xi}g_{ab} = \xi^c\nabla_c g_{ab} + g_{cb}\nabla_a\xi^c + g_{ac}\nabla_b\xi^c = \nabla_a\xi_b + \nabla_b\xi_a$. One writes for proper time (**4D**) $d\tau^2 = -ds^2 = -dx^{\mu}dx^{\nu}g_{\mu\nu}$ (thinking of signature $(-1,1,1,1)$) so for $\dot{x}^{\mu} = dx^{\mu}/d\lambda$ one writes $d\tau^2 = -\dot{x}^{\mu}\dot{x}^{\nu}g_{\mu\nu}d\lambda^2$ and a particle world-line (or geodesic) is defined by an action $I(x) = -m\int_{\lambda_a}^{\lambda_b} d\lambda\sqrt{-\dot{x}^{\mu}\dot{x}^{\nu}g_{\mu\nu}} = -m\int_{\lambda_a}^{\lambda_b}(d\tau/d\lambda)d\lambda$ (for $c = 1$ and a time-like curve of motion $\mathcal{C}$). Following [**715**] (cf. also [**839**]) the geodesic equation is (**4E**) $(\nabla_{\beta}u^{\alpha})u^{\beta} = \kappa u^{\alpha}$ where $u^{\alpha} \sim \dot{x}^{\alpha}$ is tangent to the curve $\mathcal{C}$ ($\kappa = dlog(L)/d\lambda$ where $L = \sqrt{-g_{ab}\dot{x}^a\dot{x}^b}$ here). Recall also that one uses the symbol $\sqrt{-g}$ in a Lorentz space as indicated and there are formulas like

$$\nabla_aA^a = \frac{1}{\sqrt{-g}}\partial_a\left(\sqrt{-g}A^a\right);\ \Gamma^{\mu}_{\mu a} = \frac{1}{\sqrt{-g}}\partial_a\sqrt{-g} \quad (4.3)$$

Recall also that one can write the Schwartzschild metric (for $G = c = 1$) as

$$ds^2 = -\left(1 - \frac{2M}{r}\right)dt^2 + \left(1 - \frac{2M}{r}\right)^{-1}dr^2 + r^2d\Omega^2 \quad (4.4)$$

and if $r = R(t)$ on the surface we have

$$(4.5) \qquad ds^2 = \left[ \left( \left(1 - \frac{2M}{R}\right) - \left(1 - \frac{2M}{R}\right)^{-1} \dot{R}^2 \right] dt^2 + R^2 d\Omega^2$$

We recall also the idea of Fermi normal coordinates given via

$$(4.6) \qquad g_{tt} = -1 - R_{tatb}(t)x^a x^b + O(x^3);\ g_{ta} = -\frac{2}{3}R_{tbac}x^b x^c + O(x^3);$$

$$g_{ab} = \delta_{ab} - \frac{1}{3}R_{abcd}x^c x^d + O(x^3)$$

In the preface to [**715**] it is pointed out that a most important aspect of black-hole space-times is that they contain an event horizon, a null hypersurface that marks the boundary of the black hole, and on this hypersurface there is a network (or conguence) of non-intersecting null geodesics called null generators of the horizon. We will try model a description of this following [**715**], along with extraction of other apparently standard information about relativity which I did not fully appreciate before; the lecture notes in [**839**] are also very informative for a novice in relativity and the 1984 book in [**860**] is of course justly famous for coverage and pedagogy. Thus one deals with null surfaces $\Sigma : \Phi(x^\alpha) = 0$ with parametric representations $x^\alpha = x^\alpha(y^a)$ where the $y^a$, $a = 1, 2, 3$ are intrinsic to the surface. The vector $\partial_\alpha \Phi$ is normal to $\Sigma$ and a unit normal is $n_\alpha = \epsilon \partial_\alpha \Phi / \sqrt{|g^{\mu\nu}\partial_\mu \Phi \partial_\nu \Phi|}$ where $\epsilon = \pm 1$ for timelike (resp. spacelike) $\Sigma$ ($n^\alpha$ should point in the direction of increasing $\Phi$, i.e. $n^\alpha \partial_\alpha \Phi > 0$). If however $\Sigma$ is a null surface then $g^{\mu\nu}\partial_\mu \Phi \partial_\nu \Phi = 0$ so $k_\alpha = -\partial_\alpha \Phi$ is used for normal vector (future directed when $\Phi$ increases toward the future). Since $k^\alpha$ is orthogonal to itself (i.e. $k^\alpha k_\alpha = 0$) it is also tangential to $\Sigma$ and one has a geodesic equation (**4F**) $\nabla_\beta k^\alpha k^\beta = \kappa k^\alpha$. The hypersurface is generated by null geodesics and $k^\alpha$ is tangent to the generators.

Next for $x^\alpha = x^\alpha(y^a)$, ($a = 1, 2, 3$) one writes $e^\alpha_a = \partial x^\alpha / \partial y^a$ which will be tangent to $\Sigma$ (and in the null case $e^\alpha_a k_\alpha = 0$). In the non-null case one will have $e^\alpha_a n_\alpha = 0$ where $n_\alpha$ is a unit normal and the induced metric on $\Sigma$ will be $h_{ab} = g_{\alpha\beta} e^\alpha_a e^\beta_b$. Also in the null case one can use coordinates $y^a = (\lambda, \theta^A)$ ($A = 1, 2$) with a two tensor $\sigma_{AB} = g_{\alpha\beta} e^\alpha_A e^\beta_B$ where e.g. $e^\alpha_A = (\partial x^\alpha / \partial \theta^A)_\lambda$ and $dx^\alpha = k^\alpha d\lambda$. For integration on hypersurfaces $\Sigma$ one uses $d\Sigma = |h|^{1/2} d^3 y$ when $\Sigma$ is non-null with e.g. (**4G**) $d\Sigma_\mu = \epsilon_{\mu\alpha\beta\gamma} e^\alpha_1 e^\beta_2 e^\gamma_3 d^3 y$. Here $\epsilon_{\mu\alpha\beta\gamma} = \sqrt{-g}[\mu\alpha\beta\gamma]$ where

$$(4.7) \qquad [\alpha\beta\gamma\delta] = \begin{cases} +1 & \text{if } \alpha\beta\gamma\delta \text{ is an even permutation of } 0123 \\ -1 & \text{if } \alpha\beta\gamma\delta \text{ is an odd permutation of } 0123 \\ 0 & \text{if any two indices are equal} \end{cases}$$

and (**4H**) $d\Sigma = \epsilon n_\alpha d\Sigma$. In the null case one identifies $y^1$ with $\lambda$ (the parameter for null generators) and coordinates $\theta^A$ (constant on the generators) with $e^\alpha_1 = k^\alpha, d^4 y = d\lambda d^2\theta$ and (**4I**) $d\Sigma_\mu = k^\nu dS_{\mu\nu} d\lambda$ with $dS_{\mu\nu} = \epsilon_{\mu\nu\beta\gamma} e^\beta_2 e^\gamma_3 d^2 y$. This can also be expressed as (**4J**) $d\Sigma_\alpha = -k_\alpha \sqrt{\sigma} d^2\theta d\lambda$ (with $\sigma = det[\sigma_{AB}]$ where $d\sigma$ is the 2-metric described above). Note also that (**4K**) $g^{\alpha\beta} = \epsilon n^\alpha n^\beta + h^{ab} e^\alpha_a e^\beta_b$.

Consider now intrinsic covariant derivatives via a tangent vector field (**4L**) $A^\alpha = A^a e^\alpha_a$; $A^\alpha n_\alpha = 0$; $A_a = A_\alpha e^\alpha_a$. Then the intrinsic covariant derivative of a 2-vector $A_a$ is the projection of $\nabla_\beta A_\alpha$ onto the hypersurface via (**4M**) $A_{a|b} = \nabla_\beta A_\alpha e_a^\alpha e_b^\beta$. First compute the right side of (**4M**) as

$$(4.8)\qquad \nabla_\beta A_\alpha e_a^\alpha E_b^\beta = \nabla_\beta(A_\alpha e_a^\alpha)e_b^\beta - A_\alpha(\nabla_\beta e_a^\alpha)e_b^\beta$$
$$= \partial_\beta A_a e_b^\beta - (\nabla_\beta e_{a\gamma})e_b^\beta A^c e_c^\gamma$$
$$= \frac{\partial A_\alpha}{\partial x^\beta}\frac{x^\beta}{\partial y^b} - e_c^\gamma(\nabla_\beta e_{a\gamma})e_b^\beta A^c = \partial_b A_a - \Gamma_{cab}A^c;\ \Gamma_{cab} = e_c^\gamma(\nabla_\beta e_{a\gamma})e_b^\beta$$

Equation (**4M**) then reads (**4N**) $A_{a|b} = \partial_b A_a - \Gamma^c_{ab}A_c$ which is a familiar form for the covariant derivative. One shows that (**4M**) can then be rewritten as (**4O**) $\Gamma_{a|b} = (1/2)(\partial_b h_{ca} + \partial_a h_{cb} - \partial_c h_{ab})$. Now the quantities $A_{a|b} = (\nabla_\beta A_\alpha)e_a^\alpha e_b^\beta$ are the tangential components of the vector $\partial_\beta A^\alpha e_b^\beta$ and to investigate normal components one can decompose it as in (**4K**) in the form

$$(4.9)\qquad (\nabla_\beta A^\alpha e_b^\beta = [\epsilon n^\alpha n_\mu + h^{am}e_a^\alpha e_{m\mu}](\nabla_\beta A^\mu)e_b^\beta$$
$$= \epsilon[n_\mu(\nabla_\beta A^\mu)e_b^\beta]n^\alpha + h^{am}[(\nabla_\beta A_\mu)e_m^\mu e_b^\beta]e_a^\alpha$$

One then introduces the extrinsic curvature (**4P**) $K_{ab} = (\nabla_\beta n_\alpha)e_a^\alpha e_b^\beta$ in terms of which one can write (**4Q**) $(\nabla_\beta A^\alpha)e_b^\beta = A^a{}_{|b}e_a^\alpha - \epsilon A^a K_{ab}n^\alpha$. This shows that $A^a{}_{|b}$ gives the purely tangential part of the vector field while $-\epsilon A^a K_{ab}$ represents the normal component (which vanishes if and only if the extrinsic curvature vanishes).

The Lagrangian approach to gravity involves a least action principle with variations vanishing on the boundary - but a boundary term is still necessary because of non-zero normal derivatives. Thus one considers $S_G[g] = S_H[g] + S_B[g] - S_0$ where

$$(4.10)\qquad S_H[g] = \frac{1}{16\pi}\int_V R\sqrt{-g}d^4x;$$
$$S_B[g] = \frac{1}{8\pi}\oint_{\partial V}\epsilon K|h|^{1/2}d^3y;\ S_0 = \frac{1}{8\pi}\oint_{\partial V}\epsilon K_0|h|^{1/2}d^3y$$

The matter action takes the form $S_M[\phi,g] = \int_V L(\phi,\partial_a\phi,g_{\alpha\beta})\sqrt{-g}d^4x$ and the variational principle involves $\delta g_{\alpha\beta}|_{\partial V} = 0$ (which means that $h_{ab}$ is held fixed during the variation). It is convenient to work with variations $\delta g^{\alpha\beta}$ and one arrives at (**4R**) $\delta g_{\alpha\beta} = -g_{\alpha\mu}g_{\beta\nu}\delta g^{\mu\nu}$ with (**4S**) $\delta\sqrt{-g} = -(1/2)\sqrt{-g}g_{\alpha\beta}\delta g^{\alpha\beta}$ (cf. (4.3)). There results

$$(4.11)\qquad \delta S_H = \int_V \delta\left(g^{\alpha\beta}R_{\alpha\beta}\right)d^4x$$
$$= \int_V \left(R_{\alpha\beta}\sqrt{-g}\delta g^{\alpha\beta} + g^{\alpha\beta}\sqrt{-g}\delta R_{\alpha\beta} + R\delta\sqrt{-g}\right)d^4x$$
$$= \int_V\left(R_{\alpha\beta} - \frac{1}{2}Rg_{\alpha\beta}\right)\delta g^{\alpha\beta}\sqrt{-g}d^4x + \int_V g^{\alpha\beta}(\delta R_{\alpha\beta})\sqrt{-g}d^4x$$

After some calculation one arrives at

$$(4.12)\qquad (16\pi)\delta S_H = \int_V G_{\alpha\beta}\delta g^{\alpha\beta}\sqrt{-g}d^4x - \oint_{\partial V}\epsilon h^{\alpha\beta}(\partial_\mu\delta g_{\alpha\beta})n^\mu|h|^{1/2}d^3y$$

where $G_{\alpha\beta} = R_{\alpha\beta} - (1/2)Rg_{\alpha\beta}$. Next for the variation of the boundary term one writes

$$(4.13) \quad K = \nabla_\alpha n^\alpha = (\epsilon n^\alpha n^\beta + h^{\alpha\beta})\nabla_\beta n_\alpha = h^{\alpha\beta}\nabla_\beta n_\alpha = h^{\alpha\beta}(\partial_\beta n_\alpha - \Gamma^\gamma_{\alpha\beta} n_\gamma)$$

and consequently

$$(4.14) \quad \delta K = -h^{\alpha\beta}\delta\Gamma^\gamma_{\alpha\beta} n_\gamma = -\frac{1}{2}h^{\alpha\beta}(\partial_\beta\delta_{\mu\alpha} + \partial_\alpha\delta_{\mu\beta} - \partial_\mu\delta_{\alpha\beta})n^\mu = \frac{1}{2}h^{\alpha\beta}\partial_\mu\delta g_{\alpha\beta}$$

(note tangential derivatives of $\delta_{\alpha\beta}$ vanish on $\partial V$). Hence one has **(4T)** $(16\pi)\delta S_B = \oint_{\partial V} \epsilon h^{\alpha\beta}\partial_\mu(\delta g_{\alpha\beta})n^\mu |h|^{1/2}d^3y$ which cancels out the second integral on the right in (4.12). Then since $\delta S_0 = 0$ one obtains

$$(4.15) \quad \delta S_G = \frac{1}{16\pi}\int_V G_{\alpha\beta}\delta g^{\alpha\beta}\sqrt{-g}d^4x$$

For the matter action one has

$$(4.16) \quad \delta S_M = \int_V \delta[L\sqrt{-g}]d^4x = \int_V \left(\frac{\partial L}{\partial g^{\alpha\beta}} - \frac{1}{2}Lg_{\alpha\beta}\right)\delta g^{\alpha\beta}\sqrt{-g}d^4x$$

If one defines the stress energy tensor by $T_{\alpha\beta} = -2(\partial L/\partial g^{\alpha\beta}) + Lg_{\alpha\beta}$ then the Einstein equations follow via

$$(4.17) \quad \delta S_M = -\frac{1}{2}\int_V T_{\alpha\beta}\delta g^{\alpha\beta}\sqrt{-g}d^4x; \ \delta(S_G + S_M) = 0 \Rightarrow G_{\alpha\beta} = 8\pi T_{\alpha\beta}$$

**REMARK 4.1.1.** The term $S_0 = (1/8\pi)\oint_{\partial V}\epsilon K_0|h|^{1/2}d^3y$ has no role in the equations of motion - it simply is a numerical factor in the action. It represents the gravitational action of flat space-time and makes the term $S_B - S_0$ well defined as $R \to \infty$. For a flat space-time $S_0 = 0$. ■

**REMARK 4.1.2.** For the 3+1 decomposition (as in ADM) use $t$ as a parameter with $t = c$ describing a family of nonintersecting spacelike hypersurfaces $\Sigma_t$ with coordinates $y^a$ and let $\gamma$ be a congruence of curves intersecting the $\Sigma_t$. Then $t^\alpha$ will be tangent to the congruence and $t^\alpha\partial_\alpha t = 1$ where $t^\alpha = (\partial x^\alpha/\partial t)|_{y^a}$ and one defines $e^\alpha_a = (\partial x^\alpha/\partial y^a)|_t$. Then $L_t e^\alpha_a = 0$ and one writes $n_\alpha = -N\partial_\alpha t$ so $n_\alpha e^\alpha_a = 0$ with $t^\alpha = Nn^\alpha + N^a e^\alpha_a$ where $N$ is the lapse and the 3-vector $N^a$ is called the shift. Consequently $dx^\alpha = t^\alpha dt + e^\alpha_a dy^a = (Ndt)n^\alpha + (dy^a + N^a dt)e^\alpha_a$ with

$$(4.18) \quad ds^2 = -N^2dt^2 + h_{ab}(dy^a + N^a dt)(dy^b + N^b dt)$$

where $h_{ab} = g_{\alpha\beta}e^\alpha_a e^\beta_b$ is the induced metric and $\sqrt{-g} = N\sqrt{h}$. ■

**REMARK 4.1.3.** For a Hamiltonian formulation of field theory (with one scalar field $q$) one writes $\dot{q} = L_t q$ (Lie derivative along the flow vector $t^\alpha$) and sets $p = [\partial(\sqrt{-g}L)/\partial\dot{q}]$ with $\mathcal{H}(p, q, \partial_q) = p\dot{q} - \sqrt{-g}L$ where $L$ denotes the Lagrangian. Then the Hamiltonian is defined via $H[p,q] = \int_{\Sigma_t}\mathcal{H}(p, q, q_{,a})d^3y$ (note in the coordinates $(t, y^a)$ $L_t q \sim \dot{q} \sim (\partial q/\partial t)$ and one writes for spatial derivatives $q_{,a} = \partial_\alpha q e^\alpha_a$). The action functional is then

$$(4.19) \quad S = \int_{t_1}^{t_2} dt \int_{\Sigma_t}(p\dot{q} - \mathcal{H})d^3y$$

and some calculation yields

$$\delta S = \int_{t_1}^{t_2} dt \int_{\Sigma_t} \left\{ - \left[ \dot{p} + \frac{\partial \mathcal{H}}{\partial q} - \left( \frac{\partial \mathcal{H}}{\partial q_{,a}} \right)_{,a} \right] \delta q + \left[ \dot{q} - \frac{\partial \mathcal{H}}{\partial p} \right] \delta p \right\} d^3 y \tag{4.20}$$

$$\delta S = 0 \Rightarrow \dot{p} = -\frac{\partial \mathcal{H}}{\partial q} + \left( \frac{\partial \mathcal{H}}{\partial q_{,a}} \right)_{,a} ; \; \dot{q} = \frac{\partial \mathcal{H}}{\partial p}$$

We will not try to develop the Hamiltonian theory here and refer [**715, 750, 831**] for more details. ∎

We go back now to the Schwarzschild black hole (cf. (4.4)-(4.5)). First recall the Kruskal-Szekeres coordinates and define

$$r^* = \int \frac{dr}{1-(2M/r)} = r + 2Mlog\left|\frac{r}{2M} - 1\right|; \; u = t - r^*, \; v = t + r^* \tag{4.21}$$

where $r^*(r) = (1/2)(v-u)$. Here $v$ (advanced time) and $u$ (retarded time) are "oblique" coordinates and one has a metric (**4U**) $ds^2 = -[1-(2M/r)]dudv + r^2 d\Omega^2$ which in the neighborhood of $r = 2M$ can be approximated via

$$ds^2 \simeq \mp \left(e^{-u/4M} du\right)\left(e^{v/4m} dv\right) + r^2 d\Omega^2 \tag{4.22}$$

which motivates a new set of coordinates (**4V**) $U = \mp exp[-u/4M]$ and $V = exp[v/4m]$ in terms of which

$$e^{r/2M}\left(\frac{r}{2M} - 1\right) = -UV; \; ds^2 = -\frac{32M^3}{r} e^{r/2M} dU dV + r^2 d\Omega^2 \tag{4.23}$$

which is manifestly regular at $r = 2M$. The coordinates $U$ and $V$ are called null Kruskal coordinates and one can draw an instructive diagram with perpendicular $U$, $V$ axes at 45 degrees rotation from the standard horizontal and vertical axes and directed up. Let sector I be bounded by the $+V$ axis ($r = 2M$) and the negative $U$ axis ($r = 2M$) and sector II be bounded by the positive $U$ and $V$ axes ($r = 2M$) with the black hole in the upper left $U, V$ quadrant lying betwween the hyperbola $r = 0$ and the axes $r = 2M$ ($U, V \geq 0$). The exterior $r > 2M$ is the entire upper right quadrant ($V \geq 0$ and $U \leq 0$). We do not have graphical technology on our computer but the picture should be clear (see [**715**] for details). On a Kruskal diagram all radial light rays move along curves $U = c$ or $V = c$ so light cones are oriented at 45 degrees and time-like world lines, which lie within the light cones, move with a slope larger than unity. The one-way character of the surface $r = 2M$ separating regions I and II is called the event horizon of the black hole and it is also called an apparent horizon (see [**715**] for a discussion - the concepts are different for non-stationary black holes and there is some discussion of charged rotating black holes in [**715**], e.g. Reissner-Nordström, Kerr, etc.). Now the vector $t^\alpha = \partial x^\alpha/\partial t$ is a Killing vector for the Schwarzschild space-time; it is timelike outside the black hole and space-like outside while it is null on the event horizon. One can write (**4W**) $g_{\alpha\beta} t^\alpha t^\beta = 1 - (2M/r)$ and thus the surface $r = 2M$ is also called a Killing horizon.

There is a more detailed and elaborate discussion of some of this in [**860**] (1984 book). A related picture arises in the Rindler space-time (cf. also [**743**]). Thus consider (**4X**) $ds^2 = -x^2dt^2 + dx^2$ ($-\infty < t < \infty$, $0 < x < \infty$. The singularity at $x = 0$ ($g_{ab} \to 0$ so $g^{ab}$ is singular at $x = 0$) has to be examined however, since in fact curvature behaves well at $x = 0$. The null geodesics are obtained from (**4Y**) $0 = g_{ab}k^ak^b = x^2k_t^2 + k_x^2 = 0 \Rightarrow dt/dx = \pm 1/x \Rightarrow t = \pm log(x) + c$. Write then (**4Z**) $u = t - log(x)$ and $v = t + log(x)$ to obtain ($\bullet$) $ds^2 = exp(v - u)dudv$. This does not yet extend the space-time since $-\infty < u, v < \infty$ still corresponds to $x > 0$. However ($\blacklozenge$) $U = -exp(-v)$ and $V = exp(v)$ yields $ds^2 = -dUdV$. The original Rindler space-time corresponds to $U < 0$ and $V > 0$ but there are no singularities so one can extend (à la Kruskal) to $-\infty < U, V < \infty$ with ($\bigstar$) $T = (1/2)(U + V)$, $X = (1/2)(V - U)$, and $ds^2 = -dT^2 + dX^2$ which is Minkowski space-time. The Rindler space-time is simply the wedge $X > |T|$ and corresponds to sector I of the Kruskal picture with axes $x = 0$, $t = 0$ and $x = 0$, $t = -\infty$ ($x = (X^2 - T^2)^{1/2}$ and $t = Tanh^{-1}(T/X)$). The behavior of the Schwarzschild space-time as $r \to 2M$ is analogous to Rindler space-time as $x \to 0$. Note the Kruskal metric can be written as

$$ds^2 = \frac{32M^3}{2}e^{-r/2M}(-dT^2 + dX^2) + r^2(d\theta^2 + Sin^2(\theta)d\phi^2) \tag{4.24}$$

so the congruences involve

$$\left(\frac{r}{2M} - 1\right)e^{r/2M} = X^2 - T^2;\ \frac{t}{2M} = log\left[\frac{T + X}{X - T}\right] = 2Tanh^{-1}\left(\frac{T}{X}\right) \tag{4.25}$$

**REMARK 4.1.4.** Black hole entropy and related matters will be treated in Section 5, at times in some detail, and we mention here only [**15, 139, 140, 289, 528, 675, 677, 861, 889**].

## 5. BLACK HOLES AND THERMODYNAMICS

We follow here a sequence of papers by Padmamabhan et al [**524, 523, 620, 677, 661, 662, 663, 664, 665, 666, 668, 672, 673, 674, 675, 676**] involving in particular the intriguing analogies between the gravitational dynamics of horizons and thermodynamics and supplement this with material from [**15, 16, 39, 119, 139, 140, 276, 286, 287, 289, 314, 338, 417, 421, 471, 474, 490, 524, 523, 528, 620, 634, 677, 761, 776, 858, 859, 861, 889**] (with apologies for omissions). The most attractive summary development seems to follow Padmanabhan's papers, some in collaboration with Kothawala, Mukhopadhyay, Paranjape, and Sarkar. One of the first steps was the derivation of the Einstein equations from "thermodynamics", going back to Jacobson [**474**] and extensions and variations on this theme were developed in [**661, 662, 664, 665, 675, 676, 677**] (see Section 1.2 of Chapter for material from [**664**]). We will discuss first a sketch of some subsequent work from [**676**] and follow this with a few remarks about recent connections in [**138, 139, 140, 259, 260, 286, 287, 289, 672, 675, 677, 793, 891**]. Some of the most explicit examples are in [**661, 662, 664, 665**] for spherically symmetric horizons in 4-D in which the Einstein equations can be interpreted as a thermodynamic relation $TdS = dE + PdV$ arising from virtual displacements of a horizon. This was derived from the Lagrangian $L_{EH} \propto R\sqrt{-g}$ and if gravity is indeed a long wave-length emergent phenomenon the EH action is just the first term in the expansion for the low energy effective action (see e.g.

[**676, 675**] for discussion). In [**676**] this is extended to Gauss-Bonnet corrections and Lanczos-Lovelock gravity in simple situations (and further extended in e.g. [**138, 139, 140, 259, 260, 525, 675, 677**]). In any event one finds that the identity $TdS = dE + PdV$ transcends Einstein gravity and indicates that there is a deep connection between the thermodynamics of gravitational horizons and the structure of quantum corrections to gravity.

For the EH case consider a static spherically symmetric space-time with a horizon described via

$$ds^2 = -f(r)c^2dt^2 + \frac{1}{f(r)}dr^2 + r^2 d\Omega^2 \tag{5.1}$$

with a simple zero of $f(r)$ at $r = a$ with $f'(a)$ finite so there is a horizon at $r = a$ with nonvanishing surface gravity $\kappa = f'(a)/2$. Periodicity in Euclidean time allows one to associate a temperature with the horizon as $k_B T = \hbar c\kappa/2\pi = \hbar c f'(a)/4\pi$ in normal units. The Einstein equation is $rf'(r) - (1 - f) = (8\pi G/c^4)Pr^2$, for $P$ a radial pressure evaluated at $r = a$, so that

$$\frac{c^2}{G}\left[\frac{1}{2}f'(a)a - \frac{1}{2}\right] = 4\pi P a^2 \tag{5.2}$$

Introduce now two solutions with different radii $a$ and $a + da$, multiply by $da$ in (5.2) and introduce an $\hbar$ factor <u>by hand</u> to obtain

$$\left(\frac{\hbar c f'(a)}{4\pi}\right) d\left(\frac{1}{4}4\pi a^2\right) - \left(\frac{1}{2}\frac{c^4 da}{G}\right) = Pd\left(\frac{4\pi}{3}a^3\right) \tag{5.3}$$

Consequently one reads off

$$S = \frac{1}{L_P^2}(4\pi a^2) = \frac{1}{4}\frac{A_H}{L_P^2};\ E = \frac{c^4}{2G}a = \frac{c^4}{G}\left(\frac{A_H}{16\pi}\right)^{1}/2 \tag{5.4}$$

where $A_H$ is the horizon area and $L_P^2 = G\hbar/c^3$. Thus the Einstein equations can be cast as a thermodynamic identity. One notes

(1) $TdS$ is completely classical and independent of $\hbar$ but $T \propto \hbar$ and $S \propto 1/\hbar$. This is analogous to statistical mechanics where $S \propto k_B$ and $t \propto 1/k_B$.
(2) In spite of superficial similarity (5.3) is different from the conventional first law of black hole (BH) thermodynamics (and from some previous derivations of relations between gravity and thermodynamics such as [**474**]) due to the presence of the $PdV$ term. This term is more attuned to Thorne's membrane paradigm for black holes.
(3) In standard thermodynamics one considers two equilibrium states differing infinitesimally inextensive variables like $S$, $E$, and $V$ while having the same values for intensive variables like $T$ and $P$. Then $TdS = dE + PdV$ in a connection between two quasistatic equilibrium states (cf. [**662, 676**] for more discussion).

Now take $\hbar = c = G = 1$ (natural units) and a generalization of the EH Lagrangian is determined via (**5A**) $L^D = \sum_{m=1}^K c_m L_m^{(D)}$ with

$$L_m^{(D)} = \frac{1}{16\pi}2^{-m}\delta^{a_1b_1\ldots a_mb_m}_{c_1d_1\cdots c_md_m} R^{c_1d_1}_{a_1b_1}\cdots R^{c_md_m}_{a_mb_m} \tag{5.5}$$

where $R^{ab}_{cd}$ is the $D$-dimensional Riemann tensor and the generalized ("determinant") tensor is totally antisymmetric in both sets of indices ($D = 2m$). For $D = 2m$, $16\pi L_m^{(2m)}$ is the Euler density of the $2m$-dimensional manifold and for $L_0 = 1/16\pi$, $c_0 \propto$ the cosmological constant. The EH Lagrangian is a special case of (**5A**) only when $c_1 \neq 0$ and such Lagrangians are free of ghosts and quasi-linear (cf. [**912**]). As an example consider the first correction term $L_2$ (Gauss-Bonnet Lagrangian); in 4-D this term is a total derivative with no higher order corrections so take $D > 4$. Then (A denotes action)

$$A = \int d^D x\sqrt{-g}\left[\frac{1}{16\pi}(R + \alpha L_{GB}\right] + A_M \tag{5.6}$$

where the GB Lagrangian has the form (**5B**) $L_{GB} = R^2 - 4R_{ab}R^{ab} + R_{abcd}R^{abcd}$ (corresponding to a superstring situation with $\alpha$ a positive definite inverse string tension). Thus the second term in (5.6) corresponds to a correction to Einstein gravity (all results will trivially generalize in the presence of a cosmological constant). The equation of motion for the semi-classical action in (5.6) is (**5C**) $G_{ab} + \alpha H_{ab} = 8\pi T_{ab}$ where

$$G_{ab} = R_{ab} - \frac{1}{2}g_{ab}R;\ H_{ab} = 2[RR_{ab} - 2R_{aj}R^j_b - 2R^{ij}R_{aibj} + R_a^{ijk} + R_a^{ijk}R_{bijk} - \frac{1}{2}g_{ab}L_{GB} \tag{5.7}$$

Consider again a static spherically symmetric solution of the form (**5D**) $ds^2 = -f(r)dt^2 + f^{-1}(r)dr^2 + r^2 d\Omega^2_{D-2}$ where $d\Omega^2_{D-2}$ denotes the metric of a $(D-2)$ dimensional space of constant curvature $k$ (one takes $k = 1$ for simplicity). The E-M tensor can be written (cf. [**837**]) $T^t_t = T^r_r = \epsilon(r)/8\pi$ (cf. [**599**]) and the equation of motion determining the only nontrivial metric component is

$$rf' - (D-3)(1-f) + \frac{\bar{\alpha}}{r^2}(1-f)[2rf' - (D-5)(1-f)] = \frac{2\epsilon(r)}{D-2}r^2 \tag{5.8}$$

where $\bar{\alpha} = (D-3)(D-4)\alpha$ (note that $D = 4$ and $\alpha = 0$ correspond to EH gravity (**5H**)). The horizon is obtained from the zeros of $f(r)$ and assuming only one zero at $r = a$, the temperature at this horizon being $T = \kappa/2\pi = f'(a)/4\pi$, one evaluates (5.8) to obtain

$$f'(a)\left[a + \frac{2\bar{\alpha}}{a}\right] - (D-3) - \frac{\bar{\alpha}(D-5)}{a^2} = \frac{2\epsilon(a)}{D-2}a^2 \tag{5.9}$$

Now one introduces a factor $dV$ and tries to find an expression $TdS = dE + PdV$. One multiplies both sides of (5.9) by $(D-2)A_{D-2}a^{D-4}da/16\pi$ where $A_{D-1} = 2\pi^{D/2}/\Gamma(D/2)$ is the area of a unit $(D-1)$ sphere. Identifying $P = T^r_r$ and $V = A_{D-2}a^{D-1}/(D-1)$ one obtains

$$\frac{\kappa}{2\pi}d\left(\frac{A_{D-2}}{4}a^{D-2}\left[1 + \left(\frac{D-2}{D-4}\right)\frac{2\bar{\alpha}}{a^2}\right]\right) - d\left[\frac{(D-2)A_{D-2}a^{D-3}}{16\pi}\left(1 + \frac{\bar{\alpha}}{a^2}\right)\right] = PdV \tag{5.10}$$

The first term on the left has the form $TdS$ with horizon entropy

$$S = \frac{A_{D-2}}{4}a^{D-2}\left[1 + \left(\frac{D-2}{D-4}\right)\frac{2\bar{\alpha}}{a^2}\right] \tag{5.11}$$

which is precisely the entropy for Gauss-Bonnet gravity calculated by several authors (cf. [**676**]) by more sophisticated methods. Further one can interpret the second term on the left in (5.10) as $dE$ where

$$E = \frac{(D-2)A_{D-2}a^{D-3}}{16\pi}\left(1 + \frac{\bar{\alpha}}{a^2}\right) \tag{5.12}$$

which also matches with the correct expression of energy for GB gravity without a cosmological constant.

We refer now back to (5.5) and look at

$$L_m^{(D)} = \frac{1}{16\pi}2^{-m}\delta^{a_1a_2\dots a_{2m}}_{b_1b_2\cdots b_{2m}}R^{b_1b_2}_{a_1a_2}\cdots R^{b_{2m-1}b_{2m}}_{a_{2m-1}a_{2m}} \tag{5.13}$$

Assume $D \geq 2K+1$ and ignore the cosmological constant for simplicity. These Lagrangians have the peculiar property that their variation leads to equations of motion equivalent to the ordinary partial derivatives of the Lagrangian density with respect to the metric components $g^{ab}$, namely

$$E^a_b = \sum_1^K c_m E^a_{b(m)} = \frac{1}{2}T^a_b;\ E^a_{b(m)} = \frac{1}{\sqrt{-g}}g^{ai}\frac{\partial}{\partial g^{ib}}(L_m^{(D)}\sqrt{-g}) \tag{5.14}$$

One considers the near horizon structure of the $E^t_t$ equation for a spherically symmetric metric of the form (**5D**) and will show that this can (also) be represented as $TdS = dE + PdV$. Thus consider the Rindler limit (cf. [**665**]) of such a metric by which one means to study the metric (**5D**) near the horizon at $r = a$ and bring it to the Rindler form

$$ds^2 = -N^2dt^2 + \frac{dN^2}{\kappa^2 + \mathcal{O}(N)} + \sigma_{AB}dy^Ady^B \tag{5.15}$$

This arises by using a coordinate system in which the level surfaces of the metric component $g_{00}$ (which vanishes at the horizon) define the spatial coordinate $N$ ($\kappa$ in the $g_{11}$ term will be the surface gravity of the horizon). Capital letters refer to the transverse coordinates on the $t = const.$, $N = const.$ surfaces of dimension $D-2$ and $\sigma_{AB}$ is the metric on these surfaces. Denoting the extrinsic curvature of these $(D-2)$ surfaces by $K_{AB}$ one can see that the Rindler limit gives

$$\sigma_{AB} = \sigma_{(1)AB} + \frac{N^2}{\kappa a}\sigma_{(1)AB} + \mathcal{O}(N^4); \tag{5.16}$$

$$\sigma^{AB} = \sigma^{AB}_{(1)} - \frac{N^2}{\kappa a}\sigma^{AB}_{(1)} + \mathcal{O}(N^4);\ K_{AB} = -\frac{N}{a}\sigma_{(1)AB} + \mathcal{O}(N^2)$$

where $\sigma_{(1)AB} = a^2\bar{\sigma}_{(1)AB}$ where $\bar{\sigma}_{(1)AB}$ is the metric on a unit $(D-2)$ sphere and $\sigma^{AC}_{(1)}\sigma_{(1)CB} = \delta^A_B$.

Next for the (near-horizon) structure of the $D$-dimensional Riemann tensor (dropping the $D$ for $D$-dim quantities), it turns out the components of the form

$R^{ti}_{jk}$ and $F^{jk}_{ti}$ will not contribute to the $E^t_t$ equation and the remaining components of $R^{ij}_{k\ell}$ are

$$R^{BC}_{NA} = g_{NN}(K^{C:B}_A - K^{B:C}_A) = \mathcal{O}(N);\ R^{AB}_{CD} = {}^{D-2}R^{AB}_{CD} + \mathcal{P}O(N^2); \tag{5.17}$$

$$R^{NA}_{NB} = \kappa\partial_N K^A_B + \mathcal{O}(N^2) = -\frac{\kappa}{a}\delta^A_B + \mathcal{O}(N^2);\ R^{NA}_{BC} = K^A_{C:B} - K^A_{B:C} = \mathcal{O}(N)$$

(here the colon : denotes a covariant derivative using the $(D-2)$-dimensional metric $\sigma_{AB}$). Note also that via maximal symmetry one has (**5E**) ${}^{(D-2)}R^{AB}_{CD} = (1/a^2)(\delta^A_C\delta^B_D - \delta^B_C\delta^A_D)$. Now to analyze the $E^t_t$ equation one need only analyze the terms

$$E^t_{t(m)} = \frac{1}{\sqrt{-g}}g^{tt}\frac{\partial}{\partial g^{tt}}(L^{(D)}_m\sqrt{-g}) \tag{5.18}$$

Since $R^{ij}_{k\ell} = g^{ja}R^i_{ak\ell}$ the derivatives with respect to $g^{tt}$ can be calculated and using symmetries, along with $(\partial\sqrt{-g}/\partial g^{tt}) = -(1/2)\sqrt{-g}g_{tt}$ and $g_{0N} = g_{0A} = 0$ for the static Rindler metric one arrives at

$$E^t_{t(m)} = \frac{1}{16\pi}\frac{m}{2^m}\delta^{a_1a_2\cdots a_{2m}}_{tb_2\cdots b_{2m}}R^{tb_2}_{a_1a_2}\cdots R^{b_{2m-1}b_{2m}}_{a_{2m-1}a_{2m}} - \frac{1}{2}L^{(D)}_m \tag{5.19}$$

One can show that the summations in (5.19) are cancelled by terms in $L^{(D)}_m$. Thus denote the terms in $L^{(D)}_m$ where the index $t$ appears at least once by $\{T\}$ and the others by $\{\bar{T}\}$ so that $L^{(D)}_m = \{T\} + \{\bar{T}\}$. For example in standard EH gravity $16\pi L^{(D)}_1 = R = 2R^t_t + R^{\alpha\beta}_{\alpha\beta}$ and evidently $2/r^t_t = \{T\}$ since $G^t_t = R^t_t - (1/2)R$ providing cancellation. Then one shows (cf. [**676**] for details)

$$\{T\} = \frac{2}{16\pi}\frac{m}{2^m}\delta^{a_1a_2\cdots a_{2m}}_{tb_2\cdots b_{2m}}R^{tb_2}_{a_1a_2}\cdots R^{b_{2m-1}b_{2m}}_{a_{2m-1}a_{2m}} \tag{5.20}$$

Comparison with (5.19) yields (**5F**) $Et_{t(m)} = -(1/2)\{\bar{T}\}$ and one notes that $\{\bar{T}\}$ is not a priori a null set since one assumes $D \geq 2m+1$. Putting this all together (cf. [**676**]) yields then

$$E^t_{t(m)} = \frac{\kappa m}{16\pi}\frac{1}{a^{2m-1}}(\delta^{A_1\cdots A_{2-1}}_{B_1\cdots B_{2m-1}}\delta^{B_1}_{A_1}\cdots\delta^{B_{2m-1}}_{A_{2m-1}} - -\frac{1}{2}L^{(D-2)}_m + \mathcal{O}(N) \tag{5.21}$$

and rearranging one has

$$\frac{\kappa m}{8\pi}\frac{D-2m}{a^{2m-1}}(\delta^{A_2\cdots A_{2m-1}}_{B_2\cdots B_{2m-1}}\delta^{B_2}_{A_2}\cdots\delta^{B_{2m-1}}_{A_{2m-1}} = 2E^t_{t(m)} + L^{(D-2)}_m + \mathcal{O}(N) \tag{5.22}$$

This leads to (cf. [**676**] for details)

$$\frac{\kappa}{2\pi}d\left(\sum_1^K\frac{m}{4}c_mA_{D-2}a^{D-2m}(\delta^{A-2\cdots A_{2m-1}}_{B_2\cdots B_{2m-1}}\delta^{B_2}_{A_2}\cdots\delta^{B_{2m-1}}_{A_{2m-1}}\right) \tag{5.23}$$

$$= PdV + \sum_1^K c_mA_{D-2}a^{D-2}L^{(D-2)}_m da$$

Here $\kappa/2\pi$ is the temperature $T$ one must identify the quantity inside the parentheses on the left side as the entropy $S$. A little argument using (**5E**) then leads to an entropy $S = \sum_1^K S^{(m)}$ with

$$S^{(m)} = 4\pi m c_m A_{D-2} a^{D-2} L_{m-1}^{(D-2)} = 4\pi m c_m \int_{\mathcal{H}} L_{m-1}^{(D-2)} \sqrt{\sigma} d^{D-2} y \tag{5.24}$$

Remarkably this is precisely the entropy of the horizon in Lanczos-Lovelock entropy (see e.g. [**477**]). Finally one computes

$$\sum_1^K c_m A_{D-2} a^{D-2} L_m^{(D-2)} da = d\left(\sum_1^K c_m E_{(m)}\right); \tag{5.25}$$

$$E_{(m)} = \frac{1}{16\pi} A_{D-2} a^{D-(2m-1)} \prod_2^{2m} (D-j)$$

(cf. [**676**] for details). Again the identification of $E = \sum_1^K c_m E_{(m)}$ as the energy associated with the horizon agrees with other calculations for spherically symmetric LL gravity (cf. [**151**]). In particular one can say that the energy associated with the horizon originates in the transverse geometry of the horizon. Consequently one has proved for the spherically symmetric case that the equation of motion can be recast in the form (**5G**) $(\kappa/2\pi)dS = dE + PdV$ with the differentials arising due to a change in radius of the horizon.

The emergence of a correct relation $TdS = dE + PdV$ from these procedures resonates well with an alternative perspective developed in [**675**]. This views semi-classical gravity as based on a generic Lagrangian of the form $L = Q_a^{\ bcd} R^a_{\ bcd}$ with $\nabla_b Q_a^{\ bcd} = 0$. The expansion of $Q_a^{\ bcd}$ in terms of derivatives of the metric tensor determines the structure of the theory uniquely, with zero order term the EH action and first order correction the Gauss-Bonnett action. Any such Lagrangian can be decomposed into surface and bulk terms as $\sqrt{-g}L = \sqrt{-g}L_{bulk} + L_{sur}$ whwere

$$L_{bulk} = 2Q_a^{\ bcd} \Gamma^a_{dk} \Gamma^k_{bc};\ L_{sur} = \partial_c[\sqrt{-g}V^c];\ V^c = 2Q_a^{\ bcd} \Gamma^a_{bd} \tag{5.26}$$

Both $L_{sur}$ and $L_{bulk}$ contain the same information in terms of $Q_q^{\ bcd}$ and hence can always be related to each other (cf. mypd,pdbn); it can be checked (cf. [**676**]) that

$$L = \frac{1}{2} R^a_{\ bcd} \left(\frac{\partial V^c}{\partial \Gamma^a_{bd}}\right);\ L_{bulk} = \sqrt{-g}\left(\frac{\partial V^c}{\partial \Gamma^a_{bd}}\right) \Gamma^a_{dk} \Gamma^k_{bc} \tag{5.27}$$

Further one can show that the surface term leads to the Wald entropy in spacetimes with horizon (cf. [**661, 664, 620**]) and there seem to be some as yet unresolved questions concerning Wald entropy and generalizations thereof (cf. [**138, 139, 140, 259, 260, 286, 287, 289, 675, 677, 793**]); some ambiguities are by-passed in [**675**] (0903.1254) by defining a specific Noether current.

We continue now with material from [**675**] (0807.2356) which considers gravity as an emergent phenomenon and suggests several items requiring answers. Thus

(1) Horizons cannot have temperature without the space-time having a microstructure.

(2) Einstein's equations reduce to a thermodynamic identity for virtual displacements of a horizon
(3) Einstein-Hilbert action is holographic with a surface term that encodes the same information as the bulk. In this connection one has a formula

$$\sqrt{-g}L_{surface} = -\partial_a\left(g_{ij}\frac{\partial\sqrt{-g}L_{bulk}}{\partial(\partial_a g_{ij})}\right) \tag{5.28}$$

More remarkable is the fact that one can also obtain the Einstein equations withough treating the metric as a dynamical variable.
(4) Why does the surface term in the EH action give the horizon entropy?
(5) Why do all these results hold for the much wider class of Lanczos-Lovelock gravities?

The following extract from a table of implications is also suggested and we refer to [**675**] details and discussion.

(1) The principle of equivalence
(2) $\Rightarrow$ Gravity is described by the metric $g_{ij}$
(3) $\Rightarrow$ The existence of local Rindler frames (LRF) with horizons H around any event
(4) $\Rightarrow$ The virtual displacements of H allow for the flow of energy across a hot horizon hiding an entropy $dS = dE/T$ as perceived by a given observer
(5) $\Rightarrow$ The local horizon must have an entropy $S_{grav}$
(6) $\Rightarrow$ The dynamics should arise from maximizing the total entropy of the horizon $S_{grav}$ plus matter $S_M$ for all LRF's without varying the metric
(7) $\Rightarrow$ The field equations are those of LL gravity with Einstein gravity emerging as the lowest order term
(8) $\Rightarrow$ The theory is invariant under the shift $T_{ab} \to T_{ab} + \Lambda g_{ab}$ allowing the bulk cosmological constant to be "gauged" away.

We refer to [**380, 471, 677, 861**] for discussions of the Wald entropy and gravity (cf. also [**39, 135, 139, 140, 212, 300, 328, 338, 380, 395, 528, 544, 858**]). First we recall the idea of conserved charge à la Noether (cf. [**251, 416, 544, 732, 860, 861**]. Thus, following [**860**] to put matters into context, consider a smooth Lagrangian $\mathcal{L}(\phi, \partial_a\phi)$ and recall the Euler-Lagrange equations (**5H**) $\partial\mathcal{L}/\partial\phi = \partial_a(\partial\mathcal{L}/\partial(\partial_a\phi))$ (see e.g. [**251, 416**]). Now going to e.g. [**860**] given a smooth one-parameter family $(\phi_\lambda, (\eta_\lambda)_{ab})$ of fields and flat metrics one has

$$\delta\mathcal{L} = \frac{d\mathcal{L}}{d\lambda} = \frac{\partial\mathcal{L}}{\partial\phi}\delta\phi + \frac{\partial\mathcal{L}}{\partial(\partial_a\phi)}\delta(\partial_a\phi) + \frac{\partial\mathcal{L}}{\partial\eta_{ab}}\delta\eta_{ab} \tag{5.29}$$

Consider the variations of $\mathcal{L}$ produced by a 1-parameter family of diffeomorphisms generated by a vector field $\xi^a$ leading to

$$\delta\mathcal{L} = L_\xi\mathcal{L} = \xi^a\partial_a\mathcal{L} = \frac{\partial\mathcal{L}}{\partial\phi}L_\xi\phi + \frac{\partial\mathcal{L}}{\partial(\partial_a\phi)}L_\xi\partial_a\phi + \frac{\partial L}{\partial\eta_{ab}}L_\xi\eta_{ab} \tag{5.30}$$

Taking $\xi^a$ to be a Killing field the last term in (5.11) vanishes and in the second term $L_\xi\partial_a\phi = \partial_a(L_\xi\phi)$ so using (5.10) there results

$$\partial_a\left[\frac{\partial\mathcal{L}}{\partial(\partial_a\phi)}L_\xi\phi - \xi^a\mathcal{L}\right] = 0 \tag{5.31}$$

This is called Noether's theorem as applied to the Poincaré group of symetries; in particular its validity for all translational Killing fields gives (**5I**) $\partial_a[(\partial\mathcal{L}/\partial(\partial_a\phi))-\xi^a\mathcal{L}]=0$ (note the $\xi^a$ are constant here) and this can be rewritten as (**5J**) $S^{ab}=[\partial\mathcal{L}/\partial(\partial_a\phi)]\partial^b\phi-g^{ab}\mathcal{L}=constant$ (i.e. $\partial_aS^{ab}=0$) where $S^{ab}$ is called the canonical energy-momentum tensor.

The Wald entropy is defined as a Noether charge entropy and its description was mainly carried out in differential form language by Wald et al; this is somewhat cumbersome and we refer to [**675, 677**] for a more Lagrangian approach. In [**677**] for example one adopts a simplified version of the formalism of [**162**]. Thus consider a generally covariant Lagrangian $L$ that depends on the Riemann tensor but does not contain derivatives of this tensor. Under the diffeomorphism $x^a\to x^a+\xi^a$ the metric changes via $\delta g_{ab}=-\nabla_a\xi_b-\nabla_b\xi_b$. By diffeomorphism invariance the change in the action when evaluated on-shell is given only by a surface term, leading to a conservation law $\nabla_aJ^a=0$ for which one can write $J^a=\nabla_bJ^{ab}$ where $J^{ab}$ defines (not uniquely) the antisymmetric Noether potential associated with the diffeomorphism $\xi^a$. For a Lagrangian $L=L(g_{ab},R_{abcd})$ direct computation shows that (cf. [**162**])

$$J^{ab}=-2P^{abcd}\nabla_c\xi_d+4\xi_d(\nabla_cP^{abcd});\ P^{abcd}=\frac{\partial L}{\partial R_{abcd}} \tag{5.32}$$

The Noether charge associated with a rigid diffeomorphism $\xi^a$ is defined via (**5K**) $Q=\int_S^{ab}dS_{ab}$ for a closed spacelike surface $S$. When $\xi^a$ is a time-like Killing vector (vanishing at the Killing horizon) it turns out from [**471, 861**] that the corresponding Noether charge describes the entropy $S$ associated with the horizon where $\kappa$ is the surface gravity of the black hole horizon. The integral for this "Wald entropy" can be evaluated over any space-like cross section of the Killing horizon (cf. [**476**]) and two special cases of Wald entropy are then, for $f(r)$ gravity and Gauss-Bonnett gravity respectively

$$S_f=f'(r)\frac{A}{4};\ S_{GB}=\frac{1}{4}\int d^{D-2}x\sqrt{\sigma}(1+2c_2{}^{(D-2)}R) \tag{5.33}$$

where ${}^{(D-2)}R$ is the scalar curvature of the cross-section of the horizon. One sketches in [**677**] a derivation of the equations of classical gravity via variation of these entropies plus the Clausius relation $\delta Q=T\delta S$ (cf. also [**139, 675**]).

CHAPTER 7

# ON TIME AND THE UNIVERSE

We begin with time and suddenly find ourselves immersed in the Chamseddine-Connes-Kreimer-Marcolli (CCKM) universe of ideas. A decision is made to try for understanding so we offer a sketch of some ideas and will give references for the main themes.

## 1. TIME

In [**77**] Barbour suggests that time, or more precisely duration, is redundant as a fundamental concept. Duration and the behavior of clocks emerges from a timeless law governing change. This all goes back to Poincare [**712**] and more recently the uncomfortable idea of a static universe led for example to deWitt's geometrodynamics (cf. [**262**]) and the famous Wheeler-deWitt equation discussed in Chapter 5 (see also below). Now astronomers have traditionally defined time so as to ensure that the laws of astronomy hold and there is a way to "see" this time. Mutually gravitating bodies have potential energy (**1A**) $V = -G\sum_{i<j}[m_i m_j/r_{ij}]$ where $G$ is Newton's gravitational constant. The kinetic energy $T$ can be written approximately as (**1B**) $T = \sum_i (m_i/2)(\delta d_i/\delta t)^2$ where $v_i \sim \delta d_i/\delta t$. Then conservation of energy dictates that (**1C**) $E = V + \sum_i (m_i/2)(\delta d/\delta t)^2$ is constant. The increment of time between successive "snapshots" is

$$\delta t = \sqrt{\frac{\sum m_i(\delta d_i)^2}{2(E-V)}} \tag{1.1}$$

and this is called ephemeris time following Einstein [**284**]. Accordingly the instantaneous velocity of a particle is

$$v_i = \frac{\delta d_i}{\delta t} = \sqrt{\frac{2(E-V)}{\sum m_i(\delta d)^2}} \tag{1.2}$$

so the elusive time variable has been eliminated. Now the configuration space of the universe is the key concept. In a universe of $N$ particles, each with 3 coordinates, one has a configuration space $U$ of dimension $3N$. The remarkably simple and beautiful least action principle of Jacobi singles out the special curves in $U$ for which Newton's laws hold. Thus pick 2 points in $U$ and divide trial curves joining them into short segments having action

$$\delta A = \sqrt{2}\sqrt{\left(E - \sum_{i<j}\frac{m_i m_j}{r_{ij}}\right)\sum m_i(\delta d_i)^2} \tag{1.3}$$

Now the action for a trial curve is the sum of such $\delta A$ terms and for one of the trial curves the action will be smaller than any other. For this extremal curve (and in general for no other such curve) the particles will obey Newton's laws with an emergent time defined via (1.1). There is also an interpretation of GR in which Newton's laws and time arise in much the same manner (cf. [**80**]).

Now we go to an approach whose elaboration eventually fits in with a grand sweeping mathematical theory embracing the standard model of physics (cf. [**230, 231, 233**]). "Relational time" has become popular in studies of quantum gravity (cf. [**751, 753, 754, 755**] and other (related) themes can be found in [**77, 78, 79**] (cf. also [**33, 34, 80, 160, 234, 305, 434, 559, 614, 727, 739, 812**]). In [**751**] one speaks of the disappearance of time and in [**434**] it is the emergence of time. These ideas in gravity are naturally related to the Wheeler-deWitt equation (cf. [**171, 175, 262, 353, 354, 355**] where in the Glinka version in Chapter 5, Section 7 the superspace metric is eliminated in favor of its determinant). We will extract here from [**234**] which will also serve as an introduction to von Neumann algebra techniques in covariant quantum theories, which is an important vehicle in generating the beautiful theories in [**231, 233**]. In any event there is now a great deal of background material needed for the development of non-commutative geometry and its connections to physics. It would take too long to recall all of the definitions so we will simply follow the exposition in [**234**] here and refer to [**175, 205, 206, 207, 208, 209, 230, 231, 232, 235, 236, 237, 505, 535, 562, 576, 853**] for the rest. We will emphasize the physics in keeping with the nature of this book; the mathematics, which is very complicated at times, can be found as needed in the references cited.

The cocycle Radon-Nikodym theorem (cf. [**230**]) shows that given a von Neumann algebra $M$, the 1-parameter group of automorphisms of $M$ (where $\sigma_t^\phi \in Aut(M)$, $\forall t \in \mathbf{R}$), associated by the Tomita-Takesaki (TT) theorem (cf. [**454, 827**]) to a faithful normal state $\phi \in M$, is independent (modulo inner automorphisms) to the choice of $\phi$. This means that the non-commutative von Neumann algebras are dynamical objects and inherit a natural time evolution from their non-commutativity. This is related to the association between equilibrium states and time flow. D'après Haag (cf. [**397**]) the relation between the TT construction on one side and QFT statistical mechanics in the KMS form (cf. [**762**]) represents a beautiful example of "prestabilized harmony". In [**752**] the problem of time in generally covariant theory led to a proposal of a state - dependent definition of physical time (which is shown in [**234**] to be exactly the semiclassical analogue of the map $\phi \to \sigma_t^\phi$ of the TT theorem). In quantum mechanics there are two equivalent ways of describing the time flow; namely the Schrödinger picture (flow in state space) or as a 1-parameter group of automorphisms of the algebra of observables (Heisenberg picture). Thus either one works with a flow $\alpha_t^S : \Gamma \to \Gamma$ ($\Gamma \sim$ phase space) or a 1-parameter group of automorphisms $\alpha_t f(s) = f(\alpha_t^S s)$ of the algebra of observables. In generally covariant theories such as GR there is however no preferred time flow and this has produced much discussion and various theories. A "radical" solution to this problem is based on the idea that one can extend the notion of time flow depending on the thermal state of the system and this approach was adopted in [**752**]. Thus let $\rho$ be a classical thermal state (namely a smooth positive normalized function on $\Gamma$) which defines a statistical distribution in the sense of Gibbs. In a conventional (non-gravitational) theory a Hamiltonian $H$ is given and the equilibrium thermal states are the Gibbs states $\rho = exp(-\beta H)$; the information on the time flow is encoded into the Gibbs state $\rho$ (as well as in the Hamiltonian) and the time flow $\alpha_t$ can be recovered from $\rho$ (up to the $\beta$ factor). Due to the Poisson structure of phase space the Gibbs distribution function $\rho$ naturally determines a flow (the symplectic flow of $-log(\rho)$) and this is precisely the time flow. In a generally covariant theory in which no preferred dynamics and no preferred Hamiltonian are given, a flow $\alpha_t^\rho$ (thermal time of $\rho$) is determined by any thermal state $\rho$ and hence one postulates that the thermal time $\alpha_t^\rho$ determines the physical time (cf. [**752**]). When the system is not generally covariant this will reduce to the Hamilton equations. Concrete examples show that this postulate leads to a realistic identification in a variety of cases (cf. [**752**]); in particular it works for a Robertson-Walker cosmological model and corresponds to the conventional FRW time (cf. [**752**]).

In a classical context the relation between a thermal state and a time flow is precisely the semiclassical analogue of the TT construction and, upon using the Radon-Nikodym (cocycle version), which yields the existence and uniqueness of the $\sigma_t$ in a general covariant context, one interprets this "natural" time evolution of the non-commutative algebras as the root of physical time. Thus states are positive linear functionals over the $C^*$ algebra of observables. Given a state the TT theorem provides a 1-parameter group of automorphisms $\alpha_t$ of the weak closure of the algebra (the modular group) and one postulates that the physical time is the modular flow of the thermal state. This yields a state-dependent definition of physical time flow in the context of a generally covariant QFT and the state independent characterization of time is given via the co-cycle Radon-Nikodym theorem so that thermal time is unique up to inner automorphisms. In more detail a concrete $C^*$ algebra is a linear space $\mathcal{A}$ of bounded linear operators on a Hilbert space H, closed under multiplication, adjoint conjugation $*$, and in the operator norm topology. A concrete vN algebra is a $C^*$ algebra closed in the weak topology. A positive operator $\omega$ with unit trace on H (in quantum mechanics a density matrix or physical state) defines a normalized positive linear functional over $\mathcal{A}$ via (**1D**) $\omega(A) = Tr[A\omega]$ for $A \in \mathcal{A}$. If $\omega$ is the projection operator on a "pure state" $\psi \in H$, i.e. if $\omega = |\psi><\psi|$, then (**1E**) $\omega(A) = <\psi|A|\psi>$. An abstract $C^*$ algebra and an abstract vN algebra (or $W^*$ algebra) are given by a set on which addition, multiplication, adjoint conjugation, and a norm are defined, satisfying the same algebraic relations as their concrete counterparts (cf. [**266**]), and a state $\omega$ over an abstract $C^*$ algebra is a normalized positive linear functional over $\mathcal{A}$. Given such a state the Gelfand-Naimark-Segal (GNS) construction provides a Hilbert space $H$ with a preferred state $|\psi_0>$ and a representation $\pi$ of $\mathcal{A}$ as a concrete algebra of operators on $H$ such that (**1F**) $\omega(A) = <\psi_0|\pi(A)|\psi_0>$ ($\pi(A)$ will be denoted by A in the following). The set of all states $\rho$ over $\mathcal{A}$ that can be represented as (**1G**) $\rho(A) = Tr[A\rho]$ where $\rho$ is a positive trace-class operator in $H$, is denoted as the folium (or leaf) determined by $\omega$. One considers now an abstract $C^*$ algebra $\mathcal{A}$ and a preferred state $\omega$. A von Neumann algebra $\mathcal{R}$ is then determined as the closure of $\mathcal{A}$ in the weak topology determined by the folium of $\omega$.

One considers 1-parameter groups of automorphisms $\alpha_t : \mathcal{R} \to \mathcal{R}$ of the vN algebra $\mathcal{R}$ ($t$ real). Fix a concrete vN algebra $\mathcal{R}$ on a Hilbert space $H$ and a cyclic and separating vector $|\psi> \in H$. Consider the operator $\mathfrak{S}$ defined by **(1H)** $\mathfrak{S}A|\psi> = A^*|\psi>$ for all $A \in \mathcal{R}$. One can show that this admits a polar decomposition **(1I)** $\mathfrak{S} = J\Delta^{1/2}$ where $J$ is antiunitary and $\Delta$ is a self adjoint positive operator. The TT theorem (cf. [**454, 827**]) asserts that the map $\alpha_t$ defined via **(1J)** $\alpha_t A = \Delta^{-it} A \Delta^{1/2}$ leaves $\mathcal{R}$ globally invariant and hence defines a 1-parameter group of automorphisms of the algebra $\mathcal{R}$. This is called the group of moduolar automorphisms or modular group of the state $\omega$ given in (**1E**). An automorphism $\alpha_{inner}$ of $\mathcal{R}$ is inner if there is a unitary element $U \in \mathcal{R}$ such that **(1K)** $\alpha_{inner} = U^*AU,\ \forall A \in \mathcal{R}$. Recall that two automorphisms $\alpha'$ and $\alpha''$ are equivalent when they are related by an inner automorphism $\alpha_{inner}$, i.e. **(1L)** $\alpha'(A)U = U\alpha''(A)$ for some $U \in \mathcal{R}$; the resulting equivalence classes of automorphisms are called outer automorphisms and the group of outer automorphisms of $\mathcal{R}$ is called $Out(\mathcal{R})$. In general the modular group $\alpha_t$ of (**1J**) is not a group of inner automorphisms and thus $\alpha_t$ projects down to a non-trivial 1-parameter group $\tilde{\alpha}_t \in Out(\mathcal{R})$. The co-cycle Radon-Nikodym theorem (cf. [**230**]) states that two modular automorphisms defined by two states of a vN algebra are inner-equivalent and consequently all states of a vN algebra determine the same 1-parameter group in $Out(\mathcal{R})$, i.e. $\tilde{\alpha}_t$ does not depend on $\omega$ and a vN algebra possesses a canonical 1-parameter group of outer automorphisms.

Now the idea is to ascribe to thermodynamics not so much the direction of time flow but rather the time flow itself. In the context of a conventional non-generally covariant QFT states are described by the KMS (Kubo-Martin-Schwinger) condition (cf. [**762**]). Thus let $\mathcal{A}$ be an algebra of quantum operators $A$ and consider the 1-parameter family of automorphisms of $\mathcal{A}$ defined via **(1M)** $\gamma_t A = exp(itH/\hbar)Aexp(-itH/\beta)$ with $H$ the Hamiltonian. Putting $\hbar = 1$ one says that a state $\omega$ over $\mathcal{A}$ is a KMS state at inverse temperature $\beta = 1/k_B T$ with respect to $\gamma_t$ if the function **(1N)** $f(t) = \omega(B(\gamma_t A))$ is analytic in the strip $0 < \Im t < \beta$ and **(1O)** $\omega((\gamma_t A)B) = \omega(B_{t+i\beta}A))$ for all $A, B \in \mathcal{A}$. In [**398**] it is shown that this KMS condition reduces to the well known Gibbs condition **(1P)** $\omega = Nexp(-\beta H)$ in the case of systems with a finite number of degrees of freedom, and this can be extended to systems with an infinite number of degrees of freedom. It is then postulated that the KMS condition represents the correct physical extension of the Gibbs condition (**1P**) to infinite dimensional quantum systems. There is an interesting connection of this formalism and the TT theorem. Thus any faithful state is a KMS state (with $\beta = 1$) with respect to the modular automorphism $\alpha_t$ that it generates (cf. [**266**]). Therefore in a non-generally covariant theory an equilibrium state is a state whole modular automorphism group is the time translation group (with time measured in units of $\hbar/k_B T$). Thus in the context of classical mechanics an equilibrium quantum thermal state contains all the information on the dynamics which is contained in the Hamiltonian (except for the constant factor $\beta$ which depends on the time unit). Given that the universe around us is in a state of non-zero temperature this means that the information about the dynamics can be fully replaced by the information about the thermal state. Now look at a generally covariant quantum theory given by an algebra $\mathcal{A}$ of generally covariant physical operators and a set of states $\omega$ over $\mathcal{A}$. By making a small number of physical observations one can determine a (generally) impure state $\omega$ in which the system lies. The hypothesis suggested is then

- The physical time depends on the state and given $\omega$ the physical time is given by the modular group $\alpha_t$ of $\omega$.

(cf. (**1J**) for definition of the modular group). The time flow defined on the algebra of observables by the modular group is called the thermal time. The fact that the time is determined by the state, and therefore that the system is always in an equilibrium state with respect to thermal time flow, does not imply that evolution is frozen or that we cannot detect any dynamical change. In a quantum system with an infinite number of degrees of freedom what one generally measures is the effect of small perturbations around a thermal state. In a QFT one can extract all the information in terms of vacuum expectation values of products of field operators (i.e. by means of a single quantum state $|0>$. For example if $\phi$ is a scalar field the propagator (at a fixed space point $x$) is given by

$$F(t) = <0|\phi(x,t)\phi(x,0)|0> = \omega_0(\gamma_t(\phi(x,0))\phi(x,0)) \tag{1.4}$$

where $\omega_0$ is the vacuum state over the field algebra and $\gamma_t$ is the time flow (**1M**). For a generally covariant QFT, given the quantum algebra of observables $\mathcal{A}$, and a quantum state $\omega$, the modular group of $\omega$ gives us a time flow $\alpha_t$ and then the theory describes physical evolution in the thermal time in terms of amplitudes of the form (**1Q**) $F_{A,B}(t) = \omega(\alpha_t(B)A)$ for $A, B \in \mathcal{A}$.

**REMARK 1.1.1.** In a theory where a geometrical definition of time is present the Gibbs states are the states for which the two time flows are proportional with proportionality constant $T$. Eddington in [**281**] speculated that any clock is necessarily a thermodynamical clock. ■

If one applies the thermal time hypothesis to a non-generally covariant system in thermal equilibrium there results immediately some known physics. Consider e.g. a system with a large but finite number of degrees of freedom with quantum state then given by the Gibbs density matrix (**1R**) $\omega = Nexp(-\beta H)$ where H is the Hamiltonian and (**1S**) $N^{-1} = Tr[exp(-\beta H)]$. The modular flow of $\omega$ is (**1T**) $\alpha_t A = exp(i\beta tH)Aexp(-i\beta tH)$ (with a rescaling $t \to \beta t$ (this can be directly proven - cf. [**234**]). Alternatively one can explicitly construct the modular flow from $\omega$ in terms of the operator $\mathfrak{S}$ as in (**1H**)-(**1J**). This is worth describing since it reveals the meaning of the mathematical objects in the TT theorem. $\omega$ is not given by a vector of the Hilbert space so in order to apply the TT construction one has to develop a new representation of the observable algebra in which $\omega$ is given by a vector $|0>$ using the GNS construction. A shortcut comes via [**398**] by considering the set $\mathcal{K}$ of Hilbert-Schmidt density matrices on H, namely the operators $k$ such that (**1U**) $Tr[k^*k] < \infty$. One constructs a (reducible) representation of the quantum theory on a subspace of $\mathcal{K}$ with the thermal Gibbs state given by a pure vector. Thus the operator $k_0 = \omega^{1/2}$ is in $\mathcal{K}$ and one writes it as $|k_0>$. If $A$ is a quantum operator defined on $H$ then write (**1W**) $A|k>=|Ak>$ which is again in $\mathcal{K}$. Taking $|k_0>$ as the cyclic state and acting on it with all the operators A in the observable algebra one obtains a new representation with $|k_o>$ playing the role of a vacuum state - but one can either increase or decrease its energy as required for a thermal state. Note that $|k_o>$ is time invariant since (**1X**) $U_{\mathcal{K}}(t)k_0>=|k_0>$ and a generic state $|Ak_0>$ is time translated via

$$U_{\mathcal{K}}|Ak_0>=|\gamma_t(A)k_0>=|e^{itH}Ae^{-itH}k_0>=|e^{itH}Ak_0e^{-itH}> \tag{1.5}$$

where $\gamma_t$ is given via (**1M**). Therefore in general (**1Y**) $U_{\mathcal{K}}|k>=|e^{ith}ke^{-itH}>$ and all of this still exists in the thermodynamic limit where the number of degrees of freedom tends to infinity. Although (**1R**) loses meaning the $\mathcal{K}$ representaation still exists and includes all the physical states that are formed by finite excitations around (over and below) the thermal state $|k_0>$. Now the modular group depends on the operator $\mathfrak{S}$ defined in (**1I**) and one has

$$\mathfrak{S}A|k_0>=A^*|k_0>=|A^*k_0>;\ \mathfrak{S}Ae^{-\beta H/2}=A^*e^{-\beta H/2} \tag{1.6}$$

since $k_0=\omega^{1/2}=N^{1/2}exp[-\beta H/2]$. Then $\mathfrak{S}=J\Delta^{1/2}$ is given via (**1Z**) $J|k>=|k^*>$ and $\Delta^{1/2}|k>=|exp[-\beta/2H]kexp[\beta/2H]>$. In fact

$$\mathfrak{S}A|k_0>=J\Delta^{1/2}|AN^{1/2}e^{-\beta H/2}>=J[e^{-\beta/2H}AN^{1/2}e^{-\beta H/2}e^{\beta/2H}> \tag{1.7}$$

$$=J[e^{-\beta/2H}AN^{1/2}>=|N^{1/2}A^*e^{-\beta/2H}>=A^*|k_0>$$

The operator J exchanges creation operators with annihilation operators around the thermal vacuum. The modular automorphism group $\alpha_t$ is then given via (**1J**) as

$$\alpha_tA|k_0>=\Delta^{-it}A\Delta^{it}|k_0>=\Delta^{-it}A|k_0> \tag{1.8}$$

$$=|e^{i\beta tH}Ak_0e^{-i\beta tH}>=|e^{i\beta tH}Ae^{-i\beta tH}k_0>$$

namely (♦) $\alpha_tA=exp(i\beta tH)Aexp(-i\beta tH)$ so that (★) $\alpha_t=\gamma_{\beta t}$. Thus the modular group of the Gibbs state is the time evolution flow up to a rescaling $\beta$ of the time unit and applying the thermal time postulate to the Gibbs state (**1R**) gives a definition of physical time which is proportional to the standard non-relativistic time. An interesting fact here (discussed in [**234, 752**]) is that in a special relativistic system a thermal state breaks Lorentz invariance.

**REMARK 1.1.2.** Regarding the classical limit one returns to a fully generally covariant system and considers a state $\omega$ and an observable $A$. From the definition of the modular group one has

$$\alpha_tA=e^{-itlog(\Delta)}Ae^{itlog(\Delta)}\Rightarrow\dot{A}=\frac{d}{dt}\alpha_tA|_{t=0}=i[A,log(\Delta)] \tag{1.9}$$

(note $[A,\Delta]=[A,\omega]$). In the classical limit one replaces observables and density matrices with functions on the phase space and commutators are replaced by

Poisson brackets. One thinks of $A \sim A$ and $\omega \sim (f,\omega)$ leading to

$$\frac{d}{dt}f = \{-log(\rho), f\} \sim \frac{d}{dt}f = \{H, f\};\ H = -log(\rho) \sim \rho = e^{-H} \tag{1.10}$$

Thus one obtains a classical Hamiltonian $H$ that generates the Hamiltonian evolution and the state $\rho$ is related to $H$ by the Gibbs relation. Note that (1.10) is contained in the TT relation (**1J**). These relations hold in general for a generally covariant theory but one recalls that $H$ plays a very different role than in a non-relativistic theory; it does not determine the Gibbs state but rather it is determined by any thermal state. ∎

## 2. SPACETIME, MATHEMATICS, AND PHYSICS

Since we have "obligatorily" entered the CCKM universe of ideas it seems necessary to expand on this. However this subject and its ramifications go to the frontiers of modern mathematics and the material becomes so rich (and rarified) so as to be digestible only in small quantities. The survey papers in [**205, 207**] (hep-th 0812.0165 and 0901.0577) and the books [**232, 853**] proceed at a rapid pace in an elevated manner so we only sketch a few background ideas and development from [**206, 207, 209, 237**]. There is some (overly dense) exposition in [**175**] of the definitions and a lovely discussion in [**898**] with calculations and examples; we will begin here with extractions from [**898**] (cf. also [**581**]).

Thus one recalls that the classical atom can be characterized by a set of positive real numbers, $\{\nu_i\}$ called fundamental frequencies. The atom radiates with intensity (**2A**) $I_n \propto |<\nu, n>|^4$ where $<\nu, n> d\sum n_i\nu_i$ ($n_i \in \mathbf{Z}$) and there is a group $\Gamma = \{<n,\nu>;\ n_i \in \mathbf{Z}\}$ (evidently given $<\nu,n>$ and $<\nu,n'>$ one has $<\nu, n+n'> = \sum(n_i+n_i')\nu_i$). The algebra of classical observables of this atom can be obtained as the convolution algebra of $\Gamma$. To see this write out a phase space function as an almost periodic function

$$f(q,p,t) = \sum f(q,p,n)e^{2\pi i<n,\nu>t};\ (n = (n_1,\cdots,n_k)) \tag{2.1}$$

There results (note $G$ is an Abelian group)

$$(f\bigstar g)(q,p,t,n) = \sum_{n_1+n_2=n} f(q,p,t,n_1)g(q,p,t,n_2) \tag{2.2}$$

where $f(q,p,t,n) = f(q,p,n)exp[2\pi i<n,\nu>t)$ leading to

$$fg(q,p,t) \equiv f(q,p,t)g(q,p,t) = \sum(f\bigstar g)(q,p,t,n) \tag{2.3}$$

The Ritz-Rydberg combination principle says that (♣): Frequencies in the spectrum are labeled with two indices and (♠) : $(i,j) = (i,k)\circ(k,j)$ and there is a composition law (**2B**) $\nu_{ij} = \nu_{ik} + \nu_{kj}$ written as $(i,j) = (i,k)\circ(k,j)$ so the frequencies and not parametrized by the group $\Gamma$ but by the groupoid $\Delta$ of pairs $(i,j)$ (with $\circ$ associative). The quantum group (cf. [**175**]) of observables is then the convolution algebra of $\Delta$ and this is a non-commutative matrix algebra as seen from rewriting (2.2) as

$$(F_1F_2)_{(i,j)} = \sum_{(i,k)\circ(k,j)} = (F_1)_{(i,k)}(F_2)_{(k,j)} \tag{2.4}$$

To implement the matrix aspect one should replace $f(q,p,t,n) = f(q,p,n)e^{2\pi i<n,\nu>t}$ by

$$F(Q,P)_{(i,j)} = F(Q,P)_{(i,j)}e^{2\pi i\nu_{ij}t} \tag{2.5}$$

**DEFINITION 2.1.1.** We recall that an operator T on a Hilbert space is compact if it can be approximated in norm by finite rank operators, i.e. for all $\epsilon > 0$ there exists a finite dimensional space $E \subset H$ with $\|T_{E^\perp}\| < \epsilon$. Alternatively compact operators can be defined via possession of a uniformly norm convergent expansion $T = \sum_{n\geq 0} \mu_n(T)|\psi_n >< \phi_n|$ where $0 \leq \mu_{i+1}(T) \leq \mu_i(T)$ while $\{|\psi_n >\}$ and $\{|\phi_n >\}$ are orthonormal sets. In particular the size of a compact operator is governed by the rate of decay of the $\mu_n(T)$ sequence as $n \to \infty$. One can also write $T = U|T|$ where $|T| = \sqrt{T^*T}$ with $\mu_n(T)$ (essentially) an eigenvalue of $|T|$ built on a $|\phi_n >$ expansion. ■

The characteristic values $\mu_n(T)$ satisfy (for $T_i$ compact and $T$ bounded)

$$\mu_{n+m}(T_1+T_2) \leq \mu_n(T_1) + \mu_m(T_2);\ \mu_n(T_1T) \leq \|T\|\mu_m(T_1); \tag{2.6}$$
$$\mu_{m+n}(T_1T_2) \leq \mu_n(T_1)\mu_m(T_2);\ \mu_n(TT_1) \leq \|T\|\mu_n(T_1)$$

For $n = m = 0$ such inequalities can be shown by using the fact that

$$\mu_0(T) = sup\{\|T|\chi >\| :\ |\chi \in H,\ \||\chi|\| \leq 1\} = \|T\| \tag{2.7}$$

(the latter being the operator norm). One says now that a compact operator $T$ is of order $\alpha \in \mathbf{R}^+$ if and only if there exists $C < \infty$ with $\mu_n(T) \leq Cn^{-\alpha}$, $\forall n \geq 1$; this is equivalent to $\mu_n = O(n^{-\alpha})$ as $n \to \infty$. On a $d$-dimensional manifold $M$ the Dirac operator $D = D^+ = |D|$ has eigenvalues (Weyl formula)

$$\mu_j(D) = 2\pi\left(\frac{d}{\Omega_d vol(M)}\right)^{1/d} j^{1/d} \tag{2.8}$$

(for large $j$). Hence $D^{-1}$ is infinitesimal of order $1/d$ and $D^{-d}$ is infinitesimal of order 1. For the trace one writes

$$\sigma_N(T) = Tr(T)_N = \sum_{n=0}^{N-1} \mu_n(T) \leq c log(N) + \frac{c'}{N} + finite\ terms \tag{2.9}$$

Thus the ordinary trace is at most logarithmcally divergent and should be replaced by

$$Tr(T) \to Lim_{N\to\infty}\gamma_N(T) :\ \gamma_N(T) = \frac{\sigma_N(T)}{log(N)} = \frac{1}{log(N}\sum_0^{N-1}\mu_N(T) \tag{2.10}$$

One sees that $\gamma_N$ is not linear and replaces (2.10) by (**2C**) $Tr_w(T) = Lim_w\gamma_N(T)$ (Dixmier trace) where $Lim_w$ is a linear form on the space of bounded sequences $\{\gamma_N\}$. This "trace" is not unique and is mainly just a way in which to extract the coefficient of the logarithmic divergence (see [**230, 535, 898**] for discussion). Note that infinitesimals of order 1 are in the domain of the trace and those of order higher than 1 have vanishing trace. This is discussed in terms of the quantum mechanics of non-standard analysis in [**235**] (math.QA 0111093); for PSDO T of order one with kernel $k(x,y) = -a(x)log|x-y| + O(1)$ near the diagonal one writes $⨍T = \int_M a(x)$ for the NC residue.

**REMARK 2.1.1.** We recall from [**184, 175, 230, 535**] that given a commutative $C^*$ algebra $C$ with unit one can find Hausdorff space $M$ such that $C$ is isometrically *-isomorphic to the algebra of complex continuous functions $C(M)$. Indeed one looks at the Gelfand structure space $\hat{C}$ as the equivalence classes of irreducible representations of $C$ (or maximal ideal space). Such representations are 1-dimensional and hence non-zero multiplicative conjugate linear functionals $\phi : C \to \mathbf{C}$ with $\phi(I) = 1\ \forall \phi \in \hat{C}$ (hence $\hat{C}$ is the space of characters of $C$). The Gelfand topology on $\hat{C}$ is that of pointwise convergence on $C$ and one could write for $c \in C,\ \phi(c) = \hat{c}(\phi)$ yielding an isometric isomorphism $C \leftrightarrow C(\hat{C})$. Given then a suitable (say locally compact) topological space M there is a natural $C^*$ algebra $C(M)$ and one could ask for the relation of $C(M)$ to $\widehat{C(M)}$; in fact these two spaces can be identified both set-wise and topologically (cf. [**535**]). ■

Now going back to [**898**] one can define a space $X$ by a spectral triple (or K-cycle) $(A, H, D)$ where A is an involutive algebra of bounded operators on the Hilbert space H and $D = D^+$ satisfies

(2.11)
$$D^{-1} \sim compact\ operator\ on\ H_\perp;\ [D,a] \sim bounded\ operator\ for\ any\ a \in A$$

(here $H_\perp$ is the orthogonal complement of the finite dimensional kernel of $D$). A point in $X$ above (non-commutative (NC) space) is a state of the $C^*$ algebra $A$, in other words a linear functional $\psi : A \to \mathbf{C}$ where

$$\psi(a^*a) \geq 0\ \forall a \in A;\ \|\psi\| = sup\{|\psi(a)|;\ \|a\| \leq 1\} \tag{2.12}$$

One sees that $\|\psi\| = \psi(1) = 1$ and the set $S(A)$ of all states is a convex space, i.e.: Given any two states $\psi_1$ and $\psi_2$ and a real $\lambda$ with $0 \leq \lambda \leq 1$ there results $\lambda\psi_1 + (1-\lambda)\psi_2 \in S(A)$. The spectral triple $(A, H, D)$ is even (otherwise odd) if there is a $\mathbf{Z}_2$ grading $\Gamma$ of $H$ satisfying

$$\Gamma^2 = 1;\ \Gamma^+ = \Gamma;\ \{\Gamma, D\} = [\Gamma, a] = 0\ \forall a \in A \tag{2.13}$$

The triple $(A, H, D)$ is real (otherwise complex) if there is an antilinear isometry $J :\ h \to H$ satisfying

$$J^2 = \epsilon(d)1;\ jD = \epsilon'(d)DJ;\ J\Gamma = i^d\Gamma J;\ J^+ = J^{-1} = \epsilon(d)J \tag{2.14}$$

The mod 8 periodic functions $\epsilon(d)$ and $\epsilon'(d)$ are given by

$$\epsilon(d) = (1,1,-1,-1,-1,-1,1,1);\ \epsilon'(d) = (1,-1,1,1,1,-1,1,1) \tag{2.15}$$

If $X$ is a Riemannian spin manifold $M$ then the real structure is the CP operation (**2D**) $J\psi = C\bar{\psi}$ where $C$ is the charge conjugation operator (more on this later). The Dirac operator of the K-cycle $X = (A, H, D)$ can be used to define a distance formula on the space $S(A)$ of states. Given two states (points) on $A$ (of $X$), $\psi_1$ and $\psi_2$, the distance between them is given via

$$d(\psi_1,\psi_2) = sup_{a\in A}\{|\psi_1(a) - \psi_2(a)|;\ \|[D,a]\| \leq 1\} \tag{2.16}$$

Thus $D$ contains all of the metric information about it X.

**EXAMPLE** 2.1. One checks now that (2.16) will reduce to the ordinary distance when $X$ is an ordinary manifold $M$. In that case $A = C^\infty(M)$, $D = \gamma^\mu\partial_\mu$, and the space of states is the space of characters $M(C^\infty)$ which can be identified with the manifold itself as follows. Thus

$$x \in M \to \psi_x \in M(C^\infty);\ \psi_x(f) = f(x)\ \forall f \in C^\infty(M) \tag{2.17}$$

(it takes perhaps a moment of thought to digest). The distance (2.16) then takes the form

$$d(x_1,x_2) = sup_{f\in C^\infty(M)}\{|f(x_1) - f(x_2)|;\ \|[D,f]\| \leq 1\} \tag{2.18}$$

Then since $[D,f] = \gamma^\mu\partial_\mu f$ one has $\|[D,f]\| = sup_{x\in M}\|\vec{\partial}f\|$ leading to

$$|f(x_1) - f(x_2)| = \left|\int_{x_1}^{x_2} \vec{\partial}f\cdot d\vec{x}\right| \leq \int_{x_1}^{x_2} |\vec{\partial}\cdot d\vec{x}| \tag{2.19}$$

$$\leq \int_{x_1}^{x_2} |\vec{\partial}f|ds \leq \int_{x_1}^{x_2} \|[D,f]\|ds \leq \int_{x_1}^{x_2} ds \Rightarrow d(x_1,x_2) = inf\left(\int_{x_1}^{x_2} ds\right)$$

One assumed here that the functions $f \in C^\infty(M)$ were real valued but a similar proof works for complex valued functions; the only difference is that one finds the norm of the bounded operator $[D,f]$ to be equal to the Lipschitz norm of $f$, namely

$$\|[D,f]\| = \|f\|_{Lip} = sup_{x_1\neq x_2}\frac{|f(x_1) - f(x_2)|}{Inf(\int_{x_1}^{x_2} ds)} \tag{2.20}$$

**EXAMPLE** 2.2. We follow [**898**] in giving another example to illustrate the connection of the Dirac operator $i\gamma^\mu\partial_\mu$ on a $d$-dimensional spin manifold to the metric. In particular using the Connes trace theorem

$$Tr_\omega f|D|^{-d} = \int_M f(x)\sqrt{det[g(x)]}dx^1\wedge\cdots\wedge dx^d \tag{2.21}$$

It is standard to observe that the Dirac operator $D$ is a first order elliptic pseudodifferential operator (PSDO). Pseudodifferential here means in particular that it is an operator between two Hilbert spaces $H_1$ and $H_2$ of sections of Hermitian vector bundles over $M$ which can be written in local coordinates as

$$D\psi(x) = \frac{1}{(2\pi)^d}\int e^{ip(x-y)}a(x,p)\psi(y)d^dyd^ep;\ a(x,p) = -p_\mu\gamma^\mu \tag{2.22}$$

Here $H_1 = H_2 = L^2(M,S)$ (square integrable sections of the irreducible spinor bundle over $M$ - cf. [**175, 230, 562**]) and $a(x,p)$ is called the principal symbol for $D$. It follows that the principal symbol for $D^2$ will be given by (**2E**) $p^2I = \eta^{\mu\nu}p_\mu p_\nu$ in a locally flat metric. Then

$$D^{-2}\psi(y) = \frac{1}{(2\pi)^d}\int e^{ip(y-x)}a(x,p)\psi(x)d^dxd^dy;\ a(x,p) = \frac{1}{|p|^2}I \tag{2.23}$$

This operator is of order $-2$ and one can thus envision $|D|^{-d} = (D^2)^{-d/2}$ as a PSDO of order $-d$ with principal symbol $|p|^{-d}I$. Now given $f \in C^\infty(M)$ it acts as a bounded multiplicative operator on the Hilbert space so $f|D|^{-d}$ is a PSDO of order $-d$ with principal symbol $f(x)|p|^{-d}I$. The identity $I$ here is an $N \times N$ unit matrix acting on the vector bundle (VB) so $N = 2^{d/2}$ (resp. $N = 2^{(d-1)/2}$) for even (resp. odd) dimensional $M$. The next step involves the famous Connes trace theorem stating that the Dixmier trace of a PSDO (of order $-d$) over a $d$-dimensional Riemannian manifold is proportional to the Wodzicki residue (cf. [**230, 535**])

$$(2.24)\quad Tr_w A = \frac{1}{d(2\pi)^d} Wres(A) = \frac{1}{(2\pi)^d}\int_{S^*M} Tr[a(x,p)]\sigma_p dx^1 \wedge \cdots \wedge dx^d;$$

$$\sigma_p = \sum_1^d (-1)^j p_j dp_1 \wedge \cdots \wedge \widehat{dp_j} \wedge \cdots \wedge dp_d;\ S^*M = \{(x,p) \in T^*M;\ |p| = 1\}$$

It is then fairly clear that
(2.25)

$$Tr_w f|D|^{-d} = \frac{1}{d(2\pi)^d} Wres(f|D|^{-d} = \frac{2}{(2\pi)^d}\int_{S^*M} Tr[f(x)|p|^{-d}I]\sigma_p dx^1 \wedge \cdots \wedge dx^d$$

$$= \frac{N}{d(2\pi)^d}\int_{S^{d-1}}\sigma_p \int_M f(x)dx^1 \wedge \cdots \wedge dx^d = \frac{N\Omega_{d-1}}{d(2\pi)^d}\int_M f(x)dx^1 \cdots \wedge dx^d$$

**REMARK 2.1.2.** We go to [**235**] (hep-th 9603053) for some background for the axioms of NC geometry. One can write metric data on a Riemannian manifold via (**2F**) $ds^2 = g_{\mu\nu}dx^\mu dx^\nu$ which allows one to write (**2G**) $d(x,y) = Inf \int_x^y ds$. The NC geometry is built in a similar but rather dual manner on the pair $(A, ds)$ of the algebra of coordinates and the infinitesimal length element $ds$ (note $ds$ does not commute with the coordinates, i.e. with $f \in A$). In the simplest case where A is commutative one has (**2H**) $[[f, ds^{-1}]g] = 0\ \forall f, g \in A$. When A is commutative it has a spectrum, namely the space of algebra preserving maps $A \to \mathbf{C}$ (cf. Remark 1.2.2)

$$(2.26)\quad \chi:\ A \to \mathbf{C}:\ \chi(a+b) = \chi(a) + \chi(b);\ \chi(ab) = \chi(a)\chi(b);\ \chi(\lambda a) = \lambda\chi(a)$$

When $A$ is the algebra of functions on a space $M$ the space of such maps, called characters of $A$, is identified with $M$. To each $x \in M$ corresponds the character $\chi_x$ (cf. (2.17) - $\chi_x \sim \psi_x$) (**2I**) $\chi_x(f) = f(x)$. While relations such as (**2H**) between $A$ and $ds$ hold at the universal level, a specific geometry will be specified as a unitary representation of the algebra generated by $A$ and $ds$. In general one deals with complex valued functions so there is an involution (•) $f \to f^*$ which in a model case are (♦) $f^*(x) = \overline{f(x)}$. The length element will be self-adjoint, i.e. (★) $ds^* = ds$ so $(ds)^2$ is automatically positive. Unitarity here means that (♣) $\pi(a^*) = \pi(a)^*\ \forall a \in (A, ds)$). Given then a unitary representation $\pi$ of $(A, ds)$ one measures distance via

$$(2.27)\qquad d(x,y) = Sup\{|f(x) - f(y)|;\ f \in A,\ \|[f, ds^{-1}]\| \le 1\}$$

(where the $\pi$ has been dropped but is used in an important manner in defining the norm of $[f, ds^{-1}]$. To show what model corresponds to a Riemannian geometry take $A = C^\infty(M)$ and $ds$ corresponds to a propagator for fermions denoted by (♠) $x - x$.

Thus one represents $A$ in the Hilbert space $L^2(M,S)$ of square integrable sections of the spinor bundle on $M$ via (**2J**) $(f\xi)(x) = f(x)\xi(x)\ \forall f \in A = C^\infty(M)$; $\forall \xi \in H = L^2(M,S)$ while (**2K**) $ds = D^{-1}$; $D = (1/i)\gamma^\mu\nabla_\mu$ with $D$ the Dirac operator (one ignores the ambiguity of (**2K**) on the kernel of $D$). Evidently now $[D,f]$ for $f \in A = C^\infty(M)$ is the Clifford multiplication by $\nabla f$ so that in the operator norm in H is (**2L**) $\|[D,f]\| = Sup_{x\in M}\|\nabla f\|$. It then follows by integration along the path from $x$ to $y$ that $|f(x) - f(y)| \leq$ path length, provided (**2L**) is bounded by 1; the relation between (2.27) and (**2G**) follows. Note that while $ds$ has the dimension of a length $(ds)^{-1}$ represented by $D$ has the dimension of mass. The formula (2.27) is dual to (**2G**); in the usual Riemannian case it gives the same answer but being dual it does not use the arcs connecting $x$ with $y$ but rather functions from $M$ to **C**. This allows one to treat spaces with a finite number of points on the same footing as the continuum. Another virtue of (2.27) is that it continues to make sense when the algebra $A$ is NC. ∎

## 3. NON-COMMUTATIVE GEOMETRY

In order to write down a more of less updated version of the theory as applied to physics one still needs to use a certain number of algebraic and topological ideas (e.g. K-theory, projective modules, spin structures, Clifford algebras, elementary category theory, etc.) and there are excellent presentations of the necessary information in [**230, 235, 535, 562, 853**] for example (cf. also [**175**] for a condensed version of some matters). Hence we sketch here from Chamseddine [**205**] (hep-th 0901.0577), Connes [**235**] (math.QA 0011006 and hep-th 0003006), Chamseddine and Connes [**207**] (hep-th 0812.0165), Connes and Marcolli [**232**] (math.QA 0601054), and Varilly [**853**]. Following now [**205**] one notes first that the natural group of invariance for gravity plus gauge symmetries is $G \rtimes Diff(M)$ where

$$U = C^\infty(M, U(1) \times SU(2) \times SU(3)) \tag{3.1}$$

At energies well below the Planck scale

$$M_P = \sqrt{\frac{1}{8\pi G}} = \frac{1}{\kappa} = 2.43 \times 10^{18}\ GEV \tag{3.2}$$

gravity can be safely considered as a classical theory. One conjectures now that at some energy level, space-time is the product of a continuous 4-D manifold times a discrete space $F$. Now the basic NCG will involve a real, even spectral triple defined via

(1) $A$ is an associative algebra with unit 1 and involution $*$
(2) $H$ is a complex Hilbert space carrying a faithful representation $\pi$ of $A$
(3) $D$ is a self-adjoint operator on $H$ with compact resolvent $(D-\lambda)^{-1}$ $(\lambda \in \mathbf{R})$
(4) $J$ is an anti-unitary operator which is a real structure (charge conjugation)
(5) $\gamma$ is a unitary operator on $H$ (chirality)

The following axioms are then stipulated

(1) $J^1 = \epsilon$, ($\epsilon = 1$ in zero dimensions and $\epsilon = -1$ in 4 dimensions)

(2) $[a, b^0] = 0 \; \forall a, b \in A$ where $b^0 = Jb^*J^{-1}$. This is needed to define the right action on elements of $H$, namely $\zeta b = b^0\zeta$, and is a statement that left action and right action commute
(3) $DJ = \epsilon' JD$, $J\gamma = \epsilon'' \gamma J$, $D\gamma = -\gamma D$ where $\epsilon$, $\epsilon'$, $\epsilon'' \in \{-1, 1\}$ (see table below). The reality conditions resemble the conditions governing the exisence of Majorana (real) fermions.
(4) $[[D, a], b^0] = 0 \; \forall a, b \in A$ (first order condition)
(5) $\gamma^2 = 1$ and $[\gamma, a] = 0 \; \forall a \in A$ (this permits the decomposition $H = H_L \oplus H_R$
(6) $H$ is endowed with an $A$ bimodule structure $a\zeta b = ab^0\alpha$
(7) The notion of dimension is governed by the growth of eigenvalues and may be fractal or complex

Now $A$ has a well defined unitary group **(3A)** $U = \{u \in A;\; uu^* = u^*u = 1\}$ and the natural adjoint action of $U$ on $H$ is given by $\zeta \to u\zeta u^* = uJuJ^*\zeta \; \forall \zeta \in H$. Then **(3B)** $< \zeta, D\zeta >$ is not invariant under the above transformation since **(3C)** $(uJuJ^*)D(uJuJ^*)^* = D + u[D, u^*])J^*$ and $A = A^*$ is self adjoint. This is similar to the appearance of the intersection term for the photon with electrons, namely **(3D)** $i\bar{\psi}\gamma^\mu\partial_\mu\psi \to i\bar{\psi}\gamma^\mu(\partial_\mu + ieA_\mu)\psi$ needed for invariance under variations $\psi \to exp[ie\alpha(x)]\psi$. The properties listed above for $J$ are characteristic of a real structure of KO dimension $n \in \mathbf{Z}/8$ on the spectral triple $(A, H, D)$ (see below for KO theory). The numbers $\epsilon$, $\epsilon'$, $\epsilon'' \in [-1, 1]$ are a function of $n$ mod 8 given via

(3.3)

| $n$ | 0 | 1 | 2 | 3 | 4 | 5 | 6 | 7 |
|---|---|---|---|---|---|---|---|---|
| $\epsilon$ | 1 | 1 | −1 | −1 | −1 | −1 | 1 | 1 |
| $\epsilon'$ | 1 | −1 | 1 | 1 | 1 | −1 | 1 | 1 |
| $\epsilon''$ | 1 | . | −1 | . | 1 | . | −1 | . |

**REMARK 3.1.1.** We refer to [**738**] for a discussion of KO theory (cf. also [**50**]). Thus a real algebra is a ring $A$ which is also an $\mathbf{R}$ vector space such that $\lambda(xy) = (\lambda x)y = x(\lambda y)$ for all $\lambda \in \mathbf{R}$ and $x, y \in A$. A real $C^*$ algebra is a real algebra $A$ with a linear involution $* : A \to A$ such that $(xy)^* = y^*x^*$ and a real Banach algebra satisfies $\|xy\| \leq \|x\|\|y\|$. A unital algebra implies $\|1\| = 1$ and a real $C^*$ algebra satisfies $\|x^*\| = \|x\|$ with $1 + x^*x$ invertible for all $x \in A$. In general any complex $C^*$-algebra can be regarded as a real vector space and with the same operations is a real $C^*$ algebra. Now a KO-cycle on a finite CW complex $X$ is a triple $(M, E, \phi)$ where

(1) $M$ is a compact manifold without boundary
(2) $E$ is a real vector bundle over $M$
(3) $\phi : M \to X$ is a continuous map

Two KO cycles $(M_i, E_i, \phi_i$ are isomorphic if (i) There is a diffeomorphism $h$ preserving the spin structures, (ii) $H^*(E_2) \simeq E_1$ as real vector bundles, and (iii) $\phi_2 \circ h = \phi_1$. The set of isomorphism classes of KO cycles on $X$ is denoted by $\Gamma O(X)$. Two KO cycles $(M_i, E_i, \phi_i)$ are spin bordant if there is a compact spin manifold $W$ with boundary, a real vector bundle $E \to W$ and a continuous map $\phi : W \to X$ such that the two KO cycles ($\bigstar$) $(\partial W, E_{\partial W}, \phi_{\partial W})$ and

$(M_i \coprod(-M_2), E_1 \coprod E_2, \phi \coprod \phi_2)$ (disjoint union) are isomorphic (where $-M_2$ denotes $M_2$ with spin structure on $TM_2$ reversed). The triple $(W, E, \phi)$ is called a spin bordism and $KO(X)$ is the Abelian group obtained by quotienting $\Gamma O(X)$ by the equivalence relation generated by

- Spin bordism
- Direct sum $E = E_1 \oplus E_2 \Rightarrow (M, E, \phi) \sim (M, E_1, \phi) \coprod (M, E_2, \phi)$
- Real vector bundle modification

(there is a nice discussion of spin manifolds in [**740**]).

Going back to [**205**] one takes $A$ to be a tensor product which geometrically amounts to a product space; and many results are listed (we refer to [**205**] for details and references). The spectral geometry of $A$ is given by the product rule $A = C^\infty(M) \otimes A_F$ where the algebra $A_F$ is finite dimensional and (**3E**) $H = L^2(M, S) \otimes H_F$; $D = D_M \otimes 1 + \gamma_5 \otimes D_F$ where $L^2(M, S)$ is the Hilbert space of $L^2$ spinors, and $D_M$ is the Dirac operator of the Levi-Civita spin connection on M, $D_M = \gamma^\mu(\partial_\mu + \omega_\mu)$. The Hilbert space $H_F$ is taken to include the physical fermions, the chirality operator is $\gamma = \gamma_5 \otimes \gamma_F$, and the reality operator is $J = C \otimes J_F$ where $C$ is the charge conjugation matrix. In order to avoid the fermion doubling problem, the space must be taken to be of K-theoretic dimension 6 (where $(\epsilon, \epsilon', \epsilon'') = (1, 1, -1)$ which imposes the condition $J\zeta = \zeta$). This makes the total K-theoretic dimension of the noncommutative space 10 and allows nice things to happen (this suggests perhaps a relation to supersymmetry). A number of complicated formulas are written out in detail and the spectral action principle is stated to be that the physical action depends only of the spectrum $\Sigma$ of $D$. The existence of Riemannian manifolds which are isospectral but not isometric shows that this is stronger than the usual diffeomorphism invariance of the action for general relativity (GR). The dynamics for all interactions including gravity is given via the spectral action

$$Tr\left[f\left(\frac{D_A}{\Lambda}\right)\right] + \frac{1}{2} < J\psi, D_A\psi > \tag{3.4}$$

where $f$ is a positive function, $\Lambda$ a cutoff scale needed to make $D_A/\Lambda$ dimensionless, and $\psi$ a fermionic Grassmann variable; further details about $f$ and $\Lambda$ are indicated in [**205**]. The bosonic part of the spectral action gives an action which unifies gravity with $SU(2) \times U(1) \times SU(3)$ Yang-Mills gauge theory, with a Higgs doublet H and spontaneous symmetry breaking, and a real cosmological scalar field (cf. [**208**]); this is written out in [**205**] (hep-th 0901.0577), along with many other physical equations. Thus under the main assumptions:

- Space-time is a product of a continuous 4-D manifold times a finite space
- One of the algebras $M_4(\mathbf{C})$ is subject to symplectic symmetry reducing it to $M_2(\mathfrak{H})$ ($\mathfrak{H} \sim$ quaternion field)
- The commutator of the Dirac operator with the center of the algebra is non trivial (i.e. $[D, Z(A)] \neq 0$)
- The unitary algebra $U(A)$ is restricted to $SU(A)$

one can make a number of conclusions, among which are

- The number of fermions is 16

- The algebra of the finite space is $\mathbf{C} \oplus \mathfrak{H} \oplus M_3(\mathbf{C})$
- The correct representations of the fermions with respect to $SU(3) \times SU(2) \times U(1)$ are derived
- The Higgs doublet appears as part of the inner fluctuations of the metric and a spontaneous symmetry breaking mechanism appears naturally with the negative mass term without any tuning

We have omitted much of the fascinating detail for lack of space, time, and understanding. Note that the spectral action delivers the standard model with "minimally coupled" Einstein gravity; for details and further connections to number theory, algebraic geometry, etc. we refer to [**205, 207, 208, 231, 232, 576**].

## 4. MORE ON FUNDAMENTAL MATTERS

The language of strings is generally known today and we have occasionally referred to strings in this book. Moreover it increasingly seems that strings and $M$ theory really contain many yet to be imagined deep structures for physics (and mathematics). After all the 26-dimensional Lorentz space, Leech lattice, monster Lie algebra and associated moonshine seem to be related to string theory and suggest many strange features to be revealed some day. Hence we sketch here some recent material relating quantum mechanics (QM), gravity (GR), NC geometry, strings, and all that.

**4.1. NC GEOMETRY, GRAVITY, AND STRINGS.** We go first to [**810**]-2 where one looks for a fundamental formulation of QM which does not refer to a classical spacetime. Singh is aware of the $C^*$ algebra approach discussed in Sections 7.1-7.3 but examines a somewhat simpler point of view leading to a nonlinear Schrödinger equation (SE) near the Planck energy scale. Thus (referring to [**810**] for philosophy and further details) consider a system of QM particles with a total mass-energy much less than Planck mass $m_{Pl}$ with no external classical spacetime manifold available. Since Planck mass scales inversely with the gravitational constant one neglects the gravitational field. To describe the dynamics using NC geometry consider a particle $m << m_{Pl}$ in a 2-D NC spacetime with coordinates $(\hat{x}, \hat{t})$ and introduce the non-Hermitian flat metric

$$\hat{\eta}_{\mu\nu} = \begin{pmatrix} 1 & 1 \\ -1 & -1 \end{pmatrix} \tag{4.1}$$

with line element (**4A**) $ds^2 = \hat{\eta}_{\mu\nu} d\hat{x}^\mu d\hat{x}^\nu = d\hat{t}^2 - d\hat{x}^2 + d\hat{t}d\hat{x} - d\hat{x}d\hat{t}$ ($d\hat{s}$ is also used later). One wants to generalize general covariance to the NC situation and NC dynamics will involve a velocity $\hat{u}^i = d\hat{x}^i/ds$ which via (**4A**) satisfies

$$1 = \hat{\eta}_{\mu\nu} \frac{d\hat{x}^\mu}{ds} \frac{d\hat{x}^\nu}{ds} = (\hat{u}^t)^2 - (\hat{u}^x)^2 + \hat{u}^t \hat{u}^x - \hat{u}^x \hat{u}^t \tag{4.2}$$

which is invariant under a general Lorentz transformation (cf. [**810**]-1). The generalized momentum is $\hat{p}^i = m\hat{u}^i$ so (**4B**) $\hat{p}^\mu \hat{p}_\mu = m^2$ (where $\hat{p}_\mu = \hat{\eta}_{\mu\nu}\hat{p}^\nu$). Written out this is

$$(\hat{p}^t)^2 - (\hat{p}^x)^2 + \hat{p}^t \hat{p}^x - \hat{p}^x \hat{p}^t = m^2 \tag{4.3}$$

For the dynamics one uses a complex action $S(\hat{x},\hat{t})$ and when an external classical universe with a classical manifold $(x,t)$ becomes available (see below) one defines (**4C**) $p^t = -\partial_t S$ and $p^x = \partial_x S$ leading to

$$(4.4)\qquad (\hat{p}^t)^2 - (\hat{p}^x)^2 + \hat{p}^t\hat{p}^x - \hat{p}^x\hat{p}^t = (p^t)^2 - (p^x)^2 + i\hbar\frac{\partial p^\mu}{\partial x^\mu} = m^2$$

in order to relate the NC dynamics to standard QM. In terms of the complex action the right side of (4.4) is

$$(4.5)\qquad (\partial_t S)^2 - (\partial_x S)^2 - i\hbar\left(\frac{\partial^2 S}{\partial t^2} - \frac{\partial^2 S}{\partial x^2}\right) = m^2$$

and setting $\psi = exp(iS/\hbar)$ one obtains a KG equation

$$(4.6)\qquad -\hbar^2\left(\frac{\partial^2}{\partial t^2} - \frac{\partial^2}{\partial x^2}\right)\psi = m^2\psi$$

as desired for the classical situation. The commutation relations for the NC space-time are (cf. [**810**]-1)

$$(4.7)\qquad [\hat{t},\hat{x}] = iL_{Pl}^2;\ [\hat{p}^t,\hat{p}^x] = iP_{Pl}^2$$

Here the momentum space noncommutativity distinguishes this approach of NC geometry to quantum gravity. There is a natural "grid" which when projected on the $(\hat{x},\hat{p})$ plane has an area $L_{Pl}\times P_{Pl} = \hbar$. If the mass-energy of the particle is not negligible compared to Planck mass the metric (4.1) is modified to

$$(4.8)\qquad \hat{h}_{\mu\nu} = \begin{pmatrix} \hat{g}_{tt} & \hat{\theta} \\ -\hat{\theta} & -\hat{g}_{xx} \end{pmatrix}$$

and the previous equations are replaced by

$$(4.9)\qquad ds^2 = \hat{h}_{\mu\nu}d\hat{x}^\mu d\hat{x}^\nu = \hat{g}_{tt}d\hat{t}^2 - \hat{g}_{xx}d\hat{x}^2 + \hat{\theta}[d\hat{t}d\hat{x} - d\hat{x}d\hat{t}]$$

$$(4.10)\qquad \hat{h}_{\mu\nu}\hat{p}^\mu\hat{p}^\nu = m^2;\ \hat{g}_{tt}(\hat{p}^t)^2 - \hat{g}_{xx}(\hat{p}^x)^2 + \hat{\theta}[\hat{p}^t\hat{p}^x - \hat{p}^x\hat{p}^t] = m^2$$

$$(4.11)\qquad \hat{g}_{tt}(\hat{p}^t)^2 - \hat{g}_{xx}(\hat{p}^x)^2 + \hat{\theta}[\hat{p}^t\hat{p}^x - \hat{p}^x\hat{p}^t] = g_{tt}(p^t)^2 - g_{xx}(p^x)^2 + i\hbar\theta\frac{\partial p^\mu}{\partial x^\mu}$$

along with ($\bullet$) $g_{tt}(p^t)^2 - g_{xx}(p^x)^2 + i\hbar\theta(\partial p^\mu/\partial x^\mu) = m^2$. In the macroscopic limit $m >> m_{Pl}$ the antisymmetric component $\theta \to 0$ and ($\bullet$) reduces to classical dynamics. To make contact with string theory take $\psi = exp(iS/\hbar)$ and use (**4C**) to find in the non-relativistic limit a nonlinear SE (Doebner-Goldin equation)

$$(4.12)\qquad i\hbar\partial_t\psi = -\frac{\hbar^2}{2m}\frac{\partial^2\psi}{\partial x^2} + \frac{\hbar^2}{2m}(1-\theta)\left[\frac{\partial^2\psi}{\partial x^2} - \left(\frac{\partial(log\psi)}{\partial x}\right)^2\psi\right]$$

When one considers the quantum dynamics of $D^0$ branes an effective nonlinear SE is obtained which is in fact the Doebner-Goldin equation above (cf. [**596**]). From this Singh proposes that automorphism invariance as described above is one of the fundamental symmetries of string theory and that the NC dynamics above is actually the quantum gravitational dynamics of $D^0$ branes.

Singh continues this theme in [**810**]-3 and discusses the weakly-quantum strongly gravitational dynamics of a relativistic particle whose mass is much greater than Planck mass as "dual" to the strongly quantum weakly gravitational dynamics of another particle whose mass is much less than Planck mass. The masses of the two particles are inversely related and the product of the masses is the square of the Planck mass. Recall first that the Schwarzschild radius $R_S = 2mG/c^2$ of a particle of mass $m$ can be written in Planck units as $R_{SP} = R_S/L_{Pl} = 2m/m_{Pl}$ where $L_{Pl}$ denotes the Planck length and $m_{Pl} \sim 10^{-5}$ gm. Recall also that according to relativistic quantum mechanics that (1/2) of the Compton wave length $R_C = \hbar/mc$ in Planck units becomes $R_{CP} = m_{Pl}/m$ so that $R_{SP}R_{CP} = 1$ is a universal constant. It cannot however be explained in general relativity (wherein $\hbar = 0$) nor in QM (wherein $G = 0$). Thus one motivates the theme of the constancy of this product as a consequence of a quantum-classical duality and this leads to an "explanation" of the size of the cosmological constant. Thus, using again the reformulation of QM which does not refer to an external classical spacetime (cf. [**810**]-1 for more detail) we have (**4A**) (with $\hat{s} \sim s$) and (**4D**) $[\hat{t}, \hat{x}] = iL_{Pl}^2$ with $[\hat{p}^t, \hat{p}^x] = iP_{Pl}^2$ (note that noncommutative products are understood as star products). Dynamics is described by velocities $\hat{u}^i = (d\hat{x}^i/d\hat{s})$ and $\hat{p}^i = m\hat{u}^i$ with momenta defined as gradients of a complex action $\hat{S}$ and (**4E**) $(\hat{p}^t)^2 - (\hat{p}^x)^2 + \hat{p}^t\hat{p}^x - \hat{p}^x\hat{p}^t = m^2$. One assumes that the line element and commutation relations are invariant under transformations of noncommuting spacetime coordinates. If an external classical spacetime $(x, t)$ becomes available the KG equation of standard linear QM is recovered via

$$(\hat{p}^t)^2 - (\hat{p}^x)^2 + \hat{p}^t\hat{p}^x - \hat{p}^x\hat{p}^t = (p^t)^2 - (p^x)^2 + i\hbar\frac{\partial p^\mu}{\partial x^\mu} \tag{4.13}$$

with wave function $\psi = exp(iS/\hbar)$. When the mass of the particle is comparable to the Planck mass the NC line element (**4A**) is modified to the curved NC line elemet (4.9) and the Casimir relation (**4E**) becomes the second equation in (4.10) with the correspondence rule (4.13) in the form (4.11). This leads in the simplest case where $\theta$ is a function of $m/m_{Pl}$ to the equation of motion

$$\left(\frac{\partial S}{\partial t}\right)^2 - \left(\frac{\partial S}{\partial x}\right)^2 - i\hbar\theta\frac{m}{m_{Pl}}\left[\frac{\partial^2 S}{\partial t^2} - \frac{\partial^2 S}{\partial x^2}\right] = m^2 \tag{4.14}$$

which is equivalent to a nonlinear KG equation. The NC metric is assumed to obey a NC generalization of the Einstein equations with $\theta(m/m_{Pl}) \to 1$ for $m << m_{Pl}$ (leading to standard linear QM) and $\theta \to 0$ for $m >> m_{Pl}$ (leading to classical mechanics). In the mesoscopic domain where $\theta$ is away from these limits and the mass is comparable to Planck mass both quantum and gravitational features can be simultaneously determined and some new physics emerges. We refer to [**810**] for further discussion of relations to string theory, the cosmological constant, holography, etc.

**4.2. TIME AND GEOMETRY.** Next consider some fascinating recent papers by Jejjala, Kavic, Minic, and Tze [**482, 483, 484, 485, 608**] using geometric ideas (cf. also[**30, 31, 49, 93, 222, 223, 405, 435, 604**]. There is a well known equivalence principle in classical GR which says that one cannot distinguish between gravity and acceleration. This is claimed to be at the root of the dual nature of energy and the concept of a vacuum and one wants to implement this kind of principle at the quantum mechanical level and hence the authors aim at a consistent gauging of the geometric structure of canonical quantum theory (cf. also [**49, 93, 97, 222, 223, 297**]). We think these ideas are very important and will sketch first the background following [**608**]. A few of the speculations involving the cosmological constant and big bang from [**482, 483, 484, 485**] are not covered here and we refer to these papers for interesting speculations.

Thus pure states in QM are points of an infinite dimensional Kähler manifold $P(H)$, the complex projective space of a Hilbert space $H$. Equivalently $P(H)$ is a real manifold with an integrable almost complex structure $J$ and as such it has a Kähler metric $<\psi|\phi>$ (Hermitian inner product of two states $<\psi|$ and $|\phi>$) and a Riemannian metric (**4A**) $g(\psi,\phi) = g(J\psi, J\phi) = 2k\Re(<\psi|\phi>)$ (Fubini-Study (FS) metric) with the associated symplectic 2-form (**4B**) $\omega(\psi,\phi) = 2k\Im(<\psi|\phi>)$ (with $k = \hbar = 2/c$ and $c$ the constant holomorphic sectional curvature of $P(H)$). While the symplectic form is well known in phase space the metric information provided by the Riemannian metric on $P$ embodies information about purely quantum properties (e.g. time evolution, uncertainty relations, entanglement, and measurement). An observable $A = <\hat{A}>$ is a real valued differentiable function on $P$ belonging to a special class of Kählerian functions. The evolution of states, i.e. the SE, is given by the symplectic flow generated by a Hamiltonian $H$ and states are written as $\psi = \sum e_a\psi_a$ where the $\psi_a$ are coefficients in an orthonormal eigenbasis $e_a$ The symplectic structure on $P$ is defined via (**4C**) $\omega^2 = dp_a \wedge dq^q$ with $d\omega^2 = 0$ and the Poisson bracket is (**4D**) $\{f,g\} = (\partial f/\partial p_a)(\partial g/\partial q^a) - (\partial f/\partial q^a)(\partial p_a)$; the $X^a = (p_a, q^a)$ form a set of canonical coordinates. The SE, with $h =<\hat{H}>$ is then simply the Hamilton equations (**4E**) $(dp_q/dt) = \{h, p_a\}$ and $(dq^a/dt) = \{h, q^a\}$ where (**4F**) $h = (1/2)\sum[(p^a)^2 + (q_a)^2]\omega_a$ for energy eigenvalues $\omega_a$. An observable $O$ will then evolve via (**4G**) $(dO/dt) = \{h, O\}$.

The Born rule $\psi^*\psi = 1 = (1/2\hbar)\sum[(p_a)^2 + (q^a)^2] = 1$ means that $\psi$ and $exp(i\alpha)\psi$ are to be identified. The space of rays in $H = C^{n+1}$ is $CP(n)$ which can be identified with (**4H**) $CP(n) = U(n+1)/U(n) \times U(1)$ with fiber $U(1)$ (phases). Thus QM can be viewed as a classical Hamiltonian system with phase space $CP(n)$ $(n \to \infty)$ and $U(n)$ the unitary group of quantum canonical transformations. The unique Riemannian metric on $CP(n)$, induced from the inner product of H, is the FS metric, (**4I**) $ds_{12}^2 = (1 - |<\psi_1|\psi_2>|^2) = 4(<d\psi|d\psi> - <d\psi|\psi><\psi|d\psi>) = 4[Cos^{-1}|\psi_1|\psi_2>|]^2$. The Heisenberg uncertainty relations arise from such a metric of $CP(n)$ (cf. [**31, 483**]) and the probabilistic interpretation of QM is thus hidden in the metric properties of $P(H)$. The unitary time evolution is related to the metric structure with the SE in the guise of a geodesic equation on $CP(n)$, namely (**4J**) $(du^a/ds) + \Gamma^a_{bc}u^bu^c = (1/2\Delta E)Tr(HF^a_b)u^b$ for the FS metric $g^{FS}_{ab}$ with the canonical curvature 2-form $F_{ab}$ taking values in the holonomy group $U(n)\times U(1)$ (cf. [**31**]). Here $\Delta E^2 =<H^2> - <H>^2$ and $u^a = dz^a/ds$ where $z^a$ denotes the complex coordinates on $CP(n)$ and $\Gamma^a_{bc}$ is the FS connection. As understood in [**31**] time measurement in the evolution reduces to distance on $CP(n)$ and in particular (**4K**) $\hbar ds = \Delta E dt$, giving a relational interpretation of time in QM and a connection of the geodesic distance with "time". Here $H$ appears as the "charge" of an effective particle moving with a velocity $u^a$ in the background of the Yang-Mills field $F_{ab}$. Given a curve $\Gamma$ the geometric Berry phase is given via **(4L)** $\int_{\Sigma} dp_a \wedge dq^a$ where $\Sigma$ has $\Gamma$ as boundary.

Now in [**608**]-1 this leads to coset spaces $Diff(\infty, C)/Diff(\infty - 1, C) \times Diff(1, C)$ as a minimal phase space but this is still too rigid and one goes go [**608**]-2 for a more flexible phase space. First consider the Landsman axioms for QM (cf. [**535**])

(1) The laws of physics are to be invariant under the choice of experimental setup. Thus one asserts the existence of a natural closed symplectic 2-form $\Omega$ with $d\Omega = 0$ so the state space is an even dimensional symplectic Poisson manifold.
(2) Every quantum observation or event is irreducibly statistical in nature and such events form points of a statistical (information) metric space. There is then a natural unique statistical distance function on this space of quantum events, namely the Fisher distance of information theory (cf. [**29, 170**]). More precisely via [**887**] one forms a space with the Riemannian metric $ds^2 = \sum(dp_i^2/p_i) = \sum dX_i^2$ where the $p_i = X_i^2$ denote individual probabilities leading to $ds_{12}^2 = Cos^{-1}\sum(\sqrt{p_{1i}}\sqrt{p_{2i}}$

Note that the identification of the FS and Fisher metrics here (up to a constant) immediately determines $\hbar$ (cf. [**608, 887**]) and saturates the Born rule $\sum p_i = \sum X_i^2 = 1$. To obtain QM from the above (i.e. from $\Omega$ from (1) and $g$ from (2)) one can invoke the consequent existence of an integrable complex structure **(4M)** $J = g^{-1}\Omega$ with $J^2 = -1$ for compatibility. Now one expounds in [**608**] about the inflexible kinematics of standard QM (leading to $CP(n)$ with its complex structure) and the triad of structures $g$, $\Omega$, $J$ within the state space. One can recover flexibility (and retain compatibility) however by introducing an almost complex structure $J$ leading to a coset space of $Diff(C^{n+1})$ which locally looks like $CP(n)$ and allows compatibility of the metric and symplectic structures (cf. [**347, 405, 415, 545**]). The nonlinear Grassmannian

$$Gr(C^{n+1}) = Diff(C^{n+1})/Diff(C^{n+1}, C^n \times \{0\}) \tag{4.15}$$

with $n \to \infty$ satisfies the above requirements. One notes that this is different from $Diff(\infty, C)/Diff(\infty-1, C)\times Diff(1, C)$ (cf. [**405**]). Note also that $Gr(C^{n+1})$ is only a strictly almost complex manifold and hence one only knows the existence of local time and a local metric on the space of quantum events. The local temporal equation is

$$\frac{du^a}{d\tau} + \Gamma^a_{bc}u^b u^c = \frac{1}{2E_P}Tr(HF^a_b)u^b \tag{4.16}$$

where $\hbar d\tau = 2E_P dt$ with $E_P$ the Planck energy and $F_{ab}$ is a general curvature 2-form in the holonomy group $Diff(C^{n+1}, C^n \times \{0\})$.

Nonlinear Grassmanians are Frechet spaces (cf. [**184**]), i.e. locally convex and complete topological vector spaces defined usually by a countable family of seminorms. There is generally no norm-like distance between two points although (an infinite number of) metrics can be defined. For example in [**604**] an infinite one-parameter family of non-zero geodesic distance metrics are found. Since $Gr(C^{n+1})$ is the diffeomorphism invariant counterpart of $CP^n$ the simplest and most natural topological metric to consider is the analogue of the FS metric. In [**604**] a vanishing theorem was found which states that the generalized FS metric induces on $Gr(C^{n+1})$ a vanishing geodesic distance. This arises because the curvatures are unbounded and positive in certain directions which causes the space to curl up so tightly on itself that the infimum of path lengths between any two points collapses to zero. In [**482**] this result is taken to imply in particular that the space of states out of which spacetime emerges provides an initial state in which the universe exists at a single point, namely a cosmological singularity. This suggests that a statistical notion of time may apply close to such a cosmological singularity. One recalls that in both standard geometric QM and its extension the Riemannian structure encodes the statistical nature of the theory. The geodesic distance is a measure of change in the system, e.g. via Hamiltonian time evolution. By means of the FS metric and the energy dispersion $\Delta E$ the infinitesimal distance in phase space is given by (**4K**) and time reveals its statistical quantum nature. As shown already in [**887**] what $ds$ measures on $CP^n$ is the optimal distinguishability of nearby pure states; if the states are hard to resolve experimentally then they are close to each other in the metric sense. Thus statistical distance seems to be completely fixed by the size of fluctuations. Indeed a measure of the uncertainty between two neighboring states is given by computing the volume of a spherical ball $B$ of radius $r$ as $r \to 0$ around a point $p$ of a $d$-dimensional manifold $M$. This is given by

$$\frac{Vol(B_p(r))}{Vol(B_e(1))} = r^d \left(1 - \frac{R(p)r^2}{6(d+1)} + o(r^2)\right) \tag{4.17}$$

where the left side is normalized by the volume of the d-dimensional unit sphere $Vol(B_e(1))$. $R(p)$, the scalar curvature of $M$ at $p$, can be interpreted as the average statistical uncertainty of any point $p$ in the state space (cf. [**700**]). Since $2/\hbar$ is the sectional curvature of $CP^n$ it follows that $\hbar$ can be seen as the mean measure of quantum fluctuations and (4.17) indicates that, depending on the signs and values of the curvature, the metric distance gets enlarged or shortened and may even vanish. The vanishing geodesic distance under the weak FS metric on $Gr(C^{n+1})$ is completely an effect of extremely high curvatures (cf. [**604**]). Because the space is extremely folded onto itself any two points are indistinguishable (i.e. the distance between them is zero). This is an exceptional locus in the Frechet space of all metrics on $Gr(C^{n+1})$ and a purely infinite dimensional phenomenon. We refer to [**482**] for further discussion and extrapolation.

CHAPTER 8

# GRAVITY AND THE QUANTUM POTENTIAL

## 1. INTRODUCTION

Despite the fact that Bohmian trajectories are basically unrelated to classical trajectories (cf. [**593, 594, 595**]) it nevertheless stands that deBroglie-Bohm mechanics (we write $dBB \equiv BM$) is an important alternative rendering of quantum mechanics as defined by the Schrödinger equation (SE) and we refer to e.g.[**13, 14, 114, 115, 116, 120, 133, 170, 177, 247, 267, 274, 275, 297, 310, 311, 312, 318, 332, 342, 365, 392, 436, 440, 441, 557, 589, 627, 628, 632, 629, 639, 635, 654, 679, 713, 774, 843, 844, 881, 890, 892**]) for various points of view (see [**170**] for more references). In fact from a trajectory point of view it is more realistic to consider the viewpoint of Floyd and Matone (cf. [**177, 187, 297, 310, 311, 312, 589, 713, 890**] and look at Floydian microstates (this is called the trajectory represention). In this approach Floyd looks at stationary situations for the SE $i\hbar\psi_t = -(\hbar^2/2m)\psi'' + V\psi$ where $S(x,t) = W(x) - Et$ and the quantum stationary Hamilton Jacobi equation (HSQJE) arises via Bohmian theory for $\psi = Rexp(iS/\hbar)$. Thus the SE implies

$$(1.1)\qquad \frac{1}{2m}\left(\frac{\partial W}{\partial x}\right)^2 + V + Q = E;\ (R^2W')' = 0$$

where $Q = -(\hbar^2 R''/2mR)$ will be the standard Bohmian quantum potential. The modified potential is defined as (**1A**) $U = V + Q$ and one can write for momentum (**1B**) $p(x) = \partial_x W = \pm\sqrt{2m(E-U}dx$. This can all be viewed in a different light via the quantum equivalence principle (QEP) stating that the QSHJE should have the same form under point transformations $x \to \hat{x}$. This leads to a general form of the QSHJE, namely

$$(1.2)\qquad \frac{1}{2m}(W')^2 + \mathfrak{W}(x) + Q(x) = 0;\ Q = \frac{\hbar^2}{4m}\{W, x\}$$

where $\{f, x\} = (f'''/f') - (3/2)(f''/f')^2$ is the Schwartzian derivative. Thus $\mathfrak{W} = V - E$ and one notes that (1.2) is a third order equation for $W$ (the equation (1.2) contains more information than the SE and the quantum potential (QP) in (1.2) is not in general the same as that in (1.1) - cf. [**297, 589**] and note that a general solution of the stationary SE involves a sum $\psi = (1/\sqrt{W'})[Aexp(-iW/\hbar) + Bexp(iW/\hbar)]$). Thus one could conclude that the SE is not a description of quantum mechanics, whereas the generalized HJ equation is, and the QEP seems to provide a good foundation for quantum mechanics (cf. Chapter 2). The form of (1.2) is extremely enticing of course because of the Schwartzian derivative which is known to arise in a number of important mathematical situations (cf. [**177, 297**]). It calls to mind such matters as Möbius transformations, quadratic differentials, modular forms, automorphic functions, etc. and this has relations to number theory (cf. [**97, 98, 177, 297, 458, 540, 589, 590, 591, 592**] for more on this). There seems to be a serious problem of arithmetic here with respect to quantum mechanics, e.g. Bohmian trajectories follow the probability flow and seem to have no visible relation to eigenvalues, etc. For stationary problems where there is an eigenvalue $E$ for the SE one will have generally 3 initial conditions for a microstate and there is no pilot wave effect in general; moreover no arithmetic is visible. Bohmian trajectories cannot cross but microstates can cross and can also form caustics. The trajectory represention is deterministic (but the underlying quantum equivalence principle of Faraggi-Matone [**97, 297, 589, 590**] has a vast theoretical scope into quantum mechanics and sophisticated mathematical techniques - cf. Section 8.5 for quantization). Bohmian mechanics purports to be stochastic and compatible with the Born probability amplitude, and the SE is involved with both the trajectory representation and Bohmian mechanics. Note also that the SE can be derived from many unquantum disciplines such as hydrodynamics, diffusion, etc. The eigenvalues for the SE lead to number theoretic constructions which are related to the Bohmian theory via QEP (with restrictions on the SE) and are partially visible in semi-classical theory via the Gutzwiller trace formula. There does not seem to be any arithmetic in the Bohmian trajectories as such but quantization arrives in the quantum stationary Hamilton Jacobi equation (QSHJE) (see [**590**] and Section 8.5). On the other hand working in the context of noncommutative geometry and $C^*$ algebras, etc. there is an abundance of arithmetic in the Chamsedine, Connes, Marcolli (CCM) version of the standard model with minimally coupled gravity (cf. [**205, 206, 207, 208, 209, 230, 231, 232, 233, 234, 235, 236, 237**] and see [**581**] and Chapter 7 for a "minimal" sketch of some of this (see also [**536, 537**]). Thus it seems that quantum mechanics must be connected to arithmetic (cf. also [**272**] for $p$-adic ideas about mathematical physics) but the SE does not (and should not) display all this since it does not only represent QM. Similarly the Heisenberg uncertainty relations (HUC) can be obtained in various statistical theories (including information theory) and they alone cannot represent QM (cf. [**170**]). In view of all this it seems that the QEP of Faraggi-Matone presented in the broad manner of [**97, 297, 589, 590**] is a natural way to view QM - it is not statistical, it contains more information than the SE, it is elegant and it contains much information about fundamental constants, etc.; hence we regard QEP or CCM as basic "definitions" of QM, and mention also the approach in Chapter 7, Section 4.2 of Jejjala, Kavic, Minic, and Tze as being highly fascinating.

## 2. BACKGROUND

In this section we will discuss time and gravity in the QEP perspective following [**97, 177, 187, 297, 310, 311, 312, 589**] and in Section 3 we will try to relate this to work in [**2, 170, 172, 456, 462**]. Thus we pick up the QSHJE

(1.1)-(1.2) and note that (using $q \sim x$ for the space variable)

$$\mathfrak{W}(q) = -\frac{\hbar^2}{4m}\left[exp\left(\frac{2iW}{\hbar}\right), q\right]; \; Q(q) = \frac{\hbar^2}{4m}\{W, q\} \tag{2.1}$$

(note $Q$ is the same as before and $R^2W' = c$ from (1.1)-(1.2) will give (2.1) with $Q = -(\hbar^2/4m)(R''/R)$). Further $\mathfrak{W} = V - E$ we recall that Floyd observed that $p = W' = m\dot{q}$ is not generally true and the correct version is (cf. [**297**])

$$m(1 - \partial_E Q)\dot{q} = W' \equiv m_Q\dot{q} = W' \equiv m\partial_\tau q = W'; \tag{2.2}$$

$$m_Q = m(1 - \partial_E Q); \; \tau - \tau_0 = m\int_{q_0}^{q}\frac{dx}{W'}$$

This leads to

$$t - t_0 = \partial_E \int_{q_0}^{q} W' dx = \left(\frac{m}{2}\right)^{1/2}\int_{q_0}^{q}\frac{(1 - \partial_E Q)}{\sqrt{E - U}}dx \tag{2.3}$$

and $d\tau/dt = 1/(1 - \partial_E Q)$. Thus $t = t(E)$ and one can expand upon this aspect of the theory. Note also that general solutions of the SE should be taken in the form (**2A**) $\psi = (W')^{-1}[Aexp(-iW/\hbar) + Bexp(iW/\hbar)]$ and $p \sim \partial_q W = W'$ is the generic form for $p$ corresponding to momentum in a quantum mechanical Hamiltonian $H = (1/2m)p^2 + V \sim (1/2m)(-i\hbar\partial_q)^2 + V$. Thus $p \sim -i\hbar\partial_q$ and this does not in general correspond to $m\dot{q}$.

We note also that $t \sim \partial_E W$ since a general stationary state involves $S \sim W - Et = W(q, E) - Et$ so that $\partial_t S = -E$ and $t = t(E)$. Setting $\mathfrak{S} = -S$ with $\partial_t\mathfrak{S} = E$ one can write (**2B**) $W = Et - \mathfrak{S} = t\partial_t\mathfrak{S} - \mathfrak{S}$ in Legendre form. Then given $W = W(E, q)$ with $q$ fixed one has $\partial_E W = t + E\partial_E t - \partial_t\mathfrak{S}\partial_E t = t$ so (**2C**) $\mathfrak{S} = E\partial_E W - W$ gives the dual Legendre relation. Thus the constructions above automatically entail the Legendre transformation relations (**2B**) and (**2C**) involving $\mathfrak{S} = -S$ and $W$.

**REMARK 8.2.1.** One knows that there is no self adjoint operator T acting as a translation generator of time (since $H$ should be bounded below - cf. [**9, 10, 95, 137, 145, 612**]) and various other points of view are possible (cf. [**10, 187, 234, 751, 870**]). In [**10**] for example one shows that for a closed system and internal time $T$, the time required to measure the total energy $E$ with accuracy $\Delta E$ satisfies $T\Delta E > \hbar$. ■

**REMARK 8.2.2.** More generally the nature of time itself is explored in [**234, 751**] (cf. also [**581**]) and we discuss this again briefly (cf. [**205, 206, 207, 208, 230, 231, 209, 230, 231, 232**] for proofs). A von Neuman algebra $\mathfrak{A}$ (acting on a Hilbert space $\mathfrak{H}$) comes equipped with a canonical 1-parameter group of outer automorphisms (modular group)

$$s \to \sigma^s \in Aut(\mathfrak{A}); \; \sigma^s(a) = \Delta^{is}a\Delta^{-is} \tag{2.4}$$

where $\Delta$ is given by the polar decomposition of the closure of the operator $S$ : $a\Omega \to a^*\Omega$ for $\Omega$ a cyclic and separating vector for the action of $\mathfrak{A}$. The cocycle Radon-Nikodym theorem states that $\sigma^s$ depends on $\Omega$ only, modulo inner automorphisms, so there is a unique 1-parameter family of outer automorphisms associated to $\mathfrak{A}$ via the modular theory. Fix then one $\sigma$ in this equivalence class so that $\sigma$ satisfies with respect to $\Omega$ the same properties as the time evolution $\alpha$ with respect to a thermal equilibrium state $\omega$ at inverse temperature $\beta$, namely the KMS condition (**2D**) $\omega(A\alpha^t(B)) = \omega(\alpha^{t+i\beta}(B)A)$ (cf. [**234, 397, 581, 827**]). Here A and B are observables of a thermodynamical system with Hamiltonian H, $\omega$ is a Gibbs state $\omega(A) = Z^{-1}Tr(Ae^{-\beta H})$ (with partition function Z) and $\alpha^t(A) = e^{-iHt}Ae^{iHt}$. In fact one has (**2E**) $< \Omega, a\sigma^s b\Omega > = < \Omega, (\alpha^{s-i}b)a) >$ which yields the KMS condition (**2D**) upon putting (**2F**) $\sigma^s = \alpha^{-\beta t}$. Hence an equilibrium state at inverse temperature $\beta$ is a state such that its modular group $\sigma^s$ coincides with the time flow $\alpha^t$ where (**2G**) $s = -\beta t$. Thus the modular group flow has been suggested by Connes and Rovelli [**234**] as a time flow and this demands that the ratio of the rates of the two flows (geometrical and modular) be identified as the temperature of the state. One notes that an example where the thermal time hypothesis is realized is the Unruh effect which says that the vacuum state $\Omega$ of a QFT on Minkowski space looks like a thermal equilibrium state for a uniformly accelerated observer with acceleration $a$ (namely $T_U = \hbar a/2\pi k_B c$ where $k_B$ is the Boltzman constant - cf. [**581, 582**]). ■

**REMARK 8.2.3.** We go now to Wesson [**868, 869, 870**] and sketch a way in which 5-dimensional Kaluza-Klein theory leads to a Heisenberg uncertainty principle (HUP) in 4-D (see [**868, 870**] and Chapter 4 but especially [**868**]). Induced matter theory views mass as a direct manifestation of an unconstrained fifth dimension whereas membrane theory views matter as confined to a surface in a 5-D world (cf. [**586, 718, 868, 870**]). Both theories predict that (i) Massive particles travelling on timelike geodesics in 4-D ($ds^2 > 0$) can be regarded as travelling on null geodesics in 5-D ($dS^2 = 0$), and (ii) Massive particles travelling along s-parametrized paths in 4-D in general change their mass via $m = m(s)$ and this is connected to the existence of a fifth force. Using only the induced matter theory one shows now how to arrive at the HUP. Let L scale the canonical metric $dS^2 = (1/L^2)ds^2 - d\ell^2$ in 5-D with $\Lambda = 3/L^2$ the cosmological constant. Here $x^A = (x^\alpha, \ell)$ and the 4-D interval is $ds^2 = g_{\alpha\beta}dx^\alpha dx^\beta$ for $\alpha, \beta = 0, 123$ for time and space. Usually one takes $x^4 = \ell = Gm/c^2$ in terms of the rest mass of a particle (since this is the particles Schwarzschild radius this is called the Einstein gauge) and some small argument leads one to write $\ell = h/mc$ (Wesson uses $h$ here and not $\hbar$). Then (**2H**) $dS^2 = (L^2/\ell^2)ds^2 - (L^4/\ell^4)d\ell^2$ and the corresponding Lagrangian $\mathcal{L} = (dS/ds)^2$ has 5-momenta

$$P_\alpha = \frac{\partial \mathcal{L}}{\partial(dx^\alpha/ds)} = \frac{2L^2}{\ell^2}g_{\alpha\beta}\frac{dx^\beta}{ds}; \; P_\ell = \frac{\partial \mathcal{L}}{\partial(d\ell/ds)} = -\frac{2L^4}{\ell^4}\frac{d\ell}{ds} \tag{2.5}$$

Define then a 5-D scalar

$$\int P_A dx^A = \int (P_\alpha dx^\alpha + P_\ell d\ell) = \int \frac{2L^2}{\ell^2}\left[1 - \left(\frac{L}{\ell}\frac{d\ell}{ds}\right)\right] ds \tag{2.6}$$

This is zero for $dS^2 = 0$ since (**2H**) gives then (**2I**) $\ell = \ell_0 exp[\pm(s/L)]$ with $d\ell/ds = \pm(\ell/L)$ with $\ell_0$ a constant. Note the particle under consideration has finite energy in 4-D but zero energy in 5-D since $\int P_A dx^A = 0$. Now

$$\int p_\alpha dx^\alpha = \int m u_\alpha dx^\alpha = \int \frac{hds}{c\ell} = \pm \frac{hL}{c\ell} \tag{2.7}$$

Putting $L/\ell = n$ (2.7) says that (**2J**) $\int mcds = nh$ (so the conventional action of particle physics in 4-D follows from a null line element in 5-D. Next in the same spirit one has

$$dp_\alpha dx^\alpha = \frac{h}{c}\left(\frac{du_\alpha}{ds}\frac{dx^\alpha}{ds} - \frac{1}{\ell}\frac{d\ell}{ds}\right)\frac{ds^2}{\ell} \tag{2.8}$$

(note that $dx^\alpha$ transforms as a tensor but $x^\alpha$ does not). The first term in the brackets is zero if $a = 0$ or if the scalar product of $a$ with the velocity is zero (as in conventional 4-D dynamics (cf. [**868**]). The anomalous contribution has magnitude

$$|dp_\alpha dx^\alpha| = \frac{h}{c}\left|\frac{d\ell}{ds}\right|\frac{ds^2}{\ell^2} = \frac{h}{c}\frac{ds^2}{L\ell} = n\frac{h}{c}\left(\frac{d\ell}{\ell}\right)^2 \tag{2.9}$$

using (**2I**) and $n = L/\ell$ (which implies $dn/n = -d\ell/\ell = dK_\ell/K_\ell$ where $K_\ell = 1/\ell$ is the wavenumber for the extra dimension). Then (2.9) is a Heisenberg type relation (**2K**) $|dp_\alpha dx^\alpha| = (h/c)(dn^2/n)$ and we refer to [**868**] for arguments supporting this. Note from (**2J**)-(**2K**), or directly from $dS^2 = 0$, for $n = 1$ one defines a fundamental mass $m_0 = h/cL$ which may mean that the 5-D null path determines a quantum mass (in addition to a HUP). There is much more on all of this in [**870**] (cf. also Chapter 4). ■

Now returning to microstates it is pointed out in [**177, 187**] that (following [**297**]) in solving the third order equation (1.2) one finds that $p = W'$ is determined by constants $(\ell_1, \ell_2)$ (or $\ell = \ell_1 + i\ell_2$) and

$$e^{(2i/\hbar)W(\delta)} = e^{i\alpha}\frac{w + i\ell}{w - i\bar{\ell}} \tag{2.10}$$

where $w = \psi^D/\psi$ and $\delta \sim (\alpha, \ell)$ with $\alpha$ an integration constant ($\psi$ and $\psi^D$ are linearly independent solutions of the stationary SE and $w \in \mathbf{R}$ can be arranged as in [**297**] - IJMPA, p. 1959). Thus 2 real constants (in $\ell$) will specify $p = W'$ and $\alpha$ is needed to determine W (cf. also [**310**]). Moreover for $\Omega = \psi'\psi^E - \psi(\psi^D)'$ the Wronskian one can show also (cf. [**297**])

$$p = W' = -\frac{\hbar\ell_1\Omega}{|\psi^D - i\ell\psi|^2} = f(\ell_1, \ell_2, x) \tag{2.11}$$

Now let $p$ be determined exactly with $p = p(q, E)$ via the Schrödinger equation. Then $\dot{q} = (\partial_E p)^{-1}$ is also exact so $\Delta q = (\partial_E p)^{-1}(\hat{t})\Delta t$ for some $\hat{t}$ with $t \leq \hat{t} \leq t + \Delta t$ is exact (up to knowledge of $\hat{t}$). Thus given the wave function $\psi$ satisfying the stationary SE with two boundary conditions at $q = 0$ say to fix uniqueness, one can create a probability density $|\psi|^2(q, E)$ and the function $W'$. This determines $p$ uniquely and hence $\dot{q}$ by (2.2). The additional constant needed for $W$ appears in (2.10) and we can write $W = W(\alpha, q, E)$ since from (2.10) one has

$$W - (\hbar/2)\alpha = -(i\hbar/2)log(\beta) \tag{2.12}$$

and $\beta = (w+i\bar{\ell})/(w-i\ell)$ with $w = \psi^D/\psi$ is to be considered as known via a determination of suitable $\psi$, $\psi^D$. Hence $\partial_\alpha W = -\hbar/2$ and consequently **(2L)** $\Delta W \sim \partial_\alpha W\delta\alpha = -(\hbar/2)\Delta\alpha$ measures the indeterminacy or uncertainty in $W$.

Now note that the determination of constants necessary to fix $W$ from the QSHJE is not usually the same as that involved in fixing $\ell$, $\bar{\ell}$ in (2.10). In paricular differentiating in $q$ one gets $p$ as in (2.11) (cf. [**170, 187, 297**]). and consequently

$$W' = -\frac{\hbar\ell_1\Omega}{|\psi^D - i\ell\psi|^2} \tag{2.13}$$

We see that e.g. $W'(x_0) = i\hbar\ell_1\Omega/|\psi^D(x_0) - i\ell\psi(x_0)|^2 = f(\ell_1, \ell_2, x_0)$ and $W'' = g(\ell_1, \ell_2, x_0)$ determine the relation between $(p(x_0), p'(x_0))$ and $(\ell_1,\ \ell_2)$ but they are generally different numbers. In any case, taking $\alpha$ to be the arbitrary unknown constant in the determination of $W$, we have $W = W(q, E, \alpha)$ with $q = q(S_0, E, \alpha)$ and $t = t(S_0, E, \alpha) = \partial_E S_0$ (emergence of time from the wave function). One can then write e.g. via **(2L)**

$$\Delta q = (\partial q/\partial W)(\hat{W}, E, \alpha)\Delta W = (1/p)(\hat{q}, E)\Delta W = -(1/p)(\hat{q}, E)(\hbar/2)\Delta\alpha \tag{2.14}$$

(for intermediate values $(\hat{W},\ \hat{q})$ in $[W, W = \Delta W]$ and $[q, q + \Delta q]$ leading to (cf. [**170, 187**]) the local inequality **(2M)** $\Delta p\Delta q = O(\hbar)$ which resembles the Heisenberg uncertainty relation. Similarly **(2N)** $\Delta t = (\partial t/\partial W)(\tilde{W}, E, \alpha)\Delta W$ for some intermediate value $\tilde{W}$ and hence, as before, locally **(2O)** $\Delta E\Delta t = O(\hbar)$ ($\Delta E$ being specified, independent of $\hbar$).

**REMARK 8.2.4.** It is of possible interest to note that given $E = (1/2m)p^2 + V(q) + Q(q)$ there results formally $\delta E/\delta p \sim p/m$ and from $p = m_Q\dot{q}$ follows $\delta q = (p/m_Q)\delta t$. Hence for $\delta p\delta q \geq \hbar/2$ we have

$$\delta p\delta q \sim \frac{m}{p}\delta E p\frac{\delta t}{m_Q} = \frac{m}{m_Q}\delta E\delta t \Rightarrow \delta E\delta t \geq (1 - \partial_E Q)\frac{\hbar}{2} \tag{2.15}$$

(cf. [**170, 187**]). ■

## 3. QEP AND GRAVITY

We begin with 3 papers of Matone [**589**] which contain material on connections of the QEP and QSHJE to gravity. In particular in the case of two free particles it is shown that the QP, which is attractive, may generate the gravitational potential. In [**97**] it was shown that the QEP extends to higher dimensions and to the relativistic case, suggesting that QM and general relativity (GR) may be facets of the same coin and this viewpoint is further enhanced in [**589**]. From Section 1 we recall the general term $\mathfrak{W} = V - E$ in the QSHJE. This term is of a purely quantum nature and a related aspect concerns the appearance of fundamental constants in the QSHJE (cf. [**297**]) and in the implementation of the QEP leads to the introduction of universal length scales. One knows also that the QP is always nontrivial (cf. [**297**]) and recalls that the QP here is distinguished from the Bohmian QP. In fact even in the case of $\mathfrak{W}^0$ the QP is non-trivial ($Q = (\hbar^2/4m)\{W, x\}$ from (1.2)) and we recall also that the quantum Hamiltonian characteristic function $W$

is always non-trivial; in particular (**3A**) $W \neq 0$ results from the QEP (cf. [**297**]). We recall that the QEP says that (**3B**) For each pair $\mathfrak{W}^a$, $\mathfrak{W}^b$ there is a so-called $v$-transformation such that $\mathfrak{W}^a(q) \to \mathfrak{W}^{av}(q^v) = \mathfrak{W}^b(q)$ (here $v$-transformation take $q^a \to q^b = v(q)$ and one is writing $q^v$ for $q^b$). In order to see how fundamental constants arise look at (1.2) with general solution (2.10) (note $\ell_1 \neq 0$ which is equivalent to $W \neq constant$). There is a simple reason why fundamental constants should be hidden in $\ell$ (cf. [**297**]). Thus consider the SE in the trivial case $\mathfrak{W}^0(q^0) = 0$, i.e. $\partial^2_{q^0}\psi = 0$. Two linearly independent solutions are $\psi = 1$ with $\psi^D = q^0$ and one recalls the Legendre duality between such functions (cf. here [**177, 297, 589**]). Thus via linearity of the SE one has solutions $\psi = \psi^D + i\ell\psi$ so $\psi = q^0 + i\ell$ is a solution and this suggests that $\ell_0 = \ell$ has dimensions of a length. Indeed $\psi^D/\psi$ is a Möbius transformation of the trivializing map transforming any state to $\mathfrak{W}^0$ (cf. [**297**]) so $w$ and hence $\ell$ has dimensions of a length. Since $\ell_0$ enters the QSHJE with $\mathfrak{W}^0 = 0$ one introduces some fundamental lengths. Thus $W_0 \sim \mathfrak{W}^0$ involves

$$e^{(2i/\hbar)W(\delta)} = e^{i\alpha}\frac{q^0 + i\bar{\ell}_0}{q^0 - i\ell_0};\ p_0 = \pm\frac{\hbar(\ell_0 + \bar{\ell}_0)}{2|q^0 - i\ell_0|^2} \tag{3.1}$$

Now $p_0$ vanishes only for $q^0 \to \pm\infty$ and $|p_0|$ attains its maximum at $q^0 = -\Im(\ell_0)$, namely (**3C**) $|p_0(-\Im(\ell_0)| = \hbar/\Re(\ell_0)$. Since $\Re(\ell_0) \neq 0$, $p_0$ is always finite (thus $\Re(\ell_0) \neq 0$ provides a sort of ultraviolet cutoff and this property extends to arbitrary states). Indeed

$$p = \frac{\hbar\Omega(\ell + \bar{\ell})}{2|\psi^D - i\ell\psi|^2} \tag{3.2}$$

where $\Omega = \psi'\psi^D = (\psi^D)'\psi$ is the Wronskian. Since $\Omega \neq 0$ it follows that $\psi^D$ and $\psi$ cannot have common zeros and via $\Re(\ell) \neq 0$ one sees that $p$ is finite for all $q \in \mathbf{R}$. Hence the QEP implies an ultraviolet cutoff on the conjugate momentum. In [**297**] it has been shown that fundamental constants also arise in considering the classical limit. In particular (**3D**) $lim_{\hbar\to 0}p_0 = 0$ and we note that $\Im(\ell_0)$ in (3.1) can be absorbed by a shift of $q^0$. Hence in (**3D**) one can set $\Im(\ell_0) = 0$ and distinguish the cases $q^0 \neq 0$ and $q^0 = 0$. From (**3D**) one has then

$$(p_0)\underset{\hbar\to 0}{\sim}\begin{cases} \hbar^{\gamma+1} & q_0 \neq 0 \\ \hbar^{1-\gamma} & q_0 = 0 \end{cases} \tag{3.3}$$

where $-1 < \gamma < 1$ with $\gamma$ defined via $\Re(\ell_0)\underset{\hbar\to 0}{\sim}\hbar^\gamma$. One notes that a fundamental length satisfying this condition on the power of $\hbar$ is the Planck length $\lambda_P = \sqrt{\hbar G/c^3}$ while the Compton length is excluded by the condition $\gamma < 1$. Further (as one can see below, in consideing the $E \to 0$ and $\hbar \to 0$ limits for the free particle of energy $E$) the natural choice here is the Planck length. With this choice of $\Re(\ell_0)$ the maximum of $|p_0|$ is (**3E**) $|p_0(-\Im(\ell_0)| = \sqrt{(c^3\hbar/G}$. Setting $\Im(\ell_0) = 0$ and $\Re(\ell_0) = \lambda_P$ the quantum potential associated to the trivial state $\mathfrak{W}^0$ is

$$Q^0 = \frac{\hbar}{4m}\{W^0, q^0\} = -\frac{\hbar^3 G}{2mc^3}\frac{1}{|q^0 - i\lambda_P|^4} \tag{3.4}$$

There are two basic aspects here, namely $G$ results from ensuring consistency with the classical limit and $Q$ is negative definite. Thus one can begin to see how the gravitational interaction arises (and the same features emerge also in 3-D). Also even though we are in a non-relativistic scenario the velocity of light $c$ arises in the integration constants of the QSHJE. The appearance of fundamental constants can also be seen by considering the $\hbar \to 0$ and $E \to 0$ limits for the conjugate momentum of a free particle of energy $E$

$$p_E = \pm \frac{\hbar(\ell_E + \bar{\ell}_E)}{2|k^{-1}Sin(kq) - i\ell_E Cos(kq)|^2} \tag{3.5}$$

where $k = \sqrt{2mE}/\hbar$. From [**297, 310, 311**] one should require **3F**) $lim_{\hbar\to 0}p_E = \pm\sqrt{2mE}$ and

$$lim_{E\to 0}p_E = p_0 = \pm \frac{\hbar(\ell_0 + \bar{\ell}_0)}{2|q - i\ell_0|^2} \tag{3.6}$$

However the term $\ell_E Cos(kq)$ is ill-defined in the $\hbar \to 0$ limit (cf. [**310, 311**]) so some condition on $\ell_E$ is required, which is developed in [**589**] yielding a somewhat long formula for $p_E$.

Define now

$$J_{ki} = \frac{\partial q_i}{\partial q_k^v};\ (p^v|p) = \frac{\sum_k p_k^{v^2}}{\sum_k p_k^2} = \frac{p^T J^T J_p}{p^T p} \tag{3.7}$$

The only possibility to reach another state $\mathfrak{W}^v \neq 0$ starting from $\mathfrak{W}^0$ is that it transforms with an inhomogeneous term (cf. [**297**])

$$\mathfrak{W}^v(q) = (p^v|p^a)\mathfrak{W}(q^a) + (q^q; q^v);\ Q^v(q^v) = (p^v|p^a)Q^a(q^a) - (q^a; q^v) \tag{3.8}$$

Here $(q^a; q^v)$ is a "cocycle" satifsying the cocycle condition

$$(q^a; q^c) = (p^c|p^b)[(q^a; q^b) - (q^c; q^b)] \tag{3.9}$$

which is a basic ingredient in the QEP (cf. [**297**]). In particular (3.9) implies a basic Möbius invariance of $(q^a; q^b)$ and one has $\mathfrak{W}^v(q^v) = (q^0; q^v)$ so that in general (**3G**) $\mathfrak{W}(q) = (q^0; q)$ and according to (**3B**) all states correspond to the inhomogeneous part in the transformation of the $\mathfrak{W}$ state induced by some $v$-map. Since the inhomogeneous part has a purely quantum origin we see that interactions have a purely quantum origin. For the quantum potential (QP) one sees that QEP implies (**3H**) $\mathfrak{W}^v(q^v) + Q^v(q^v) = (p^v|p)(\mathfrak{W}(q) + Q(q))$ and taking $\mathfrak{W} = \mathfrak{W}^0 = 0$ this gives (omitting the superscript $v$) (**3I**) $\mathfrak{W}(q) = (p|p^0)Q^0(q^0) - Q(q)$ (showing that any potentials can be expressed in quantum terms).

**REMARK 8.3.1.** We extract from [**589**] the observation that the $SL(2, \mathbf{C})$ symmetry

$$e^{2iW/\hbar} \to \frac{Ae^{2iW/\hbar} + B}{Ce^{2iW/\hbar} + D} \tag{3.10}$$

of the equation $\{exp[2iW/\hbar], q\} = -4m\mathfrak{W}/\hbar^2$ is equivalent to the QSHJE. Changing the integration constant $\ell$ in the SE corresponds to a Möbius transformation leaving $\mathfrak{W}$ unchanged but mixing the QP and the kinetic term. Thus the QP is parametrized by $SL(2, \mathbf{C})$ transformations in which the constants $A$, $B$, $C$, $D$ depend on fundamental constants. This all generalizes to 3 and 4 dimensions. ■

Thus in some sense the QP is at the origin of interactions and one considers the 2-particle model with masses $m_1$, $m_2$ satisfying

$$\sum \frac{1}{2m_i}(\nabla_i W)^2 - E - \sum_i \frac{\hbar^2}{2m_i}\frac{\Delta_i R}{R} = 0 \tag{3.11}$$

$$\sum_i \frac{\nabla_i \cdot (R^2 \nabla_i W)}{m_i} = 0 \tag{3.12}$$

Then set

$$r = r_1 - r_2,\ r_{cm} = \frac{m_1 r_1 + m_1 r_2}{m_1 + _2};\ m = \frac{m_1 m_2}{m_1 + m_2} \tag{3.13}$$

Then e.g. (3.11) takes the form

$$\frac{1}{2\sum m_i}(\nabla_{r_{cm}} W)^2 + \frac{1}{2m}(\nabla W)^2 - E - \frac{\hbar^2}{2\sum m_i}\frac{\Delta r_{cm} R}{R} - \frac{\hbar^2}{2m}\frac{\Delta R}{R} = 0 \tag{3.14}$$

where the differential symbols refer to components of $r$ (resp. $r_{cm}$). These can be decomposed into equations for the center of mass $r_{cm}$ and those for the relative motion; for the latter one has a QSHJE

$$\frac{1}{2m}(\nabla S)^2 - E - \frac{\hbar^2}{2m}\frac{\Delta R}{R} = 0;\ \nabla \cdot (R^2 \nabla W) = 0 \tag{3.15}$$

In [**310**] it has been stressed that the continuity equation implies (**3J**) $R^2 \partial_t W = \epsilon_i^{i_2 \cdots i_d} \partial_{i_2} F_{i_3 \cdots i_d}$ where $F$ is a $(d-2)$ form. In th 3-D case $R^2 \partial_i W$ is the curl of a vector $B$ with (**3K**) $\nabla W = R^{-2} \nabla \times B$. The QSHJE (3.15) reduces to the canonical form (**3L**) $j^2 = \hbar^2 R^3 \Delta R + 2mER^4$ where $j^2 = j_k j^k$ with $j = \nabla \times B$. Using the identity $(a \times b) \cdot (c \times d) = (a \cdot c)(b \cdot d) - (a \cdot d)(b \cdot c)$ (**3L**) becomes

$$\Delta B^2 - (\nabla B)^2 = \hbar^2 R^3 \Delta R + 2mER^4 \tag{3.16}$$

and we note that $j$ resembles the usual current. It should be emphasized again that $R$ and $W$ here are not in general the ones that would be obtained by identifying $Rexp(iW/\hbar)$ with the wave function. Nevertheless by construction we have that indeed $\psi = Rexp(iW/\hbar)$ does solve the SE so (**3M**) $j = (i\hbar/2)(\psi \nabla \bar{\psi} - \bar{\psi} \nabla \psi)$ can be identified with a standard QM current.

Since the QP seems strongly connected with gravity one considers now solutions of the QHSJE leading to the classical HJ equation for the classical Newton gravitational interaction. Thus given two free particles one asks whether a quantum potential (**3N**) $Q = -(\hbar^2/2m)(\Delta R/R)$ admits the form (**3O**) $Q = V_G$ where $lim_{\hbar \to 0} V_G = -Gm_1 m_2/r$. If such a solution exists then for $\hbar \to 0$ (3.15) corresponds to the HJ equation for the gravitational potential (**3P**) $(1/2m)(\nabla W^{cl})^2 - (Gm_1 m_2/r) - E = 0$. Thus the problem corresponds to finding all $R$ satisfying

$$\frac{\hbar^2}{2m}\frac{\Delta R}{R} = -V_G = G\frac{m_1 m_2}{r} + O(\hbar) \tag{3.17}$$

where $R$ and $W$ satisfy (3.15). Consider the set $\mathfrak{R}$ of solutions of (3.17) and the above problem is equivalent to finding the set $\mathfrak{B}$ of solutions of (3.16) with $R \in \mathfrak{R}$. It follows that the set of possible potentials with gravitational behavior $r^{-1}$ is given by

$$\mathcal{V}_G = \left\{ -\frac{\hbar^2}{2m}\frac{\Delta R}{R};\ R \in \mathbf{R}_G \right\} \tag{3.18}$$

where $\mathbf{R}_G = \{R : R \in \mathbf{R}; \textit{and } B \textit{ exists}\}$ (i.e. one has to find all $R$ satisfying (3.17) and then restricting to those for which there is a field $B$ satisfying (3.16)); note that the fact that higher order terms in (3.17) are not fixed so there are many solutions. Here only a special situation is considered via a reformulation avoiding the $B$ field. First note that by (3.15) and (3.17) W should satisfy the equation

$$\frac{1}{2m}(\nabla W)^2 = E + G\frac{m_1 m_2}{r} + O(\hbar) \tag{3.19}$$

Then it is convenient to solve this first since it is simpler than (3.17). A general solution would involve terms depending on $\theta$ and $\phi$ but here one looks only for a function of $r$ so that $\nabla W = \hat{r}\partial_r W(r)$ where $\hat{r}$ is a unit vector along $r$. Then (3.19) becomes

$$\frac{1}{2m}(\partial_r W)^2 = E + G\frac{m_1 m_2}{r} + O(\hbar) \tag{3.20}$$

and the continuity equation is $\nabla \cdot (R^2 \hat{r}\partial_r W) = 0$ so **(3Q)** $R = 1/r\sqrt{\partial_r W)}$. Since the radial part of the Laplacian is $r^{-1}\partial_r^2$ we see that the QSHJE (3.15) becomes

$$\frac{1}{2m}(\partial_r W)^2 - E + \frac{\hbar^2}{4m}\{W, r\} = 0 \tag{3.21}$$

Formally this is the 1-D QSHJE for a free particle on the non-negative real axis and by (3.5) one obtains

$$\partial_r W = \pm \frac{\hbar|\ell_E + \bar{\ell}_E)}{2(k^{-1}Sin(kr) - i\ell_E Cos(kr)|^2} \tag{3.22}$$

and possible solutions of this are discussed in [**589**].

**REMARK 8.3.2.** In [**297, 589**] there are profound discussions of fundamental constants related to the QEP and relations of the QEP to Schwarzian derivatives and Möbius transformations in [**97, 297, 589, 590**] suggest connections to arithmetic as in [**231, 232**]. We mention here a few items from [**590**] (to be expanded in Section 8.5). Thus

(1) The QEP (or EP) cannot be consistently implemented in classical mechanics (CM) in any dimension. If $\mathcal{H}$ is the space of all possible $\mathfrak{W} = V - E$ then for each pair $\mathfrak{W}^a$, $\mathfrak{W}^b \in \mathcal{H}$ there is a $v$-transformation such that **(3R)** $\mathfrak{W}^a(q) \to \mathfrak{W}^{av}(q^v) = \mathfrak{W}^b(q^v)$.
(2) One can assume for any pair of 1-particle states that there is a "field" $W$ such that **(3S)** $W^b(q_b) = W^a(q_a)$ is well defined and in a suitable limit $W \to W^{cl}$. Since $p_i = \partial W(q)/\partial q^i$, **(3R)** implies that $p_i^b = \Lambda_i^j p_j^a$ where $\Lambda_i^j = (\partial q_a^j/\partial q_b^i)$ (note $p_i^b dq_b^i = p_i^a dq_a^i$). One can say that **(3R)** imposes the QEP.

(3) The QEP is implemented by deformation of the classical SHJE via a QP to arrive at **(3T)** $\mathfrak{W}^v(q^v) + Q^v(q^v) = (p^v|p)[\mathfrak{W}(q) + Q(q)]$ where $(p^v|p) = p^t\Lambda^t\Lambda p/p^tp$. Further **(3U)** $Q^v(q^v) = (p^v|p^a)Q^a(q_a) - (q_a; q^v)$ and the QEP can be expressed via the cocycle condition **(3V)** $(q_a; q_c) = (p^c|p^b)[(q_a; q_b) + (q_b; q_c)]$ equivalent to **(3W)** $(q_a; q_c) = (\partial_{q^c}q_b)^2(q_a; q_b) + (q_b; q_c)$ (which is satisfied by the Schwarzian derivative). In [**97**] it is shown that the cocycle condition fixes the higher dimensional version of the Schwarzian derivative (which involves higher dimensional Möbius invariance). The relativistic version is sketched in Section 4. ■

**3.1. WEYL GEOMETRY AND THE QUANTUM POTENTIAL.** Connections between Weyl geometry and the QP evolve from at least two points of view (summarized here with some repetition from Chapter 2).

(1) The first origin is evolved from a statistical point of view (see e.g. [**53, 54, 117, 170, 179, 185, 172, 191, 192, 193, 465, 642, 770, 797, 876**]). One can think of a SE in Weyl space as in [**770**] (see also [**170, 179, 172, 191, 193**] for direct connections). Thus assume random particle motion $q^i(t,\omega)$ ($\omega \sim$ is a sample space tag) with probability density $\rho(q,t)$ and $q^i(t_0,\omega) = q_0^i(\omega)$. This can be put in a Lagrangian formulation and the geometrical structure on M involves $ds^2 = g_{ij}dq^idq^j$ with scalar curvature $R(q,t)$. In a Weyl space for $q^i \to q^i + \delta q^i$ one has $\delta A^i = \Gamma^i_{jk}A^jdq^k$ and for $\ell = (g_{ik}A^iA^k)^{1/2}$ one has $\delta\ell = \ell\phi_k dq^k$ where $\phi_k$ are the covariant components of a so-called Weyl vector. This leads to a Ricci-Weyl curvature

$$R = \dot{R} + (n-1)(n-2)|\phi|^2 - \frac{2}{\sqrt{g}}\partial_i(\sqrt{g}\phi^i) \tag{3.23}$$

where $\dot{R}$ is the Riemannian curvature built from Christoffel symbols. The Lagrangian has the form

$$L(q,\dot{q},t) = L_C(q,\dot{q},t) + \gamma\frac{\hbar^2}{m}R(q,t);\ \gamma = \frac{n_2}{6(n-1)};\ n = dim(M) \tag{3.24}$$

(cf. [**170, 179, 172, 191, 193, 770**] for more details). Setting $\hat{\rho}(q,t) = (1/\sqrt{g})\rho(q,t)$ implies **(3X)** $\phi_i(q,t) = -[1/(n-2)]\partial_i[log(\hat{\rho}(q,t))]$ exhibiting the action of the quantum force **(3Y)** $f_i = (\hbar^2/m)\partial_i R$. Then the Weyl vector is developed from the probability density flow of $\hat{\rho}$ based on **(3Z)** $\partial_t\hat{\rho} + (1/\sqrt{g})\partial_i(\sqrt{g}v^i\hat{\rho}) = 0$ (via the continuity equation $\partial_i\rho + \partial_i(\rho v^i) = 0$. In non-relativistic situations this leads to a SE

$$\partial_t S = \frac{1}{2m}g^{ik}\partial_i S\partial_k S + V - \gamma\left(\frac{\hbar^2}{m}\right)R = 0 \tag{3.25}$$

which for $n = 3$ involves a QP

$$Q = -\frac{\hbar^2}{2m}\left[\dot{R} + \frac{8}{\sqrt{\hat{\rho}g}}\partial_i\left(\sqrt{g}g^{ik}\partial_k\sqrt{\hat{\rho}}\right)\right] \tag{3.26}$$

(cf. [**170, 179**]).

(2) For the second origin we refer to [**117, 170, 185, 172, 191, 192, 797**] and extract here from [**172**] (item 5). The article [**185**] was designed to show relations between conformal general relativity (CGR) and Dirac-Weyl (DW) theory with identification of conformal mass $\hat{m}$ and quantum mass $\mathfrak{M}$ following [**117, 197, 170, 797**] and precision was added via [**642**]. However the exposition became immersed in technicalities and details and we simplify matters here. Explicitly we enhance the treatment of [**117**] by relating $\mathfrak{M}$ to an improved formula for the quantum potential based on [**642**] and we provide a specific Bohmian-Dirac-Weyl theory wherein the identification of CGR and DW is realized. Much has been written about these matters and we mention here only [**4, 24, 53, 54, 55, 104, 117, 197, 170, 185, 191, 193, 192, 249, 256, 465, 748, 770, 797, 798, 876**] and references therein. One has an Einstein form for GR of the form

$$S_{GR} = \int d^4x\sqrt{-g}(R - \alpha|\nabla\psi|^2 + 16\pi L_M) \tag{3.27}$$

(cf. [**117, 729**]) whose conformal form (conformal GR) is an integrable Weyl geometry based on

$$\hat{S}_{GR} = \int d^4x\sqrt{-\hat{g}}e^{-\psi}\left[\hat{R} - \left(\alpha - \frac{3}{2}\right)|\hat{\nabla}\psi|^2 + 16\pi e^{-\psi}L_M\right] \tag{3.28}$$

$$\int d^4x\sqrt{-\hat{g}}\left[\hat{\phi}\hat{R} - \left(\alpha - \frac{3}{2}\right)\frac{|\hat{\nabla}\hat{\phi}|^2}{\hat{\phi}} + 16\pi\hat{\phi}^2 L_M\right]$$

where $\Omega^2 = exp(-\psi) = \phi$ with $\hat{g}_{ab} = \Omega^2 g_{ab}$ and $\hat{\phi} = exp(\psi) = \phi^{-1}$ (note $(\hat{\nabla}\psi)^2 = (\hat{\nabla}\hat{\phi})^2/(\hat{\phi})^2$). One sees also that (3.28) is the same as the Brans-Dicke (BD) action when $L_M = 0$, namely (using $\hat{g}$ as the basic metric)

$$S_{BD} = \int d^4x\sqrt{-\hat{g}}\left[\hat{\phi}\hat{R} - \frac{\omega}{\hat{\phi}}|\hat{\nabla}\hat{\phi}|^2 + 16\pi L_M\right]; \tag{3.29}$$

which corresponds to (3.28) provided $\omega = \alpha - (3/2)$ and $L_M = 0$. For (3.28) we have a Weyl gauge vector $w_a \sim \partial_a\psi = \partial_a\hat{\phi}/\hat{\phi}$ and a conformal mass $\hat{m} = \hat{\phi}^{-1/2}m$ with $\Omega^2 = \hat{\phi}^{-1}$ as the conformal factor above. Then we identify $\hat{m}$ with the quantum mass $\mathfrak{M}$ of [**797**] where for certain model situations $\mathfrak{M} \sim \beta$ is a Dirac field in a Bohmian-Dirac-Weyl theory as in (3.34) below with quantum potential $\mathfrak{Q} = (\hbar^2/m^2c^2)(\Box_g|\psi|/|\psi|)$ (for $\psi$ a solution of the Klein-Gordon equation) determined via $\mathfrak{M}^2 = m^2 exp(\mathfrak{Q})$ (cf. [**170, 642, 797**] and note that $m^2 \propto T$ where $8\pi T^{ab} = (1/\sqrt{-g})(\delta\sqrt{-g}\mathfrak{L}_M/\delta g_{ab})$). We discuss a variation of this via the QEP in Section 4. Then $\hat{\phi}^{-1} = \hat{m}^2/m^2 = \mathfrak{M}^2/m^2 \sim \Omega^2$ for $\Omega^2$ the standard conformal factor of [**797**]. Further one can write ($\bullet$) $\sqrt{-\hat{g}}\hat{\phi}\hat{R} = \hat{\phi}^{-1}\sqrt{-\hat{g}}\hat{\phi}^2\hat{R} = \hat{\phi}^{-1}\sqrt{-g}\hat{R} = (\beta^2/m^2)\sqrt{-g}\hat{R}$. Recall here from [**170**] that for $g_{ab} = \hat{\phi}\hat{g}_{ab}$ one has $\sqrt{-g} = \hat{\phi}^2\sqrt{-\hat{g}}$ and for the Weyl-Dirac geometry we give a brief survey following [**170, 465**].

(a) Weyl gauge transformations: $g_{ab} \to \tilde{g}_{ab} = e^{2\lambda} g_{ab}$; $g^{ab} \to \tilde{g}^{ab} = e^{-2\lambda} g^{ab}$ - weight e.g. $\Pi(g^{ab}) = -2$. $\beta$ is a Dirac field of weight $-1$. Note $\Pi(\sqrt{-g}) = 4$.
(b) $\Gamma^c_{ab}$ is the Riemannian connection; the Weyl connection is $\hat{\Gamma}^c_{ab}$ and $\hat{\Gamma}^c_{ab} = \Gamma^c_{ab} = g_{ab} w^c - \delta^c_b w_a - \delta^c_a w_b$.
(c) $\nabla_a B_b = \partial_a B_b - B_c \Gamma^c_{ab}$; $\nabla_a B^b = \partial_a B^b + B^c \Gamma^b_{ca}$
(d) $\hat{\nabla}_a B_b = \partial_a B_b - B_c \hat{\Gamma}^c_{ab}$; $\hat{\nabla}_a B^b = \partial_a B^b + B^c \hat{\Gamma}^b_{ca}$
(e) $\hat{\nabla}_\lambda g^{ab} = -2 g^{ab} w_\lambda$; $\hat{\nabla}_\lambda g_{ab} = 2 g_{ab} w_\lambda$ and for $\Omega^2 = exp(-\psi)$ the requirement $\nabla_c g_{ab} = 0$ is transformed into $\hat{\nabla}_c \hat{g}_{ab} = \partial_c \psi \hat{g}_{ab}$ showing that $w_c = -\partial_c \psi$ (cf. [**117**]) leading to $w_\mu = \hat{\phi}_\mu / \hat{\phi}$ and hence via $\beta = m\hat{\phi}^{-1/2}$ one has $w_c = 2\beta_c/\beta$ with $\hat{\phi}_c/\hat{\phi} = -2\beta_c/\beta$ and $w^a = -2\beta^a/\beta$.

Consequently, via $\beta^2 \hat{R} = \beta^2 R - 6\beta^2 \nabla_\lambda w^\lambda + 6\beta^2 w^\lambda w_\lambda$ (cf. [**170, 185, 256, 465**]), one observes that $-\beta^2 \nabla_\lambda w^\lambda = -\nabla_\lambda(\beta^2 w^\lambda) + 2\beta\partial_\lambda \beta w^\lambda$, and the divergence term will vanish upon integration, so the first integral in (3.28) becomes

$$I_1 = \int d^4x \sqrt{-g} \left[ \frac{\beta^2}{m^2} R + 12\beta\partial_\lambda \beta w^\lambda + 6\beta^2 w^\lambda w_\lambda \right] \tag{3.30}$$

Setting now $\alpha - (3/2) = \gamma$ the second integral in (3.28) is

$$I_2 = -\gamma \int d^4x \sqrt{-\hat{g}} \hat{\phi} \frac{|\hat{\nabla}\hat{\phi}|^2}{|\hat{\phi}|^2} \tag{3.31}$$

$$= -4\gamma \int d^4x \sqrt{-\hat{g}} \hat{\phi}^{-1} \hat{\phi}^2 \frac{|\hat{\nabla}\beta|^2}{\beta^2} = -\frac{4\gamma}{m^2} \int d^4x \sqrt{-g} |\hat{\nabla}\beta|^2$$

while the third integral in (3.28) becomes (★) $16\pi \int \sqrt{-g} d^4x L_M$. Combining now (3.39), (3.31), and (★) gives then

$$\hat{S}_{GR} = \frac{1}{m^2} \int d^4x \sqrt{-g} \left[ \beta^2 R + 6\beta^2 w^\alpha w_\alpha \right. \tag{3.32}$$

$$\left. + 12\beta\partial_\alpha \beta w^\alpha - 4\gamma |\hat{\nabla}\beta|^2 + 16\pi m^2 L_M \right]$$

We will think of $\hat{\nabla}\beta$ in the form (♦) $\hat{\nabla}_\mu \beta = \partial_\mu \beta - w_\mu \beta = -\partial_\mu \beta$. Putting then $|\hat{\nabla}\beta|^2 = |\partial\beta|^2$ (1.6) becomes (recall $\gamma = \alpha - (3/2)$)

$$\hat{S}_{GR} = \frac{1}{m^2} \int d^4x \sqrt{-g} \left[ \beta^2 R + (3 - 4\alpha)|\partial\beta|^2 + 16\pi m^2 L_M \right] \tag{3.33}$$

One then checks this against some Weyl-Dirac actions. Thus, neglecting terms $W^{ab} W_{ab}$ we find integrands involving $dx^4 \sqrt{-g}$ times

$$-\beta^2 R + 3(3\sigma + 2)|\partial\beta|^2 + 2\Lambda\beta^4 + \mathfrak{L}_M \tag{3.34}$$

(see e.g. [**170, 185, 465, 797**]); the term $2\Lambda\beta^4$ of weight $-4$ is added gratuitously (recall $\Pi(\sqrt{-g}) = 4$). Consequently, omitting the $\Lambda$ term, (3.34) corresponds to (3.33) times $m^2$ for $\mathfrak{L}_M \sim 16\pi L_M$ and (♦♦) $9\sigma + 4\alpha + 3 = 0$. Hence one can identify conformal GR (without $\Lambda$) with a Bohmian-Weyl-Dirac theory where conformal mass $\hat{m}$ corresponds to quantum mass $\mathfrak{M}$.

**REMARK 8.3.3.** The origin of a $\beta^4$ term in (3.34) from $\hat{S}_{GR}$ in (3.28) with a term $2\sqrt{-\hat{g}}\hat{\Lambda}$ in the integrand seems to involve writing (★★) $2\sqrt{-\hat{g}}\hat{\Lambda} = 2\sqrt{-\hat{g}}\hat{\phi}^2\Omega^4\hat{\Lambda} = 2\sqrt{-g}\beta^4\hat{\Lambda}/m^4$ so that $\Lambda$ in (3.28) corresponds to $\hat{\Lambda}$. Normally one expects $\Lambda\sqrt{-g} \to \sqrt{-\hat{g}}\hat{\phi}^2\Lambda$ (cf. [**24**]) or perhaps $\Lambda \to \hat{\phi}^2\Lambda = \Omega^{-4}\Lambda = \hat{\Lambda}$. In any case the role and nature of a cosmological constant seems to still be undecided.∎

## 4. ON THE KLEIN-GORDON EQUATION

The Klein-Gordon (KG) equation looks like (•) $\Box\psi + m^2\psi = 0$ where $\Box \sim \eta^{ab}\partial_a\partial_b$ (d'Alembertian in Minkowski flat space) or more generally with a Riemannian metric (♦) $\Box \sim (1/\sqrt{-g})[\partial_a(\sqrt{-g}g^{ab}\partial_b)/\partial x^b]$. For a conformally invariant d'Alembertian in dimension 4 one needs (★) $\boxtimes \sim g^{ab}\nabla_a\nabla_b + (1/6)R_g$ where $R_g$ is the Ricci scalar (cf. [**117, 170, 185, 327, 388, 558, 860**]). Note then that for $\tilde{g}^{ab} = \hat{\phi}g^{ab}$ there results (••) $\Omega\,\tilde{\boxtimes}\,(\hat{\phi}^{1/2}\chi) = \boxtimes\chi$ (where $\tilde{\boxtimes}$ is formed with $\tilde{\nabla}_a$, $\tilde{\nabla}_b$). Looking at $\psi = |\psi|exp(iS/\hbar)$ in (♦) for example in [**797**] (cf. also [**170, 185, 642**]) one has a Bohmian form (♣) $\partial^\mu S\partial_\mu S = \mathfrak{M}^2c^2$ (corresponding to (♦♦) $\tilde{\nabla}^\mu S\tilde{\nabla}_\mu S = m^2c^2$) with variable $\mathfrak{M}^2 = m^2 exp(\mathfrak{Q})$ where (♠) $\mathfrak{Q} = (\hbar^2/m^2c^2)(\Box|\psi|/|\psi|)$ (see (4.14) et suite and cf. [**170, 185, 860**]). Thus $\mathfrak{Q}$ is based on an original $g_{ab}$ considered in [**97, 297**] but we continue to use the notation of (♠) (see Remark 8.4.2).

**REMARK 8.4.1.** We gather here a few formulas from various sources concerning covariant derivatives and conformal transformations (see e.g. [**54, 55, 97, 103, 170, 191, 193, 327, 388, 530, 558, 603, 729, 770, 860, 876**]). Thus with more or less standard notation we recall ($g$ or $-g$ refer to the determinant and $V$ is a Riemannian volume element)

$$\text{(4.1)}\quad \nabla^i\phi = \phi^{;i} = g^{ik}\partial_k\phi;\ \nabla_i V^i = \partial_i V^i + V^k\partial_k(\sqrt{-g} = \frac{1}{\sqrt{-g}}\partial_m(V^m\sqrt{-g});$$

$$\nabla_i\phi = \partial_i\phi;\ \Delta f = \nabla_i\nabla^i f = \frac{1}{\sqrt{-g}}\partial_j\left(g^{jk}\sqrt{-g}\partial_k f\right)$$

Under a conformal change of variables $\tilde{g}_{ij} = exp(2\lambda)g_{ij}$

$$\text{(4.2)}\qquad \tilde{\Gamma}^k_{ij} = \Gamma^k_{ij} + \delta^k_j\partial_j\lambda + \delta^k_i\partial_i\lambda = g_{ij}\nabla^k\lambda$$

(4.3)

$$\tilde{R}_{ij} = R_{ij} = (n-2)[\nabla_i\partial_j\lambda - (\partial_i\lambda)(\partial_j\lambda)] + (\Delta\lambda + (n-2)\|\nabla\lambda\|^1)g_{ij};\ d\tilde{V} = e^{n\lambda}dV$$

$$\tilde{R} = e^{-2\lambda}(R + 2(n-1)\Delta\lambda = (n-2)(n-1)\|\nabla\lambda\|^2)$$

where in (4.3) one notes that for a Riemannian metric $\Delta f = \nabla^i\partial_i f = g^{ik}\partial_k\partial_i f$ is the trace of the Hessian (via $H(f)_{ij} = H_f(X_i, X_j) = \nabla_{X_i}\nabla_{X)_j}f - \nabla_{\nabla_{X)_i}X_j}f$ and $\Delta f = \sum g^{ij}H(f)_{ij}$). For $n > 2$ in particular there results

$$\text{(4.4)}\qquad \tilde{R} = e^{-2\lambda}\left[R + \frac{4(n-1)}{n-2}e^{-(n-2)(\lambda/2)}\Delta(e^{(n-2)(\lambda/2)})\right]$$

Thus for $n = 4$, $(1/2)(n-2) = 1$ and

$$\tilde{R} = e^{-2\lambda}\left[R + 6e^{-\lambda}\Delta(e^{\lambda})\right] \tag{4.5}$$

Note also for $\tilde{g}_{ab} = e^{2\lambda}g^{ab}$ $(e^{2\lambda} \sim m^2/\mathfrak{M}^2)$ one has

$$\nabla^a\phi\nabla_a\phi = g^{ab}\nabla_b\phi\nabla_a\phi \Rightarrow e^{2\lambda}\nabla^a\phi\nabla_a\phi = \tilde{g}^{ab}\nabla_b\phi\nabla_a\phi = \tilde{\nabla}^a\phi\tilde{\nabla}_a\phi \tag{4.6}$$

since $\tilde{\nabla}^a\phi = \tilde{g}^{ab}\partial_b$ and $\tilde{\nabla}_a\phi = \nabla_a\phi = \partial_a\phi$. ∎

Next we sketch the development of [**97**] following the QEP. The QEP or equivalence principle of Faraggi-Matone is extended in [**97**] to multidimensional and relativistic systems, and reminiscent of the Einstein equivalence of GR the idea is to show that it is possible to connect all physical systems by coordinate transformations. In particular there should always be a coordinate system connecting a system with non-trivial potential $V$ and energy $E$ to the one with $V - E = 0$. This requires a quantum context in the 1-D situation and the resulting Floydian trajectories there depend on Planck length. The $p-q$ duality is related to Möbius symmetry and this fixes the QSHJE and connections to gravity arise in indicated in Section 8.3 and (refer to [**97, 297**] for more on philosophy). The multi-dimensional SE follows the format of 1-D and the cocycle $C_{a,b} = (q^a; q^b)$ is related to a higher dimension Möbius group by the rule that $C_{a,b} = 0$ if and only if $q^a$ and $q^b$ are related by a Möbius transformation. We will skip to the relativstic situation.

Given a potential $V(q,t)$ the relativistic classical HJ equation (RCHJE) is (note here that the metric is the flat Minkowski metric)

$$\frac{1}{2m}\sum_1^D(\partial_k S^{cl}(q,t))^2 + \mathfrak{W}_{rel}(q,t) = 0; \tag{4.7}$$

$$\mathfrak{W}_{rel}(q,t) = \frac{1}{2mc^2}[m^2c^4 - (V(q,t) + \partial_t S^{cl}(q,t))^2]$$

(cf. [**97**] for details). In the time independent case $S^{cl}(q,t) = S_0^{cl}(q) - Et$ and (4.7) becomes

$$\frac{1}{2m}\sum_1^D(\partial_k S_0^{cl})^2 + \mathfrak{W}_{rel} = 0;\ \mathfrak{W}_{rel} = \frac{1}{2mc^2}[m^2c^4 - (V(q) - E)^2] \tag{4.8}$$

In the latter case one repeats previous arguments to obtain

$$\frac{1}{2m}(\nabla S_0)^2 + \mathfrak{W}_{rel} - \frac{h^2}{2m}\frac{\Delta R}{R} = 0;\ \nabla\cdot(R^2\nabla S) = 0 \tag{4.9}$$

and (4.9) leads to the stationary KG equation (**4A**) $-\hbar^2c^2\Delta\psi + (m^2c^4 - V^2 + 2EV - E^2)\psi = 0$ where $\psi = Rexp(iS_0/\hbar)$. In the time dependent $(D+1)$ dimensional case the RCHJE is (**4B**) $(1/2m)\eta^{ab}\partial_a S^{cl}\partial_b S^{cl} + \mathfrak{W}'_{rel} = 0$ where $\eta \sim diag(-1,1,1,1)$ and (**4C**) $\mathfrak{W}'_{rel}(q) = (1/2mc^2)[m^2c^4 - V^2(q) - 2cV(q)\partial_0 S^{cl}(q)]$ where $q = (ct, q_1, \cdots, q_D)$. Then in order to implement the QEP we have add a term $\mathfrak{Q}$ to get (**4D**) $(1/2m)(\partial S)^2 + \mathfrak{W}_{rel} + \mathfrak{Q} = 0$. Here one needs the identification (**4E**) $\mathfrak{W}_{rel}(q) = (1/2mc^2)[m^2c^4 - V^2(q) - 2cV(q)\partial_0 S(q)]$ where $S$ now appears (rather than $S^{cl}$). Implementation of the QEP requires now that for an arbitrary $\mathfrak{W}^a$ state

$$(4.10)\quad \mathfrak{W}^b_{rel}=(p^b|p^a)\mathfrak{W}^a_{rel}(q^a)+(q^a;q^b);\ \mathfrak{Q}^b(q^b)=(p^b|p^a)\mathfrak{Q}^a(q^a)-(q^a;q^b)$$

$$(p^b|p)=\frac{\eta^{\mu\nu}p^b_\mu p^b_\nu}{\eta^{\mu\nu}p_\mu p_\nu}=\frac{p^TJ\eta J^Tp}{p^T\eta p};\ J^\mu_\nu=\frac{\partial q^\mu}{\partial q^{b^\nu}}$$

Furthermore one has the cocycle condition (**4F**) $(q^a;q^c)=(p^c|p^b)[(q;q^b)-(q^c;q^b)]$.

Next one looks at the identity ($\psi\sim Rexp(iS/\hbar)$; $\Box=-(1/c^2)\partial_t^2+\sum_1^3\partial_i^2=\eta^{ab}\partial_a\partial_b$)

$$(4.11)\qquad \alpha^2(\partial S)^2=\frac{\Box(Re^{\alpha S})}{Re^{\alpha S}}-\frac{\Box R}{R}-\frac{\alpha}{R^2}\partial\cdot(R^2\partial S)$$

Then if $\partial(R^2\cdot\partial S)=0$ (continuity condition) and one sets $\alpha=i/\hbar$ there results

$$(4.12)\qquad \frac{1}{2m}(\partial S)^2=-\frac{\hbar^2}{2m}\frac{\Box(Re^{iS/\hbar}}{Re^{iS/\hbar)}}+\frac{\hbar^2}{2m}\frac{\Box R}{R}$$

Using this one arrives at the RQHJE in the form (cf. [**97**])

$$(4.13)\qquad \frac{1}{2m}(\partial S)^2+\mathfrak{W}_{rel}-\frac{\hbar^2}{2m}\frac{\Box R}{R}=0;\ \partial\cdot(R^2\partial S)=0$$

The nonrelativistic limit is then checked, leading to the time dependent SE, and the procedure suggests that the QEP is most simply implemented once one considers the minimal coupling prescription (i.e. $\partial_\mu\to\partial_\mu+ieA_\mu$)). Thus consider $P^{cl}_\mu=p^{cl}_\mu+eA_\mu$ where $p^{cl}_\mu$ is particle momentum and $P^{cl}_\mu=\partial_\mu S^{cl}$. Then the RCHJE is (**4G**) $(1/2m)(\partial S^{cl}-eA)^2+(1/2)mc^2=0$ where $A_0=-V/ec$. Note that (**4H**) $\mathfrak{W}=(1/2)mc^2$ and the critical case corresponds to the limit situation where $m=0$. As usual to implement the QEP one adds a correction to (**4G**), namely (**4I**) $(1/2m)(\partial S-eA)^2+(1/2)mc^2+\mathfrak{Q}=0$. In addition there are transformation properties

$$(4.14)\quad \mathfrak{W}^b(q^b)=(p^b|p^q)\mathfrak{W}^a(q^a)+(q^a;q^b);\ \mathfrak{Q}^b(q^b)=(p^b|p^a)\mathfrak{Q}^a(q^a)-(q^a;q^b);$$

$$(p^b|p)=\frac{(p^b-eA^b)^2}{(p-eA)^2}=\frac{(p-eA)^TJ\eta J^T(p-eA)}{(p-eA)^T\eta(p-eA)}$$

where $J=\partial q^\mu/\partial q^{b^\nu}$. These equations imply the standard cocycle condition and it is proved that $(q^a;q^b)$ vanishes if $q^a$ and $q^b$ are related by a conformal transformation. To continue one has the relevant identities

$$(4.15)\qquad \alpha^2(\partial S-eA)^2=\frac{D^2Re^{\alpha S}}{Re^{\alpha S}}-\frac{\Box R}{R}-\frac{\alpha}{R^2}\partial\cdot(R^2(\partial S-eA));$$

$$D_\mu=\partial_\mu-\alpha eA_\mu;\ D^2=D^\mu D_\mu=\Box-2\alpha eA\partial+\alpha^2e^2A^2-\alpha e(\partial A)$$

Since (4.10) holds for any $R$, $S$, and $\alpha$ one can require $\partial\cdot(R^2(\partial S-eA))=0$ and set $\alpha=i/\hbar$ to obtain

$$(4.16)\qquad (\partial S-eA)^2=\hbar^2\left(\frac{\Box R}{R}-\frac{D^2(Re^{iS/\hbar})}{Re^{iS/\hbar}}\right)$$

There is no loss of generality in considering (4.11) since $(\partial\cdot(R^2(\partial S - eA)) = 0$. One can show now that (recall $\mathfrak{W}\sim(1/2)mc^2$)

$$\mathfrak{W} = \frac{\hbar^2}{2m}\frac{D^2(Re^{iS/\hbar})}{Re^{iS/\hbar}} \tag{4.17}$$

(cf. [**97**]) and then, using (**4I**) and (4.11), there results (★★) $\mathfrak{Q} = -\frac{\hbar^2}{2m}\frac{\Box R}{R}$. The RQHJE now reads

$$(\partial S - eA)^2 + m^2c^2 - \hbar^2\frac{\Box R}{R} = 0;\ \partial\cdot(R^2(\partial S - eA)) = 0 \tag{4.18}$$

Note that the same result can be directly obtained from (4.7) since (**4B**) coincides with (**4G**) after setting $\mathfrak{W}_{rel} = mc^2/2$ and replacing $\partial_\mu S^{cl}$ with $\partial_\mu S^{cl} - eA_\mu$. One can also check that (4.13) implies the KG equation

$$(i\hbar\partial + eA)^2\psi + m^2c^2\psi = 0;\ \psi = Re^{iS/\hbar} \tag{4.19}$$

If one considers $\psi = Rexp(-iS/\hbar)$ one would have the conjugate equation (**4J**) $(i\hbar\partial - eA)^2\psi + m^2c^2\psi = 0$. In the time independent limit $A_\mu = (-(V/ec), 0, \cdots, 0)$ with $\partial_t V = 0$ and (4.13) reduces to the stationary KG equation (**4A**).

**REMARK 8.4.2.** In the case of the RCHJE the fixed point $\mathfrak{W}^0 \equiv 0$ corresponds to $m = 0$ and hence the QEP implies that all other masses can be generated by coordinate transformations. Thus masses correspond to the inhomogeneous term in transformation of the $\mathfrak{W}^0$ state or (**4K**) $(1/2)mc^2 = (q^0; q)$. Furthermore via (4.8) masses are expressed in terms of the quantum potential (**4L**) $(1/2)mc^2 = (p|p^0)\mathfrak{Q}^0(q^0) - \mathfrak{Q}(q)$. In particular following [**297**] the role of the QP as a sort of intrinsic self-energy has a realization in (**4L**). We consider now (4.13) for $A = 0$ and compare with (**4I**); then

$$(\partial S)^2 + m^2c^2 - \frac{\hbar^2\Box R}{R} = 0 \equiv \frac{1}{2m}(\partial S)^2 + \frac{1}{2}mc^2 + \mathfrak{Q} = 0 \tag{4.20}$$

It follows that (**4M**) $\mathfrak{Q} = -\frac{\hbar^2\Box R}{2mR}$ as in (★★) and we can compare this with (♣) and (♠) at the beginning of Section 4 from the Shojai theory (with a flat Minkowski metric) where $\mathfrak{Q} = \mathfrak{Q}_S = (\hbar^2/m^2c^2)(\Box R/R)$ arises from using a quantum mass $\mathfrak{M}$ in the equation (**4N**) $\partial^\mu S\partial_\mu S = \mathfrak{M}^2c^2$ with $\mathfrak{M} = mexp(\mathfrak{Q}/2)$ which is identified with (**4O**) $\tilde{\nabla}^\mu S\tilde{\nabla}_\mu S = m^2c^2$ (cf. [**97, 170, 297, 797**]). Note for $V = 0$ one has $\mathfrak{W}_{rel} = m^2c^4/2mc^2 = (1/2)mc^2$ from (**4E**) and dividing (4.18) by $1/2m$ yields $\mathfrak{Q}_{BFM} = -(\hbar^2/2m)(\Box R/R)$ in the quantum potential position. Then changing the signature (cf. Remark 4.1.1)) and comparing (4.20) above with (4.22) and (**4O**) in Chapter 2 shows that (†) $(\nabla S)^2 = m^2c^2[1 + (\hbar^2/m^2c^2)(\Box R/R)] = m^2c^2(1+Q) \sim \mathfrak{M}^2c^2$. ■

**4.1. ENHANCEMENT.** The apparent discrepency indicated in Remark 4.2 betwen $\mathfrak{Q}_S$ and $\mathfrak{Q}_{BFM}$ suggests taking a closer look at the KG equation with some embellishments based on [**170, 642, 800**]. Thus in nonrelativistic deBroglie-Bohm theory the quantum potential is $Q = -(\hbar^2/2m)(\nabla^2|\Psi|/|\Psi|)$. The particles trajectory can be derived from Newton's law of motion in which the quantum force $-\nabla Q$ is present in addition to the classical force $-\nabla V$. The enigmatic quantum behavior is attributed here to the quantum force or quantum potential (with $\Psi$ determining a "pilot wave" which guides the particle motion). Setting $\Psi = \sqrt{\rho}exp[iS/\hbar]$ one has

$$\frac{\partial S}{\partial t} + \frac{|\nabla S|^2}{2m} + V + Q = 0; \ \frac{\partial \rho}{\partial t} + \nabla \cdot \left(\rho \frac{\nabla S}{m}\right) = 0 \tag{4.21}$$

The first equation in (4.21) is a Hamilton-Jacobi (HJ) equation which is identical to Newton's law and represents an energy condition $E = (|p|^2/2m) + V + Q$ (recall from HJ theory $\nabla S = p$). The second equation represents a continuity equation for a hypothetical ensemble related to the particle in question. For the relativistic extension one could simply try to generalize the relativistic energy equation $\eta_{\mu\nu}P^\mu P^\nu = m^2c^2$ to the form

$$\eta_{\mu\nu}P^\mu P^\nu = m^2c^2(1+Q) = \mathcal{M}^2c^2; \ Q = (\hbar^2/m^2c^2)(\Box|\Psi|/|\Psi|) \tag{4.22}$$

$$\mathcal{M}^2 = m^2\left(1 + \alpha\frac{\Box|\Psi|}{|\Psi|}\right); \ \alpha = \frac{\hbar^2}{m^2c^2} \tag{4.23}$$

(as in Section 2). This can be derived by setting $\Psi = \sqrt{\rho}exp(iS/\hbar)$ in the Klein-Gordon (KG) equation and separating the real and imaginary parts, leading to the relativistic HJ equation $\eta_{\mu\nu}\partial^\mu S\partial^\nu S = \mathfrak{M}^2c^2$ (as in (4.21) - note $P^\mu = -\partial^\mu S$) and the continuity equation is $\partial_\mu(\rho\partial^\mu S) = 0$. The problem of $\mathcal{M}^2$ not being positive definite here (i.e. tachyons) is serious however and in fact (4.22) is not the correct equation (see e.g. [**170, 642**]). Thus one considers a Klein-Gordon (KG) equation and a wave function $\psi = Rexp(iS/\hbar)$ from the viewpoint that a correct Bohmian representation involves Poincaré invariance with a probability interpretation and no acausal Bohmian traectories. This leads to $\mathfrak{M}^2 = m^2exp(Q)$ via the following argument. Equation (4.22) can be rewritten as (cf. also pp. 2-9 and 2.10)

$$\frac{dx^\mu}{d\tau} = \frac{\partial^\mu S(x^\nu(\tau))}{\mathfrak{M}} \tag{4.24}$$

where $\tau$ denotes proper time and this leads to

$$\mathfrak{M}(x^\gamma(\tau))\frac{d^2x^\mu(\tau)}{d\tau^2} = \left(c^2\eta^{\mu\nu} - \frac{dx^\mu}{d\tau}\frac{dx^\nu}{d\tau}\right)\partial_\nu\mathfrak{M}(x^\gamma(\tau)) \tag{4.25}$$

In the non-relativistic limit $\tau \to t$ with $x^\mu(t) \sim (ct, x^\alpha(t))$ and $|\dot{x}^\alpha| << c$ the geodesic equation becomes

$$O\left(\frac{\dot{x}(t)}{c}\right)^2 = c\dot{x}^\alpha\partial_\alpha\mathfrak{M}; \ \mathfrak{M}\ddot{x}^\alpha = c^2\partial^\alpha\mathfrak{M} + \dot{x}^\alpha\dot{x}^\beta\partial_\beta\mathfrak{M} - \dot{x}^\alpha\partial_t\mathfrak{M} \tag{4.26}$$

The second equation only has a good non-relativistic limit provided that $\partial_\beta\mathfrak{M} \sim \mathfrak{M}/c$ which implies that the first equation is identically satisfied. It is then reasonable to think of $\partial_t\mathfrak{M} \sim \mathfrak{M}/c^2$ and hence (**4P**) $\mathfrak{M}\ddot{x}^\alpha = c^2\partial^\alpha\mathfrak{M}$. One recalls here from [**170, 642, 797**] that the non-relativistic equation involves

$$m\ddot{x}^\alpha = -\partial_\alpha\left(mc^2log[\mathfrak{M}/\mu]\right) = -\partial_\alpha Q_{cl} \tag{4.27}$$

where $\mu$ is any mass scale (this is a quantum force equation). Consequently, for $\mu = m$, $\mathfrak{M} = mexp(Q_{cl}/mc^2)$ and this suggests that the relativistic mass satisfies $\mathfrak{M} = mexp(Q/2)$ for $Q = Q_{rel}$ as in (4.22). An important aspect here is that (4.25) is the geodesic equation associated to the space time metric $g_{\mu\nu} = (\mathfrak{M}/m)^2\eta_{\mu\nu}$

(this is discussed in Section 3.1). Note also that (4.22) is equivalent to (**4O**) or (♦♦), namely $g^{\mu\nu}\partial_\mu S\partial_\nu S = m^2c^2$ and this suggests that one should see the quantum effects on the motion of the relativistic particle as being given by the above conformal transformation of the space metric (this point of view is developed in [**770**] for example and in Section 8.3.1).

**REMARK 8.4.3.** Evidently the quantum potential (QP) as in (**1B**) of p. 1-1 does not "characterize" quantum mechanics (QM) since the QP arises from fluctuations, thermodynamics, hydrodynamics, etc. as well as from QM (cf. [**170**]). The connection of QP to "quantum mass" and conformal GR as well as to the QEP and the QHJE however indicates a fundamental (quantum) importance for the QP via the QEP in quantum situations. Similarly the emergence of fundamental constants in e.g. [**97, 297, 589**] indicates a deep connection of the QEP to Möbius transformations, Schwarzian derivatives, etc. and thence to number theory. Many connections of QM to number theory have been spelled out at great length in e.g. [**231**] via noncommutative geometry and this seems to confirm the need for arithmetic in QM. Such arithmetic is surely connected to atomic eigenvalue behavior but one does not see eigenvalues in hydrodynamics for example where the SE also arises à la Madelung et al (cf. [**843**]). The nature of the QP suggests that it measures momentum fluctuations and information (cf. [**170**]). Perhaps the fluctuations (and hence the QP) are really to be thought of as metric fluctuations and some systems (e.g. hydrodynamics or information theory) create their own metric without having atomic properties (cf. [**197, 843**]). This also suggests that diminishing the temperature might freeze certain systems into having atomic properties (see e.g. [**859**]) although this doesn't seem compatible with the QEP. ■

## 5. QUANTIZATION

We go now to [**590**] where it is shown that energy quantization is a direct consequence of the existence of the QSHJE and the QEP under duality transformations (cf. also [**97, 98, 297, 591, 592**]). This is a very important result since it brings the eigenvalue picture (and hence number theory) into QM via the QEP and we will reproduce various arguments from [**590**] and previous sections. Thus (with some repetition) assume that for any pair of 1-particle states there is a field $S_0$ such that (**5A**) $S_0^b(q_b) = S_0^a(q_a)$ is well defined. Assume also that in a suitable limit $S_0$ reduces to $S_0^{cl}$. Thus (**5A**) is the "scalar" hypothesis and since the conjugate momentum is defined via (**5B**) $p_i = (\partial/\partial q^i)S_0(q)$ it follows that (**5C**) $p_i^b = \Lambda_i^j p_j^a$ where $\Lambda_i^j = \partial q_a^j/\partial q_b^i$ which implies (**5D**) $p_i^b dq_b^i = p_i^a dq_a^i$ (with $Det(\Lambda(a)) \neq 0$). Thus (**5A**) implies that two 1-particle states are always connected by a coordinate transformation and we can consider this as imposing the equivalence postulate (EP). In arbitrary dimensions the coordinate transformation is given by imposing (**5C**) and in the 1-D situation one has (**5E**) $q_b = (S_0^b)^{-1} \circ S_0^a(q_a)$. As consequences of (**5A**) one denotes by $H$ the space of all possible $\mathfrak{W} = V - E$ (as in earlier Sections) and let $v$-transformation denote transformations leading from one system to another. Then (**5A**) is equivalent to saying that for each pair $\mathfrak{W}^a$, $\mathfrak{W}^b \in H$ there is a $v$-transformation such that (**5F**) $\mathfrak{W}^a(q) \to \mathfrak{W}^{av}(q^v) = \mathfrak{W}^b(q^v)$. This implies that there always exists a trivializing coordinate $q_0$ for which $\mathfrak{W}^0(q_0) \equiv 0$

and this means that implementation of (**5F**) requires that $\mathfrak{W}$ states transform inhomogeneously (as mentioned earlier) and in particular the EP cannot be implemented in classical mechanics. For this one must deform the classical HJ equation (CHJE) and as in [**297**] one define a Legendre transform of the reduced action via (**5G**) $T_0(p) = q^k p_k - S_0(q)$ with $S_0(q) = p_k q^k - T_0(p)$. Here $T_0(p)$ is a coordinate generating function (**5H**) $p_k = \partial S_0/\partial q^k$ and $p^k = \partial T_0/\partial p_k$. Then $S_0$ should satisfy a differential equation having the structure (**5I**) $\mathfrak{F}(\nabla S_0, \Delta S_0, \cdots) = 0$ which can be written as (cf. [**97, 297, 590**] for discussion)

$$\frac{1}{2m}\sum_1^D (\partial_{q^k} S_0(q))^2 + \mathfrak{W}(q) + Q(q) = 0 \tag{5.1}$$

The transformation properties of $\mathfrak{W} + Q$ are determined via

$$\frac{1}{2m}\sum_1^D \left[\frac{\partial_{q^{kv}} S_0^v(q^v))^2}{2m}\right] + \mathfrak{W}(q) + Q(q) = 0 \tag{5.2}$$

so that

$$\mathfrak{W}^v(q^v) + Q(q^v) = (p^v|p)[\mathfrak{W}(q) + Q(q)] \tag{5.3}$$

In particular (**5J**) $lim_{\hbar\to 0} = 0$ must hold leading to

$$\mathfrak{W}^v(q^v) = (p^v|p^a)\mathfrak{W}^a(q_a) + (q_a; q^v);\ Q^v(q^v) = (p^v|p^a)Q^a(q_a) - (q_q; q^v) \tag{5.4}$$

(cf. (5.3) and discussion in [**590**]). The $\mathfrak{W}^0$ state plays a special role and from (5.4) follows $\mathfrak{W}^v(q^v) = (q_0; q^v)$. Hence according to (**5F**) all of the states correspond to the inhomogeneous part in the transformation of $\mathfrak{W}^0$ induced by some $v$-map. Comparing

$$\mathfrak{W}^b(q_b) = (p^b|p^a)\mathfrak{W}^a(q_a) + (q_a; q_b) = (q_0; q_b) \tag{5.5}$$

and the same formula with $a \leftrightarrow b$ one has (♦) $(q_b; q_a) = -(p^a|p^b)(q_a; q_b)$ so in particular $(q; q) = 0$ and

$$\mathfrak{W}^b(q_b) = (p^b|p^c)\mathfrak{W}^c(q_c) = (p^b|p^a)\mathfrak{W}(q_a) + (p^b|p^c)(q_a; q_c) + (q_c; q_b) \tag{5.6}$$

leading via (5.5) to

$$(q_a; q_c) = (p^c|p^b)[(q_a; q_b) + (q_b; q_c)] \tag{5.7}$$

which expresses the essence of the EP. In the 1-D case one has (**5K**) $(q_a; q_c) = (\partial_{q^c} q_b)^2 (q_a; q_b) + (q_b; q_c)$ which is satisfied by the Schwarzian derivative and is essentially the unique solution. It follows that the coboundary term must be zero so (cf. [**297**]) (**5L**) $(q_a; q_b) = -(\beta^2/4m)\{q_a, q_b\}$ where $\{f(q), q\} = (f'''/f') - (3/3)(f''/f')^2$ is the Schwarzian derivative (with $\beta \sim \hbar$). Consequently $S_0$ satisfies the QSHJE

$$\frac{1}{2m}\left(\frac{\partial S_0(q)}{\partial q}\right)^2 + V(q) - E + \frac{\hbar^2}{4m}\{S_0, q\} = 0 \tag{5.8}$$

Here $\psi = (S_0')^{-1/2}(Ae^{-iS_0/\hbar} + Be^{iS_0/\hbar})$ solves the SE (**5M**) $(= (\hbar^2/2m)\partial_q^2 + V)\psi = E\psi$ and the ratio $w = \psi^D/\psi$, where $\psi$ and $\psi^D$ are two real linearly independent solutions of (**5M**), is the trivializing map transforming any $\mathfrak{W}$ to $\mathfrak{W}_0 \equiv 0$ in analogy with uniformization theory (this formulation extends to the relativistic theory as well - cf. [**97, 297**]). Then the crucial theorem here is

**THEOREM.** If

$$V(q) = E - E \geq \begin{cases} P_-^2 > 0 & q < q_- \\ P_+^2 > 0 & q > q_+ \end{cases} \tag{5.9}$$

then $w$ is a local self-homeomorphism of $\hat{\mathbf{R}} = \mathbf{R} \cup \{\infty\}$ if and only if (**5M**) has an $L^2(\mathbf{R})$ solution.

The consequence is that, since the QSHE is defined if and only if $w$ is a local self-homeomorphism of $\hat{\mathbf{R}}$, it follows that the QSHJE by itself implies energy quantization (and this does not use any probabilistic interpretation of the wave function. To prove this theorem (from [**590**]) recall that the QSHJE is a consequence of the EP which imposes some constraints on $w = \psi^D/\psi$; in particular the existence of the QSHJE (5.7) or equivalently of (**5N**) $\{w, q\} = -(4m/\hbar^2)\mathfrak{W}(q)$ is required. This means that implementation of the EP imposes that $\{w; q\}$ exist so that (**5O**) $w \neq const.$, $w \in C^2(\mathbf{R})$, and $\partial_q^2 w$ is differentiable on $\mathbf{R}$. Moreover implementing the EP involves satisfying the transformation and symmetry properties of the Schwarzian derivative and this involves a non-trivial inversion. Thus (**5N**) must entail also (**5P**) $\{w^{-1}, q\} = -\frac{4m}{\hbar^2}\mathfrak{W}(q)$ and there is also a duality property for the Schwarzian derivative, namely (**5Q**) $\{w, q\} = -(\partial w/\partial q)^2\{q, w\}$. Thus $\{w, q^{-1}\} = q^4\{w, q\}$ so that (**5N**) can be written in an equivalent form (**5R**) $\{w, q^{-1}\} = -(4m/\hbar^2)q^4\mathfrak{W}(q)$. In other words starting from the EP one arrives at either (**5N**) or (**5R**). This means that under (**5S**) $q \to (1/q)$ the point $0^\pm$ map to $\pm\infty$ and one must be able to extend (**5O**) to the point at $\infty$ so that (**5O**) should hold on the extended real line $\hat{\mathbf{R}} = \mathbf{R} \cup \{\infty\}$. This is related to the fact that Möbius transformations map circles to circles and the existence of the QSHJE forces one to impose smooth joining conditions even at $\pm\infty$, i.e. **5O**) must be extended to (**5T**) $w \neq constant$, $w \in C^2(\hat{\mathbf{R}})$, and $\partial_q^2$ is differentiable on $\hat{\mathbf{R}}$. One checks that $w$ is a Möbius transformation of the trivializing map (cf. [**297**]) and (**5Q**), requiring invertibility of $w(q)$ (i.e. $\partial_q w \neq 0$ on $\mathbf{R}$ is a consequence of the cocycle condition (5.7)). By (**5R**) the local univalence must be extended to $\hat{\mathbf{R}}$ and hence a joining condition is also stipulated, namely

$$w(-\infty) = \begin{cases} w(\infty) & w(-\infty) \neq \pm\infty \\ -w(\infty) & w(-\infty) = \pm\infty \end{cases} \tag{5.10}$$

One sees that the EP implies the QSHJE and, in addition to this equation implying the SE, it also involves aspects of the canonical variables which are not directly related to the SE. One shows in [**590**] that energy quantization is also contained in this approach. In fact the continuity conditions correspond to an existence theorem for the SE and strictly speaking the continuity conditions come from the continuity of a probability density $\rho = |\psi|^2$ which should also satisfy the continuity equation (★) $\partial_t \rho + \partial_q j = 0$ where $j = i\hbar(\psi\partial_q\bar{\psi} - \bar{\psi}\partial_q\psi)/2m$. Since for stationary states $\partial_t\rho = 0$ it follows that $j = const.$ and it is just this interpretation of the wave function in terms of probability amplitude, with the consequent meaning of $\rho$ and $j$, that provides the physical motivation for imposing the continuity of the wave function and its first derivative. In the QEP = EP approach the continuity conditions arise from the QSHJE and (**5T**) (i.e. the QEP) implies (**5U**) continuity of $\psi^D$, $\psi$ with $\partial_q\psi^D$ and $\partial_q\psi$ differentiable. If e.g. $V(q) > E\ \forall q \in \mathbf{R}$ there are no solutions such that the ratio $\psi^D/\psi$ corresponds to a local self-homeomorphism of $\hat{\mathbf{R}}$. Thus if $V > E$ at $\pm\infty$ a physical situation requires that there are at least two points where $V = E = 0$ and if the potential is not continuous $V(q) - E$ should have at least two turning points. Note by (5.9) one has

$$(5.11)\qquad \int_{q_-}^{-\infty} dx\kappa(x) = -\infty;\ \int_{q_+}^{\infty} dx\kappa(x) = \infty$$

where $\kappa = \sqrt{2m(V-E)}/\hbar$. One needs to show that in the case of (**5.9**) the joining condition (5.10) requires that the corresponding SE has an $L^2(\mathbf{R})$ solution. Note that(**5T**), following from the EP, can be recognized as a standard condition (**5U**) for the SE; the other condition (5.10) (following from the EP and QSHJE) is not directly recognized as pertinent to the SE. However this condition leads to energy quantization while in the standard approach to the SE another assumption is required so the QSHJE and the SE are fundamentally different. It is stressed that (**5T**) and (5.10) guarantee that $w$ is a local self-homeomorphism of $\hat{\mathbf{R}}$.

Next one shows that if the SE has an $L^2(\mathbf{R})$ solution then (5.10) can be satisfied. Let $\psi \in L^2(\mathbf{R})$ with $\psi^D$ a linearly independent solution. One can show that $\psi^D \not\propto \psi$ implies that $\psi^D \notin L^2(\mathbf{R})$. In particular $\psi^D$ is divergent at both $q = \pm\infty$. Indeed consider the real ratio (**5V**) $w = [(A\psi^D + B\psi)/(C\psi^D + D\psi)]$ where $AD - BC \neq 0$. Since $\psi \in L^2(\mathbf{R})$ one has

$$(5.12)\qquad lim_{q\to\pm\infty} w = lim_{q\to\pm\infty}\frac{A}{C}$$

so $w(\infty) = w(-\infty)$. When $C = 0$ one has

$$(5.13)\qquad lim_{q\to\pm\infty} w = lim_{q\to\pm\infty}\frac{A\psi^D}{D\psi} = \pm\epsilon\cdot\infty$$

where $\epsilon = \pm 1$ so the ratio diverges. It is then checked that the joining condition (5.10) holds via

$$(5.14)\qquad \psi^D(q) = c\psi(q)\int_{q_0}^{q} dx\psi^{-2} + d\psi(q)$$

for $c \in \mathbf{R}/\{0\}$ and $d \in \mathbf{R}$ (cf. [**590**]). Finally one shows in two different ways in [**590**] that the existence of an $L^2(\mathbf{R})$ solution of the SE is a necessary condition to satisfy the joining condition.

Thus the EP or QEP gives rise to the usual quantized spectrum for the SE via the QSHJE. This provides a connection to number theory and in a sense redeems the Bohmian theory, where trajectories are often unreliable guides. The requirements of the QEP are severe and will apparently exclude many situations where a SE arises from hydrodynamics or statistical theory. It might be interesting to develop a theory of the quantum potential for situations with no $L^2$ solutions of the SE.

# Bibliography


[1] W. Abou Salem, math-ph 0601047
[2] S. Abraham, P. deCordoba, J. Isidro, and J. Santander, hep-th 0810.2236 and 0810.2356
[3] E. Abreu, C. Neves, and W. Oliveira, hep-th 0411108
[4] R. Adler, M. Bazin, and M. Schiffer, Introduction to general relativity, McGraw Hill, 1965
[5] S. Adler, Quantum theory as an emergent phenomenon, Cambridge Univ. Press, 2004; hep-th 9703053
[6] G. Agarwal and E. Wolf, Phys. Rev. D, 2 (1970), 2161-2186, 2187-2205, and 2206-2225
[7] M. Agop, P. Nica, and M. Girtu, General Relativity Gravitation, 40 (2008), 35-55
[8] J. Aguilar, C. Romero, and A. Barros, General Relativity Gravitation, 40 (2008), 117-130
[9] Y. Aharonov, S. Massar, and S. Popescu, quant-ph 0110004
[10] Y. Aharonov and B. Reznik, quant-ph 9906030
[11] G. Allemandi, M. Capone, S. Capozziello, and M. Francaviglia, hep-th 0409198
[12] K. Ajith, E. Harikumar, V. Rivelles, and M. Sivakmar, hep-th 0801.2043
[13] V. Allori and N. Zanghi, Found. Phys., 39 (2009), 20-32
[14] V. Allori, S. Goldstein, R. Tumulka, and N. Zanghi, quant-ph 0603027
[15] M. Akbar, hep-th 0702029; gr-qc 0808.3308
[16] M. Akbar and R. Cai, hep-th 0602156 and 0609128; gr-qc 0612089
[17] M. Akbar and E. Woolgar, gr-qc 0808.3126
[18] S. Albeverio and R. Hoegh-Krohn, Mathematical theory of Feynman path integrals, Lect. Notes Math., 523 (1976)
[19] O. Alcantara Bonfim, J. Florencio, and F. Sa Barreto, Phys. Rev. E, 58 (1998), R2693-R2696
[20] G. Allemandi, M. Capone, S. Capozziello, and M. Francaviglia, hep-th 0409198
[21] A. Alexandrov, math.DS 0901.4067
[22] R. Alicki and M. Fannes, Quantum dynamical systems, Oxford Univ. Press, 2001
[23] A. Allahverdyan and D. Janzing, cond-mat 0708.1175
[24] G. Allemandi, M. Capone, S. Capozziello, and M. Francaviglia,hep-th 0309198
[25] P. Aluffi and M. Marcolli, math.AG 0811.2514 and 0901,2197
[26] J. Alvarez, H. Quevedo, and A. Sanchez, gr-qc 0801.2279
[27] J. Aman, I. Bengtsson, and N. Pidokrajt, Gen. Relativ. Grav., 35 (2003), 1733-1743; 38 (2006), 1305-1315
[28] J. Aman and N. Pidokrajt, gr-qc 0801.0016
[29] S. Amari and H. Nagaoka, Methods of information geometry, Amer. Math. Soc., 2000
[30] J. Anandan, Found. Phys., 21 (1991), 1265-1284
[31] J. Anandan and Y. Aharonov, Phys. Rev. Lett., 65 (1990), 1697

[32] A. Anderson and J. Halliwell, gr-qc 9304025
[33] E. Anderson and J. Barbour, gr-qc 0201092
[34] E. Anderson, J. Barbour, B. Foster, and N. O'Murchadha, gr-qc 0211022
[35] H. Aoki, N. Ishibashi, S. Iso, H. Kawai, Y. Kitazawa, and T. Tada, hep-th 9908141
[36] D. Appleby, Found. Phys., 29 (1999), 1863-1883 and 1885-1916
[37] G. D'Ariano, quant-ph 0807.4383
[38] R. Aros, gr-qc 0801.4591
[39] M. Arzano, hep-th 0711.3222
[40] P. Aschieri, hep-th 0608172
[41] P. Aschieri, M. Dmitrijevic, F. Meyer, and J. Wess, hep-th 0510059
[42] P. Aschieri, C. Blohmann, M. Dimitrijevic, F. Meyer, P. Schupp, and J. Wess, hep-th 0504183
[43] P. Aschieri, J. Madore, P. Manoussselis, and G. Zoupanos, hep-th 0310072, 0401200, and 0503039
[44] P. Aschieri, H. Steinacker, J. Madore, P. Manousselis, and G. Zoupanos, hep-th 0704.2880
[45] P. Aschieri and B. Jurco, hep-th 0409200
[46] P. Aschieri, L. Cantini, and B. Jurco, hep-th 0312154
[47] A. Ashtekar and B. Krishnan, Phys. Rev. D, 68 (2003), 104030
[48] A. Ashtekar, in Gravitation and quantizations, Ed. B.Julia, Elsevier, 1994
[49] A. Ashtekar and T. Schilling, gr-qc 9706069
[50] M. Atiyah, K-theory, Benjamin, 1967
[51] T. Aubin, Some nonlinear problems in Riemannian geometry, Springer, 1998
[52] T. Aubin, Nonlinear analysis on manifolds. Monge-Ampere equations, Springer, 1982
[53] J. Audretsch, Phys. Rev. D, 27 (1983), 2872-2884
[54] J. Audretsch, F. Gähler, and N. Straumann, Comm. Math. Phys., 95 (1984), 41-51
[55] J. Audretsch and C. Lämmerzahl, Class. Quant. Grav., 5 (1988), 1285-1295
[56] G. Bacciagaluppi, quant-ph 0302099
[57] S. Bahrami and S. Nasiri, SIGMA, 3 (2007), 001
[58] R. Baierlein, D. Sharp, and J. Wheeler, Phys. REv., 126 (1962), 1864-1865
[59] D. Bak and S. Rey, hep-th 9902173
[60] I. Bakas, hep-th 0511057 and 0702034
[61] A. Balachandran, T. Govindarajan, and B. Ydri, hep-th 9911087 and 0006216
[62] C. Balazs and I. Szapuudi, hep-th 0605190
[63] N. Balazs and A. Voros, Phys. Repts., 143 (1986), 109-240
[64] A. Balcerzak and M. Dabrowski, Phys. Rev. D, 77 (2008), 023524; hep-th 0804.0855
[65] R. Balian and C. Bloch, Annals Phys., 85 (1974), 514-545
[66] K. Ball, Geom. Dedicata, 41 (1992), 241-250
[67] R. Banerjee, hep-th 0409022
[68] R. Banerjee and K. Kumar, hep-th 0404110 and 0505245
[69] R. Banerjee and H. Yang hep-th 0404064
[70] R. Banerjee, C. Lee, and H. Yang, hep-th 0312103
[71] T. Banks, W. Fischler, S. Shenker, and L. Susskind, hep-th 9610043
[72] D. Bar and L. Horwitz, quant-ph 0209012
[73] O. Barabash and Y. Shtanov, astro-[h 9904144; hep-th 9807291
[74] B. Barbashov, L. Glinka, V. Pervushin, and A. Zakharov, gr-qc 0612028
[75] B. Barbashov, V. Pervushin, A. Zakharov, and V. Zinchuk, hep-th 0501242
[76] B. Barbashov, L. Glinka, V. Pervushin, S. Shuvalov, and A. Zakharov, hep-th 0612252

[77] J. Barbour, 0211021, 0309089, and 0903.3489; The end of time, Oxford Univ. Press, 1999
[78] J. Barbour, Class. Quant. Grav., 11 (1994), 2853-2873 and 2875-2897
[79] J. Barbour and L. Smolin, hep-th 9203041
[80] J. Barbour, B. Foster, and N. O'Murchadha, gr-qc 0012089
[81] J. Bardeen, B. Carter, and W. Hawking, Comm. Math. Phys., 31 (1973), 161-170
[82] J. Barrett, hep-th 0608221
[83] J. Acacio de Barros and N. Pinto-Neto, gr-qc 9611029; Inter. Jour. Mod. Phys. D, 7 (1998), 201
[84] A. Bassi, G. Ghirardi, and D. Salvetti, quant-ph 0009020 and 0707.2940
[85] H. Baumgärtel, math-ph 0410036
[86] C. Beck and F. Schlögl, Thermodynamics of chaotic systems, Cambridge Univ. Press, 1993
[87] K. Becker, M. Becker, and J. Schwartz, String theory and M theory, Cambridge Univ. Press, 2007
[88] J. Bekenstein, astro-ph 0412652 and 0701848; Phys. Rev. D, 70 (2004), 083509 (astro-ph 0403694); Phys. Rev. D, 7 (1973), 2333
[89] J. Bekenstein and R. Sanders, astro-ph 0509519
[90] B. Belchev and M. Walton, quant-ph 0810.3893
[91] C. Bender, D. Brody, and D. Hook, hep-th 0804.4169
[92] G. Benenti and G. Casati, quant-ph 0808.3243
[93] I. Bengtsson and K. Zyczkowski, Geometry of quantum states, Cambridge Univ. Press, 2006
[94] E. Beniaminov, math-ph 0812.5116
[95] G. Beretta, quant-ph 0511091
[96] M. Berry and M. Tabor, Jour. Phys. A, 10 (1977), 371
[97] G. Bertoldi, A. Faraggi, and M. Matone, Class. Quantum Gravity, 17 (2000), 3965 (hep-th 9909201)
[98] G. Bertoldi, J. Isidro, M. Matone, and P. Pasti, hep-th 0003131 and 0003200
[99] A. Bessi, Einstein manifolds, Springer, 1987
[100] V. Bianca and L. Vega, Ann. H. Poincaré, 25 (2008), 697-711
[101] E. Bibbona, L. Fatibene, and M. Francaviglia, math-ph 0608063 and 0604053
[102] J. Bisognano and E. Wichman, Jour. Math. Phys., 16 (1975), 984 and 17 (1976), 303
[103] M. Blagojevich, Gravitation and gauge symmetry, IOP Press, 2002
[104] D. Blaschke and M. Dabrowski, hep-th 0407078
[105] M. Blasone, P. Jizba, and G. Vitiello, quant-ph 0301031 and 0302011; hep-th 0007138
[106] M. Blasone, P. Jizba, and H. Kleinert, Phys. Rev. A, 71 (2005), 052507 (quant-ph 0409021), Annals Phys.; 320 (2005), 468-486 (quant-ph 0504200); quant-ph 0504047
[107] M. Blasone, P. Jizba, and F. Scardigli, quant-ph 0901.3907
[108] A. Blaut and J. Kowalski-Glikman, gr-qc 9506081, 9509040, 9607004, and 9710136
[109] R. Blume-Kohout and W. Zurek, quant-ph 0704.3615
[110] N. Bogoliubov, A. Logunov, A. Oksak, and I. Todorov, General principles of quantum field theory, Kluwer, 1990
[111] N. Bogoliubov and A. Prykarpatsky, gr-qc 0810.3303
[112] N. Bogoliubov, A. Prykarpatsky, and U. Taneri, gr-qc 0808.0871
[113] O. Bohigas, M. Giannoni, and C. Schmit, Phys. Rev. Lett., 52 (1984), 1-4
[114] D. Bohm, Phys. Rev., 84 (1951), 166; Phys. Rev., 85 (1952), 166-179 and 180-193

[115] D. Bohm and B. Hiley, Phys. Rev. Lett., 55 (1985), 2511-2514
[116] A. Bolivar, Physica A, 315 (2002), 601-615; Canad. Jour. Phys., 81 (2003), 971-976
[117] R. Bonal, I. Quiros, and R. Cardenas, gr-qc 0010010
[118] P. Bonifacio, C. Wang, J. Toto Mendonca, and R. Bingham, gr-qc 0903.1668
[119] I. Booth, gr-qc 0508107
[120] A. Bouda, quant-ph 0004044 and 0210193
[121] A. Bouda and A. Gharbi, quant-ph 0810.0826
[122] G. Bowman, Found. Phys., 35 (2003), 605-625
[123] T. Boyer, Amer. Jour. Phys., 71 (2003), 866-870
[124] M. Brack and R. Bhaduri, Semiclassical physics, Addison-Wesley, 1997
[125] C. Brans and R. Dicke, Phys. Rev., 124 (1961), 925-935
[126] L. Breen, math.CT 0802.1833
[127] L. Brenig, quant-ph 0608025 and 0610142
[128] D. Broadhurst and D. Kreimer, Phys. Lett. B, 393 (1997), 403-412
[129] D. Brody and D. Hook, cond-mat 0809.1166
[130] D. Brody, D. Ellis, and D. Holm, Jour. Phys. A, 41 (2008), 502002 (quant-ph 0808.2380); quant-ph 0901.2025
[131] D. Brody and L. Hughston, Phys. Lett. A, 245 (1998), 73-78; gr-qc 9701051, 9708032,and 0406121; quant-ph 9706030, 9711057, 9906086, and 0601020
[132] M. Brown and B. Hiley, quant-ph 0005026
[133] M. Brown, quant-ph 9703007 and 0102102; Birbeck thesis, 2006
[134] H.Brown, quant-ph 0901.1278
[135] J. Brown and J. York, Phys. Rev. D, 47 (1993), 1407-1419
[136] L. Bruneau and J. Derezinski, math-ph 0511069
[137] R. Brunetti and K. Fredenhagen, quant-ph 0207048
[138] R. Brustein, hep-th 0702108
[139] R. Brustein and M. Hadad, hep-th 0903.0823
[140] R. Brustein, D. Gorbonos, and M. Hadad, hep-th 0712.3206
[141] J. Brylinski, Loop spaces, characteristic classes, and geometric quantization, Birkhäuser, 2008
[142] M. Buric and J. Madore, hep-th 0406232, 0507064, and 0807.0960
[143] M. Buric, T. Grammatikopoulis, J. Madore, and G. Zoupanos, hep-th 0603044
[144] M. Buric, J. Madore, and G. Zoupanos, hep-th 0709.3159 and 0712.4024
[145] P. Busch, quant-ph 0105049
[146] C. Cafaro, nlin 0702029, math-ph 0810.4622, 0810.4624, 0810.4625, and 0810.4639
[147] C. Cafaro and A. Ali, nlin 0702027; math-ph 0810.4523 and 0810.4626
[148] R. Cai, Phys. Lett. B, 582 (2004), 237-242; hep-th 0109133, 0111093, 0112253, and 0311240
[149] R. Cai and S. Kim, hep-th 0501055
[150] R. Cai and L. Cao, gr-qc 0611071; Phys. Rev. D, 75 (2007), 04003; Nucl. Phys. B, 785 (2007), 135
[151] R. Cai, Phys. Lett. B, 582 (2004), 237-242; hep-th 0109133, 0111093, 0112253, and 0311240
[152] R. Cai and Y. Myung, hep-th 0210300
[153] L. Caiani, L. Casetti, C. Clementi, and M. Pettini, Phys. Rev. Lett., 79 (1997), 4361-4364
[154] J. Calmet and X. Calmet, math-ph 0403033
[155] X. Calmet and J. Calmet, cond-mat 0410452
[156] M. Campisi, P. Talkner, and P. Hänggi, cond-mat 0901.3974
[157] S. del Campo, R. Herrera, and P. Lebrana, gr-qc 0711.1559

[158] V. Canuto, P. Adams, S. Hsieh, and E. Tsiang, Phys. Rev. D, 16 (1977), 1643-1663
[159] S. Capozziello and R. de Ritis, Classical and Quantum Gravity, 11 (1994),107-117
[160] A. Carati, cond-mat 0501588
[161] G. Carcassi, quant-ph 0809.0909 and 0902.0141
[162] G. Cardoso, B. deWitt, and T. Mohaupt, hep-th 9904005
[163] M. Carfora, math.DG 0607309 and 0710.3342
[164] M. Carfora and T. Buchert, math-ph 0801.0553
[165] M. Carfora and S. Romano, hep-th 0902.2061
[166] J. Carinena, J. Clemente-Gallardo, and G. Marmo, quant-ph 0707.3539; math-ph 0701053
[167] J. Carinena, X. Gracia, G. Marmo, E. Martinez, M. Munoz-Lecanda, and N. Roman-Roy, math-ph 0604063
[168] J. Caro and L. Salcedo, Phys. Rev. A, 60 (1999), 842-852
[169] L. Caron, H. Jirari, H. Kröger, and G. Melkonyan, quant-ph 0106159
[170] R. Carroll, Fluctuations, information, gravity and the quantum potential, Springer, 2006; On the quantum potential, Arima Publ., 2007
[171] R. Carroll, Teor. i Mat. Fiz., 152 (2007), 904-914
[172] R. Carroll, math-ph 0703065 and 0710.4351 (Prog. in Phys., 4 (2007), 22-24 and 1 (2008), 21-24); Prog. in Phys., 2 (2008), 89-90; math-ph 0712.3251
[173] R. Carroll, math-ph 0807.1320, 0807.4158; Prog. in Phys., 2 (2009), 24-28 (math-ph 0808.2965)
[174] R. Carroll, Quantum theory, deformation, and integrability, North-Holland, 2000
[175] R. Carroll, Calculus revisited, Kluwer, 2002
[176] R. Carroll, Lect. Notes Phys., 502 (1998), pp. 33-56
[177] R. Carroll, quant-ph 0506075
[178] R. Carroll, gr-qc 0512146; physics 0511176 and 0602036
[179] R. Carroll, Foundations of Physics, 35 (2005), 131-154
[180] R. Carroll, Inter. Jour. Evolution Equation, 1 (2005), 23-56; Applicable Analysis, 84 (2005), 1117-1149
[181] R. Carroll, quant-ph 0309023 and 0309159
[182] R. Carroll, physics 0511176 and 0602036; gr-qc 0512046
[183] R. Carroll, gr-qc 0501045
[184] R. Carroll, Abstract methods in partial differential equations, Harper-Row, 1969
[185] R. Carroll, gr-qc 0705.3921
[186] R. Carroll, math-ph 0701077
[187] R. Carroll, Canad. Jour. Phys., 77 (1999), 319-325
[188] S. Carroll, Spacetime and geometry, Addison-Wesley, 2004
[189] L. Casetti, M. Pettini, and E. Cohen, cond-mat 9912092
[190] L. Casetti, C. Clementi, and M. Pettini, Phys. Rev. E, 54 (1996), 5969-5984 (chao-dyn 9609010)
[191] C. Castro, Foundations of Physics, 22 (1992), 569-615; Foundations of Physics Lett., 4 (1991), 81-99; Jour. Math. Phys., 31 (1990), 2633-2626
[192] C. Castro, Found. Phys., 37 (2007), 366-409
[193] C. Castro and J. Mahecha, Prog. Phys., 1 (2006), 38-45
[194] A. Caticha, quant-ph 9810007; math-ph 0008018; gr-qc 0109068; physics 0710.1068 and 0808.0012
[195] A. Caticha and C. Caforo, physics 0710.1071
[196] A. Caticha, Phys. Rev. A, 57 (1998), 1572-1582; Found. Phys., 30 (2000), 227-251; gr-qc 0109008, 0301061, and 0508108
[197] A. Caticha, cond-mat 0409175; quant-ph 0610076 and 0907.4335; physics 0710.1071

[198] A. Caticha and C. Cafaro, physics 0710.1071
[199] A. Caticha and A. Giffin, physics 0608185
[200] R. Cavalcanti, math.DG 0501406
[201] M. Celerier and L. Nottale, quant-ph 0609107 and 0609161; gr-qc 0505012; hep-th 0112213 and 0210027
[202] B. Cerchiai, G. Fiori, and J. Madore, math.QA 0002007 and 0002215
[203] B. Cerchiai, R. Hinterding, J. Madore, and J. Wen, math.QA 9807123
[204] M. Chachian, P. Kulish, K. Nishijima, and A. Tureanu, hep-th 0408069
[205] A. Chamseddine, hep-th 9806046, 0009153, 0301112, and 0901.0577
[206] A. Chamseddine and A. Connes, hep-th 0705.1786, 0706.3688, and 0706.3690
[207] A. Chamseddine and A. Connes, hep-th 9606056, and 0812.0165
[208] A. Chamseddine, A. Connes, and M. Marcolli, Adv. Theor. Math., 11 (2007), 991-1090 (hep-th 0610241)
[209] A. Chamseddine and A. Connes, Comm. Math. Phys., 186 (1997), 731-750 (hep-th 9606001)
[210] I. Chavel, Riemannian geometry - a modern introduction, Cambridge Univ. Press, 1993
[211] J. Cheeger, Problems in analysis, Princeton Univ. Press, 1970
[212] C. Chen and J. Nester, gr-qc 0001088
[213] Y. Chen and Y. Xiao, hep-th 0712.319
[214] Y. Cheng, Macroscopic and statistical thermodynamics, World Scientific, 2006
[215] N. Chetaev, The stability of motion (Ustoichevost Dvizhenia), Permagon, 1961; Theoretical mechanics, Mir-Springer, 1989
[216] S. Cho, R. Hinterding, J. Madore, and H. Steinacker, hep-th 9903239
[217] B. Chow and D. Knopf, The Ricci flow: An introduction, Amer. Math. Soc., 2004
[218] B. Chow, Peng Lu, and Lei Ni, Hamilton's Ricci flow, Amer. Math. Soc., 2006
[219] T. Roy Chowdhury and T. Padmanabhan, gr-qc 0404091
[220] T. Chrobok and H. Borzeszkowski, Gener. Relativ. Gravit., 38 (2006), 397-415
[221] D. Chruscinski and K. Mlodawski, quant-ph 0501163
[222] R. Cirelli, A. Mania, and L. Pizzocchero, Jour. Math. Phys., 31 (1990), 2891-2897 and 2898-2903
[223] R. Cirelli, M. Gatti, and A. Mania, quant-ph 0202076
[224] F. Cianfrani, A. Marrocco, and G. Montani, gr-qc 0508126
[225] T. Ciufolini and J. Wheeler, Gravitation and inertia, Princeton Univ. Press, 1995
[226] T. Clunan, S. Ross, and D. Smith, gr-qc 0402044
[227] L. Cohen, Jour. Math. Phys., 7 (1966), 781; Found. Phys., 20 (1990), 1455-1473
[228] M. Combescure, J. Ralston, and D. Robert, math-ph 9807005
[229] M. Combescure and D. Robert, Jour. Math. Phys., 36 (1995), 6596-6610; Asymptotic Anal., 14 (1997), 377-404
[230] A. Connes, Noncommutative geometry, Academic Press, 1994
[231] A. Connes and M. Marcolli, Noncommutative geometry, quantum fields, and motives, Colloq. Pub., AMS, 2008
[232] A. Connes and M. Marcolli, math.NT 0409306
[233] A. Connes, In "On space and time", Ed. S. Majid, Cambridge Univ. Press, 2008, pp. 196-237
[234] A. Connes and C. Rovelli, Class. Quant. Grav., 11 (1994), 2899-2917
[235] A. Connes, math.QA 0111093; hep-th 9603053, 0101093, and 0003006
[236] A. Connes and D. Kreimer, hep-th 9904044
[237] A. Connes, hep-th 0608226; Comm. Math. Phys., 182 (1996), 155-176
[238] M. Consoli, hep-th 0109215, 0002098, and 0306070; gr-qc 0306105

[239] F. Cooperstock, V. Faraoni, and G. Perry, gr-qc 9512025
[240] L. Cornalba, hep-th 9909081
[241] N. Costa Dias and J. Nuno Prata, Ann. Phys., 313 (2004), 110-146; Phys. Lett. A, 291 (2001), 355-366; hep-th 0504166; quant-ph 0208156
[242] A. Coutant and S. Rajeev, cond-mat 0807.4632
[243] M. Crampin and F. Pirani, Applicable differential geometry, Cambridge Univ. Press, 1986
[244] S. Creagh, J. Robbins, and R. Littlejohn, Phys. Rev. A, 42 (1990), 1907-1922
[245] J. Cresson, math.GM 0211071
[246] L. Crowell, Quantum fluctuations of spacetime, World Scientific, 2005
[247] J. Cushing, A. Fine, and S. Goldstein (editors), Bohmian mechanics and quantum theory, Kluwer, 1996
[248] P. Cvitanovic et al, ChaosBook.org, 2008
[249] M. Dabrowski, T. Denkiewicz, and D. Blaschke, hep-th 0507068
[250] U. Danielsson, Phys. Rev. D, 71 (2005), 023516
[251] A. Das, Field theory, World Scientific, 2006; Lectures on quantum field theory, World Scientific, 2008
[252] G. Date, Class. Quant. Gravity, 18 (2001), 5219-5225
[253] M. Davidovic, D. Arsenovic, M. Bozic, A. Sanz, and S. Miret-Artes, quant-ph 0803.2606
[254] M. Davidson, Physica A, 96 (1979), 465
[255] P. Davies, Jour. Phys. A, 8 (1975), 609
[256] D. Delphenich, gr-qc 0211065 and 0209091
[257] E. Deotto and G. Ghirardi, quant-ph 9704021
[258] A. Deriglazov, math-ph 0903.1328
[259] N. Deruelle and J. Katz, gr-qc 0104007 and 0512077
[260] N. Deruelle, J. Katz, and S. Ogusi, gr-qc 0310098
[261] C. Dewdney and Z. Malik, quant-ph 9506027
[262] B. DeWitt, Phys. Rev., 160 (1967), 1113-1148
[263] K. Dienes and M. Lennek, hep-th 0312173, 0312216, and 0312217
[264] M. Dimassi and J. Sjöstrand, PNLDE 21, Birkhäuser, 126-142
[265] P. Dirac, Proc. Royal Soc. London A, 209 (1951), 291-296; 212 (1952), 330-339; 332 (1973), 403-418
[266] J. Dixmier, Von Neumann algebras, North-Holland, 1982
[267] T. Djama, quant-ph 0111121, 0111142, 0201003, 0311057, 0311059, 0404175, 0404098, 0409044
[268] H. Doebner and G. Goldin, Phys. Lett. A, 162 (1992), 397
[269] M. and N. Dolarsson, Tensors, relativity, and cosmology, Elserier, 2005
[270] M. Douglas and N. Nekrasov, hep-th 0106048
[271] F. Dowker, J. Henson, and R. Sorkin, gr-qc 0311055
[272] B. Dragovich, A. Khrennikov, S. Kozyrev, and I. Volovich, math-ph 0904.4205
[273] D. Dubin, M. Hennings, and T. Smith, Mathematical aspects of Weyl quantization and phase, World Scientific, 2000
[274] D. Dürr, S. Goldstein, and N. Zanghi, quant-ph 9511016, 9512031, 0308038, and 0308039; Jour. Stat. Phys., 67 (1992), 843-907, 68 (1992), 259-270, and 116 (2004), 959-1055; Phys. Lett. A, 172 (1992), 6-12
[275] D. Dürr, S. Goldstein, R. Turulka, and N. Zanghi, quant-ph 0903.2601
[276] G. Dvali and S. Solodukhin, hep-th 0806.3976
[277] M. Dzugatov, E. Aurell, and A. Valpiani, Phys. Rev. Lett., 81 (1998), 1762-1765

[278] W. Ebeling and I. Sokolov, Statistical thermodynamics and stochastic theory of nonequilibrium systems, World Scientific, 2005
[279] K. Ecker, D. Knoff, L. Ni, and P. Topping, math.DG 0608470
[280] J. Eckmann and D. Ruelle, Rev. Mod. Phys., 57 (1985), 617-656
[281] A. Eddington, The nature of the physical world, MacMillan, 1930
[282] C. Efthymiopoulos and G. Contopoulos, Jour. Phys. A, 39 (2006), 1819-1852
[283] C. Efthymiopoulos, C. Kalapotharakos, and G. Contopoulos, quant-ph 0709.2038
[284] A. Einstein, Principe de relativité, Arch. Sci. Phys. Natur., ser. 4, 29 5 (1910)
[285] E. Eisenberg and L. Horwitz, quant-ph 9502010
[286] C. Eling, R. Guedens, and T. Jacobson, Phys. Rev. Lett., 96 (2006), 121301 (gr-qc 0602001)
[287] C. Eling and J. Bekenstein, gr-qc 0810.5255
[288] E. Elizalde, hep-th 0712.1346
[289] E. Elizalde and P. Silva, hep-th 0804.3721 and 0903.2732
[290] H. Elze, gr-qc 0307014, 0301109, 0512016, and 0704.2683; quant-ph 0306096, 0710.2765, and 08063408; hep-th 0411176, 0508095, and 0510267
[291] S. Esposito, Found. Phys., 28 (1998), 231-244; Found. Phys. Lett., 12 (1999), 165-177
[292] L. Evans, Partial differential equations, Amer. Math. Soc., 1998
[293] L. Faddeev and V. Popov, Teor. Mat. Fiz., 1 (1969), 3-18; Phys. Lett. B, 5 (1967), 29-30
[294] D. Fairlie, hep-th 9806198
[295] V. Faraoni, Cosmology in scalar-tensor gravity, Kluwer, 2004
[296] P. Falsaperla and G. Fonte, Phys. Lett. A, 316 (2003), 382-390
[297] A. Faraggi and M. Matone, Phys. Lett. B, 437 (1998), 445 (1999), 77, and 450 (1999), 369; Inter. Jour. Mod. Phys. A, 15 (2000), 1869- 2017; Phys. Lett. A, 249 (1998), 180; Phys. Rev. Lett., 78 (1997), 163; hep-th 9801033, 9705108, and 9809125
[298] L. Fatibene and M. Francaviglia, Natural and gauge foralism for classical field theories, Kluwer, 2003
[299] L. Fatibene, M. Ferraris, M. Francaviglia, and S. Mercadante, Entropy, 9 (2007), 169-185
[300] L. Fatibene, M. Ferraris, M. Francaviglia, and M. Raiteri, gr-qc 0003019
[301] A. Fatollahi, Phys. Lett. B, 665 (2008), 257-259
[302] H. Feichtinger, Annales Inst. Fourier, 40 (1990), 537-555
[303] M. Feldman, T. Ilmanen, and L. Ni, math.DG 0405036
[304] B. Felsager, Geometry, particles, and fields, Springer, 1995
[305] J. Finkelstein, quant-ph 0508103
[306] G. Fiori and J. Madore, math.QA 9904027
[307] D. Fivel, quant-ph 0104123, 0205012, and 0311145
[308] S. Flego, B. Frieden, A. Plastino, A.R. Plastino, and B. Soffer, Phys. Rev. E, 68 (2003), 016105
[309] C. Flesia and C. Piron, Helv. Phys. Acta, 57 (1984), 697-703
[310] E. Floyd, Phys. Rev. D, 26 (1982), 1339, 25 (1982), 1547, 29 (1984), 1841, and 39 (1986), 3246; Inter. Jour. Mod. Phys. A, 14 (1999), 1111; Found. Phys. Lett., 9 (1996), 489 and 13 (2000), 235; Inter. Jour. Theor. Phys., 27 (1988), 273-281; Phys. Lett. A, 214 (1996), 259-265
[311] E. Floyd, quant-ph 9705051, 9708026, and 9907092
[312] E. Floyd, quant-ph 0009070, 0206114, 0302128, 0307090, and 0605120
[313] G. Folland, Harmonic analysis in phase space, Princeton Univ. Press, 1989

[314] B. Foster, gr-qc 0509121
[315] P. diFrancesco, P. Mathieu, and D. Senechal, Conformal field theory, Springer, 1997
[316] D. Freed, hep-th 9212115
[317] D. Friedan, Phys. Rev.Lett., 45 (1980), 1057-1060; Annals of Phys., 163 (1985), 318-419
[318] B. Frieden and R. Gatenby, Exploratory data analysis using Fisher information, Springer, 2007
[319] B. Frieden, Physics from Fisher information, Cambridge Univ. Press, 1998; Science from Fisher information, Springer, 2004
[320] B. Frieden, Amer. Jour. Phys., 57 (1989), 1004-1008
[321] B. Frieden and B. Soffer, Phys. Rev. E, 52 (1995), 2274-2286; Physica A, 388 (2009), 1315-1330
[322] B. Frieden, A. Plastino, A.R. Plastino, and B. Soffer, Phys. Rev. E, 60 (1999), 48-53 and 66 (2002), 046128; Phys. Lett. A, 304 (2002), 73-78
[323] B. Frieden, Physica A, 198 (1993), 262-338; Phys. Rev. A, 41 (1990), 4265-4276
[324] H. Frisk, Phys. Lett. A, 227 (1997), 139-142
[325] A. Frolov and L. Kofman, hep-th 0212327
[326] Y. Fuji and K. Maeda, The scalar tensor theory of gravitation, Cambridge Univ. Press, 2003
[327] S. Fulling, Aspects of QFT in curved space time, Cambridge Univ. Press, 1989
[328] D. Fursaev, hep-th 9809049
[329] G. Gallavotti and E. Cohen, Jour. Stat. Phys., 80 (1995), 931-970
[330] S. Garashchuk and V. Rassolov, Jour. Chem. Phys., 118 (2003), 2482-2490
[331] M. Garay, math-ph 0705.2950
[332] P. Garbaczewski, Entropy, 7 (2005), 253-299; Jour. Stat. Phys., 123 (2007), 315-355
[333] P. Garbaczewski, cond-mat 0202463, 0211362, 0301044, 0510533, 0706248, 0604538, and 0703147; quant-ph 0408192, 0504086, 0612151 and 0805.1536; stat-mech. 0703204
[334] P. Garbaczewski, cond-mat. 0510533; quant-ph 0706.2481
[335] P. Garbaczewski, cond-mat 0202463 and 0211362
[336] G. Garcia-Calderon and M. Moshinsky, Jour. Phys. A, 13 (1980), L185- L188
[337] G. Garcia de Polavieja, Phys. Lett. A (220), 1996), 303-314; Phys. Rev. A, 53 (1996), 2059-2061
[338] D. Garfinkle and R. Mann, gr-qc 0004056
[339] X. Ge, hep-h 0703253
[340] I. Gelfand and A. Yaglom, Jour. Math. Phys., 1 (1960), 48-69
[341] M. Gellman and C. Tsallis, Nonextinsive entropy, Oxford Univ. Press, 2004
[342] G. Ghirardi, A. Rimini, and T. Weber, Phys. Rev. D, 34 (1986), 470-491
[343] G. Ghirardi and L. Marinatto, quant-ph 0206021, 0401065, 0505195, and 0509194
[344] G. Ghirardi, L. Marinatto, and R. Romano, quant-ph 0310120
[345] G. Giachetta, L. Mangiarotti, and G. Sardanashvili, Geometric and algebraic topological methods in field theory, World Scientific, 2005
[346] G. Giachetta, G. Sardanashvili, and L. Mangiarotti, New Lagrangian and Hamiltonian methods in field theory, World Scientific, 1999
[347] G. Gibbons and H. Pohle, Nuclear Phys. B, 410 (1993), 117
[348] P. Gibilisco and T. Isola, math-ph 0407007 and 0509146
[349] P. Gibilisco, D. Imparato, and T. Isola, math-ph 0701062, 0702058, 0706.0791, and 0707.1231
[350] P. Gibilisco, F. Hiai, and D. Petz, math-ph 0712.1208

[351] A. Giffin and A. Caticha, physics 0708.1593
[352] D. Gilbarg and N. Trudinger, Elliptic partial differential equations of second order, Springer, 1983
[353] L. Glinka, gr-qc 0612079; arXiv 0707.3341
[354] L. Glinka, gr-qc 0711.1380, 0803.1533 and 0804.3516
[355] L. Glinka, gr-qc 0712.1674, and 0801.4157; hep-th 0712.2769
[356] L. Glinka, V. Pervushin, and R. Kostecki, gr-qc 0703062
[357] L. Glinka, hep-th 0902.2829
[358] L. Glinka and V. Pervushin, gr-qc 0705.0655
[359] S. Goldberg, Curvature and homology, Academic Press, 1962
[360] Y Goldfarb, I. Degani, and D. Tannor, quant-ph 0604150 and 0705.2132
[361] Y. Goldfarb, J. Schiff, and D. Tannor, quant-ph 0706.3508, 0707.0117, and 0807.4659
[362] Y. Goldfarb and D. Tannor, quant-ph 0706.3507
[363] G. Goldin, quant-ph 0002013
[364] S. Goldstein and W. Struyve, quant-ph 0704.3070
[365] S. Goldstein, R. Tumulka, and N. Zanghi, quant-ph 0710.0885
[366] S. Goldstein, R. Tumulka, and N. Zanghi, quant-ph 0710.0885
[367] Y. Gong and A. Wang, hep-th 0704.0793; Phys. Rev. Lett., 991 (1997), 21130
[368] T. Gonzalez, G. Leon, and I. Quiros, astro-ph 0502383 and 0702227
[369] M. de Gosson, Maslov classes, metaplectic representations, and Lagrangian quantization, Akad. Verlag, 1997; Symplectic geometry and quantum mechanics, Birkhäuser, 2006; The principles of Newtonian and quantum mechanics, Imperial Coll. Press, 2001
[370] M. de Gosson, quant-ph 0808.2774; ESI preprint 1951 (2007); math.SG 0411453;
[371] M. de Gosson and F. Luef, Phys. Lett. A, 364 (2007), 453-457
[372] M. de Gosson, Jour. Phys. A, 31 (1998), 4239-4247; 38 (2005), L325-L329 and 9263-9287; Operator theory, 164 (2006), 121-132
[373] M. de Gosson, math-ph 0505073; math.SG 0503708 and 0504013
[374] M. de Gosson, math-ph 0505073 and 0602055; quant-ph 0808.2774; math.SG 0504013; Jour. Phys. A, 37 (2004), 7297-7314 and L325-L329
[375] M. deGosson and S. deGosson, math-ph 0508051
[376] M. deGosson, Bull. Sci. Math., 121 (1997), 301-322; Annales Inst. Poincaré, 70 (1999), 547-573; Annales Inst. Fourier, 40 (1990), 537-555
[377] M. deGosson, Found. Phys., 39 (2009), 194-214
[378] W. Graf, gr-qc 0602054; Phys. Rev. D, 67 (2003), 024002 and 38 (2005), L325-L329
[379] R. Graham, D. Rockaerts, ant T. Tel, Phys. Rev. A, 31 (1985), 3364-3375
[380] S. Gralla and R. Wald, gr-qc 0806.3293
[381] A. Granik, quant-ph 0409018 and 0801.3311; physics 0309059
[382] A. Granik and G. Chapline, quant-ph 0302013
[383] M. Grasselli, math-ph 0007039
[384] M. Green, J. Schwartz, and E. Witten, Superstring theory, Cambridge Univ. Press, 1987
[385] B. Greenbaum, K. Jacobs, and B. Sundaram, quant-ph 0705.2227
[386] K. Gröchenig, Foundations of time frequency analysis, Birkhäuser, 2001
[387] M. Gromov, Inventiones Math., 82 (1985), 307-347
[388] O. Gron and S. Hervik, Einstein's general theory of relativity, Springer, 2007
[389] F. Gronwald, F. Hehl, and J. Nitsch, physic 0506219
[390] C. Grosche and F. Steiner, Phys. Lett. A, 123 (1987), 319-328; Annals Phys., 182 (1988), 120-156

[391] D. Gross, quant-ph 0602001 and 0702004
[392] G. Grössing, Phys. Lett. A, 372 (2008), 4556-4562; Found. Phys. Lett., 17 (2004), 343-362; quant-ph 0201035, 0204070, 0205047, 0311109, 0404030, 0410236, 0508079, 0711.4954, 0806.4462, 0808.3539, and 0812.3561
[393] M. Gualtieri, math.DG 0401221 and 0703298
[394] V. Guillemin and S. Sternberg, Symplectic techniques in physics, Cambridge Univ. Press, 1984
[395] H. Guo, C. Huang, and X. Wu, gr-qc 0208067
[396] M. Gutzwiller, Chaos in classical and quantum mechanics, Springer, 1990
[397] R. Haag, Local quantum physics, Springer, 1996
[398] R. Haag, N. Hugenholtz, and M. Winnik, Comm. Math. Phys., 5 (1967), 215
[399] G. Haciberiroglu, M. Caglar, and Y. Polatoglu, nlin.CD 0706.1709
[400] M. Hall and M. Reginatto, Jour. Phys. A, 35 (2002), 3289-3303; Fortschr. Phys., 50 (2002), 646-651; quant-ph 0201084; Phys. Rev. A, 72 (2005), 062109
[401] M. Hall, K. Kumar, and M. Reginatto, Jour. Phys. A, 36 (2003), 9779-9794
[402] M. Hall, gr-qc 0408098; quant-ph 0007116; Jour. Phys. A, 37 (2004), 7799 and 9549 (quant-ph 0404123 and 0406054)
[403] M. Hall, quant-ph 0804.2505; Jour. Phys. A, 37 (2004)
[404] M. Hall and M. Reginatto, quant-ph 0509134
[405] S. Haller and C. Vizman, math.DG, 0305089
[406] J. Halliwell, Phys. Rev. Lett., 83 (1999), 2481-2485; Phys. Rev. A, 58 (1998), 105015 and 60 (1999), 105031; quant-ph 0305084, 0607132, 9902008, 9905094, and 0507136; gr-qc 0208018
[407] J. Halliwell and S. Hawking, Phys. Rev. D, 31 (1985), 1777
[408] J. Hartle and S. Hawking, Phys. Rev. D, 28 (1983), 2960
[409] I. Hamilton, R. Mosna, and L. Delle Site, physics 060918
[410] D. Han, Y. Kim, and M. Noz, Phys. Rev. A, 37 (1988), 807-814; 40 (1989), 902-912
[411] J. Hannay and A. Ozorio de Almeida, Jour. Phys. A, 17 (1984), 3429-3440
[412] E. Harikumar and V. Rivelles, hep-th 0607115
[413] J. Hartle, quant-ph 0209104; gr-qc 9404017
[414] J. Hartle, R. Laflamme, and D. Marolf, Phys. Rev. D, 51 (1995), 7007-7016
[415] Y. Hatakeyama, Tohoku Math. Jour., 18 (1966), 338
[416] B. Hatfield, Quantum field theory of point particles and strings, Westview Press, 1992
[417] S. Hawking, Comm. Math. Phys., 43 (1975), 199-220
[418] S. Hawking and G. Ellis, The large scale structure of space-time, Cambridge Univ. Press, 1973
[419] M. Hayashi, Quantum information, Springer, 2006
[420] G. Hayward, Phys. Rev. D, 43 (1991), 3861-3872; 49 (1994), 831-839
[421] S. Hayward, gr-qc 9710089 and 0004042
[422] S. Hayward, S. Mukohyama, and M. Ashworth, gr-qc 9810006
[423] Y. He, V. Jejjala, and D. Minic, math-ph 0903.4321
[424] E. Heby, Sobolev spaces on Riemannian manifolds, Lect. Notes Math., 1365, Springer, 1996
[425] E. Heby, Nonlinear analyaia on manifolds, Courant Lect. Notes Math., Vol 5, 1999
[426] F. Hehl and Y. Obukhov, Foundations of classical electrodynamics, Birkhäuser, 2003
[427] F. Hehl and G. Kerlick, Gen. Relativ. Gravit., 9 (1978), 691-710
[428] F. Hehl, P. vonder Heyde, and G. Kerlick, Rev. Mod. Phys., 48 (1976), 393-416
[429] F. Hehl, Y. Itin, and Y. Obukhov, physics 0610221

[430] F. Hehl and Y. Obukhov, gr-qc 0001010 and 0508024; physics 0404101 and 0407022
[431] B. Helffer and D. Robert, Jour. Fnl. Anal., 5 (1983), 246-268
[432] B. Helffer and J. Sjöstrand, Comm. PDE, 9 (1984), 337-408
[433] S. Helgason, Differential geometry, Lie groups, and symmetric spaces, Amer. Math. Soc., 2001
[434] M. Heller and W. Sasin, gr-qc 9711051
[435] A. Heslot, Amer. Jour. Phys., 51 (1983), 1096-1102; Phys. Rev. D, 31 (1985), 1341-1348
[436] B. Hiley, Quo vadis quantum mechanics, Springer, 2005, pp. 299-324; Foundations of Physics, 2009
[437] B. Hiley, R. Callaghan, and O. Maroney, quant-ph 0010020
[438] N. Hitchin, Quart. Jour. Math., 54 (2003), 281
[439] H. Hofer and E. Zehnder, Symplectic invariants and Hamiltonian dynamics, Birkhäuser, 1994
[440] P. Holland, The quantum theory of motion, Cambridge Univ. Press, 1997
[441] P. Holland, Phys. Rev. A, 60 (1999), 4326-4330; 67 (2003), 062105; Found. Phys., 28 (1998), 881-911; Annals Phys., 315 (2005), 505-531
[442] P. Holland, quant-ph 0305175 and 0401017; arXiv 0807.4482 and 0901.0402
[443] S. Hollands and R. Wald, gr-qc 0803.2003 and 0805.3419
[444] G. 't Hooft, hep-th 0003005, 0104080, 0104219, 0105105, 0707.4568, and 0707.4572; quant-ph 0212095 and 0604008
[445] G. 't Hooft, quant-ph 9612018; hep-th 0003005 and 0405032; gr-qc 9509050, 9903084, 0401027, and 0804.0328
[446] G. 't Hooft, Found. Phys., 38 (2008), 733-757
[447] L. Hörmander, The analysis of partial differential operators, 1-4, Springer, 1983
[448] M. Horvat and T. Prossen, quant-ph 0601165
[449] L. Horwitz, E. Eisenberg, and Y. Strauss, quant-ph 9610025
[450] L. Horwitz and C. Piron, Helv. Phys. Acta, 66 (1993), 693-711
[451] L. Horwitz, J. Levitan, M. Lewkowicz, M. Schiffer, and Y. ben Zion, physics 0701212
[452] D. Huard, H. Kröger, G. Melkonyan, and L. Nadeau, quant-ph 0406131
[453] G. Iacomelli and M. Pettini, Phys. Lett. A, 212 (1996), 29-38
[454] A. Inoue, Tomita-Takesaki theory in algebras of unbounded operators, Springer, 1998
[455] N. Ishibashi, H. Kawai, Y. Kitazawa, and A. Tsuchiya, hep-th 9612115
[456] J. Isidro, J. Santander, and P.F. de Cordoba, hep-th 0808.2351, 0808.2717, and 0902.0143; gr-qc 0804.0169
[457] J. Isidro, hep-th 0110151 and 0411015; quant-ph 0407159; math-ph 0710.3544
[458] J. Isidro, hep-th 0510075 and 0407161; math-ph 0708.0720; quant-ph 0503053
[459] J. Isidro and M. de Gosson, Jour. Phys. A, 40 (2007), 3549-3567 (hep-th 0608087); Mod. Phys. Lett. A, 22 (2007), 191-200 (quant-ph 0608093)
[460] J. Isidro, hep-th 0204178 and 0304175; quant-ph 0307172
[461] J. Isidro, hep-th 0512241 and 0612186; math-ph 0708.0720; Mod. Phys. Lett. A, 38 (2005), 2813-2918
[462] J. Isidro and F. deCordoba, math-ph 0712.1961; hep-th 0804.1060
[463] J. Isidro, hep-th 0009221; quant-ph 0105012 and 0112032
[464] J. Isidro, quant-ph 0204128, 0304143, 0310092, and 0407159
[465] M. Israelit, The Weyl-Dirac theory and our universe, Nova Science Pub., 1999; Found. Phys., 29 (1999), 1303-1322; 32 (2002), 295-321 and 945-961; Gen. Relativ. Gravit., DOI 10.1007/s10714-008-0633-5

[466] M. Israelit, Found. Phys., 28 (1998), 205-228; 29 (1999), 1303-1322; gr-qc 9608035; Found. Phys., 35 (2005), 1725-1748 and 1769-1782; 37 (2007), 1628-1642
[467] M. Israelit and N. Rosen, Found. Phys., 22 (1992), 555-568; 24 (1994), 901-915; 25 (1995), 763
[468] M. Israelit, Gen. Relativ. and Gravit., 29 (1997), 1411-1424 and 1597-1614; gr-qc 0710.3410, 0710.3690, 0710.3913, 0710.3923, and 0905.2482
[469] Y. Itin and F. Hehl, gr-qc 0401016
[470] V. Ivancevic and D. Reid, nlin.AO 0809.4069
[471] V. Iyer and R. Wald, Phys. Rev. D, 52 (1995), 4430-4439 (gr-qc 9403028)
[472] G. Izquierdo and D. Pavon, Phys. Lett. B, 633 (2006), 420-426
[473] J. Jackson, Classical electrodynamics, Wiley, 1999
[474] T. Jacobson, Phys. Rev. Lett., 75 (1995), 1260-1263 (gr-qc 9504004)
[475] T. Jacobson and R. Parentani, gr-qc 0302099
[476] T. Jacobson, G. Kang, and R. Myers, Phys. Rev. D, 49 (1994), 6587-6598 (gr-qc 9312023); Phys. Rev. D, 52 (1995), 3518-3528
[477] T. Jacobson and R. Myers, Phys. Rev. Lett., 70 (1993), 3684
[478] S. Jalalzadeh, Genera. Relativ. Gravit., 39 (2007), 387-400
[479] P. Jang and R. Wald, Jour. Math. Phys., 18 (1977), 41
[480] W. Janke, D. Johnston, and R. Kenna, cond-mat 0401092 and 0210571
[481] E. Jaynes, Probability theory, Cambridge Univ. Press, 2003
[482] V. Jejjala, M. Kavic, D. Minic, and C. Tze, gr-qc 0905.2992; hep-th 0804.3598
[483] V. Jejjala, M. Kavic, and D. Minic, hep-th 0705.4581 and 0706.2252
[484] V. Jejjala, D. Minic, and C. Tze, gr-qc 0406037
[485] V. Jejjala and D. Minic, hep-th 0605105
[486] H. Jirari, H. Kröger, X. Luo, G. Melkonyan, and K. Moriarty, hep-th 0103027; quant-ph 0108094
[487] H. Jirari, H. Kröger, X. Luo, and K. Moriarty, quant-ph 0102032
[488] M. John, quant-ph 0102087, 0109093, and 0809.5101
[489] C. Johnson, D-branes, Cambridge Univ. Press, 2003
[490] D. Johnston, W. Janke, and R. Kenna, cond-mat 0408316
[491] J. Jost, Riemannian geometry and geometric analysis, Springer, 2002
[492] A. Joseph and S. Rajeev, hep-th 0807.3957
[493] B. Jurco and P. Schupp, Euro. Phys. Jour., C14 (2000), 367-370 (hep-th 0001032)
[494] B. Jurco, L. Möller, S. Schraml, P. Schupp, and J. Wess, hep-th 0104153
[495] B. Jurco P. Schupp, and J. Wess, hep-th 0005005, 0012225, and 0102129
[496] H. Kandrup and C. Bohn, astro-ph 0108038
[497] H. Kandrup, astro-ph 9903434
[498] G. Kang, Phys. Rev. D, 54 (1996), 7483-7489
[499] G. Kaniadakis, quant-ph 0112049; physics 0901.1058
[500] G. Kaniadakis and A. Scarfone, cond-mat 0303334
[501] D. Kaplan and R. Sundrum, hep-th 0505265
[502] Z. Karkuszewski, quant-ph 0512223
[503] D. Kazakov, V. Pervushin and S. Pushkin, Jour. Phys. A, 11 (1978), 2093-2105
[504] P. Kazinski, Phys. Rev. E, 77 (2008), 041119; cond-mat 0809.0233; cond-mat. 0711.3644
[505] M. Khalkhali and M. Marcolli, An invitation to noncommutative geometry, World Scientific, 2008
[506] A. Kholodenko, gr-qc 0110064; hep-th 0303334 and 0701084
[507] A. Kholodendo and E. Ballard, gr-qc 0410029
[508] A. Kholodenko and K. Freed, Jour. Chem. Phys., 80 (1984), 900-924

[509] C. Kiefer, Quantum gravity, Oxford Univ. Press., 200
[510] Y. Kim and E. Wigner, Amer. Jour. Phys., 58 (1990), 439-448
[511] E. Kiritsis, String theory in a nutshell, Princeton Univ. Press, 2007
[512] D. Klammer and H. Steinacker, hep-th 0805.1157; gr-qc 0903.0986
[513] J. Klauder, quant-ph 0112010 and 0308049
[514] J. Klauder and B. Skagerstam, quant-ph 0612037
[515] H. Kleinert, Path integrals ..., World Scientific, 2004
[516] S. Kobayashi and K. Nomizu, Foundations of differential geometry, Vols. 1 and 2, Wiley, 1996
[517] B. Koch, quant-ph 0801.4635; hep-th 0810.2786; gr-qc 0901.4106
[518] J. Koga and K. Maeda, gr-qc 9803086
[519] I. Kolar, P. Michor, and J. Slovak, Natural operators in differential geometry, Springer, 1993
[520] M. Kondratieva and T. Osborn, quant-ph 0505066
[521] S. Konkel and A. Makowski, Phys. Lett. A, 238 (1998), 95-100
[522] M. Kontsevich, Lett. Math. Phys., 66 (2003), 157-216
[523] D. Kothawala, T. Padmanabhan, and S. Sarkar, gr-qc 0807.1481
[524] D. Kothawala, S. Sarkar, and T. Padmanabhan, gr-qc 0701002
[525] D. Kothawala and T. Padmanabhan, gr-qc 0904.0215
[526] J. Kowalski-Glikman, gr-qc 9511014
[527] V. Krasnoholovets and F. Columbus, New topics in quantum physics research, Nova Sci. Pub., 2006
[528] C. Krishnan and S. Kuperstein, hep-th 0903.2169
[529] H. Kröger, quant-ph 0106087
[530] W. Kühnel, Differential geometry, Amer. Math. Soc., 2002
[531] J. Kurchan, cond-mat 0901.1271
[532] P. Lamberti, M. Portesi, and J. Sparacino, quant-ph 0807.0583
[533] C. Lämmerzahl and F. Hehl, gr-qc 0409072
[534] G. Landi, An introduction to non-commutative spaces and geometry, Springer, 1997
[535] N. Landsman, Mathematical topics between classical and quantum mechanics, Springer, 1998; quant-ph 0506082
[536] M Lapidus, In search of the Riemann zeros, Amer. Math. Soc., 2008
[537] M. Lapidus and M. van Frankenhuysen, Fractal geometry, complex dimensions, and zeta functions, Springer, 2006
[538] V. Latera and M. Baranger, Phys. Rev. Lett., 82 (1999), 520-523
[539] V. Latera, M. Baranger, A. Rapisarda, and C. Tsallis, Phys. Lett. A, 273 (2000), 97-103
[540] P. Lax and R. Phillips, Scattering theory (revised edition), Acad. Press, 1989; Scattering theory for automorphic functions, Princeton Univ. Press, 1976
[541] R. Leacock and M. Padgett, Phys. Rev. D, 28 (1983), 2491
[542] C. Leavens and R. Sala Mayato, Phys. Lett. A, 280 (2001), 163-172
[543] C. LeBrun, math.DG 0803.3734
[544] J. Lee and R. Wald, Jour. Math. Phys., 31 (1990), 725-743
[545] L. Lempert, Jour. Diff. Geom., 38 (1993), 519
[546] P. Li and S. Yau, Proc. Symp. Pure Math., 36 (1980), 205-239
[547] I. Licata, quant-ph 0705.1173 and 0711.2973
[548] A. Liddle, A. Mazumdar, and f. Schnuck, Phys. Rev. D, 58 (1998), 061301
[549] T. Liko and P. Wesson, gr-qc 0310067 and 0505024
[550] T. Liko, J. Overduin, and P. Wesson, gr-qc 0311054

[551] J. Ling, math.DG 0406061, 0406296, 0406437, 0406562, 0407138, 0710.2574, 0710.4291, and 07104326
[552] H. Liu and B. Mashhoon, Ann. Phys. (Leipzig), 4 (1995), 565
[553] H. Liu and P. Wesson, Astrophys. Jour., 562 (2001), 1
[554] F. Lizzi, R. Szabo, and A. Zampini, hep-th 0107115
[555] R. Littlejohn, Phys. Repts., 138 (1986), 193-291; Jour. Math. Phys., 31 (1990), 2952-2977, and Jour. Stat. Phys., 68 (1992), 7-50
[556] R. Littlejohn and J. Robbins, Phys. Rev. A, 36 (1987), 2953-2961
[557] M. Lombardi and A. Matzkin, quant-ph 0506188
[558] E. Lord, Tensors, relativity, and cosmology, Tata McGraw Hill, 1976
[559] M. Lorente, gr-qc 0312119
[560] M. Lyapunov, Problème général de la stabilité du mouvement, Princeton Univ. Press, 1047
[561] M. Lyra and C. Tsallis, Phys. Rev. Lett., 80 (1998), 53-56
[562] J. Madore, An introduction to noncommutative differential geometry and its physical applications, Cambridge Univ. Press, 1995
[563] J. Madore, hep-th 9506183; gr-qc 9607065, 9611026, 9705083, 9709002, and 9906059
[564] J. Madore, T. Masson, and J. Mourad, hep-th 9411127
[565] J. Madore and J. Mourad, hep-th 9506041 and 9601169; gr-qc 9607060
[566] J. Madore, S. Schraml, P. Schupp, and J. Wess, hep-th 0001203 and 0009230
[567] R. de la Madrid and J. Isidro, quant-ph 0702111
[568] R. Maia, F. Nicacio, and R. Vallejos, Phys. Rev. Lett., 100 (2008), 184102
[569] S. Majid, In "On space and time", Cambridge Univ. Press, 2008, pp. 56-140
[570] J. Makela and A. Peltola, gr-qc 0205128, 0406032, and 0612078
[571] J. Makela, gr-qc 0605098, 0711128, 0805.3952, 0805.3955, and 0810.4910
[572] A. Makowski, P. Peploswski, and S. Dembinski, Phys. Lett. A, 266 (2000), 241-248
[573] Z. Malik and C. Dewdney, quant-ph 9506026
[574] O. Manko, V. Manko, G. Marmo, and E. Sudarshan, Phys. Lett. A, 357 (2006), 255-260
[575] P. Mannheim, Brane localized gravity, World Scientific, 2005
[576] M. Marcolli, math-ph 0804.4824
[577] G. Marmo, G. Scolarici, A. Simoni, and F. Ventriglia, hep-th 0501094
[578] J. Marsden and T. Ratiu, Introduction to mechanics and symmetry, Springer, 1999
[579] G. Marsh, Found. Phys., 38 (2008), 293-300
[580] M. Martin, A.R. Plastino, and A. Plastino, Physica A, 275 (2000), 262-271
[581] P. Martinetti, math-ph 0306046; gr-qc 0501022 and 0904.4865
[582] P. Martinetti and C. Rovelli, Class. Quant. Grav., 20 (2003), 4919-4932
[583] S. Martinez, F. Nicolas, F. Pennini, and A. Plastino, Physica A (2000), 489-502
[584] S. Martinez, F. Pennini, A. Plastino, and H. Vucetich, cond-mat 0105355
[585] B. Mashhoon and P. Wesson, Class. Quant. Gravit., 21 (2004), 3611-3620; Gen. Relativ. Gravit., 39 (2007), 1403-1412
[586] B. Mashhoon, H. Liu, and P. Wesson, Phys. Lett. B, 331 (1994), 305
[587] S. Masi, cond-mat 0611300
[588] V. Maslov and M. Fedoriuk, Semiclassical approximations in quantum mechanics, Reidel, 1981
[589] M. Matone, Brazil. Jour. Phys., 35 (2005); Found. Phys. Lett., 15 (2002), 311-328; hep-th 0212260
[590] M. Matone, hep-th 0502134
[591] M. Matone, P. Pasti, S. Shadchin, and R. Volpato, hep-th 0607133

[592] M. Matone and R. Volpato, hep-th 0506231 and 0806.4370; math.AG 0605734 and 0710.2124
[593] A. Matzkin, quant-ph 0208018, 0411084, 0607095, 0703251, 0709.2114, 0802.0613, 0806.3240, 0808.2420; Phys. Lett. B, 357 (1995), 342-348; Foundations of Physics, 39 (2009), 903-920
[594] A. Matzkin and V. Nurok, quant-ph 0609172
[595] A. Matzkin and M. Lombardi, quant-ph 0706.2978
[596] N. Mavromatos and R. Szabo, Inter. Jour. Mod. Phys. A, 16 (2001), 209-250
[597] M. Maziashvili, hep-th 0708.1472; gr-qc 0709.0898
[598] S. McDonald and A. Kaufman, Phys. Rev. Lett., 42 (1979), 1189-1191
[599] A. Medved, D. Martin, and M. Visser, Class. Quant. Grav., 21 (2004), 3111
[600] B. Mehlig and M. Wilkinson, cond-mat 0012027
[601] C. Mehta, Jour. Phys. A, 1 (1968), 385-392
[602] S. Mercuri and G. Montani, gr-qc 0401102
[603] A. Messiah, Quantum mechanics, Vols. 1 and 2, Dover, 1999
[604] P. Michor and D. Mumford, Documenta Math., 10 (2005), 217-245
[605] J. Mickelsson, math-ph 0603031
[606] T. Micklitz and A. Altland, nlin.CD 0901.3137
[607] P. Milgrom, Astrophys. Jour., 270 (1983), 365, 371, and 384; 333 (1988), 689; 302 (1986), 617; 287 (1984), 571
[608] D. Minic and C. Tze, hep-th 0305193 and 0309239
[609] S. Mirabotalebi, F. Ahmadi, and H. Salehi, Gen. Relativ. Gravit., 40 (2008), 81-92
[610] C. Misner, K. Thorne, and J. Wheeler, Gravitation, Freeman, 1973
[611] I. Moerdijk, math.AT 0212266
[612] S. Molotkov, quant-ph 0201113
[613] G. Montani, gr-qc 0412030
[614] M. Montesinos and C. Rovelli, gr-qc 0002024
[615] J. Moser, Trans. Amer. Math. Soc., 120 (1965), 286
[616] H. Motavali and M. Golshani, hep-th 0011064 and 0012119
[617] R. Mosna, I. Hamilton, and L. Delle Site, quant-ph 0504124 and 0511068
[618] R. Mrugala, J. Nulton, C. Christian Schön, and P. Salamon, Phys. Rev. A, 41 (1990), 3156-3160
[619] J. Muga, R. Sala, and R. Snider, Physica Scripta, 47 (1993),
[620] A. Mukhopadhyay and T. Padmanabhan, hep-th 0608120
[621] R. Müller, Differential Harnack inequalities and the Ricci flow, European Math. Soc., 2006
[622] F. Müller-Hoissen, gr-qc 0710.4418
[623] P. Muratore-Ginanneschi, nlin.CD 0210047
[624] M. Murray, math.DG 0712.1651
[625] B. Muthukumar, hep-th 0412069 and 0609172; Phys. Rev. D, 71 (2005), 105007
[626] Y. Myung, Y. Kim, and Y. Park, hep-th 0802.2152; Mod. Phys. Lett. A, 23 (2008), 91-98; gr-qc 0708.3145; Phys. Rev. D, 76 (2007), 104045
[627] M. Nagasawa, Schrödinger equation and diffusion theory, Birkhäuser, 1997; Stochastic processes in quantum physics, Birkhäuser, 2000
[628] S. Nasiri, SIGMA, 2 (2006), 062 )quant-ph 0511125)
[629] S. Nasiri and H. Safari, quant-ph 0505147
[630] S. Nasiri, Y. Sobouti, and F. Taati, quant-ph 0605129
[631] A. Nassimi, quant-ph 0706.0237
[632] E. Nelson, Quantum fluctuations, Princeton Univ. Press, 1985; Dynamical theory of Brownian motion, Princeton Univ. Press, 1967

[633] V. Nemytskii and V. Stepanov, Qualitative theory of differential equations, Dover, 1989
[634] A. Nielsen and M. Visser, Class. Quantum Gravity, 23 (2006), 4637-4658
[635] H. Nikolic, Found. Phys. Lett., 19 (2006), 553-566 (quant-ph 0505143); quant-ph 07072319
[636] H. Nikolic, hep-th 0512186; quant-ph 0512065, 0602024, 0603207, and 0610138
[637] H. Nikolic, Euro. Phys. Jour. C, 421 (2005), 365-374 (hep-th 0407228), gr-qc 9909035 and 0111029; hep-th 0202204 and 0601027
[638] H. Nikolic, gr-qc 0312063; hep-th 0501048; quant-ph 0603207, 0703071, 0804.4564, 0811.1905, and 0512065
[639] H. Nikolic, hep-th 0708.3962 and 0801.4471; quant-ph 0609163, 0707.2319, 0801.1905, and 0804.4563
[640] H. Nikolic, hep-th 0605250, 0702060, 0705.3452, 0708.0729, 0806.1431, and 0904.2287; gr-qc 0805.2555
[641] B. Nikolov and B. Frieden, Phys. Rev. E, 49 (1994), 4815-4820
[642] J. Noldus, gr-qc 0508104
[643] L. Nottale, Fractal space-time and microphysics: Toward a theory of scale relativity, World Scientific, 1993; physics, 0901.1270; Euro. Jour. Phys. C, 421 (2005), 365-374, 18 (2005), 123-148, 17 (2004), 63-380
[644] L. Nottale, Chaos, Solitons, and Fractals, 7 (1996), 877-938; 10 (1999), 459-468; 12 (2000), 1577-1583; 16 (2003), 539-564
[645] M. Novello, J. Salim, and F. Falciano, gr-qc 0901.3741
[646] H. Ohanian and A. Ruffini, Gravitation and spacetime, Norton, 1994
[647] M. Ohya and D. Petz, Quantum entropy and its uses, Springer, 2004
[648] Y. Okawa and H. Ooguri, hep-th 0012218
[649] B. Oksendal, Stochastic differential equations, Springer, 2003
[650] L. Olavo, Physica A, 262 (1999), 197-214 and 27 (1999), 260-302; Phys. Rev. E, 64 (2001), 036125
[651] T. Oliynyk and E. Woolgar, math.DG 0607438
[652] T. Oliynyk, V. Suneeta, and E. Woolgar, hep-th 0410001, 0510239, and 0705.0827
[653] J. Oppenheim, gr-qc0105101 and 0112001
[654] A. Orefice, R. Giovanelli, and D. Ditto, quant-ph 0705.4049 and 0706.3102
[655] J. Ortega and T. Ratiu, Momentum maps and Hamiltonian reduction, Birkhäuser, 2004
[656] J. Overduin and P. Wesson, Phys. Repts., 283 (1997), 303-378 (gr-qc 9805018)
[657] J. Overduin, P. Wesson, and B. Mashhoon, astro-ph 0707.3148
[658] A. Ozorio de Almeida and J. Hannay, Annals Phys., 138 (1982), 115-154
[659] T. Padmanabhan, Phys. Rev. Lett., 78 (1997),1854; Phys. Rev. D, 57 (1998), 6206
[660] T. Padmanabhan, Gen. Relativ. Grav., 40 (2008), 023 (arXiv 0802.0973)
[661] T. Padmanabhan, Gen. Relativ. Gravit., 34 (2002), 2029-2035; gr-qc 0202080
[662] T. Padmanabhan, Class. Quant. Gravity, 19 (2002), 5387-5408; gr-qc 0202078, 0204019, 0205090, and 0209088; hep-th 0205278
[663] T. Padmanabhan, gr-qc 0308070 and 0311036
[664] T. Padmanabhan, gr-qc 0405072; Inter. Jour. Mod. Phys. D, 14 (2005), 2263 (gr-qc 0408051); Brazil. Jour. Phys., 35 (2005), 362 (gr-qc 0412068); astro-ph 0603114
[665] T. Padmanabhan, gr-qc 0503107 and 0510015
[666] T. Padmanabhan, Gen. Relativ. Gravit., 38 (2006), 1547-1552; gr-qc 0606061
[667] T. Padmanabhan, Gen. Relativ. Gravit., 34 (2002), 2029-2935; 35 (2003), 2097-2103
[668] T. Padmanabhan, gr-qc 0706.1654

[669] T. Padmanabhan and A. Patel, hep-th 0305165 and gr-qc 0309053
[670] T. Padmanabhan, Class. Quant. Gravity, 21 (2004), 4485-4494; Phys. Rev. D, 39 (1989), 2924-2932; hep-th 0406060
[671] T. Padmanabhan and T. Singh, Class. Quant. Gravity, 7 (1990), 411-426
[672] T. Padmanabhan and A. Paranjape, gr-qc 0701003
[673] T. Padmanabhan, gr-qc 0609012
[674] T. Padmanabhan, gr-qc 0606061, 0609012, and 0706.1654
[675] T. Padmanabhan, hep-th 0705.2533 and 0903.1254; astro-ph 0603114 and 0812.2610; Gen. Relativ. Grav., 40 (2008), 2031-2036 (hep-th 0807.256)
[676] A. Paranjape, S. Sarkar, and T. Padmanabhan, Phys. Rev. D, 74 (2006), 104015 (hep-th 0607240)
[677] M. Parikh and S. Sarkar, hep-th 0903.1176
[678] M. Parikh, hep-th 0508108
[679] R. Parmenter and R. Valentine, Phys. Lett. A, 201 (1995), 1-8; 227 (1997), 5-14
[680] R. Parmenter and A. diRenzo, quant-ph 0305183
[681] R. Parwani, quant-ph 0408185 and 0412192; hep-th 0401190; quant-ph 0506005 and 0508125
[682] O. Passon, quant-ph 0412119 and 0611032
[683] M. Pawlowski and V. Pervushin, Inter. Jour. Mod. Phys. A, 16 (2001), 1715-1742
[684] P. Pearle and A. Valentini, quant-ph 0506115
[685] L. de la Pena and A. Cetto, Found. Phys., 12 (1982), 1017-1037
[686] B. Peng, K. Hunt, P. Hunt, A. Suarez, and J. Ross, Jour. Chem. Phys., 102 (1995), 4548-4562
[687] F. Pennini and A. Plastino, cond-mat 0402467, 0405033, and 0407110; Phys. Rev. E, 69 (2004), 057101; Phys. Lett. A, 326 (2004), 20-26
[688] F. Pennini and A. Plastino, Phys. Lett. A, 349 (2006), 15-20; 177 (1993), 177-179
[689] F. Pennini, A. Plastino, A.R. Plastino, and M. Casas, Phys. Lett. A, 302 (2002), 156-162
[690] F. Pennini, A.R. Plastino, and A. Plastino, Physica A, 258 (1998), 446-457
[691] F. Pennini, A. Plastino, and G. Ferri, cond-mat 0408072
[692] R. Penrose, The emperor's new mind, Oxford Univ. Press, 1989
[693] G. Perelman, math.DG 0211159, 0303109, and 0307245
[694] A. Peres, Quantum theory: Concepts and methods, Kluwer, 1995
[695] A. Peres, Annals Phys., 19 (1962), 279-286
[696] G. Perry and F. Cooperstock, gr-qc 9810045
[697] V. Pervushin and V. Zinchuk, gr-qc 0601067
[698] V. Pervushin and V. Smirichinski, Jour. Phys. A, 32 (1999), 6191-6201
[699] M. Pettini, Geometry and topology in Hamiltonian dynamics and statistical mechanics, Springer, 2007
[700] D. Petz, Jour. Phys. A, 35 (2002), 929
[701] R. Picken, Twenty years of Bialowieza, World Scientific, 2005
[702] N. Pinto-Neto, gr-qc 0410001, 0410117, and 0410225
[703] N. Pinto-Neto and E. Santini, gr-qc 0009080 and 0302112; Gener. Relativ. Gravitation, 34 (2002), 505; Phys. Lett. A, 315 (2003), 36; Phys. Rev. D, 59 (1999), 123517 (gr-qc 9811067)
[704] J. Pissondes, Chaos, Solitons, and Fractals, 9 (1998), 1115-1142; hep-th 0009169 and 0010146
[705] A. Plastino and A.R. Plastino, Phys. Lett. A, 226 (1997), 257-263
[706] A. Plastino and E. Curado, cond-mat 0412336 and 0509070; Phys. Rev. E, 72 (2005), 047103

[707] A. Plastino, A.R. Plastino, and H. Miller, Phys. Lett. A, 235 (1997), 129-134; Physica A, 235 (1997), 577-588

[708] A. Plastino, A.R. Plastino, H. Miller, and F. Khanna, Phys. Lett. A, 221 (1996), 29-33

[709] A.R. Plastino, M. Casas, and A. Plastino, Phys. Lett. A, 246 (1998), 498-504

[710] J. Plebanski and A. Krasinski, An introduction to general relativity and cosmlogy, Cambridge Univ. Press, 2006

[711] M. Pletyukhov and M. Brack, Jour. Phys. A, 36 (2003), 9949-9469

[712] H. Poincaré, Rev. Métaphys. Morale, 5 (1898), 1

[713] B. Poirier, quant-ph 0802.3472, 0803.0142, 0803.0143, and 0803.0193; Jour. Chem. Phys., 121 (2004), 4501-451

[714] B. Poirier and G. Parlant, Jour. Phys. Chem. A, 111 (2007), 10400

[715] E. Poisson, A relativists toolkit, Cambridge Univ. Press, 2004

[716] J. Polchinski, String theory, Vols. 1 and 2, Cambridge Univ. Press, 2001

[717] J. Ponce de Leon, gr-qc 0104008 and 0106020

[718] J. Ponce de Leon, Gen. Relativ. Gravit., 25 (1993), 865-880; Phys. Lett. B, 523 (2001), 311-316; Mod. Phys. Lett. A, 16 (2001), 2291-2203 (gr-qc 0111011); Inter. Jour. Mod. Phys. D, 11 (2002), 1355 (gr-qc 0105120); Gen. Relativ. Gravit., 36 (2004), 1335-1360 (gr-qc 0310078)

[719] J. Ponce de Leon, gr-qc 0105120, 0207108, 0209013, 0212058, 0305041, 0310117, 0511067, 0512067, 0703094, 0711.1004, 0712.4301, and 0802.1953

[720] V. Ponomariov and Y. Obukhov, Gen. Relativ. Gravit., 14 (1982), 309-330

[721] N. Poplawski, gr-qc 0511071, 0701176, 0702129, 0705.0251, 0706.4474, 08014797, and 0802.4453

[722] M. Portesi, M. Pennini, and A. Plastino, cond-mat 0511490 and 0510434

[723] A. Prykarpatsky, N. Bogoliubov, J. Golenia, and U. Taneri, math-ph 0902.4411

[724] H. Quevedo, physics 0604164; math-ph 0712.0868; gr-qc 0704.3102; Gen. Relat. Gravit., 40 (2008), 971-984

[725] H. Quevedo and A. Vazquez, math-ph 0712.0868

[726] H. Quevedo and A. Sanchez, hep-th 0805.3003; gr-qc 0902.3388

[727] R. Qi, quant-ph 0211082

[728] Qian Shu Li, Gong Min Wei, and Li Qiang Lu, Phys. Rev. A, 70 (2004), 022105

[729] I. Quiros, Phys. Rev. D, 61 (2000), 124026; hep-th 0009169

[730] I. Quiros, hep-th 0010146; gr-qc 9904004 and 0004010

[731] I. Quiros, R. Bonal, and R. Cardenas, gr-qc 9905071 and 0007071

[732] I. Racz and R. Wald, Class. Quant. Grav., 9 (1992), 2643-2656

[733] S. Rajeev, hep-th 0210179 and 0711.4319; math-ph 0703061; quant-ph 0701141

[734] M. Ravicule, M. Casas, and A. Plastino, Phys. Rev. A, 55 (1997), 1695-1702

[735] E. Recami and G. Salesi, Phys. Rev. A, 57 (1998), 98-105

[736] M. Reginatto, quant-ph 9909065; gr-qc 0501030; Phys. Rev. A, 58 (1998), 1775-1778; gr-qc 0501030

[737] F. Reif, Fundamentals of statistical and thermal physics, McGraw-Hill, 1965

[738] R. Reis, R. Szabo, and A. Valentino, hep-th 0610177

[739] M. Reisenberger and C. Rovelli, gr-qc 0111016

[740] A. Rennie, math-ph 9903021

[741] M. Reuter, hep-th 9804036

[742] K. vanRijsbergen, The geometry of information retrieval, Cambridge Univ. Press, 2004

[743] W. Rindler, Essential relativity, Springer, 1977; Relativity, second edition, Oxford Univ. Press, 2006

[744] V. Rivelles, Phys. Lett. B, 558 (2003), 191-196 (hep-th 0212262); hep-th 0305122
[745] M. Roberts, gr-qc 9905005
[746] H. Robertson, Phys. Rev. 34 (1929), 163-164
[747] M. Ronan, Symmetry and the monster, Oxford Univ. Press, 2006
[748] N. Rosen, Found. Phys., 12 (1982), 213-224; 13 (1983), 363-372
[749] H. Rosu, Mod. Phys. Lett. A, 18 (2003), 1-10
[750] C. Rovelli, Quantum gravity, Cambridge Univ. Press, 2004
[751] C. Rovelli, gr-qc 9503067, 9609002, and 0903.3832
[752] C. Rovelli, Class. Quant. Gravity, 10 (1993), 1549-1566 and 1567-1578
[753] C. Rovelli, Phys. Rev. D, 42 (1990), 2638-2647 and 43 (1991), 442-436
[754] C. Rovelli, Class. Quant. Gravity, 8 (1991), 297-316 and 317-331
[755] C. Rovelli, gr-qc 0111037, 0202079, and 0604045
[756] G. Ruppeiner, Rev. Mod. Phys., 67 (1995), 605-659
[757] V. Rusov, quant-ph 0804.1427
[758] V. Rusov and D. Vlasenko, quant-ph 08064050
[759] Y. Rylov, physics 0210003, 0303065, 0402068, 0405117, 0510243, 0603237, and 0604111; Jour. Math. Phys., 40 (1999), 256-278
[760] H. Sadjadi, Phys. Lett. B, 645 (2007), 108-112; Phys. Rev. D, 73 (2006), 063525
[761] A. Sakharov, Sov. phys. Dokl., 12 (1968), 1040
[762] S. Sakai, Operator algebras in dynamical systems, Cambridge Univ. Press, 1991
[763] D. Salamon, Symplectic geometry, Cambridge Univ. Press, 1993
[764] L. Salcedo, Jour. Chem. Phys., 126 (2007), 057101
[765] H. Salehi, H. Motavali, and M. Golshani, hep-th 0011062 and 0011063
[766] M. Salizzoni, A. Torrielli, and H. Yang, hep-th 0512249
[767] L. Saloff-Coste, Aspects of Sobolev type inequalities, Cambridge Univ. Press, 2002
[768] L. Saloff-Coste, Surveys Differential Geometry, 9, Inter. Press, 2004, 351-384
[769] J. Samuel and S. Roy Chawdhury, gr-qc 0711.0428 and 0711.0430
[770] E. Santamato, Phys. Rev. D, 29 (1984), 216-222; 32 (1985), 2615-2621; Jour. Math. Phys., 25 (1984), 2477-2480, Phys. Lett. A, 130 (1988), 199-202
[771] E. Santini, gr-qc 0005092
[772] A. Sanz, quant-ph 0412050
[773] A. Sanz and F. Borondo, quant-ph 0803.2581
[774] A. Sanz and S. Miret-Artes, quant-ph 0703161, 0710.2841, and 0806.2105
[775] G. Sardanishvili, Gauge theory in jet manifolds, Hadronic Press, 1993
[776] S. Sarkar and T. Padmanabhan, Entropy, 9 (2007), 100-107 (gr-qc 0607042); gr-qc 0611071
[777] S. Sasa, nlin-0006042
[778] N. Sasakura, hep-th 0001161
[779] H. Sati, hep-th 0608190
[780] D. Saunders, The geometry of jet bundles, Cambridge Univ. Press, 1989
[781] V. Sbitnev, quant-ph 0808.1245
[782] J. Schiff, Y. Goldfarb, and D. Tannor, quant-ph 0807.4659
[783] I. Schmelzer, quant-ph 0803.4657
[784] M. Schön, gr-qc 9304024
[785] D. Schuch, Phys. Lett. A, 338 (2005), 225-231; quant-ph 0805.1687
[786] D. Schuch and M. Moshinsky, Phys. Rev. A, 73 (2006), 062111; Sigma, 4 (2008), 054 (quant-ph 0807.1966)
[787] L. Schulman, Techniques and applications of path integration, Dover, 2005
[788] P. Schupp, hep-th 0110038
[789] U. Schwengelbeck and F. Faisal, Phys. Lett. A, 199 (1995), 281-286

[790] S. Seahra and P. Wesson, Class. Quantum Grav., 19 (2002),1139-1155; 20 (2003), 1321-1339; Gen. Relativ. Gravit., 33 (2001), 1731-1752; 37 (2005), 1339-1347
[791] N. Seiberg, hep-th 0601234
[792] N. Seiberg and E. Witten, JHEP, 09 (1999), 032 (hep-th 9908142
[793] A. Sen, hep-th 0506177, 0508042, and 0708.1270
[794] S. Sengupta and P. Chattaraj, Phys. Lett. A, 215 (1996), 119-127
[795] G. Sewell, Quantum mechanics and its emergent macrophysics, Princeton Univ. Press, 2002; math-ph 0509069, 0612020, 0710.1239, 0710.3315, and 0711.3295
[796] D. Shirikov and V. Kovalev, hep-th 0001210
[797] F. and A. Shojai, gr-qc 0404102 and 0306099; Inter. Jour. Mod. Phys. A, 15 (2000), 1859-1868
[798] F. Shojai and M. Golshani, Inter. Jour. Mod. Phys. A, 13 (1998), 677-693 and 2135-2144
[799] A. and F. Shojai and N. Dadhich, gr-qc 0504137
[800] A. and F. Shojai, Physica Scripta, 64 (2001), 413-416
[801] A. Shojai, Inter. Jour. Mod. Phys. A, 15 (2000), 1757-1771
[802] F. Shojai and A. Shirinifard, gr-qc 0504138
[803] A. and F. Shojai, quant-ph 0109025 and 0609109; gr-qc 0105102, 0306100 and 0311076
[804] F. and A. Shojai and M. Golshani, Mod. Phys. Lett. A, 13 (1998), 2725-2729 and 2915-2922; gr-qc 9903050
[805] F. Shojai, Phys. Rev. D, 60 (1999), 124001
[806] F. and A. Shojai, Class. Quant. Grav., 21 (2004), 1-9; gr-qc 0409020, 0105102, 0109052, and 0409036
[807] Y. Shtanov, quant-ph 9705024; Phys. Rev. D, 54 (1996), 2569-2570
[808] G. Sierra, math-ph 0510572, 0711.1063, and 0712.0705
[809] J. Silverman, The arithmetic of dynamical systems, Springer, 2007
[810] T. Singh, hep-th 0605112; gr-qc 0510042, 0901.0978, and 09052548
[811] J. Sjostrand and M. Zworski, math-ph 0108052
[812] W. Smilga, physics 0502040
[813] L. Smolin, Phys. Lett. A, 113 (1986), 408-412; quant-ph 0609109
[814] L. Smolin, hep-th 0201031, 0712.0977, 0803.2926, and 0812.3761
[815] Y. Sobouti and S. Nasiri, Inter. Jour. Mod. Phys. B, 7 (1993), 3255-3272
[816] Y. Sobouti and S. Nasiri, Inter. Jour. Mod. Phys., 7 (1993), 3255-3273
[817] C. Sochichiu, hep-th 0506086
[818] S. Solodukhin, hep-th 0709045 and 0802.3117; Phys. Lett. B, 646 (2007), 268-274
[819] T. Sotiuiou, V. Faraoni, and S. Liberati, gr-qc 0707.2748
[820] T. Sotiriou and S. Liberati, Phys. Rev. D, 74 (2006), 044016 (gr-qc 0603096); gr-qc 0703080
[821] H. Steinacker, hep-th 0708.2426, 0712.3194, 0806.2032, and 0903.1015
[822] W. Struyve and A. Valentini, quant-ph 0808.0290
[823] A. Suarez and J. Ross, Jour. Chem. Phys. (1994), 4563-4573
[824] R. Szabo, hep-th 0109062 and 0602233
[825] R. Szabo, An introduction to string theory and D-brane dynamics, Imperial Coll. Press, 2004
[826] H. Takata, 0008019 and 0010081
[827] M. Takesaki, Tomita's theory of modular Hilbert algebras and its applications, Springer, 1970
[828] X. Tang and A. Boozer, physics 9803006
[829] V. Tapia, math-ph 0903.2031

[830] S. Tasaki, E. Eisenberg, and L. Horwitz, hep-th 0310186
[831] T. Thiemann, Modern canonical quantum general relativity, Cambridge Univ. Press, 2007
[832] P. Tillman, gr-qc 0610141
[833] K. Tintarev and K. Fieseler, Concentration compactness, Imperial Coll. Press, 2007
[834] S. Tiwari, gr-qc 0612099 and 0307079; Superluminal phenomena in modern perspective, Rinton Press, 2003
[835] D. Toms, hep-th 0411233
[836] P. Topping, Lectures on the Ricci flow, Cambridge Univ. Press, 2006
[837] T. Torii and H. Maeda, Phys. Rev. D, 72 (2005), 064007
[838] G. Torres-Vega and J. Frederick, Jour. Chem. Phys., 93 (1990), 8862-8879; 98 (1993), 3103-3120
[839] P. Townsend, gr-qc 9707012
[840] C. Trahan and B. Poirier, Jour. Chem. Phys., 124 (2006), 034115 and 034116
[841] C. Tsallis, cond-mat 0010150
[842] C. Tseng and A. Caticha, cond-mat 0411625
[843] R. Tsekov, Jour. Phys. A, 28 (1995), L557-L561 and 40 (2007), 10945-10947; quant-ph 07111442, 0803.4409, 0808.0326, 0903.3644, and 0904.0723
[844] R. Tumulka, math-ph 0711.0035; quant-ph 0611283
[845] W. Unruh, Phys. Rev. D, 14 (1975), 870
[846] W. Unruh and R. Wald, Phys. Rev. D, 27 (2006), 044016 (gr-qc 0603096)
[847] A. Unterberger, Quantization and arithmetic, Birkhäuser, 2008
[848] A. Valentini, quant-ph 0104067, 0203049, 0309107, and 0812.2941
[849] A. Valentini, quant-ph 0811,0810
[850] A. Valentini and H. Westman, quant-ph 0403034
[851] I. Vancea, gr-qc 9801072; hep-th 0309214
[852] A. Vankov, Found. Phys., D)I 10.1007/as10701-008-9219-z
[853] J. Varilly, An introduction to non-commutative geometry, EMS Lecture series in mathematics, 2006
[854] A. Vazquez, H. Quevedo, and A. Sanchez, hep-th 0805.4819
[855] A. Vercin, quant-ph 0803067 and 9803067
[856] J. Vigier, Phys. Lett. A, 135 (1989), 99
[857] M. Visser, Phys. Rev. D, 48 (1993), 583-591 and 5697-5705
[858] D. Vollick, gr-qc 0712.1859
[859] G. Volovik, The universe in a helium droplet, Oxford Univ. Press, 2003
[860] R. Wald, General relativity, Univ. Chicago Press, 1984; Quantum field theory in curved spacetime, Univ. Chicago Press, 1994
[861] R. Wald, gr-qc 9307038, 9901033, and 9912119; Jour. Math. Phys., 31 (1990), 2378-2384
[862] D. Wands, gr-qc 0307034
[863] B. Wang, C. Lin, D. Pavon, and E. Abdalla, hep-th 0711.221
[864] S. Weinberg, Phys. Rev. Lett., 62 (1989), 485-488
[865] S. Weinberg, Annals Phys., 194 (1989), 336
[866] Y. Weinstein, S. Lloyd, and C. Tsallis, cond-mat 0206039
[867] J. Wess, hep-th 0408080 and 0511025
[868] P. Wesson, Space, time, matter, World Scientific, 1999
[869] P. Wesson, 5-dimensional physics, World Scientific, 2006
[870] P. Wesson, Jour. Math. Phys., 43 (2002), 2423-2438; Class. Quant. Gravity, 19 (2002), 2825; gr-qc 0302092 and 0309134; General. Relativ. Gravit., 38 (2006), 937-944; Phys. Rev. Lett., 276 (1992), 299

[871] P. Wesson and J. Ponce de Leon, Jour. Math. Phys., 33 (1992), 3883- 3887
[872] P. Wesson and H. Liu, Internat. Jour. Theor. Phys., 36 (1977), 1865-1879
[873] C. Wetterich, quant-ph 0212031 and 0809.2671
[874] J. Wheeler, Phys. Rev. D, 41 (1990), 431-441
[875] J. Wheeler, hep-th 9706215, 9708088, 0002068, and 0305017; gr-qc 9411030
[876] J. Wheeler, Scientific American Library, NY, 1990, p. 212
[877] M. Wilkinson, Jour. Phys. A, 20 (1987), 2415-2423; Jour. Phys. A, 21 (1988), 1173-1190
[878] S. Williams, D. Searles, and D. Evans, Phys. Rev. E, 70 (2004), 066113
[879] D. Wisniacki, F. Borondo, and R. Benito, Europhys. Lett., 64 (2003), 441-447 (quant-ph 0309083)
[880] D. Wisniacki, E. Pujals, and F. Borondo, Europhys. Lett., 73 (2006), 671-676 (nlin 0507015)
[881] D. Wisniacki and F. E. Pujals, Europhys. Lett., 71 (2005) 159-165 (quant-ph 0502108)
[882] D. Wisniacki, E. Vergini, R. Benito, and F. Borondo, nlin 0311052 and 0402022
[883] D. Wisniacki and F. Toscano, quant-ph 0810.5740
[884] W. Wood and G. Papini, Found. Phys. Lett., 6 (1993), 207-223
[885] N. Woodhouse, Geometric quantization, Oxford Univ. Press, 1991
[886] E. Woolgar, hep-th 0708.2144
[887] W. Wooters, Phys. Rev. D, 23 (1981), 357-362
[888] S. Wu, B. Wang, and G. Yang, hep-th 0711.1209
[889] S. Wu, B. Wang, G. Yang, and P. Zhang, hep-th 0801.2688
[890] A. Wyatt, Quantum dynamics with trajectories, Springer, 2005
[891] Li Xiang and X. Wen, gr-qc 0901.0603
[892] C. Yang, Chaos, solitons, and fractals, 30 (2006), 342-362 and 33 (2007), 1073-1092; Annals Phys., 321 (2006), 2876-2926
[893] H. Yang, hep-th 0402002, 0512215, 0611174, and 0704.0929
[894] H. Yang, hep-th 0612231, 0711.0234, 0809.4728, and 0902.0035
[895] H. Yang and M. Salizzoni, hep-th 9512215
[896] H. Yang, Inter. Jour. Mod. Phys. A, 23 (2008), 2181-2183; Mod. Phys. Lett. A, 22 (2007), 1119-1132
[897] S. Yau, Annales Sci. Ecole Norm. Sup., 8 (1975), 487-507
[898] B. Ydri, hep-th 0110006
[899] J. York, Phys. Rev. Lett., 28 (1972), 1082-1085 and 26 (1971), 1656-1658; Jour. Math. Phys., 14 (1973), 456-464 and 13 (1972), 125-130; gr-qc 0405005 and 9307022; Found. Phys., 16 (1986), 249-257
[900] D. Youm, Phys. Rev. D, 62 (2000), 084002; Mod. Phys. Lett. A, 16 (2001), 2371-2380
[901] C. Zachos, D. Fairlie, and T. Curtright, Quantum mechanics in phase space, World Scientific, 2005
[902] P. Zanardi, P. Giorda, and M. Cozzini, quant-ph 0701061
[903] T. Zhu, J. Ren, and M. Li, hep-th 0811.0212
[904] W. Zimdahl and A. Balakin, Phys. Rev. D, 63 (2000), 023507 and Entropy, 4 (2002), 49-64
[905] W. Zimdahl, astro-ph 9702070; gr-qc 9711081
[906] W. Zimdahl and D. Pavon, astro-ph 0005352
[907] J. Zinn-Justin, Path integrals in quantum mechanics, Oxford Univ. Press, 2005
[908] M. Zirpel, quant-ph 0903.3878
[909] A. Zotov, Mod. Phys. Lett. A, 16 (2001), 615

[910] Y. Zunger, hep-th 0002074
[911] W. Zurek, quant-ph 0707.2832 and 0903.5082
[912] B. Zweibach, A first course in string theory, Cambridge Univ. Press, 2004

# Index